D0573880

HEALTH SCIENCES CHEMISTRY

HEALTH SCIENCES CHEMISTRY

Gershon Shugar, B.S., M.A., Ph.D.
Professor, Essex County College, Newark, N.J.

Ronald A. Shugar, B.S., M.D.
Menlo Park Medical Group, Edison, N.J.

Lawrence Bauman, B.S., D.D.S.
Former Associate Professor, New York University,
School of Dentistry, New York, N.Y.

Rose Shugar Bauman, B.S.
Science Writer, Watchung, N.J.

F.A. DAVIS COMPANY, Philadelphia

Second Printing 1978

Printed in the United States of America.

Library of Congress Cataloging in Publication Data

Health sciences chemistry.

Includes index.
1. Chemistry. 2. Biological chemistry.
I. Shugar, Gershon J., 1918-
QD31.2.H39 540 78-1065
ISBN 0-8036-7835-5

This book is dedicated with all our love to
Gregory Phillip Shugar
by his family, the authors: grandfather, father, uncle and aunt.

Gershon J. Shugar
Ronald A. Shugar
Lawrence Bauman
Rose Shugar Bauman

PREFACE

Careers in the field of health care are expanding as a result of the team concept of total health care delivery. This has necessitated the development of specialized health careers, and students must learn the theories, principles, and concepts of the natural scientific laws that govern our universe. Furthermore they must be able to apply what they have learned in order to stand as a professional in their chosen field.

Health care personnel who make up the team concept of total health care delivery to the community must have knowledge in the field of chemistry, but not the extensive knowledge that is required of a professional chemist. They must have the basic chemistry that relates to the health care field.

This textbook in health sciences chemistry has been written for you, the student, in the language that you understand, with references that relate to everyday experiences. Particular attention has been made to relate chemical concepts to clinical situations. We believe that this text includes the fundamentals of chemistry that students must know in order to assure a successful performance in their professional fields.

ACKNOWLEDGEMENT

We wish to express our sincere appreciation and thanks to Mr. Anthony Zuppardi, Associate Professor, Essex County College, for editing this book. Mr. Zuppardi has been deeply involved in teaching chemistry to allied health personnel for the past eight years. He is currently pursuing his Ph.D. in Biochemistry at the College of Medicine and Dentistry in Newark, N.J. His suggestions, criticisms and contributions in the biochemical area were especially valuable and helpful.

CONTENTS

HEALTH SCIENCES CHEMISTRY

OBJECTIVES

When you have completed this chapter you will be able to:

1. State why health care personnel must know the fundamentals of chemistry.
2. Give the reason why conversion between the different measurement systems is important to health care personnel.
3. Distinguish between determinate and indeterminate errors.
4. Draw a diagram showing how to read a liquid level and discuss how a too high or too low volume reading can be obtained.
5. Describe how the dimensional factor conversion method is applied to measurement units.
6. Recognize the two fractions which can be obtained from an equality.
7. Determine which fraction should be used to convert from one measurement system to another.
8. Demonstrate that you have selected the correct fraction to be used in the conversion.
9. Convert lengths, weights, and volumes in the metric system.
10. Convert length, weights and volumes in the English system.
11. Interconvert units of measurements between the English and metric systems.

UNITS AND MEASUREMENTS

You, the prospective health care employee, will be intimately involved in the delivery of health care at community centers, hospitals, doctor's offices, private residences, and in the field, when and wherever the need or emergency arises. You will be in contact with people, dispensing medications, administering prescribed treatments, analyzing blood and body fluids, preparing solutions and mixtures, and handling various kinds of metal and plastic equipment.

All matter is composed of the basic building blocks of nature called *elements.* There are 107 different elements, but only 88 of them occur naturally; the rest are made by man. Every substance and object is composed of one or more of these elements joined together in different shapes and arrangements (Fig. 1-1). Some are relatively simple and consist of a single element (such as silver, gold or lead); others, such as the hemoglobin of the blood, are fantastically complex; while still others, such as some substances in the living body, may contain arrangements of as many as a million particles joined together.

These elements are part of life, death, disease, and recovery and you will be able to do a more professional job if you understand the chemistry of the elements. Chemistry is the study of these elements, how they behave, how they react, and what makes them "tick." Chemistry is involved with the composition of substances, their structure, and how they are changed or transformed to other substances.

The elements cannot think or reason. They behave as they do because of the conditions to which they are exposed, subjected or treated. They behave the same way every time they are handled if they are treated identically. This means identical weights, temperatures, volumes, lengths, radiation, or lengths of time. This necessary precision requires the use of some form of measurement involving length, area, volume, or time. In order to determine these values, instruments for scientific measurement must be used (Fig. 1-2).

Measurements are complicated because there are a number of different systems in the world, just as there are different currencies with different values in the various countries. Yet, businessmen can trade and engage in commerce

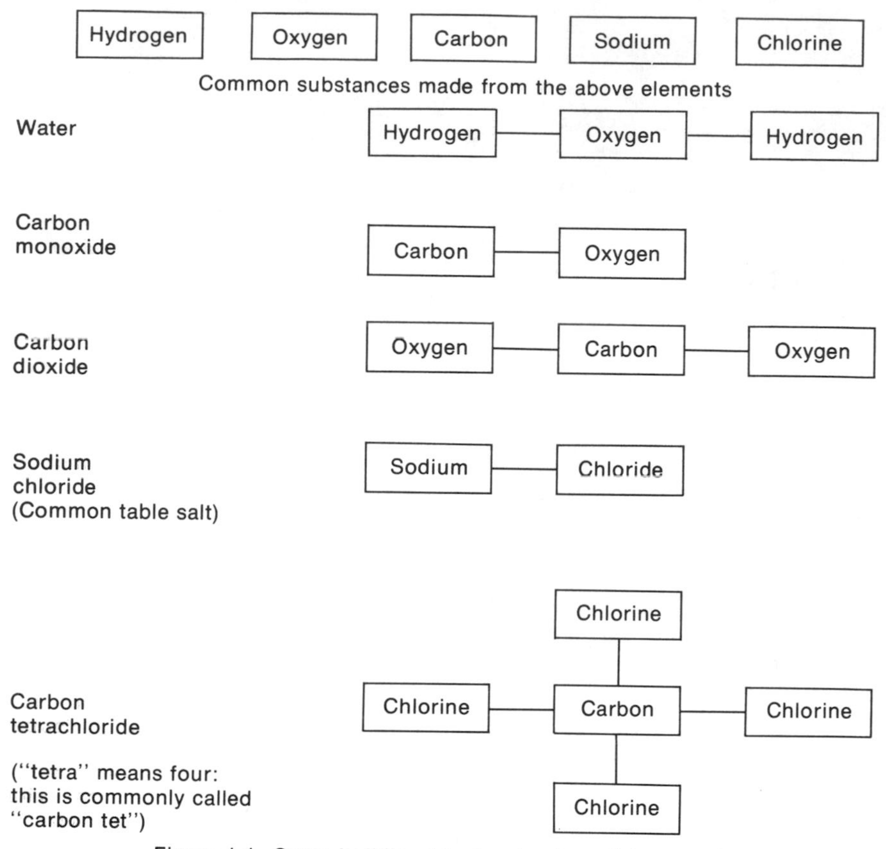

Figure 1-1. Some building blocks of nature, "elements."

because they can determine the *equivalent value* of one country's currency in another country's currency. *What is equal is equal.* One system of measurement can be substituted for another, and health care personnel must be able to convert one system into another, and possibly into a third system of equal value. There are times when a particular measurement system is not available. In that case you can either push the panic button or can use your professional training to convert that quantity into the units that you do have at your disposal.

ACCURACY IN MEASUREMENT

Accuracy of measurements and readings is important, but accuracy should not be carried to impractical extremes. For example, patients are weighed to the nearest kilogram; medications to the nearest gram or grain. You should use the scientific system of measurement to the accuracy that the situation warrants. By the same reasoning, it is not sufficient to record a patient's temperature to the nearest whole degree, because omission of the fractional degree would sacrifice significant information to the diagnostician. Even if the thermometer used would read temperature to the *thousandth* of a degree, only the nearest

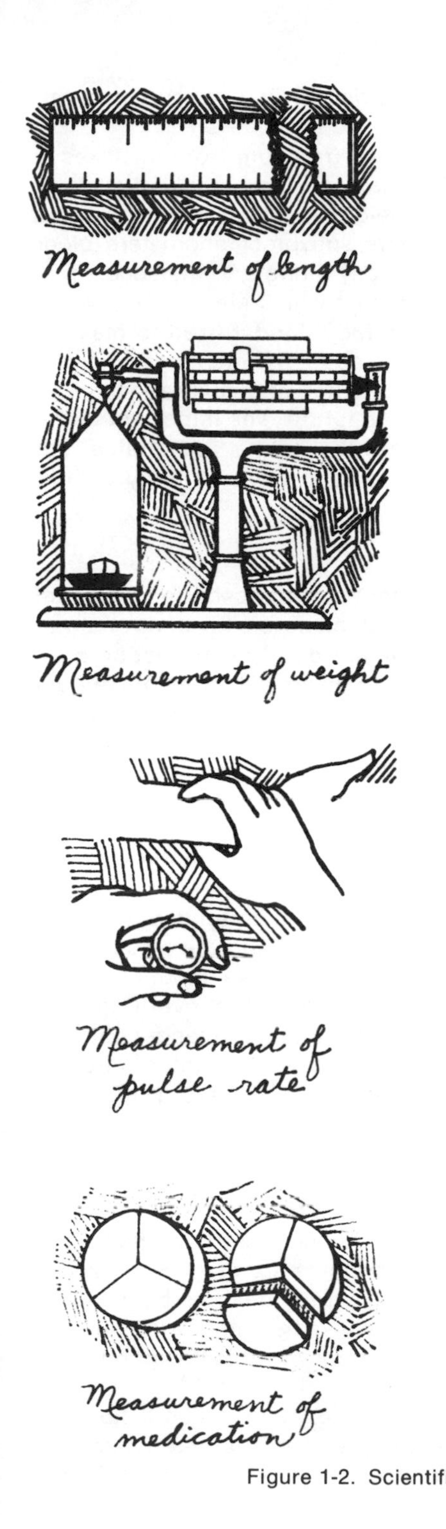

Figure 1-2. Scientific measuring instruments.

tenth of a degree would be recorded. *Only those figures in a measurement that are reasonable, trustworthy, and meaningful should be reported.*

Measurements are no better than the accuracy (freedom from mistakes or error) of the instruments. A balance that does not work properly, that sticks or gives erratic readings should not be used for weighing. This is common sense. Yet, many professionals blindly use thermometers, sphygmomanometers (blood pressure apparatus), or balances without giving any thought to the accuracy of the apparatus. Generally, most pieces of equipment which are used in the scientific and medical fields are fairly well calibrated (standardized to maximal accuracy). There is nothing positive and absolute about the markings on equipment, because mistakes have been made in markings and calibrations in the past on instruments and they will be made in the future. The important fact to remember is *to think, to reason* and *to judge* if the data is reasonable. You should always examine instruments and never report data blindly without checking results.

Error

Errors are always made. In fact, no measurement is ever absolutely and completely accurate. Regardless of the care exercised the possibility of error always exists. Some errors are the fault of the instrument due to malfunction or incorrect calibration. These are called determinate errors. Other errors are due to the human factor, and these are called indeterminate errors, caused by error in interpretation or reading of the instrument. In all cases, use care, and read your instruments accurately.

Parallax Errors

Deviations can be found in readings when there is space between the scale and the pointer or indicator. This difference in readings is due to parallax. This principle must be considered when you read the level of liquid in a burette or syringe, because the attractive force between the liquid and the glass causes the liquid to assume a concave form. For accuracy, your eye should be at the same level as that of the liquid, never above or below that level (Fig. 1-3). Parallax error is well illustrated when a driver and front seat passenger compare their readings of the speedometer. The driver's eye sees the pointer in front of the dial, whereas the passenger reads the speed with his eye at an angle, thereby reading a lower speed.

Measuring Systems

Systems of measurement used throughout the world for length, area, volume and weight are the metric and English systems. For temperature, the Celsius (centigrade), Fahrenheit and Kelvin (Absolute) systems are used. Each system has its own unique unit of measure with specific relationships between the units. The English system has evolved over many years with various uneven relationships between the units; whereas the metric system is based on multiples of ten (the decimal system).

It is almost impossible for anyone to remember every conversion factor used to convert every unit of one system into another, and tables that list specific conversion numbers are large and cumbersome. There is, however, a simple,

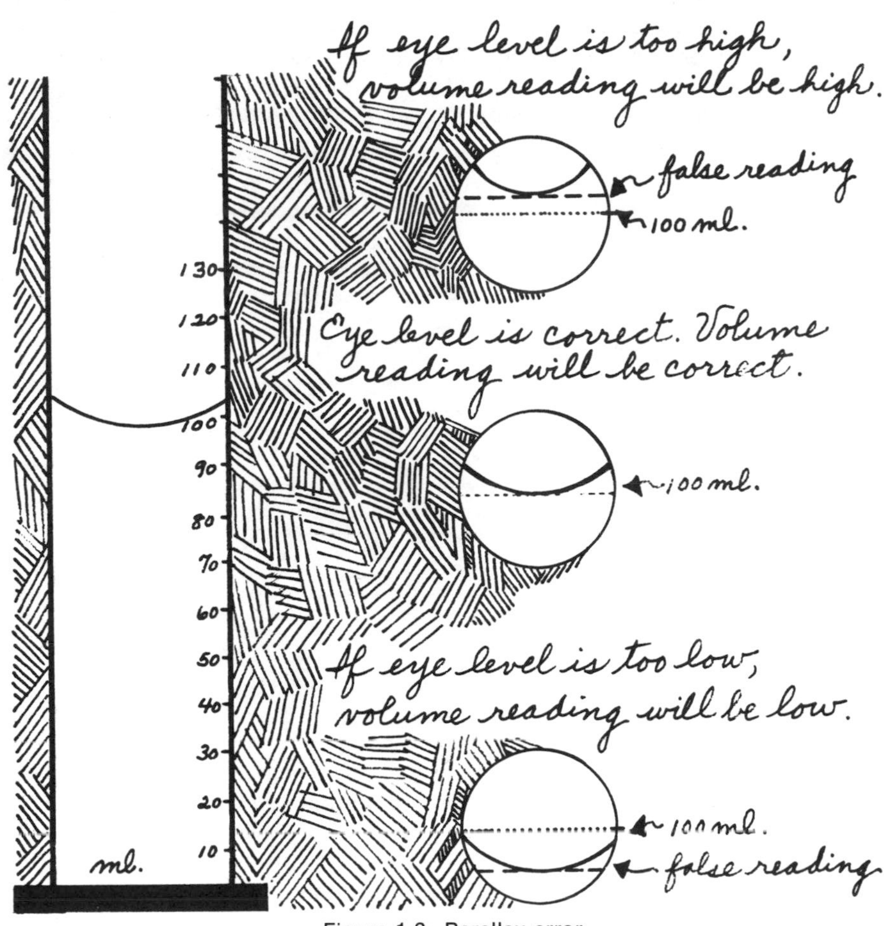

Figure 1-3. Parallax error.

easy to learn and easy to use method which will enable you to convert any measurement in a given system to any other system. It is called the *dimensional factor conversion method.*

THE DIMENSIONAL FACTOR METHOD FOR CONVERTING MEASUREMENTS

Many professionals attempt to memorize formulas to convert their measurements, and many times these professionals make mistakes such as using the wrong formula, multiplying when they should be dividing, or solving the problem but getting the wrong answer. If you wish to avoid these common mistakes and *always* get the right answer without pain or strain, you can master the procedure called dimensional factor conversion. What this scientific sounding name actually means is that you proceed in a logical step-by-step fashion, using basic arithmetical methods in a reasoning process to get what you want.

An equality means that *things are equal,* such as an equation: 12 rolls = 1 dozen. Twelve is a dozen, and one dozen amounts to 12, so each side is equal in value to the other side, and the equality is represented by the "=" sign. From basic arithmetic you should remember that dividing *each* side of an equation by the *same* number *does not* change the value of the equation (the equality).

Twelve is equal to 1 dozen: This is an EQUALITY.

Take the equality: $12 = 1 \text{ doz}$

Divide each side by *12*: $\frac{\cancel{12}}{\cancel{12}} = \frac{1 \text{ doz}}{12}$

And cancel the 12s on the left side

Answer: $1 = \frac{1 \text{ doz}}{12}$

OR

Divide each side by *1 doz*: $\frac{12}{1 \text{ doz}} = \frac{\cancel{1 \text{ doz}}}{\cancel{1 \text{ doz}}}$

And cancel the 1 doz on the right side

Answer: $\frac{12}{1 \text{ doz}} = 1$

NOTE: The two answers are merely *reciprocals of the same fraction* and BOTH are equal to ONE. *When any fraction is multiplied by* ONE, *the value of the fraction is not changed.*

THE KEY TO SOLVING CONVERSIONS IS TO SELECT THE RIGHT EQUALITY, having the units wanted for the answer in the *numerator* of the fraction and the original units (which are to be converted) in the denominator of the fraction. Therefore, the correct procedure is to:

1. Know the equalities.
2. Select the proper equality to give the right units.
3. Multiply (clearing and canceling) fractions.

Before attempting to work with unfamiliar metric units, you should first get the confidence that successful experience gives by converting commonly encountered units and measures.

PROBLEM

There are 36 rolls of bandage. How many dozen are there?

The equality that is needed is 1 dozen = 12 rolls. Therefore, according to the procedure there are two possible fractions:

$$1 = \frac{1 \text{ doz}}{12 \text{ rolls}} \text{ or } 1 = \frac{12 \text{ rolls}}{1 \text{ doz}}$$

WHICH FRACTION SHOULD BE SELECTED?

The correct one to select is the one which has the *desired units* (in this case, 1 dozen) in the *numerator*.

Set up equation $36 \text{ rolls} \times \frac{1 \text{ doz}}{12 \text{ rolls}} =$

cancel $36 \cancel{\text{rolls}} \times \frac{1 \text{ doz}}{12 \cancel{\text{rolls}}} =$

divide $\overset{3}{\cancel{36}} \text{ rolls} \times \frac{1 \text{ doz}}{\underset{1}{\cancel{12}} \text{ rolls}} =$

clear $3 \times 1 \text{ doz} = 3 \text{ doz}$ (answer)

If the wrong fraction is chosen, then the answer will be incorrect and the fact that a wrong choice was made is evident in the answer. It will have the *wrong unit labels* in it.

For example, in the previous problem, if you had chosen to multiply the 36 rolls by the fraction $\frac{12 \text{ rolls}}{1 \text{ doz}}$ then the answer would look like this:

$$36 \text{ rolls} \times \frac{12 \text{ rolls}}{1 \text{ doz}} = \frac{432 \text{ rolls}^2}{\text{doz}}$$

(Note that *no units cancel,* producing a squared rolls/dozen unit)

Obviously this answer is wrong. *If no units cancel out,* then your choice of the fraction was incorrect. *Units must cancel out to give you the desired units.*

PROBLEM

How many seconds are there in one week?

The equalities are:	The fractions are:
60 sec = 1 min	$\frac{1 \text{ min}}{60 \text{ sec}}$ & $\frac{60 \text{ sec}}{1 \text{ min}}$
60 min = 1 hr	$\frac{1 \text{ hr}}{60 \text{ min}}$ & $\frac{60 \text{ min}}{1 \text{ hr}}$
24 hr = 1 day	$\frac{1 \text{ day}}{24 \text{ hr}}$ & $\frac{24 \text{ hr}}{1 \text{ day}}$
7 days = 1 week	$\frac{1 \text{ week}}{7 \text{ days}}$ & $\frac{7 \text{ days}}{1 \text{ week}}$

NOTE: Select the equality which will leave only the desired units in the answer.

To convert 60 min (1 hr) to sec, use $\frac{60 \text{ sec}}{1 \text{ min}}$

$$\frac{60 \cancel{\text{min}}}{1 \text{ hr}} \times \frac{60 \text{ sec}}{1 \cancel{\text{min}}} = \frac{3{,}600 \text{ sec}}{1 \text{ hr}}$$

to convert 24 hr (1 day) to sec, use

$$\frac{3600 \text{ sec}}{1 \cancel{\text{hr}}} \times \frac{24 \cancel{\text{hr}}}{1 \text{ day}} = \frac{86{,}400 \text{ sec}}{1 \text{ day}}$$

to convert one week (7 days) to sec, use

$$\frac{86{,}4000 \text{ sec}}{1 \not{\text{day}}} \times \frac{7 \not{\text{day}}}{1 \text{ week}} = \frac{604{,}800 \text{ sec}}{1 \text{ week}}$$

METRIC SYSTEM

In the metric system every unit is either ten, a hundred, a thousand, or a million times larger or smaller than another unit. There are no unusual odd equalities such as 5,280 feet in a mile or 16 ounces in a pound. Each unit is designated by a PREFIX which tells you how much smaller or larger the unit is than the standard. The prefixes are:

smaller than one	micro, which means *one-millionth* of
	milli, which means *one-thousandth* of
	centi, which means *one-hundredth* of
	deci, which means *one-tenth* of
larger than one	kilo, which means one *thousand* times

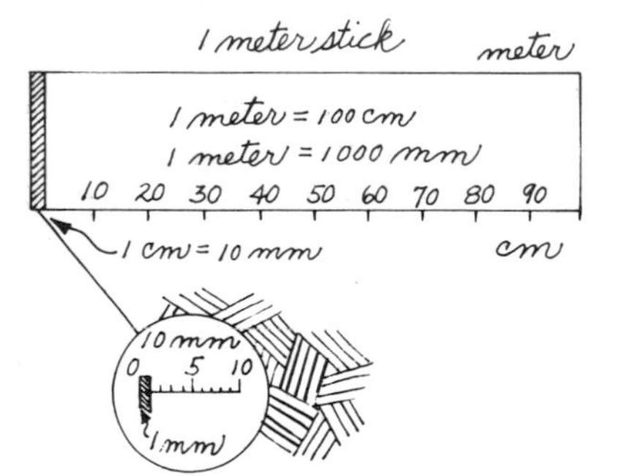

Figure 1-4. Linear units.

Length

Length is the distance between two points and it is measured in *linear* units. The standard unit is the meter; other units are (Fig. 1-4):

*milli*meter	*one-thousandth* of a meter (0.001)	: mm
*centi*meter	*one-hundredth* of a meter (0.01)	: cm
*deci*meter	*one-tenth* of a meter (0.1)	: dm
*kilo*meter	One *thousand* meters (1,000)	: km

The equalities are:	The fractions are:
1,000 mm = 1 m	$\frac{1 \text{ m}}{1{,}000 \text{ mm}}$ & $\frac{1{,}000 \text{ mm}}{1 \text{ m}}$
100 cm = 1 m	$\frac{1 \text{ m}}{100 \text{ cm}}$ & $\frac{100 \text{ cm}}{1 \text{ m}}$
10 mm = 1 cm	$\frac{1 \text{ cm}}{10 \text{ mm}}$ & $\frac{10 \text{ mm}}{1 \text{ cm}}$
1,000 m = 1 km	$\frac{1 \text{ km}}{1{,}000 \text{ m}}$ & $\frac{1{,}000 \text{ m}}{1 \text{ km}}$

NOTE: a meter is slightly *longer* than 1 yard and is approximately 40 inches long.

PROBLEM

The x-ray table is 2 m long. What is the length in cm?

$$\text{Equality: } 100 \text{ cm} = 1 \text{ m}$$

$$\text{The two possible fractions are: } 1 = \frac{100 \text{ cm}}{1 \text{ m}} \quad \text{or} \quad \frac{1 \text{ m}}{100 \text{ cm}}$$

$$\text{Units desired are cm, therefore select: } 1 = \frac{100 \text{ cm}}{1 \text{ m}}$$

$$\text{Multiply and cancel: } 2 \cancel{\text{m}} \times \frac{100 \text{ cm}}{1 \cancel{\text{m}}} = 200 \text{ cm (answer)}$$

Weight

The gram (g) is the standard unit of weight measurement (Fig. 1-5). Other units are broken down as follows:

*micro*gram	*one-millionth* gram (0.000001)	: μg
*milli*gram	*one-thousandth* gram (0.001)	: mg
*centi*gram	*one-hundreth* gram (0.01)	: cg
*deci*gram	*one-tenth* gram (0.1)	: dg
*kilo*gram	one *thousand* grams (1,000)	: kg

Figure 1-5. One gram.

Metric weight equivalents are shown in Figure 1-6.

The equalities are:	The fractions are:
1,000,000 μg = 1 g	$\frac{1 \text{ g}}{1{,}000{,}000 \ \mu\text{g}}$ & $\frac{1{,}000{,}000 \ \mu\text{g}}{1 \text{ g}}$
1,000 mg = 1 g	$\frac{1 \text{ g}}{1{,}000 \text{ mg}}$ & $\frac{1{,}000 \text{ mg}}{1 \text{ g}}$
100 cg = 1 g	$\frac{1 \text{ g}}{100 \text{ cg}}$ & $\frac{100 \text{ cg}}{1 \text{ g}}$
10 dg = 1 g	$\frac{1 \text{ g}}{10 \text{ dg}}$ & $\frac{10 \text{ dg}}{1 \text{ g}}$
1,000 g = 1 kg	$\frac{1 \text{ kg}}{1000 \text{ g}}$ & $\frac{1000 \text{ g}}{1 \text{ kg}}$

PROBLEM

How many mg of penicillin K are there in 100 g?

$$\text{Select the needed equality: } 1 \text{ g} = 1000 \text{ mg}$$

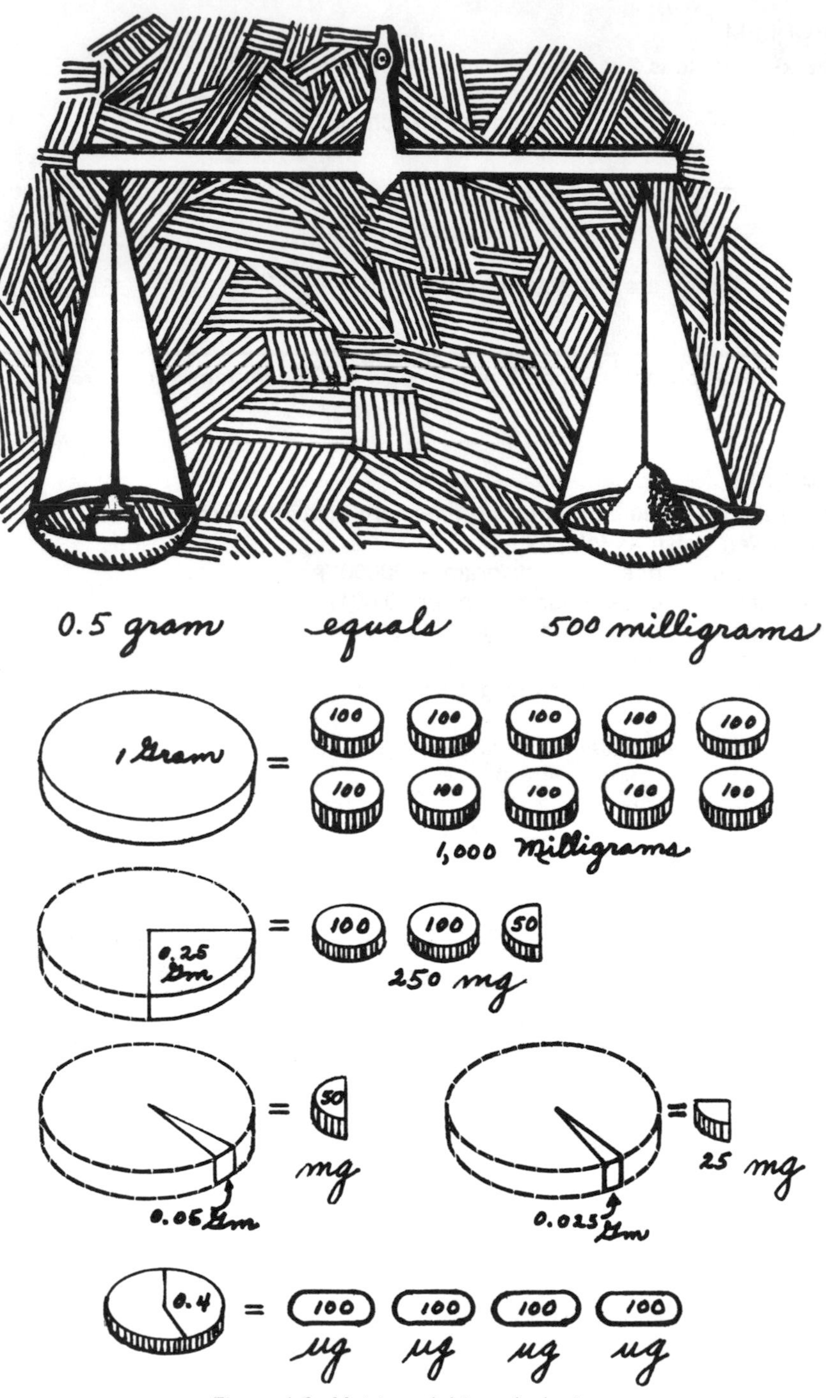

Figure 1-6. Metric weight equivalents.

The two possible fractions are:

$$1 = \frac{1000 \text{ mg}}{1 \text{ g}} \quad \text{or} \quad \frac{1 \text{ g}}{1000 \text{ mg}}$$

Units desired are mg; select $1 = \frac{1000 \text{ mg}}{1 \text{ g}}$

Multiply and cancel: $100 \cancel{g} \times \frac{1000 \text{ mg}}{1 \cancel{g}} = 100{,}000 \text{ mg}$ (answer)

Volume

Volume is measured in the metric system by the liter, which is the standard unit of measure, or a fraction of the liter called a milliliter (Fig. 1-7). The liter is slightly larger than a quart. The units of volume are:

*micro*liter	*one-millionth* liter (0.000001)	: μl
*milli*liter	*one-thousandth* liter (0.001)	: ml
*centi*liter	*one-hundredth* liter (0.01)	: cl
*deci*liter	*one-tenth* liter (0.1)	: dl
*kilo*liter	one *thousand* liters (1,000)	: kl

The equalities are:	The fractions are:
1 L = 1,000,000 μl	$\frac{1 \text{ L}}{1{,}000{,}000\ \mu\text{l}}$ & $\frac{1{,}000{,}000\ \mu\text{l}}{1 \text{ L}}$
1 L = 1,000 ml	$\frac{1 \text{ L}}{1{,}000 \text{ ml}}$ & $\frac{1{,}000 \text{ ml}}{1 \text{ L}}$
1 L = 100 cl	$\frac{1 \text{ L}}{100 \text{ cl}}$ & $\frac{100 \text{ cl}}{1 \text{ L}}$
1 L = 10 dl	$\frac{1 \text{ L}}{10 \text{ dl}}$ & $\frac{10 \text{ dl}}{1 \text{ L}}$
1,000 L = 1 kl	$\frac{1 \text{ kl}}{1{,}000 \text{ L}}$ & $\frac{1{,}000 \text{ L}}{1 \text{ kl}}$

PROBLEM

How many ml are contained in 0.5 L of I.V. saline solution?

Equality: 1,000 ml = 1 L

The two fractions are: $\text{L} = \frac{1000 \text{ ml}}{1 \text{ L}} \quad \text{or} \quad \frac{1 \text{ L}}{1{,}000 \text{ ml}}$

Units desired are ml; select $\frac{1000 \text{ ml}}{1 \text{ L}}$

Multiply and cancel: $0.5 \cancel{L} \times \frac{1{,}000 \text{ ml}}{1 \cancel{L}} = 500 \text{ ml}$ (answer)

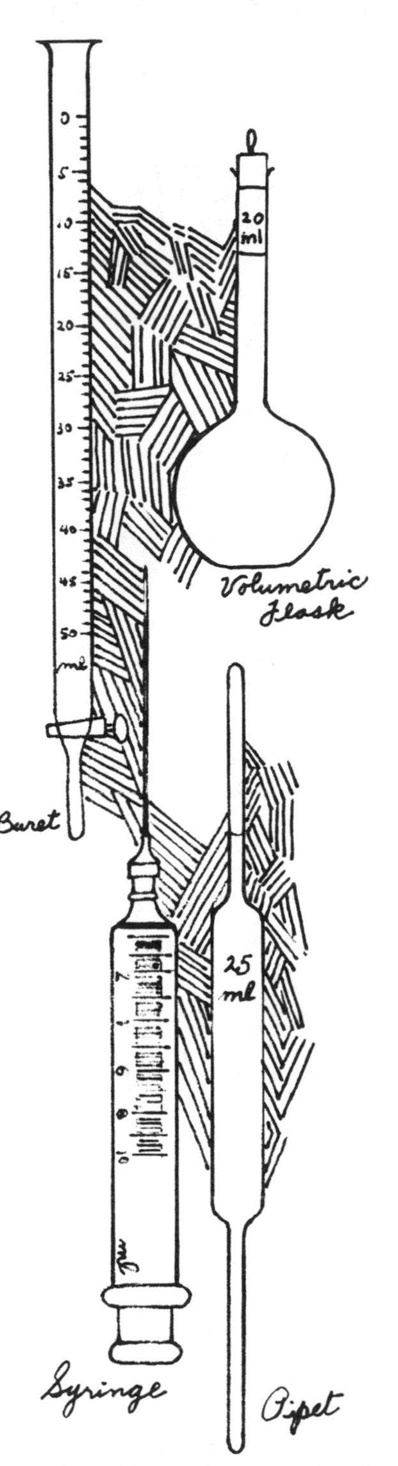

Figure 1-7. Volumetric measurement apparatus.

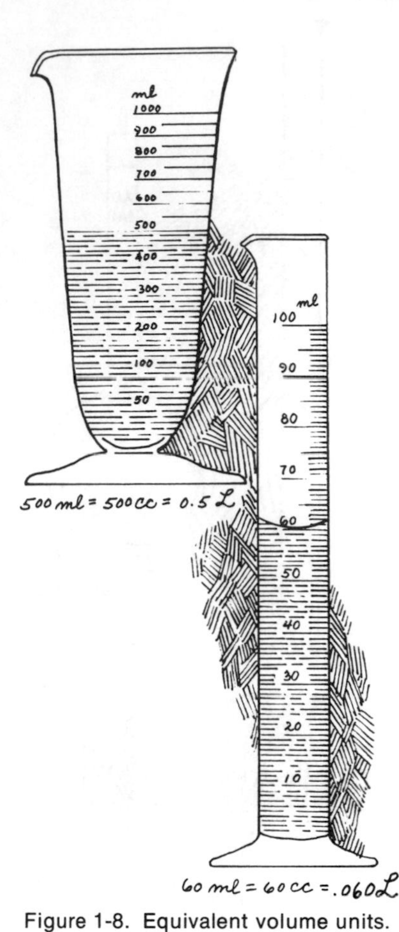

Figure 1-8. Equivalent volume units.

Another unit of volume used in the metric system is the cubic centimeter, which represents the volume of a cube which has a length, width, and height of 1 centimeter on each edge. One cubic centimeter (1 cc) is equal to 1 milliliter, therefore these units are interchangeable (Fig. 1-8).

250 cc = 250 ml
750 cc = 750 ml

Consequently, 1,000 ml which is equal to 1 liter can also be represented as 1000 cc = 1 liter.

ENGLISH SYSTEM OF MEASUREMENT

Length

In the English system length is measured by the following units (Fig. 1-9): inches (in); feet (ft); yards (yd).

The equalities are:	The fractions are:
12 in = 1 ft	$1 = \frac{12\text{ in}}{1\text{ ft}}$ or $\frac{1\text{ ft}}{12\text{ in}}$
3 ft = 1 yd	$1 = \frac{3\text{ ft}}{1\text{ yd}}$ or $\frac{1\text{ yd}}{3\text{ ft}}$

PROBLEM

A roll of adhesive tape is marked 108 in. How many ft does it contain? How many yards?

First equality: 12 in = 1 ft

Fraction desired: $\frac{1\text{ ft}}{12\text{ in}}$

Multiply and cancel: $\overset{9}{\cancel{108\text{ in}}} \times \frac{1\text{ ft}}{\cancel{12\text{ in}}} = 9\text{ ft}$ (answer)

Second equality: 3 ft = 1 yd

Fraction desired: $\frac{1\text{ yd}}{3\text{ ft}}$

Multiply and cancel: $\overset{3}{\cancel{9\text{ ft}}} \times \frac{1\text{ yd}}{\cancel{3\text{ ft}}} = 3\text{ yd}$ (answer)

Weight

Weight in the English system is measured in units called ounces (oz), pounds (lb), and tons.

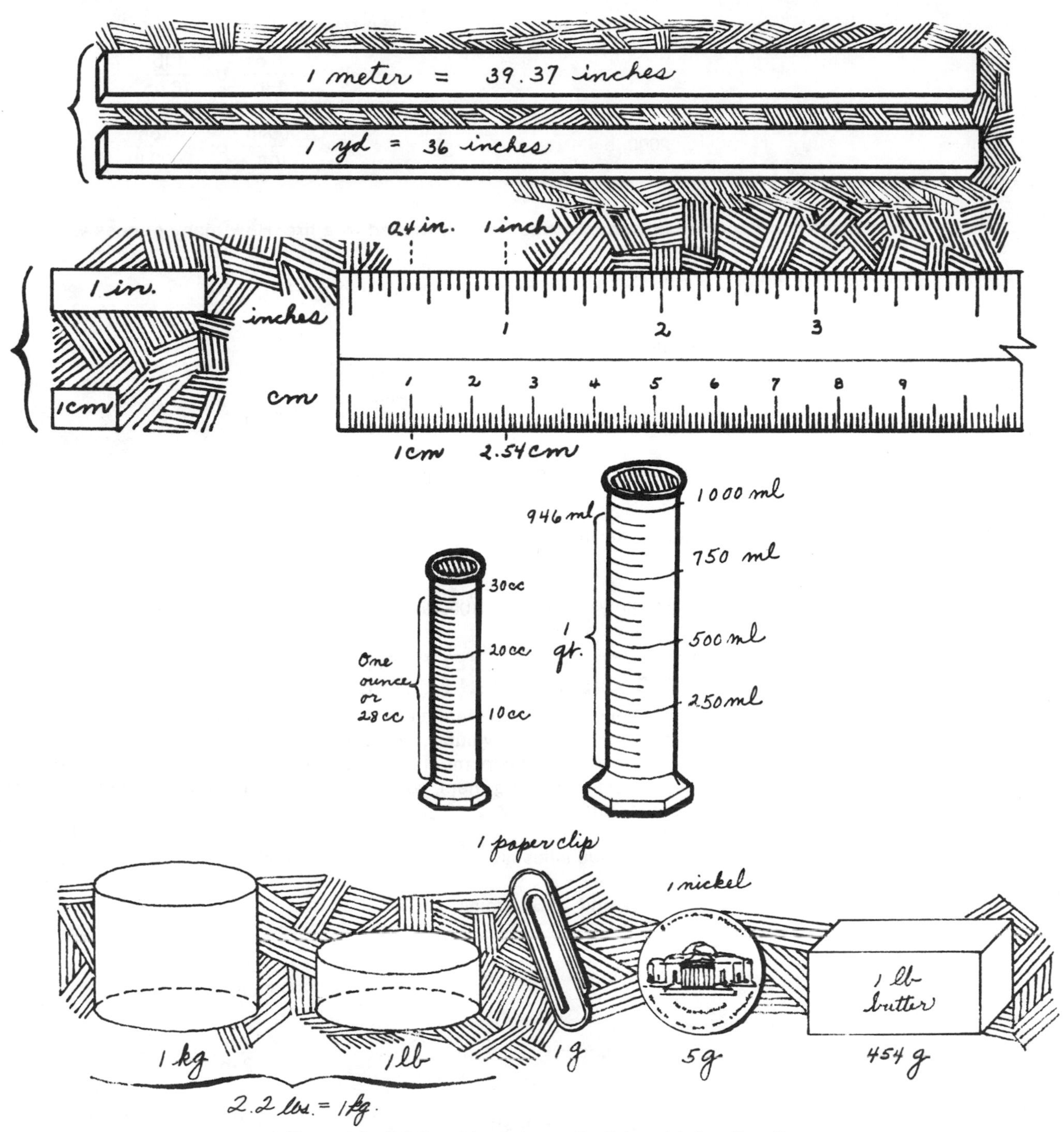

Figure 1-9. Relationships between English-metric length units.

The equalities are: The fractions are:

$$16 \text{ oz} = 1 \text{ lb} \qquad 1 = \frac{16 \text{ oz}}{1 \text{ lb}} \text{ or } \frac{1 \text{ lb}}{16 \text{ oz}}$$

$$2000 \text{ lb} = 1 \text{ ton} \qquad 1 = \frac{2000 \text{ lb}}{1 \text{ ton}} \text{ or } \frac{1 \text{ ton}}{2000 \text{ lb}}$$

PROBLEM

One-half ton of drugs were destroyed in a fire. How many ounces were destroyed?

First equality: 2000 lb = 1 ton

Fraction desired: $\frac{2000 \text{ lb}}{1 \text{ ton}}$

Multiply and cancel: $0.5 \cancel{\text{ton}} \times \frac{2000 \text{ lb}}{1 \cancel{\text{ton}}} = 1000 \text{ lb}$

Second equality: 16 oz = 1 lb

Fraction desired: $\frac{16 \text{ oz}}{1 \text{ lb}}$

Multiply and cancel: $1000 \cancel{\text{lb}} \times \frac{16 \text{ oz}}{1 \cancel{\text{lb}}} = 16{,}000 \text{ oz}$ (answer)

INTERCONVERSION BETWEEN THE METRIC AND ENGLISH SYSTEMS

Now that you understand the dimensional factor method it should be easy to convert any length, area, volume, or weight from one system to another. What you need to know at this point are the equalities which relate the English and metric systems. Knowing them, you merely select the appropriate fraction of the equality and multiply.

The equalities are:	The fractions are:
1 pound = 454 grams	$\frac{1 \text{ lb}}{454 \text{ g}}$ & $\frac{454 \text{ g}}{1 \text{ lb}}$
1 inch = 2.54 cm	$\frac{1 \text{ in}}{2.54 \text{ cm}}$ & $\frac{2.54 \text{ cm}}{1 \text{ in}}$
2.2 pounds = 1 kilogram	$\frac{1 \text{ kg}}{2.2 \text{ lb}}$ & $\frac{2.2 \text{ lb}}{1 \text{ kg}}$
946 ml = 1 quart	$\frac{1 \text{ qt}}{946 \text{ ml}}$ & $\frac{946 \text{ ml}}{1 \text{ qt}}$

The key to easy conversion is to change the measurement units according to the following table:

	in METRIC SYSTEM change to	in ENGLISH SYSTEM change to
Length	centimeters	inches
Weight	grams	pounds
Volume	milliliters	quarts

Once these units are derived, the equalities listed above are used as the appropriate fraction and multiplied.

Weight: English to Metric

PROBLEM

Eight ounces of a drug will yield how many 10 mg doses?

The oz can be changed to lb:

$$\cancel{8\ \text{oz}} \times \frac{1\ \text{lb}}{\underset{2}{\cancel{16\ \text{oz}}}} = 0.5\ \text{lb}$$

The lb is then converted to g:

$$\frac{1\ \cancel{\text{lb}}}{0.5} \times \frac{454\ \text{g}}{1\ \cancel{\text{lb}}} = 277\ \text{g}$$

The g are converted to mg:

$$277\ \cancel{\text{g}} \times \frac{1{,}000\ \text{mg}}{1\ \cancel{\text{g}}} = 27{,}7000\ \text{mg}$$

Multiplying the number by $\frac{1\ \text{dose}}{10\ \text{mg}}$ will give the answer:

$$\overset{2{,}770}{\cancel{27{,}700\ \text{mg}}} \div \frac{1\ \text{dose}}{\cancel{10\ \text{mg}}} = 2{,}770\ \text{doses (answer)}$$

Length: Metric to English

PROBLEM

There are 7.62 meters of tubing in a roll. How many feet does it contain?

First convert meters to centimeters:

$$7.62\ \cancel{\text{m}} \times \frac{100\ \text{cm}}{1\ \cancel{\text{m}}} = 762\ \text{cm}$$

Then change metric to English, cm to in:

$$762\ \cancel{\text{cm}} \times \frac{1\ \text{in}}{2.54\ \cancel{\text{cm}}} = 300\ \text{in}$$

Finally change in to ft:

$$\overset{25}{\cancel{300\ \text{in}}} \times \frac{1\ \text{ft}}{\cancel{12\ \text{in}}} = 25\ \text{ft (answer)}$$

SUMMARY

All matter in the world is composed of building blocks called elements, and chemistry is the study of how they behave under various conditions. Measurements can be made in the metric or English systems, and health care personnel must be able to interconvert measurement units between these systems. The measurements must be accurate and meaningful. One error commonly encountered in measurement is called parallax.

The dimensional factor method of converting measurements requires knowledge of the equalities between the various units and how to convert the equality into the needed fraction. When the proper fraction is used, the answer will always have the correct numerical value and unit.

EXERCISE

1. What are two kinds of errors found in scientific measurements? Describe them.
2. What is the effect of the position of your eye on the accuracy of reading the volume of liquid in a burette?
3. How accurately should a patient's body temperature reading be reported?
4. A patient is 152 cm tall. What is his height in ft and in?
5. A roll of adhesive tape is marked 300 in. Express the measurement in cm and in m.
6. A patient excreted 1500 ml of urine in one day. Express this in liters and in quarts.
7. A sample weighs 0.2 g. How many mg does it weigh?
8. How many micrograms are there in 0.5 g?
9. How many milligrams are in 0.5 lb?
10. A patient weighs 150 lb. Express his weight in kg.
11. A patient weighing 132 pounds is to be given a glucose tolerance test. At the rate of 1 g of glucose for each kg of body weight, how many ounces of glucose should be given? (Hint: equalities, 2.2 lb = 1 kg; 454 g = 1 lb ; 16 oz = 1 lb).
12. The measuring rule on the instrument reads 0.75 m. Express this value in mm, cm, in, and ft and in.

OBJECTIVES

When you have completed this chapter you will be able to:

1. List three major divisions of chemistry and the characteristics of each division.
2. Distinguish between mass and weight and explain how they behave in outer space.
3. Distinguish between the three states of matter and list the major differences between them.
4. Define a physical change, what state causes that change, and give an example of a substance which can exist in three different states.
5. Define a chemical change and how it differs from a physical change.
6. Write the formula for density calculations and give the units that each term is expressed in.
7. Calculate the density, weight, and volume of substances, given sufficient data.
8. Define specific gravity and explain why it has a numerical value but no units.
9. Explain why substances apparently weigh less when they are under water.
10. Describe a urinometer and state why it is used in the hospital.
11. Define the law of conservation of mass and give an example.

Matter

GOALS AND METHODS

To you, the beginning student in the health care field, the necessity of studying chemistry may not be obvious, as your interests are in the fields of biology and medical science. However, as time goes on, you will find that the dividing line between chemistry and biology has almost disappeared. In all of the many health care fields, which are rapidly increasing in number, an understanding of the basic principles of chemistry and of the chemistry of the living body will help you perform more efficiently and more effectively. It will also help you to understand more fully the functions and duties of your fellow professionals on the health care team. It will give you a new perspective on life and the interdependence of all people concerned with the goal of universal good health and physical well being.

Good health is the result of good body chemistry. Sickness is the result of a malfunction of the normal body chemistry. The objectives of the health care team are to alleviate pain and suffering, to counteract the effects of illness, and to work in aiding progress toward curing disease.

The subject of chemistry is vast. Professional chemists continue their studies throughout their careers merely to keep abreast of advances in their specialized fields. In this course, you will receive the fundamentals of chemistry, which will serve as a foundation for a better understanding of the chemistry of the living body.

The major divisions of chemistry are: general chemistry, the study of matter, its composition and transformations; organic chemistry, the study of carbon containing substances, their names, classifications, properties, uses, transformations, and their effect on the living body; and biochemistry, the chemistry of substances in the living body, the changes these substances undergo and the resulting effects on the living body.

MASS AND WEIGHT

You, your body, food, water, clothing, possessions, air, the things you handle such as cars, instruments or drugs are all *matter.* You can describe matter as *anything that takes up room* (that occupies space) and *has mass* (weight as we term it on earth). The terms *mass* and *weight* are used interchangeably by us, however, there *is* a difference between them. *Mass* is defined as the *amount of matter* in a substance or object. *Weight* is defined as the Earth's attraction for that matter. The force of attraction of the earth is what we call the force of gravity. An astronaut weighs 200 pounds on Earth, because that is the amount of the Earth's attraction on the mass of the astronaut's body. The same astronaut in space has no weight. He has the same mass, the same amount of matter as he had on Earth. On a spring balance he would not show any weight. This is why objects and astronauts in the spacecraft float around freely. The moon is smaller than the Earth. Its gravitational pull is about $^1/_6$ that of the Earth. This is why the astronauts seemed as though they were walking in water, a soft bouncy walk, when they walked on the moon; they weighed $^1/_6$ of their Earth weight (Fig. 2-1).

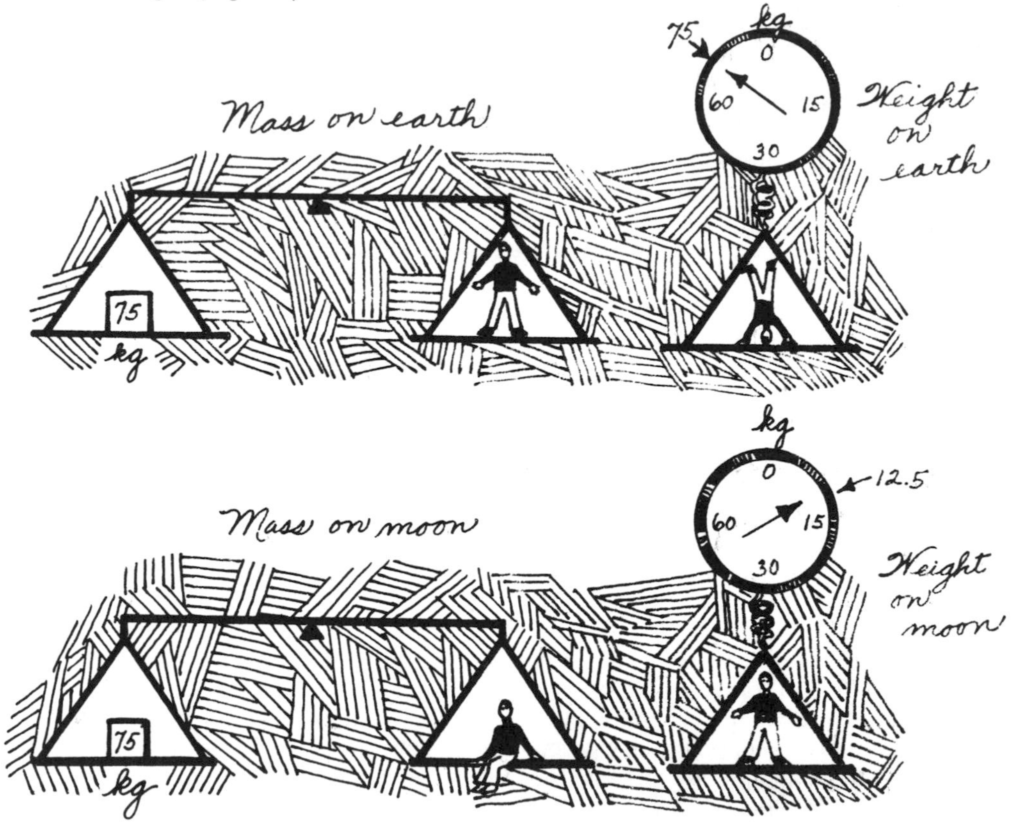

Figure 2-1. Comparison between mass and weight on the earth and moon. Mass on earth is equal to mass on the moon. Weight on earth is greater than the weight on the moon because of the earth's larger gravitational attraction.

STATES OF MATTER

All matter can be divided into three classes or states: solids, liquids and gases. Each of these have special characteristics which describe them.

Solids

Solids have a definite volume and shape (Fig. 2-2). They can be formed, bent, distorted, but they cannot be compressed into a smaller volume. *Solids cannot be compressed.*

Liquids

Liquids have a definite volume, but do not have a definite shape (Fig. 2-3). Liquids will adapt themselves to conform to the shape of the container, whether it be square, round, oblong, rectangular or irregular. Liquids may not fill the container, because they do have a definite volume. Furthermore, liquids have the same characteristic as solids; *they cannot easily be compressed into smaller volumes.*

Gases

Gases, on the other hand, do not have a definite volume or a definite shape. Gases can expand indefinitely to fill whatever space they are put into (Fig. 2-4). You can easily verify this by opening an odoriferous chemical bottle at one end of the room. Shortly, the gaseous vapor will have filled the room. And, just as gases can expand, *they can be compressed into smaller volumes.* You probably know this from seeing the steel cylinders of compressed oxygen in hospitals and from inflating your tires with compressed air when they are flat.

Physical Change

Matter can change from one state to another state and still remain the same material. *No new substance is produced:* the state may change, the color may change, and the change that occurs is usually caused by a *change in temperature.* Take water as an example. Water can freeze and become ice, remelt to become liquid water, evaporate to become water vapor or steam, or condense back to liquid water (Fig. 2-5). Yet water remains water. There has been no change whatsoever in its composition. All that happened was that it changed its physical state. *Physical change is a change where a substance is unaltered in its composition; it is a reversible change.* Many substances can exist in all three states, while others exist in two.

Chemical Change

When a chemical change occurs, such as the smoking of a cigarette, the burning of gasoline, the cooking of food, or the rusting of iron, substances are *completely changed into new and different substances* (Fig. 2-6). The tobacco and cigarette paper change to smoke and ashes. Gasoline burns with oxygen to form carbon monoxide, carbon dioxide and water. Pure iron combines with the oxygen of the air to form iron oxide (rust). The characteristics of the reacting substances are completely changed. For example, an iron nail is shiny, metallic, strong, durable, malleable, and is attracted by a magnet. On the other hand, iron oxide is a red powder, crumbles, has no strength, and is not attracted to

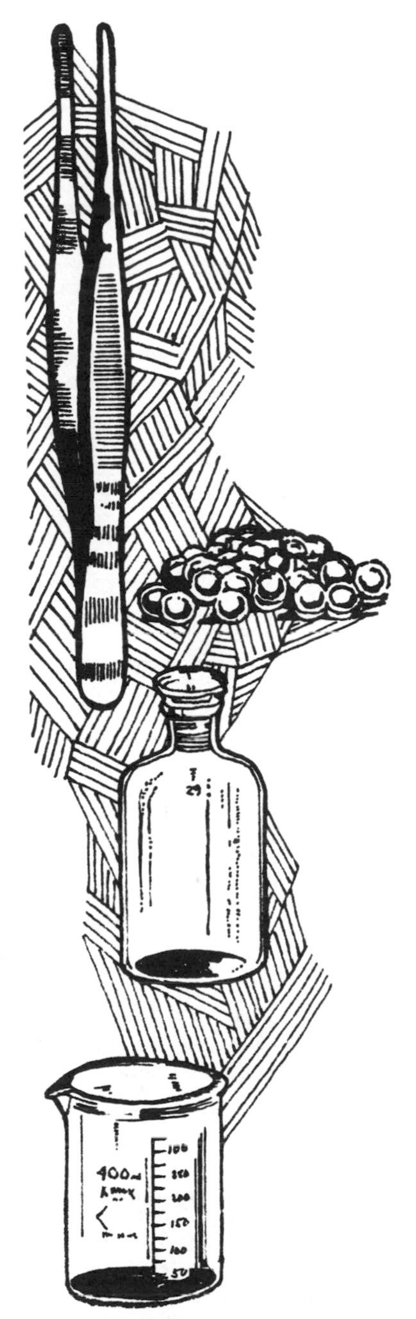

Figure 2-2. Solids have a definite volume and shape.

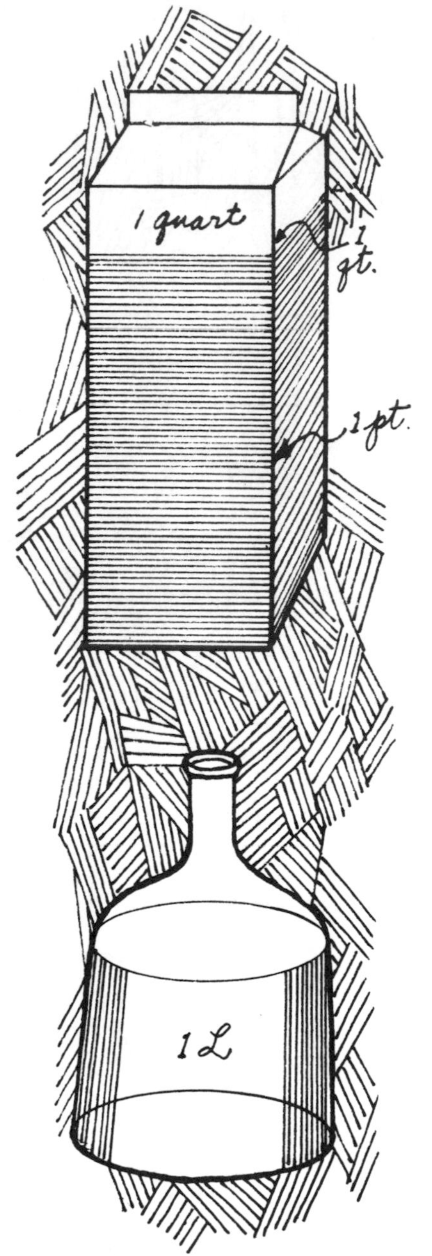

Figure 2-3. Liquids have a definite volume, but they assume the shape of the container.

Figure 2-4. Gases have neither shape nor definite volume. They can expand or be compressed.

Figure 2-5. Physical change. The form in which water exists depends upon its temperature.

a magnet. The iron has undergone a chemical change, and, although it was not noticeable, heat energy was given off. In fact, in either physical or chemical changes, heat energy is either absorbed or is given off. In chemical change you cannot unburn the cigarette or gasoline nor uncook food by cooling, whereas in physical change, you can resolidify butter or change liquid water to ice by cooling.

Figure 2-6. Chemical changes.

Density

One of the most important physical characteristics of matter is density, which relates the weight of a substance to the volume it occupies. From your everyday experience you would say that aluminum is lighter than lead or that solid gold jewelry is heavier than an identical piece of costume jewelry. There *is* a difference in the weights of these substances, but what you really are saying is that the *same volume* of gold is heavier than the same volume of copper, and that the same volume of aluminum is lighter than the same volume of lead. Each substance has its own characteristic density which is related to two measurements: the weight and the volume of that substance. Density is expressed in the terms grams per cubic centimeter, g/cm^3, and the relationship, therefore, is:

$$\text{Density (D)} = \frac{\text{Weight (W)}}{\text{Volume (V)}} = \frac{\text{grams}}{\text{cubic centimeters}} = \frac{\text{g}}{\text{cm}^3}$$

Dividing the weight in grams by the volume in cubic centimeters yields the density in g/cm^3 or g/ml. A cubic centimeter is a cube which is 1 cm long by 1 cm wide by 1 cm high. Its volume is equal to its length times its width times its height:

$$V = 1\text{ cm} \times 1\text{ cm} \times 1\text{ cm} = 1\text{ cm}^3$$

Also

$$1\text{ cm}^3 = 1\text{ cc} = 1\text{ ml in volume}$$

In the health care field you will be working with all three states of matter, and the substances will range in density from about one ten-thousandth of a gram per cubic centimeter (hydrogen gas) to about 19 g/cm^3 (gold). Water has a density of 1 g/cm^3. Ether (diethyl ether, an anesthetic) has a density of 0.7 g/cm^3, while carbon tetrachloride is heavier than water, having a density of 1.6 g/cm^3. Physical therapists would use, where applicable, aluminum appliances rather than iron, because the weight of the iron appliance would be almost three times as heavy. The following table lists some of the densities of substances that health care personnel may come in contact with, expressed to two decimal places.

Table 2-1 Densities in g/cm^3.

Substance	*Density*
Blood plasma	1.03
Aluminum	2.70
Iron	7.90
Lead	11.34
Gold	19.30
Copper	8.92
Alcohol	0.79
Ethyl Ether	0.71
Carbon Tetrachloride	1.59
Mercury	13.55
Gasoline	0.70

CALCULATIONS INVOLVING DENSITY:

There are three terms involved in the formula for density calculations, and if you know any two of them, you can calculate the third (Fig. 2-7). As you have learned, the density formula is:

$$\text{Density} = \frac{\text{Weight}}{\text{Volume}}$$

NOTE: The weight must be expressed in grams and the volume in cubic centimeters for the density value to be in the term g/cm^3.

PROBLEM:

A piece of metal weighs 79 grams and its volume is 10 cm^3. What is the density and what is the metal?

$$\text{Weight} = 79 \text{ g}$$
$$\text{Volume} = 10 \text{ cm}^3$$

Substitute in formula and solve:

$$D = \frac{79 \text{ g}}{10 \text{ cm}^3} = 7.9 \text{ g/cm}^3 \text{ (answer).}$$

Examine Table 2-1: the metal is iron.

PROBLEM

A 50 cubic centimeter beaker is filled with mercury. Could you weigh it on a balance which had a *maximum capacity of 500 g*? (Neglect the weight of the beaker).

$$\text{Density of mercury} = 13.55 \text{ g/cm}^3$$
$$\text{Volume} = 50 \text{ cm}^3$$

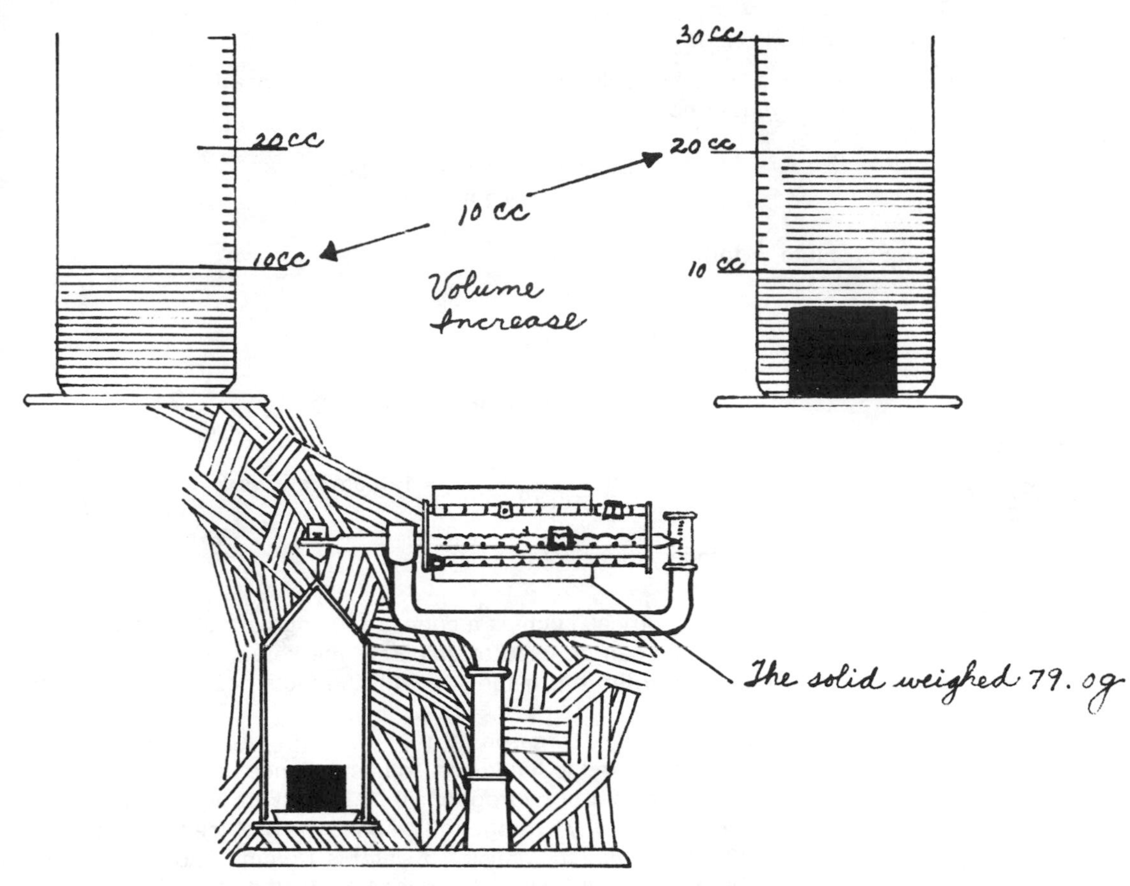

Figure 2-7. Density determination. The solid displaced 10 cc of liquid, therefore its volume is 10 cc. The density of the solid $= \frac{\text{weight}}{\text{volume}} = \frac{79.0 \text{ g}}{10 \text{ cc}} = 7.90$ g/cc.

Substitute in the formula and solve:

$$13.55 \text{ g/cm}^3 = \frac{\text{Weight}}{50 \text{ cm}^3}$$

Solve for weight:

$$\text{Weight} \times \frac{13.55 \text{ g}}{\cancel{\text{cm}^3}} \times 50 \cancel{\text{cm}^3} = 677.50 \text{ g (answer)}$$

When you multiply, the terms "cm^3" cancel out, leaving "grams" as the unit of the answer.

No. The weight of the mercury would be too heavy for the balance.

PROBLEM

You need 94 g of ethyl alcohol (alcohol) to prepare a medication. Can a *100 ml graduated* cylinder hold this amount?

$$\text{Density alcohol} = 0.79\ \text{g/cm}^3$$
$$\text{Weight} = 94\ \text{g}$$

Substitute in formula and solve:

$$0.79\ \text{g/cm}^3 = \frac{94\ \text{g}}{\text{Volume}}$$

$$\text{or } 0.79\ \text{g/cm}^3 \times \text{Volume} = 94\ \text{g}$$

$$\text{and Volume} = \frac{94\ \cancel{\text{g}}}{0.79\ \cancel{\text{g}}/\text{cm}^3} = 120\ \text{cm}^3\ \text{(answer).}$$

No. The graduated cylinder is too small.

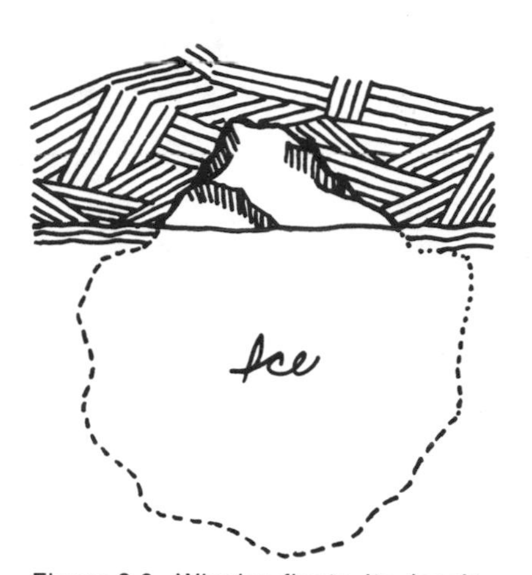

Figure 2-8. Why ice floats. Its density is less than that of water.

If the object is symmetrical, such as a cube, a rectangular solid, or a sphere, mathematical formulas can be used to calculate the volume, and the weight can be determined in order to provide you with the facts needed to calculate the density.

Gas densities are usually expressed in grams/*liter* (L) because they are extremely small when compared to liquid and solid substance densities. The density of air is 1.3 g/L, and that of hydrogen is 0.09 g/L. Hydrogen has a smaller density than air, and that is why hydrogen-filled balloons rise. Ice floats because its density is 0.92 g/cm^3, while the density of water is 1 g/cm^3. About 90 percent of ice is submerged, leaving 10 percent exposed above the water (Fig. 2-8).

Specific Gravity

In the health care field you will encounter the term specific gravity more often than the term density. Actually *specific gravity has exactly the same numerical value as density, but it has no units.* It is only a number. This is because the specific gravity is the *ratio (the comparison) of the weight of a substance to the weight of an equal volume of water:* this is actually the ratio of the densities.

$$\text{Specific gravity} = \frac{\text{Density of the substance}}{\text{Density of water}}$$

In this formula, where the density of the substance, expressed in g/cm^3 is divided by the density of water, which is 1 g/cm^3, the units cancel out, leaving the number value only, which is the specific gravity.

$$\text{Specific gravity of alcohol} = \frac{0.79\ \cancel{\text{g/cm}^3}}{1\ \cancel{\text{g/cm}^3}} = 0.79$$

Therefore, you can see, by referring back to Table 2-1, that the specific gravity of aluminum is 2.70, and that of iron is 7.90. The specific gravity tells you *how many times heavier* or *how many times lighter a volume of a substance is* than *the weight of an equal volume of water.* This is actually a comparison of densities, because the density is the weight of 1 cm^3 of a substance.

Bouyancy

Have you ever wondered why some things float in water, while others sink? Have you ever pulled someone out of a swimming pool and found that as they emerged from the water they seemed to get heavier? This phenomenon is due to buoyancy, which is the upward force exerted by a liquid or a gas upon a body which is within or upon it. When that upward force is greater than the weight of the object, the object floats. That upward force depends upon the density of the liquid, and we use this principle to determine specific gravity.

Floating objects will displace a volume of water which has the *same weight as the object.* This is the principle of the hydrometer, a floating tubular instrument, whose indicator scale will give the specific gravity reading of a liquid. In pure water, the hydrometer reads 1.00, in alcohol, 0.79, and in carbon tetrachloride, 1.59.

In the hospital specific gravity readings are made on body fluids, such as urine, because the readings are an important diagnostic tool. The specific gravity of normal urine should range between 1.005 and 1.025. Should the specific gravity of a urine sample read 1.001, this would mean that the patient is excreting almost pure water, and that the kidneys are not functioning properly to concentrate the urine. The urinometer is simply a hydrometer which is calibrated to determine the specific gravity of urine specimens (Fig. 2-9).

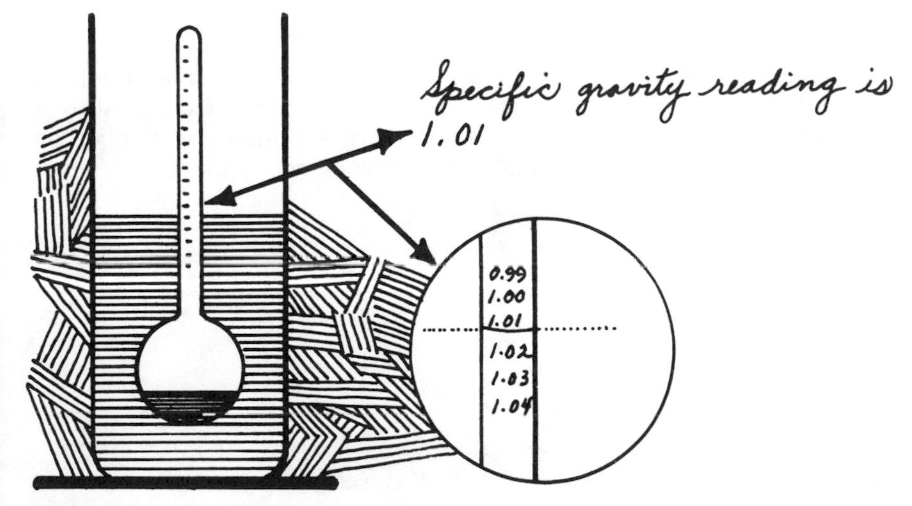

Figure 2-9. A urinometer is a hydrometer used to determine the specific gravity of urine specimens.

CONSERVATION OF MASS

Scientists have shown that matter cannot be created or destroyed in ordinary chemical reactions. What this means is that the weight of all of the reacting substances is equal to the weight of all the products formed in the reaction. If a flash cube is weighed before and after being used, the weights will be the same. Of course, the flash cube cannot be reused because the silvery threads inside the cube have now changed to a white powder. A chemical change has taken place, and the new substance formed has properties which are entirely different from the original substances. The form and composition has been changed, but no matter was created or destroyed. Pure carbon is a black, solid material and will burn completely in the presence of oxygen, a colorless gas, to form carbon dioxide, a colorless gas. This is the identical gas which bubbles out of soda water, carbonated beverages, beer and champagne. Yet, when the carbon was burned, that weight when added to the weight of oxygen consumed, would equal the weight of the carbon dioxide formed. The law of conservation of mass holds true in all chemical reactions except radioactivity.

SUMMARY

Mass is the amount of matter in a substance and weight is the gravitational attraction on that matter. The greater the attractive force, the greater the weight. Matter can be classified into solids, liquids, or gases, and each has specific characteristics. Solids have a definite volume and shape, liquids have a definite volume, and gases have no volume or shape, and can be compressed or expand indefinitely. Matter can change from one state to another and still remain the same identical material, this being a physical change caused by temperature. When new and different substances are formed, the change is a chemical one, where characteristics of the reacting substance are completely changed. An important characteristic of matter is density, the weight of one cubic centimeter (cm^3), and the formula is: $\text{Density} = \frac{\text{Weight}}{\text{Volume}}$. Another important term, specific gravity, has the same numerical value as density, but has no unit label; it is an important diagnostic tool in the hospital. Matter cannot be created or destroyed in normal chemical reactions, except radioactivity, and the mass of all of the reacting matter equals the mass of all of the products formed.

EXERCISE

1. Name the three major divisions of chemistry, and briefly state their scope.
2. Distinguish the difference between mass and weight.
3. Why does the same astronaut weigh less on the moon than on the earth?
4. Give the three states of matter, the characteristics which distinguish them from each other, and two examples of each.
5. Which of the following substances can be compressed into smaller volumes: water, air, oxygen gas, iron, oil, glass, stainless steel?
6. Is there such a thing as compressed water? Why?
7. What is a physical change and what can cause it?
8. Give 3 examples of a physical change.
9. Distinguish between a physical change and a chemical change.
10. Can a chemical change be reversed by lowering the temperature?
11. State four characteristics of metallic iron and four characteristics of iron oxide that prove to you a chemical and not a physical reaction took place.
12. Define density, and state the formula.
13. Give the units that the three components of the density formula are expressed in.
14. Milk has a density of 1.03 g/cm^3. How much will 1 liter weigh?
15. The density of blood plasma is 1.027 g/cm^3. What volume would 500 g occupy?
16. One hundred cubic centimeters of an unknown liquid weighs 89 g. What is the density?
17. Define specific gravity and explain why it is only a number without units.
18. How does a hydrometer work? Give an example of how it is used in diagnostic procedures in the hospital.
19. Explain the law of conservation of mass and state the exception to the law.

OBJECTIVES

When you have completed this chapter you will be able to:

1. Define the term energy.
2. List five types of energy and give examples of each.
3. State the law of conservation of energy and give the exception to the law.
4. Illustrate the transformations of energy, beginning with the generation of electrical energy from various sources.
5. Explain what the term temperature means.
6. Describe how a thermometer works.
7. Distinguish between the three temperature measurement scales.
8. Interconvert temperature readings between Celsius, Fahrenheit, and Kelvin scales.
9. Distinguish between the intensity and the quantity of heat energy.
10. Define the term calorie and Calorie.
11. Draw a calorimeter and describe what it is used for.
12. Define specific heat and make calculations using the formula.
13. Define heat of vaporization and summarize why steam burns may be severe.

ENERGY

Energy is something that can produce changes in matter; it is the ability to do work. Heat energy changes liquid water to steam. Light energy fades clothing or forms an image on photographic film. Electrical energy runs motors to do work, silver-plates dinnerware, lights electric bulbs, and transmits voices and pictures from space by television. Chemical energy can heat homes by the combustion (burning) of coal, wood, oil, or natural gas and can move cars by burning gasoline. The food that you eat contains the chemical energy which is needed by the body to move about and perform work. Two relatively new sources of energy are nuclear energy, which can power electrical generating plants, ships, and pacemakers, and radiant energy (solar energy), that energy given off by the sun.

LAW OF CONSERVATION OF ENERGY

Energy, like matter, cannot be created or destroyed in ordinary chemical reactions, except in radioactive chemical reactions. However, energy can be transformed from one form to another. Chemical energy contained in foods is stored in the body and transformed into mechanical energy to move the body. Chemical energy stored in coal or natural gas is changed to heat energy to drive the boilers at the electrical generating plant which, in turn, is changed to electricity (electrical energy). The electricity is transmitted to homes, where again it is changed to light energy, heat energy, mechanical energy, and chemical energy (Fig. 3-1). Some appliances do not change energy to another form efficiently. An electric bulb is not very efficient, because all of the electrical energy is not changed to light energy; some is lost as heat energy (Fig. 3-2). The body cells are very efficient and change most of the chemical energy to mechanical energy. In every instance the *sum* of the input and output energies are equal because of the law of the conservation of energy: energy is neither created nor destroyed in ordinary chemical reactions.

You have already learned that the physical state of a substance can be changed by raising or lowering the temperature. What really happens is that

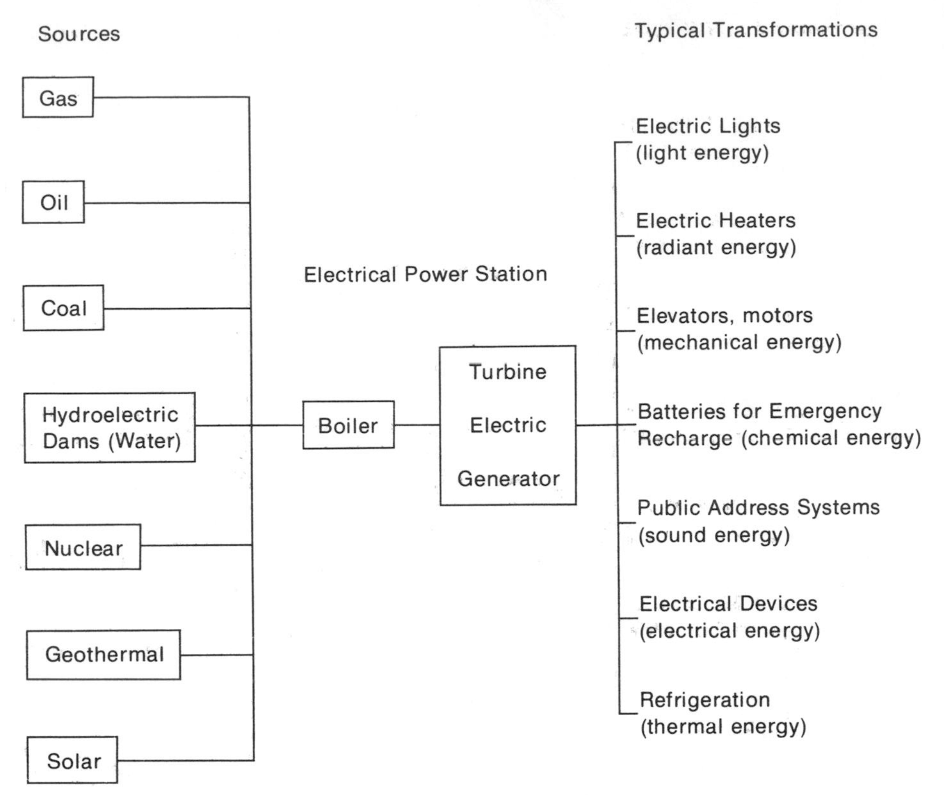

Figure 3-1. Transformations of energy.

Figure 3-2. Efficiency in energy transformation. An incandescent light bulb gives off a great deal of heat energy, and is not very efficient. A fluorescent tube is cool, more of the electrical energy is converted to light. It is more efficient than the incandescent bulb.

heat energy is either added to or removed from the substance. For example, you have already seen that heat energy was given off when gasoline burned. As far as your study is concerned, most of the reactions that you will observe will be those in which heat energy is generated.

TEMPERATURE

The temperature of the body is an important indicator. When a patient has fever, you can measure the intensity of the heat of the body with a thermometer, and follow the patient's progress toward recovery by regular observations of the temperature (Fig. 3-3). Temperature plays an important part in the behavior of solids, liquids, and gases. It is *not* heat energy, but it is a measure of the *intensity* of heat energy. When a body absorbs heat energy, the temperature rises. When a body loses heat energy, the temperature falls.

A thermometer is a commonly used temperature measuring device, although today the trend is to use electronic indicators. In the normal glass thermometer, a thin glass bulb that is attached to a thick-walled glass capillary tube is filled with mercury or alcohol (Fig. 3-4). When heated, liquids expand, and the mercury or alcohol rises in the calibrated capillary tube. The height to which the liquid rises is a measure of the temperature of the patient.

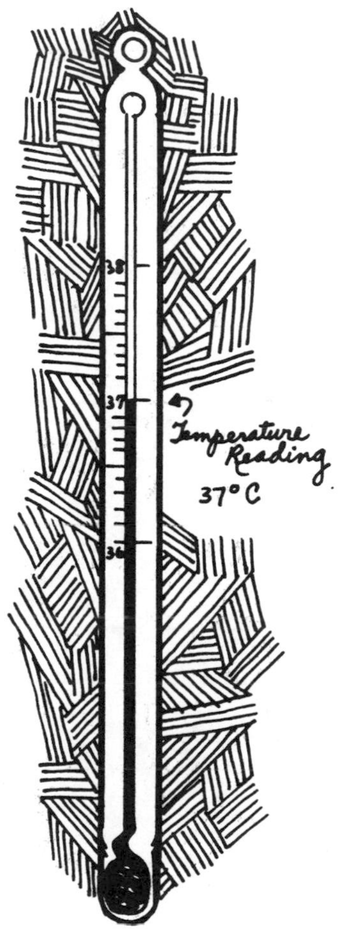

Figure 3-4. A thermometer.

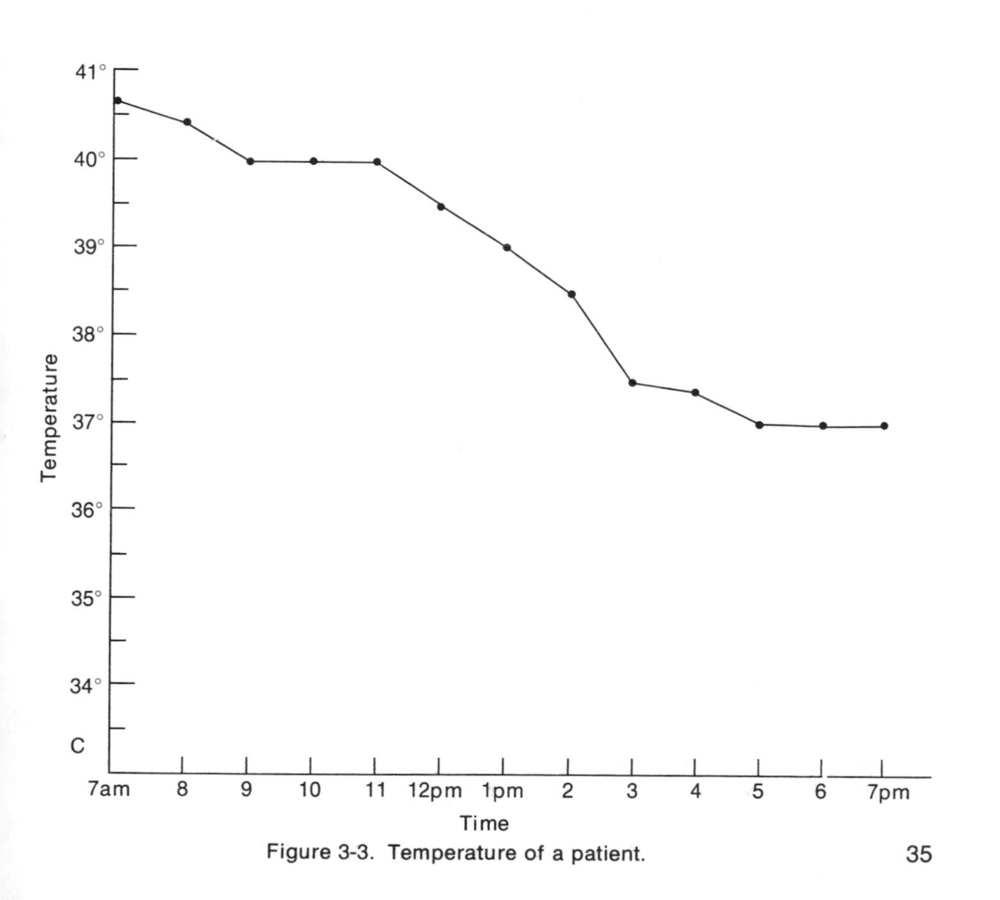

Figure 3-3. Temperature of a patient.

CAUTION: Never immerse a clinical thermometer in very hot water. The expansion of the liquid will force the liquid to the top of the capillary and break the thermometer.

TEMPERATURE MEASUREMENT SCALES

Temperature is measured according to three scales: Celsius (Centigrade), Fahrenheit, and Kelvin (Absolute). Previously, only the scientific community used the Celsius scale, while the medical community used Fahrenheit. Today there is a trend toward the use of the Celsius scale everywhere, except where calculations involved with gases, as in inhalation therapy, are made. In the case of these calculations, natural scientific laws require the use of the Kelvin scale.

Just as you are now able to interconvert metric and English system measurement units, you also can easily interconvert these three scales (Fig. 3-5).

NOTE: The degrees Fahrenheit are recorded by °F.
The degrees Celsius (Centigrade) by °C.
The degrees Kelvin (or Absolute) by °A or °K

PROBLEM

The patient's temperature is 98.6°F. What is the temperature in °C?

Formula: 5/9 (°F −32)

Procedure:	
Write the °F:	Temperature = 98.6°F
Subtract 32 from °F:	98.6 − 32 = 66.6
Multiply result by 5/9:	5/9 × 66.6 = 37
	or you can cancel,
	5 × 7.4 = 37
You have the answer:	37 °C

PROBLEM

The room temperature is 20°C. What is the temperature in °F?

Formula: 9/5 C + 32

Procedure:	
Write the °C temperature:	T = 20°C
Multiply by $\frac{9}{5}$:	$\frac{9}{5} \times 20 = \frac{180}{5} = 36$
Add 32:	36 + 32 = 68
You have the answer:	68°F

PROBLEM

What is the boiling point of water in °K if it is 100°C?

Formula: °C + 273 = °K

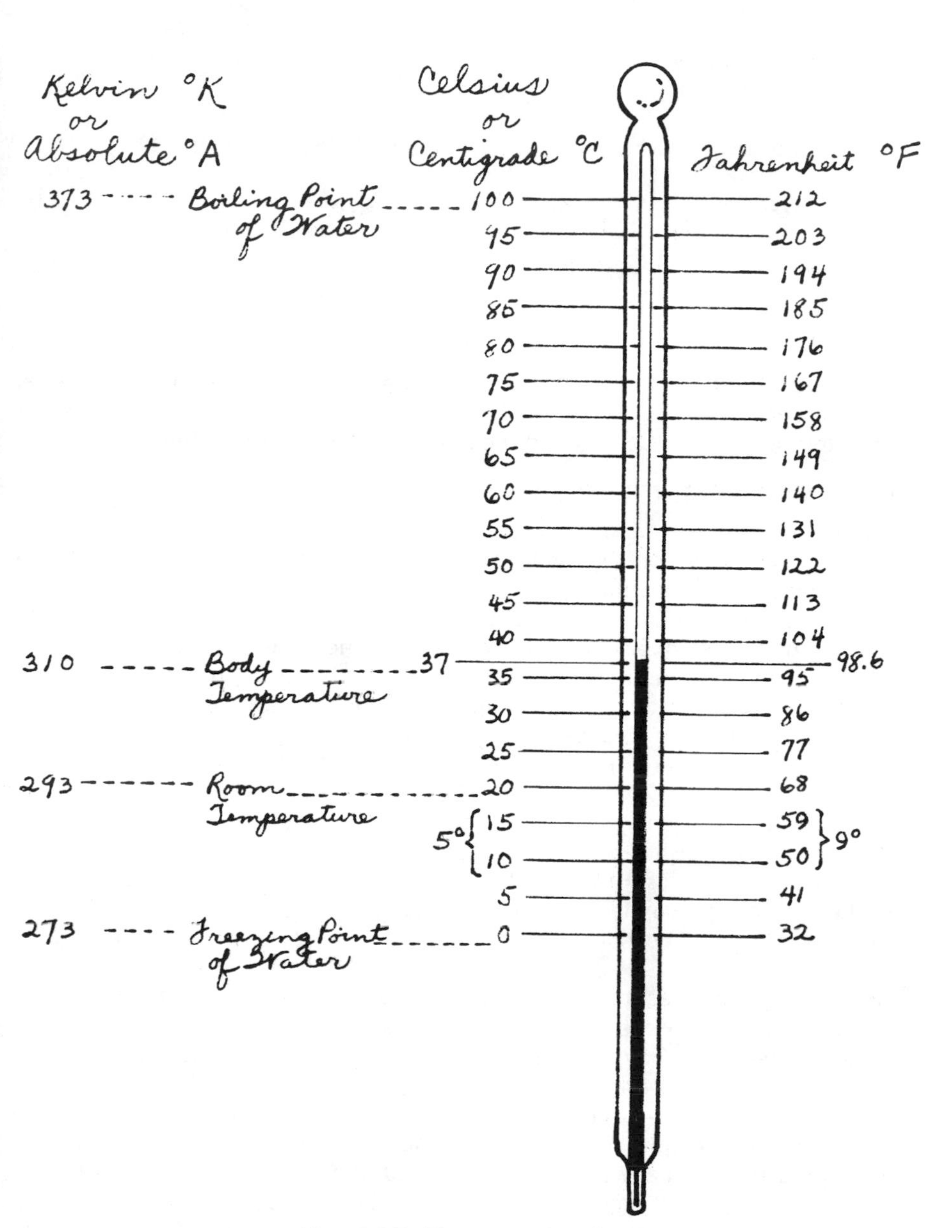

Figure 3-5. Temperature scales.

Procedure:

Write the Celsius temperature:	T = 100°C
Add 273 to °C:	100 + 273 = 373
You have the answer:	373°K

PROBLEM

The freezing point of water is 273°K. What is the temperature in °C?

Formula: °C = °K − 273

Write the Kelvin temperature:	273°K
Subtract 272 from °K:	273 − 273 = 0
You have the answer:	0°C

To change Fahrenheit to Kelvin, change the Fahrenheit temperature to Celsius, and then convert Celsius to Kelvin. To change Kelvin to Fahrenheit, change the Kelvin temperature to Celsius and then convert this to Fahrenheit.

HEAT AND TEMPERATURE

As you know, the thermometer measures the intensity of heat, not the amount of heat. It requires more heat energy to heat a large quantity of a substance than to heat a small quantity of the same substance. In fact, the amount of heat energy required to heat any one substance depends on the amount and the mass of that substance. For example, to raise the temperature of 1,000 g of water 1°C requires ten times the heat energy to raise 100 g of water 1°C. Even though the thermometer of each reads the same, there is ten times the heat energy contained in the 1,000 g of water than in the 100 g. Therefore, we can reason that the heat energy content depends upon how much of the substance is involved.

We measure heat energy by the calorie, which is defined as the amount of heat energy required to raise the temperature of 1 g of water 1°C. This term is written with a small "c." However, this quantity of heat is so small that scientists and health care personnel use the large Calorie, written with a large "C," as the standard unit of measure. This is actually a kilocalorie, and is equal to 1,000 calories.

Heat energy is measured in Calories. The body requires energy in order to function, but when we eat more food than is required to supply the energy requirements of the body, that energy is usually stored as fat, the technical term being adipose tissue. Should a diet contain fewer Calories than the body requirements, the stored fat is used to supply body energy requirements, and the body then loses weight. Today, with the current emphasis on reducing diets and caloric intake, many people are conscious of their diets. Therefore, considering normal body chemistry, the safe and sure way to lose weight is to restrict your caloric intake while still maintaining a balanced diet (Fig. 3-6). You will see later in the book what a balanced diet consists of, but in all cases dieting should be done in conjunction with medical supervision.

The caloric value of substances is determined in an instrument called a calorimeter (Fig. 3-7), which actually burns a weighed amount of the substance in

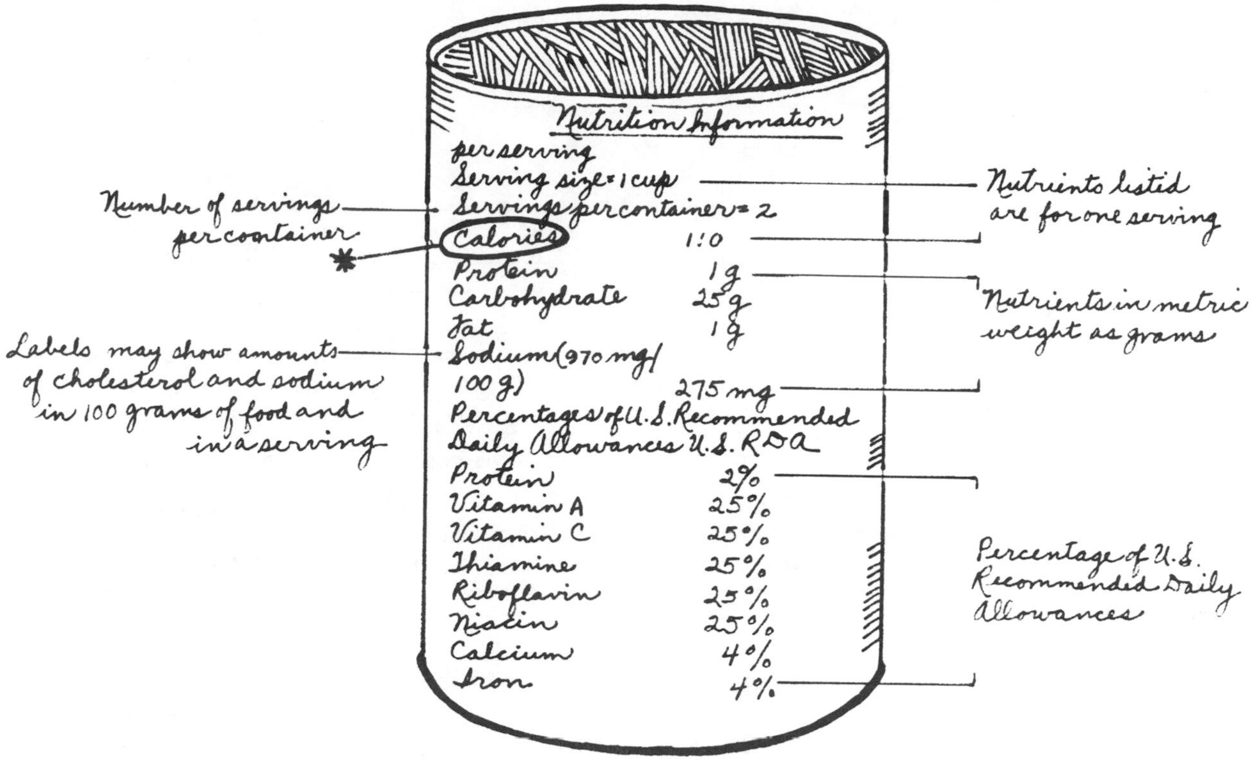

Figure 3-6. Nutritional data label.

a closed container. The heat generated raises the temperature of the water in the calorimeter, the rise being dependent upon the calories of heat evolved. The same number of calories are released when the substance is burned in the calorimeter or in the body, so nutritionists can actually measure the caloric value of foods. One slice of bread contains about 100 Calories (100,000 calories) and this is enough energy to raise the temperature of 1,000 g of water from 0°C to 100°C. So you can see that it takes only a little cheating on your diet to acquire excess adipose tissue.

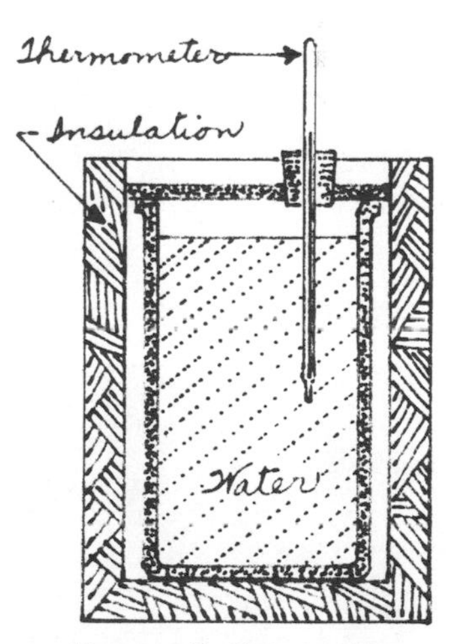

Figure 3-7. Calorimeter.

SPECIFIC HEAT

Some substances hold more heat than other substances, and thus it takes longer to raise their temperatures than identical objects made of other substances. Water requires the input of more heat energy than any other substance, and because it can contain more heat energy, it takes longer for water to cool than any other substance. This is why geographical areas surrounded by water have a more even temperature than land-locked areas, water being the natural air conditioner. It is also the reason why hot water bags are used in health care; they retain heat for a long time.

Water is the standard by which the temperature of other substances are meas-

Table 3-1. Specific heat.

Substance	*Specific heat in cal/g/°C*
Water	1.00
Aluminum	0.22
Copper	0.09
Gold	0.03
Alcohol	0.65
Chloroform	0.23
Lead	0.03
Iron	0.11

ured. It requires 1 calorie of heat to raise the temperature of 1 g of water 1°C. Table 3-1 lists the specific heats of various substances. You can see that copper is preferred to aluminum cookware, because the low specific heat of copper (about $\frac{1}{2}$ that of aluminum) allows the copperware to heat up in half the time. Food products have different specific heats and therefore require the removal of more heat energy in order to refrigerate or cool them. The greater the water content, the longer it takes to cool or to freeze. Dried foodstuffs cool very quickly due to their low water content.

Calculations Involving Specific Heat

The number of calories which are required to heat a substance to a certain temperature depends upon these factors:

1. The weight of the substance.
2. The specific heat of the substance.
3. The temperature rise, that is the difference between the initial temperature and the final temperature in degrees Celsius.

The formula for calculating the number of calories required is:

Number of calories = (weight in g) (specific heat) (temperature rise)

There are four items: calories, weight, specific heat, and number of degrees change. If you know any three, you can calculate the fourth by substituting in the formula.

PROBLEM

(1) How many calories are required to raise the temperature of 100 g of water from 50°C to 60°C?
(2) How many for 1000 g of water?

Calories = weight × specific heat × change in temperature
Calories = 100 × 1 × 10 = 1,000 calories
Calories = 1,000 × 1 × 10 = 10,000 calories

PROBLEM

Which would provide more calories of heat: 500 g of water or 500 g of aluminum, if they both were at 90°C and both cooled to 80°C?

Facts: The weight of each substance = 500 g
The specific heat of water = 1 cal/g/°C
The specific heat of aluminum = 0.22 cal/g/°C
The temperature drop for each = 10°C

Substituting in formula:
For water: calories = $500 \times 1 \times 10 = 5000$ calories
For aluminum: calories = $500 \times 0.22 \times 10 = 1100$ calories

Answer: The water has a higher specific heat, therefore it provided almost 5 times the number of calories as the aluminum.

HEAT OF VAPORIZATION

You will be able to apply your knowledge of this characteristic of liquids in many ways. From a safety standpoint, you will see why steam causes such severe burns, and from health care, why an alcohol rub is more effective than a water rub.

It requires heat energy to vaporize liquids to gases, and the quantity of heat energy to vaporize 1 g of a substance in calories is called the heat of vaporization. Naturally, when the same quantity of vapor condenses (changes back to liquid) that heat energy must be given off. The calories absorbed to vaporize must be given off to condense the substance. By looking at Table 3-2 you can easily see the reason why certain substances evaporate more rapidly than others: they have a lower heat of vaporization.

Use extreme care when working around steam in the laboratory, with sterilizers, and in inhalation therapy. The very large amount of heat evolved when steam condenses on the skin is the reason why steam burns may be quite severe. Use protective grease generously when exposed to steam vapors for your patient and yourself.

Table 3-2. Heat of vaporization of substances (at the boiling point).

Substance	*g-cal/g*
Acetone	124
Carbon tetrachloride	46
Chloroform	59
Ethanol	204
Ethyl acetate (nail polish solvent)	88
Ethyl ether	83
Water	540

SUMMARY

Energy is defined as the capacity to do work, and is found as heat, chemical, light, radiant, electrical, nuclear and solar energy. It can be changed from one form to another, and it cannot be created or destroyed. The *intensity* of heat energy is measured by the temperature on 3 scales: Celsius, Fahrenheit, and Kelvin. The readings on the 3 scales can be interconverted. The *quantity* of heat energy is measured by the calorie, the amount of heat required to raise the temperature of 1 g of water 1°C. The kilocalorie or *Calorie* equals 1,000 calories and is the measure used for determining the caloric content of food. Some substances require fewer calories than others to register a rise in temperature, and the number of calories required to raise the temperature of one gram of water by 1°C is called specific heat. Water has the highest specific heat, 1 calorie per gram per °C. This is why it takes longer to heat water than other substances and why water holds its heat longer than other substances. To change from a liquid to a vapor state requires heat energy, which is called the heat of vaporization. Each substance has its specific value, water having 540 calories per gram. When steam vapor condenses heat energy is released; this is why steam burns may be more severe than hot water burns.

EXERCISE

1. Define the term energy.
2. Name four kinds of energy and give two examples of each.
3. State the law of the conservation of energy and explain what it means.
4. Explain what is meant by transformation of energy and give an example.
5. What is the exception to the law of the conservation of energy?
6. Are all energy transformations totally efficient? If not, give an example where all available energy is not transformed as desired.
7. Define temperature.
8. What common device is used to measure temperature?
9. What basic principle does the thermometer work upon?
10. Name the three temperature scales normally used.
11. In the hospital the recommended temperature is 68°F. What is the corresponding Celsius temperature?
12. Normal body temperature is 98.6°F. What is the corresponding Celsius temperature?
13. A patient's temperature is 40.0°C. What is his temperature in Fahrenheit??
14. Chloroform boils at 61.0°C. What is the temperature on the Fahrenheit scale?
15. Change the following temperatures as indicated:
    a. −50°C to °K
    b. −40°C to °F
    c. 100°C to °K
    d. 273°K to °C; to °F
    e. 212°F to °K
16. Define a calorie and a Calorie. What is the relationship between the two?
17. How is heat content measured?
18. What is adipose tissue?
19. What instrument is used by technicians to measure the caloric content of foods? What is the basic principle of the instrument?
20. Define specific heat.
21. What substance has the highest specific heat and what is its value?
22. How many calories of heat are required to raise the temperature of 250 g of the following substances from 20°C to 50°C?
    a. water   b. copper   c. aluminum
23. Define heat of vaporization.
24. Which would give a more severe burn, condensing steam or alcohol? Why?

OBJECTIVES

When you have completed this chapter you will be able to:

1. State the four ideas of John Dalton which are the foundation of modern chemistry.
2. Define an atom.
3. Mathematically equate scientific notation with fraction expressions.
4. Explain why a single atom cannot be weighed.
5. Locate the atomic weights of any element.
6. Write the symbols for the common elements.
7. Describe the composition of the atom and draw a simple representation of it.
8. List the components of the atom, give their location, their relative mass, and electrical charge.
9. Describe an electron cloud and explain its existence.
10. State the number of electron levels and in which levels the electrons have the least and most energy.
11. Define atomic number, and compare the concept to identifying elements and people.
12. Draw the atoms of the first 20 elements, giving the proton number in the nucleus and the electron configuration.
13. Calculate atomic masses of elements knowing their composition.
14. Define isotopes and explain why their masses differ.
15. Name three isotopes of hydrogen, and draw their structure.
16. Distinguish between an atom and a molecule.

THE STRUCTURE OF MATTER

17. Write the formulas of simple chemical compounds, and list the number and type of atoms contained in them.
18. Calculate molecular weights.
19. Define Avogadro's number.
20. Explain the law of definite composition and give examples.
21. Calculate the percentage composition of a molecule.

The modern concept of the atom was proposed by an Englishman named John Dalton, and he is credited with the development of the atomic theory, which is the foundation of modern chemistry. His concept was:

1. Elements are substances that cannot be broken down into any simpler substances by ordinary chemical means, and are composed of submicroscopic particles called atoms.
2. Atoms of any *one* element are identical in *size, weight*, and *chemical properties*.
3. Atoms of *different* elements have *different sizes, weights,* and *chemical properties.*
4. Atoms of two or more different elements may combine in more than one ratio to form more complex particles, which are called molecules.

ATOMS AND ATOMIC WEIGHTS

Most of Dalton's ideas concerning the atom still hold true today, even though recent research by scientists has revealed additional facts concerning the nature of atoms. For example, atoms can be split into what are called subatomic particles, and therefore the atom is much more complex than Dalton realized.

An atom is defined as the smallest particle of an element that retains the chemical identity of that element. It is extremely small, taking over 100 million of them laid end to end to cover one centimeter, about 0.4 in. Obviously it would be impossible to weigh one atom on a balance, but by weighing a definite number of atoms, the weight of one atom can be calculated. For example, scientists have weighed a definite number of hydrogen atoms and found the weight of a single hydrogen atom to be 1.67×10^{-24}* g $\left(\frac{1.67}{1000000000000000000000000}\text{ g}\right)$. Other atoms weigh in the 10^{-24} to 10^{-22} g range.

*SCIENTIFIC NOTATION. This is an efficient method to express numbers by writing them with exponents (the small numeral written to the upper right of the number). *The exponent tells you how many times that number is being multiplied by itself.*

EXAMPLES: $10^2 = 10 \times 10 = 100$
$10^3 = 10 \times 10 \times 10 = 1{,}000$
$10^4 = 10 \times 10 \times 10 \times 10 = 10{,}000$

A negative exponent, an exponent which has a minus sign in front of it, can be changed into a positive exponent by taking the reciprocal of the numerical expression.

$$10^{-6} = \frac{1}{10^6} = \frac{1}{10\times10\times10\times10\times10\times10} = \frac{1}{1{,}000{,}000} = 0.000001$$

Therefore you can easily see why 1.67×10^{-24} is such an infinitesimally small number.

Neither the chemist nor the health care personnel are concerned with the weight of the individual atom because of its small size. We are more concerned with the *relative* weight of an atom with respect to other atoms. This relationship is known as the *atomic weight* of an atom. Atomic weights are determined for each element on the basis of an arbitrary but helpful choice, the carbon-12 atom, which has a value of 12.00 atomic mass units (amu) assigned to it. This means that if a carbon atom has an atomic weight of 12, then the hydrogen atom will have an approximate atomic weight of 1, as the carbon atom weighs about 12 times as much as the hydrogen atom. Similarly, the sulfur atom weighs about 2.67 times as much as the carbon atom and therefore its atomic weight is about 32. The precise atomic weights of the elements are listed in Table 4-1. The gram-atomic weight of an element is the weight of the element expressed in grams.

You should not attempt to memorize atomic weights of elements, because atomic weight charts and tables are always available. However, knowing the *approximate* atomic weights of a few of the most commonly encountered ones will save you time.

Hydrogen: 1
Carbon : 12
Nitrogen : 14

Oxygen : 16
Sulfur : 32
Chlorine: 35

Table 4-1. Table of atomic masses (based on carbon-12).

Name	Symbol	Atomic No	Atomic Weight	Name	Symbol	Atomic No	Atomic Weight
Actinium	Ac	89	(227)	Mercury	Hg	80	200.59
Aluminum	Al	13	26.98	Molybdenum	Mo	42	95.94
Americium	Am	95	(243)	Neodymium	Nd	60	144.24
Antimony	Sb	51	121.75	Neon	Ne	10	20.18
Argon	Ar	18	39.95	Neptunium	Np	93	(237)
Arsenic	As	33	74.92	Nickel	Ni	28	58.71
Astatine	At	85	(210)	Niobium	Nb	41	92.91
Barium	Ba	56	137.34	Nitrogen	N	7	14.01
Berkelium	Bk	97	(247)	Nobelium	No	102	(254)
Beryllium	Be	4	9.01	Osmium	Os	76	190.2
Bismuth	Bi	83	208.98	Oxygen	O	8	16.00
Boron	B	5	10.81	Palladium	Pd	46	106.4
Bromine	Br	35	79.90	Phosphorus	P	15	31.00
Cadmium	Cd	48	112.40	Platinum	Pt	78	195.09
Calcium	Ca	20	40.08	Plutonium	Pu	94	(242)
Californium	Cf	98	(251)	Polonium	Po	84	(210)
Carbon	C	6	12.01	Potassium	K	19	39.10
Cerium	Ce	58	140.12	Praseodymium	Pr	59	140.91
Cesium	Cs	55	132.90	Promethium	Pm	61	(147)
Chlorine	Cl	17	35.45	Protactinium	Pa	91	(231)
Chromium	Cr	24	52.00	Radium	Ra	88	(226)
Cobalt	Co	27	58.93	Radon	Rn	86	(222)
Copper	Cu	29	63.55	Rhenium	Re	75	186.2
Curium	Cm	96	(247)	Rhodium	Rh	45	102.90
Dysprosium	Dy	66	162.50	Rubidium	Rb	37	85.47
Einsteinium	Es	99	(254)	Ruthenium	Ru	44	101.07
Erbium	Er	68	167.26	Samarium	Sm	62	150.35
Europium	Eu	63	151.96	Scandium	Sc	21	44.96
Fermium	Fm	100	(253)	Selenium	Se	34	78.96
Fluorine	F	9	19.00	Silicon	Si	14	28.09
Francium	Fr	87	(223)	Silver	Ag	47	107.87
Gadolinium	Gd	64	157.25	Sodium	Na	11	23.00
Gallium	Ga	31	69.72	Strontium	Sr	38	87.62
Germanium	Ge	32	72.59	Sulfur	S	16	32.06
Gold	Au	79	196.97	Tantalum	Ta	73	180.95
Hafnium	Hf	72	178.49	Technetium	Tc	43	(99)
Helium	He	2	4.00	Tellurium	Te	52	127.60
Holmium	Ho	67	164.93	Terbium	Tb	65	158.92
Hydrogen	H	1	1.01	Thallium	Tl	81	204.37
Indium	In	49	114.82	Thorium	Th	90	232.04
Iodine	I	53	126.90	Thulium	Tm	69	168.93
Iridium	Ir	77	192.2	Tin	Sn	50	118.69
Iron	Fe	26	55.85	Titanium	Ti	22	47.90
Krypton	Kr	36	83.80	Tungsten	W	74	183.85
Lanthanum	La	57	138.91	Uranium	U	92	238.03
Lawrencium	Lr	103	(257)	Vanadium	V	23	50.94
Lead	Pb	82	207.19	Xenon	Xe	54	131.30
Lithium	Li	3	6.94	Ytterbium	Yb	70	173.04
Lutetium	Lu	71	174.97	Yttrium	Y	39	88.90
Magnesium	Mg	12	24.31	Zinc	Zn	30	65.37
Manganese	Mn	25	54.94	Zirconium	Zr	40	91.22
Mendelevium	Md	101	(256)				

SYMBOLS FOR THE ELEMENTS

You learned earlier that there are over 100 building blocks of nature which we call elements, and even though you may have never studied chemistry, you have probably heard of some of them. The scientist uses a shorthand method to designate these elements by means of abbreviations or symbols from their English names. A few are derived from Latin.

SYMBOLS FOR THE ELEMENTS

Element	Symbol	
Aluminum	Al	
Argon	Ar	
Arsenic	As	
Barium	Ba	
Beryllium	Be	
Calcium	Ca	
Carbon	C	
Chlorine	Cl	
Chromium	Cr	
Cobalt	Co	
Copper	Cu	(Latin: cuprum)
Fluorine	F	
Gold	Au	(Latin: aureum)
Helium	He	
Hydrogen	H	
Iodine	I	
Iron	Fe	(Latin: ferrum)
Lead	Pb	(Latin: plumbum)
Magnesium	Mg	
Mercury	Hg	(Latin: hydrargyrum, silver water)
Neon	Ne	
Nickel	Ni	
Nitrogen	N	
Oxygen	O	
Phosphorus	P	
Potassium	K	(Latinized Arabic, kalium)
Radium	Ra	
Silicon	Si	
Silver	Ag	(Latin: argentum)
Sodium	Na	(Latin: natrium)
Sulfur	S	
Tin	Sn	(Latin: stannum)
Zinc	Zn	

THE ATOM AND IT'S COMPOSITION

The smallest particle of an element that can exist and still retain the chemical properties of that element is called an atom. To help you understand more about the structure of the atom, we can start off by comparing it to our solar

Table 4-2. Properties of the particles making up the atom.

Particle	*Symbol*	*Electrical charge*	*Relative mass*	*Location*
Proton	p	Plus 1:+1 Positive 1	1 amu	Nucleus
Neutron	n	0	1 amu	Nucleus
Electron	e	Minus 1 −1; Negative 1	0*	Outside nucleus

*For our purposes the relative mass is considered to be zero, even though its mass = 1/1836 that of the proton. Because it is so small, we neglect it completely.

system. The atom has a heavy nucleus, comparable to our sun, with particles revolving around it in definite orbits. We know that the atom is extremely small and it is composed of a number of electrical particles, the most important being the proton, neutron, and electron. Each of these particles differ from the others by its weight and electrical charge, and each has its special space location in the atom (Table 4-2). The *electrons* are always located in orbit *outside* the nucleus, and the *protons* and *neutrons* are always located within the nucleus (Fig. 4-1).

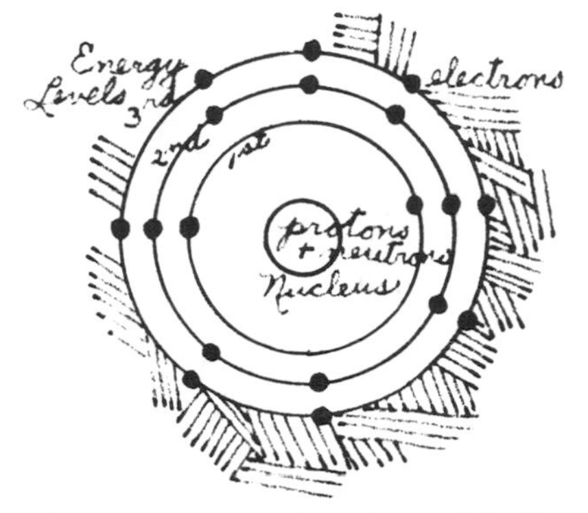

Figure 4-1. Atom (not to scale). Nucleus is composed of protons and neutrons. Electrons orbit the nucleus (not all electron orbits are shown).

All of these particles are composed of electricity and they obey the laws of electricity. Consequently, one plus charge (+) neutralizes 1 minus (−) charge and one minus (−) charge neutralizes 1 plus (+) charge. Therefore, when you have an equal number of positive and negative charges present, the electrical condition is neutral, zero charge. *Every atom is electrically neutral*; therefore *in every atom*, the *number of protons* (which are positively charged) must be equal to the *number of the electrons* (which are negatively charged). The electrons orbit the nucleus containing the protons and neutrons in a "fuzzy cloud" and they can be compared to the blades in an electric fan which, when rotating fast, appear to be everywhere at the same time. We therefore consider the electrons to form an electron cloud around the nucleus, at specified distances from the nucleus in energy levels, electron shells, or orbits.

The Energy Levels-Electron Shells

There are seven energy levels surrounding the nucleus, with the first level nearest it and containing the least energy. Level number 7 is the furthest from the nucleus and has the most energy, with energy levels 2 through 6 in between (Fig. 4-2). Health care personnel are only concerned with the first three levels and our discussions will be limited to these levels (Fig. 4-3).

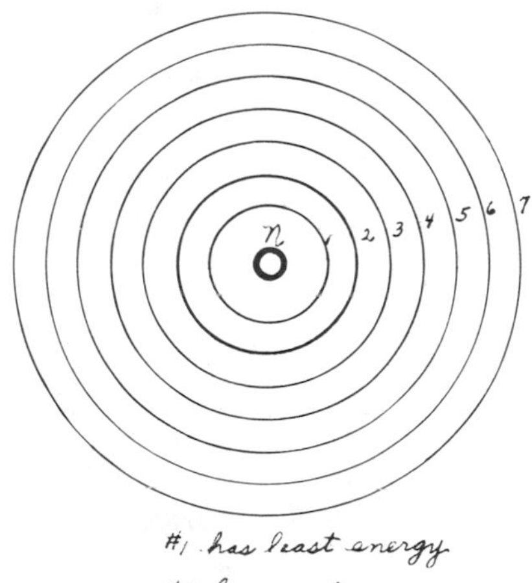

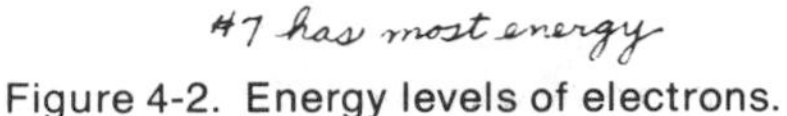

Figure 4-2. Energy levels of electrons.

FACTS TO REMEMBER

First energy level can contain a maximum of TWO electrons.
Second energy level can contain a maximum of EIGHT electrons.
Third energy level can contain a maximum of EIGHT electrons.

The first energy level *must be filled* before electrons can start filling the second energy level, and the second energy level *must be filled* before electrons can begin filling the third energy level. This procedure holds true for the first 20 elements, but it varies after that. *It is the number of electrons in the shells and their location that determines the chemical characteristics of the element.*

Atomic Number

Just as every person in the world can be *positively* identified by his fingerprints, so can each element be positively identified by its *atomic number.* This atomic number is the *number of protons* in the nucleus of the atom (Fig. 4-4). It is also the number of electrons in orbit around the nucleus, because in any atom which is electrically neutral *the number of protons always equals the number of electrons.*

THE ATOMIC NUMBER IDENTIFIES THE ELEMENT

Table 4-3 lists the first twenty elements in order of their increasing atomic number, starting with one. The element which has 1 proton in its nucleus (and 1 electron in orbit) is hydrogen. ANY atom that has *one proton in its nucleus is a hydrogen atom* regardless of where it was found or made (Fig. 4-5). The next element has 2 protons in its nucleus (also 2 electrons in its orbit), and that element is called helium. Notice that the two electrons are in the first energy level (Fig. 4-6). Following helium (atomic number 2) is lithium (Fig. 4-7) with an atomic number of 3, and now the third electron is found in the second energy level because the first energy level holds a maximum of 2 electrons. As the atomic number increases to 10, the electrons continue to fill the second energy

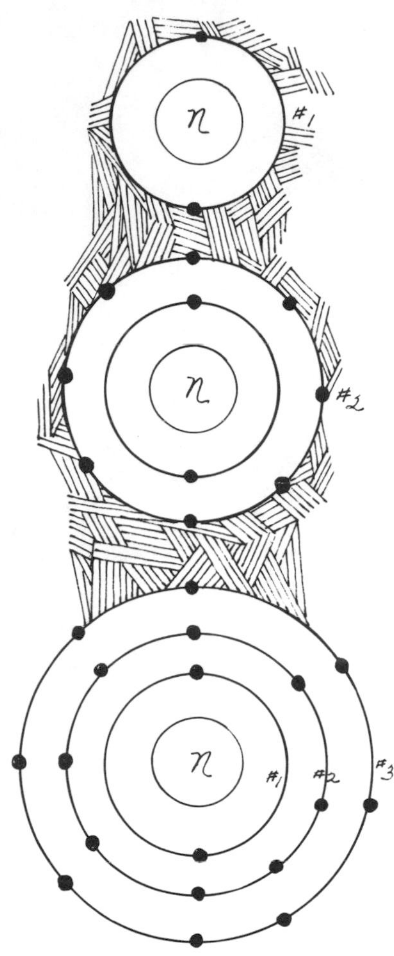

Figure 4-3. Electron capacity of the first three energy shells.

Figure 4-4. Nucleus. 13 protons means an atomic number of 13: this means that the element is aluminum.

Table 4-3. Electron arrangement of the first twenty elements.

			Electron arrangement			
Element	*Symbol*	*Atomic number*	*First energy level*	*Second energy level*	*Third energy level*	*Fourth energy level*
Hydrogen	H	1	1			
Helium	He	2	2			
Lithium	Li	2	2	1		
Beryllium	Be	4	2	2		
Boron	B	5	2	3		
Carbon	C	6	2	4		
Nitrogen	N	7	2	5		
Oxygen	O	8	2	6		
Fluorine	F	9	2	7		
Neon	Ne	10	2	8		
Sodium	Na	11	2	8	1	
Magnesium	Mg	12	2	8	2	
Aluminum	Al	13	2	8	3	
Silicon	Si	14	2	8	4	
Phosphorus	P	15	2	8	5	
Sulfur	S	16	2	8	6	
Chlorine	Cl	17	2	8	7	
Argon	Ar	18	2	8	8	
Potassium	K	19	2	8	8	1
Calcium	Ca	20	2	8	8	2

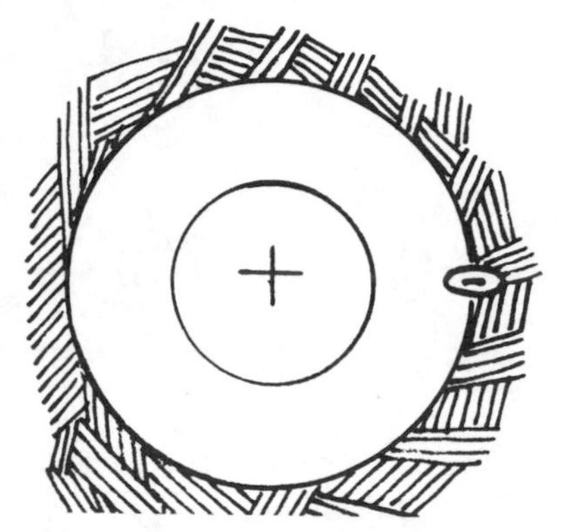

Figure 4-5. Hydrogen's atomic number is 1.

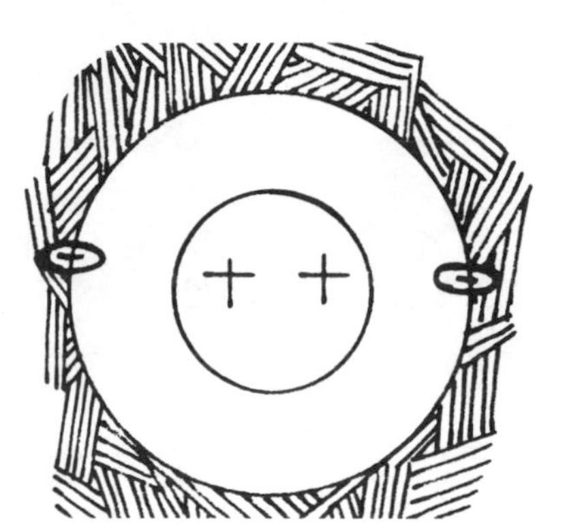

Figure 4-6. Helium's atomic number is 2.

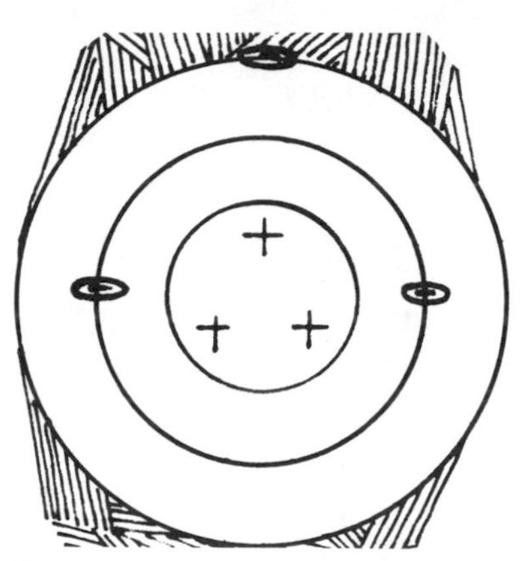

Figure 4-7. Lithium's atomic number is 3.

level. When the atomic number is 10, with the element neon (Fig. 4-8), both the first energy level (maximum of 2 electrons) and the second energy level (maximum capacity of 8 electrons) are filled to capacity. The next higher atomic number element having one more electron is sodium. Its electron structure would be 2 electrons in first level, 8 in second, and one in third level (Fig. 4-9). Calcium (Fig. 4-10), atomic number of 20, would therefore have 2 electrons in first, 8 in second, 8 in third, and 2 electrons in the fourth energy level. Up through element 20, calcium, the electrons fill the shells in the 2-8-8 configuration, but beginning with element 21, electrons shuttle back and forth between the third and fourth shells. The first 20 elements include almost all of the elements of greatest importance to health care personnel, so that is why we stop at element 20 (Fig. 4-11).

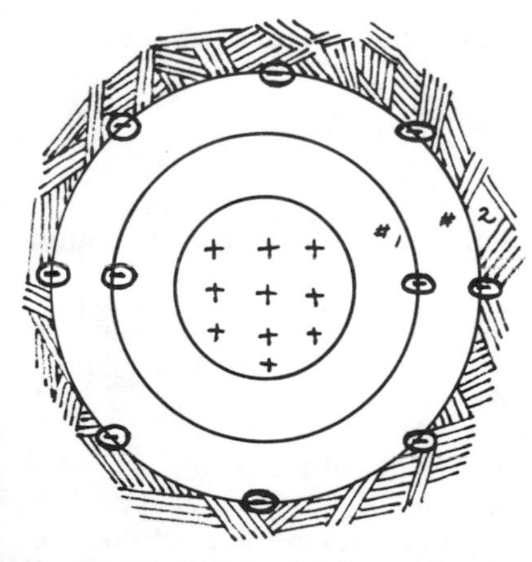

Figure 4-8. Neon's atomic number is 10.

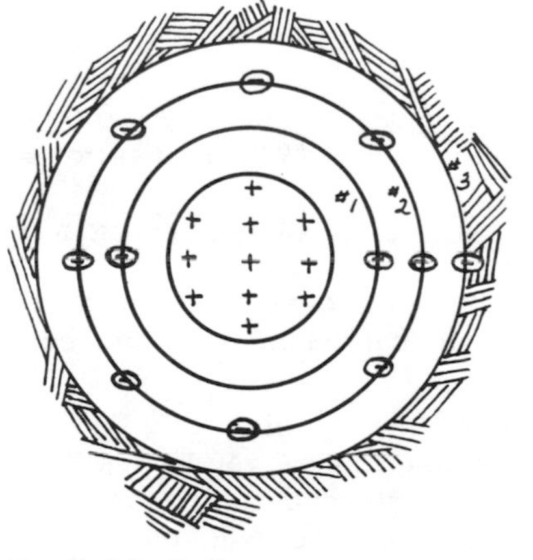

Figure 4-9. Sodium's atomic number is 11.

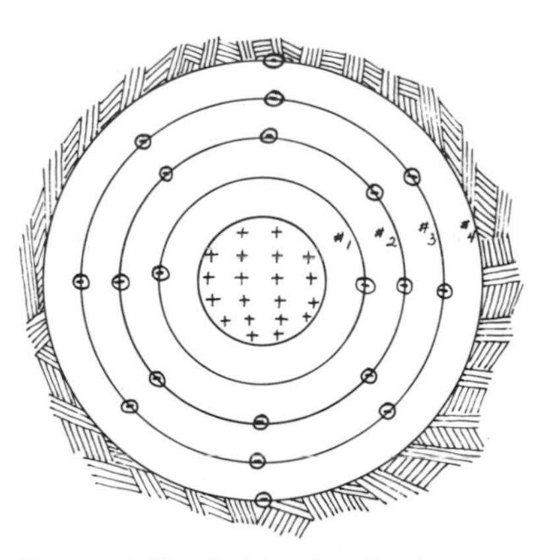

Figure 4-10. Calcium's atomic number is 20.

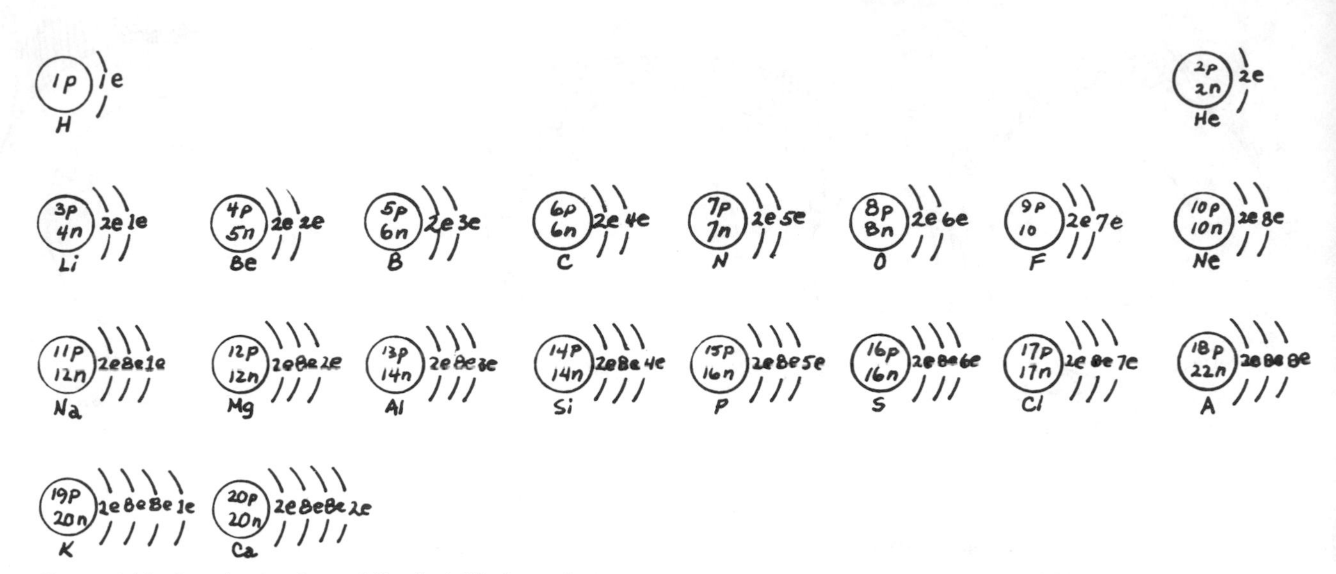

Figure 4-11. Atomic structure of the first 20 elements.

Atomic Weight

The atom is composed of protons, neutrons, and electrons, but only the protons and the neutrons have mass (weight). We previously considered the mass of the electron to be zero because it is so negligible. Therefore, the only factors which contribute to the weight of the atom are the protons and the neutrons. You might relate this to weighing yourself on a scale as you are, and then weighing yourself after removing your watch. The difference in weight would not be detectable, therefore we neglect the masses of the electrons when we consider the mass of the atom.

$$\text{Mass Number of the atom} = \text{Mass of protons} + \text{mass of neutrons}$$

But, since

$$\text{Mass of a proton} = 1 \text{ amu}$$
$$\text{Mass of a neutron} = 1 \text{ amu}$$

Then:

$$\text{Mass of the atom} = \text{number of protons} + \text{number of neutrons}$$

NOTE: All protons are identical, all neutrons are identical, and all electrons are identical regardless of their source.

Isotopes

According to what you have just learned, the weights of the atoms should be *whole numbers,* because there are no fractional neutrons or fractional protons, and the weight of the atom is the sum of the protons and the neutrons. Yet, when you examine the periodic table you do not find whole number atomic weights. The few elements having atomic weights that come close to whole

Table 4-4. The three isotopes of hydrogen.

Name	*Symbol**	*Protons*	*Neutrons*	*At No*	*At mass*
Protium	$^{1}_{1}H$	1	0	1	1 + 0 = 1
Deuterium	$^{2}_{1}H$	1	1	1	1 + 1 = 2
Tritium	$^{3}_{1}H$	1	2	1	1 + 2 = 3

*The superscript = atomic mass
The subscript = atomic number

numbers are helium, carbon, oxygen and nitrogen. All of the rest have atomic weights that are not even close to whole numbers; for example, chlorine has an atomic weight of 35.453.

The answer to this question has been found in the nucleus of atoms. Apparently atoms of the same element can contain different numbers of neutrons. Those atoms all contain the *same number of protons*, which identifies them as being that *element*, but because they contain different numbers of neutrons, they have a different weight. The neutrons contribute only to the weight of the atom but do not affect the atomic number because they are neutrally charged. Atoms of the same element which contain different numbers of neutrons are called *isotopes*. Their atomic numbers are the same (the number of protons in the nucleus are identical), but the number of neutrons is different, therefore their weights are different. There are two isotopes of chlorine (Fig. 4-12), each having the 17 protons which identifies the atom as being chlorine, but one isotope contains 18 neutrons while the other contains 20 neutrons. The atomic mass of the 18-neutron isotope is 35 amu (17 protons + 18 neutrons) and that of the 20-neutron isotope is 37 amu (17 protons + 20 neutrons). Thus ordinary chlorine can be considered to be about 25 percent of the 37-isotope and 75 percent of the 35-isotope, which averages the atomic weight of chlorine to 35.453. The isotopes are written as the symbol with the atomic mass as the superscript and the atomic number as the subscript: $^{35}_{17}Cl$ and $^{37}_{17}Cl$. Most of the elements have several isotopes. Hydrogen has three isotopes (Fig. 4-13): protium, deuterium, and tritium.

The atomic weight of hydrogen as found in the periodic table is 1.0080, which represents the *average weight of the isotopes of hydrogen* in the world. Similarly, the atomic weight of the other elements are fractional numbers because of the existance of the isotopes. Therefore, because of the presence of isotopes, a more precise definition of an element is: *a kind of matter in which all of the elements have the same atomic number.* The three isotopes of hydrogen all have the same chemical properties; for example, they all react with oxygen to form water. However they do have different masses, contributed by the neutrons. Isotopes are extremely important to health care personnel because certain isotopes of elements are radioactive and are useful in the diagnosis and treatment of certain cancers and other diseases.

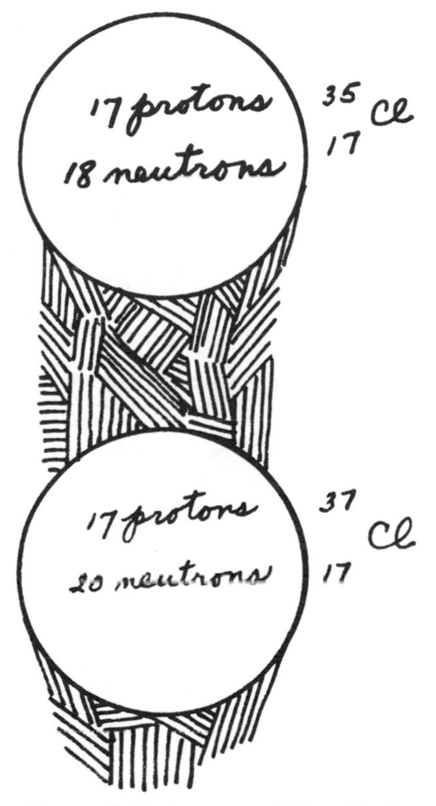

Figure 4-12. Isotopes of the chlorine nuclei.

PHYSICAL AND CHEMICAL PROPERTIES OF ISOTOPES

The chemical properties of isotopes of the same element are identical because of two factors: (1) Isotopes have the same number of protons in the nucleus, and (2) isotopes have the same number of electrons in orbit. Isotopes of the same element have the same atomic number but different atomic masses. It is the difference in atomic masses that results in a difference in the physical properties.

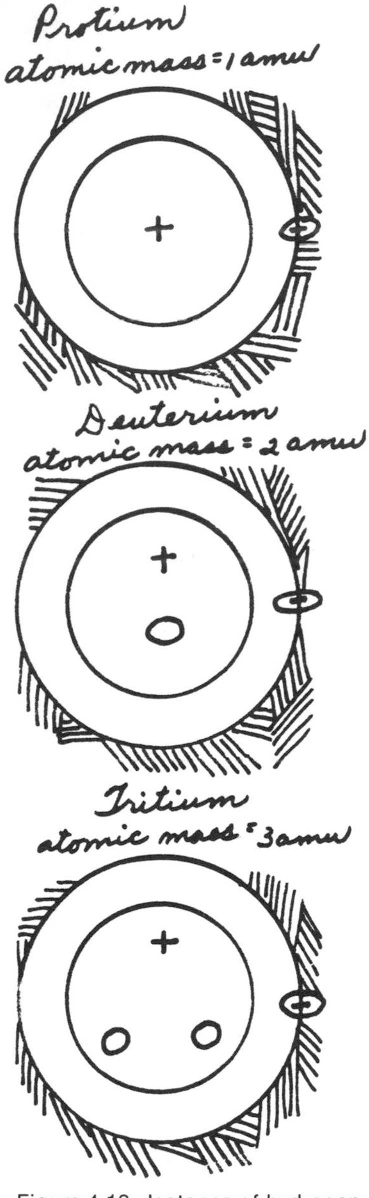

Actually you can relate the concept of isotopes to yourself. You are the same person day after day because your fingerprints positively identify you. Yet, your mass changes daily and even during the day due to intake and excretion of liquids and solids, the utilization of body fats for energy, and the general metabolism of the body. The loss and gain of mass (body fluids and fat) has not changed your identity. An analogy exists with the elements. Their identity is fixed by their atomic number, but their mass is dependent upon the sum of the protons and neutrons in the nucleus, the neutrons contributing to the mass but not changing the identity of the atom. Another analogy would be a set of identical twins who are identical in personality, mentality, fingerprints, coloring, and features, except that one weighs a few pounds more than the other.

The number of neutrons in any atom can be found by subtracting the atomic number from the atomic mass:

Atomic mass	= number of protons + number of neutrons
(−) Atomic number	= number of protons
Number of neutrons	

Example: $^{35}_{17}Cl$ 35 (protons and neutrons), subtract 17 protons, this leaves 18 neutrons.
$^{37}_{17}Cl$ 37 (protons and neutrons), subtract 17 protons, this leaves 20 neutrons.

Protium
atomic mass = 1 amu

Deuterium
atomic mass = 2 amu

Tritium
atomic mass = 3 amu

Figure 4-13. Isotopes of hydrogen.

THE MOLECULE

You already know that the atom is the smallest particle that truly represents an element. When atoms of different elements combine, they form an entirely *new substance* called a *compound*. The smallest particle of that compound that can exist and still retain the chemical properties of that compound is called a *molecule.*

Just as a symbol is used to express an element, a *formula* is used to represent a chemical compound. The formula tells you the *number and kind of atoms of which the molecule is composed.* For example, the substance water has the formula, H_2O, which expresses the fact that 2 atoms of hydrogen have chemically combined with 1 atom of oxygen to form the substance water. Most of the compounds found in nature are much more complex than water, being composed of many different kinds and numbers of atoms, and having molecules composed of hundreds of thousands of atoms. The formula tells you the number of each type of atom present, using the chemical symbols of the elements. When more than one atom of the same element is in the molecule, a subscript to the right of the symbol designates how many atoms of that particular element are contained in the molecule. If there is no subscript, only one atom of the element is contained in the molecule.

Table 4-5. Compounds: Formulas and atomic analysis.

Compounds	*Formula*	*Atom analysis*
Carbon monoxide	CO	1 carbon: 1 oxygen
Carbon dioxide	CO_2	1 carbon: 2 oxygens
Hydrochloric acid	HCl	1 hydrogen: 1 chlorine
Sulfuric acid	H_2SO_4	2 hydrogens: 1 sulfur: 4 oxygens
Nitric acid	HNO_3	1 hydrogen: 1 nitrogen: 3 oxygens
Aluminum hydroxide	$Al(OH)_3$	1 aluminum: 3 oxygens: 3 hydrogens
Calcium phosphate	$Ca_3(PO_4)_2$	3 calciums: 2 phosphorus: 8 oxygens

When two or more elements are surrounded by a parentheses which has a *subscript outside* the parenthesis (see aluminum hydroxide and calcium phosphate in Table 4-5), each of the element subscripts must be multiplied by the parenthesis subscript to find the number of atoms of the element in the molecule.

EXAMPLE: analysis of aluminum hydroxide and calcium phosphate.

$Al(OH)_3$ = 1 aluminum atom
$(OH)_3$ means that there are:
3 (OH), therefore there are:
3×1 oxygen or 3 oxygens and
3×1 hydrogen or 3 hydrogens.

$Ca_3(PO_4)_2$ = 3 calciums
$(PO_4)_2$ means that there are:
2 (PO_4), therefore there are:
2×1 phosphorus or 2 phosphorus and
2×4 oxygens or 8 oxygens.

Molecular Weights

The individual molecule is so small that it is impossible to weigh a single molecule, just as it is impossible to weigh one atom of an element. However, the weight of the molecule, which is the combined weight of all of the atoms which compose the molecule, can be calculated by adding the atomic weights of every atom in the molecule. The molecular weight is expressed in grams and is called the gram-molecular weight or *mole*. Therefore, the gram-molecular weight or mole is equal to the sum of the weights of all of the atoms of the molecule, that is, equal to the sum of all the atomic weights which are expressed in grams. The atomic weight of every element can be found in the periodic chart or in the periodic table. To calculate the molecular weight, multiply the atomic weight of each element by the number of atoms of that element in the molecule and add them together. Here are several representative molecular weight calculations:

Hydrogen chloride HCl:

1 hydrogen	$= 1 \times 1.008$		$= 1.008$
1 chlorine	$= 1 \times 35.453$		$= 35.453$
		molecular weight	$= 36.461$

Sulfuric Acid H_2SO_4:

2 hydrogens	$= 2 \times 1.008$		$= 2.016$
1 sulfur	$= 1 \times 32.06$		$= 32.06$
4 oxygens	$= 4 \times 15.999$		$= 63.996$
		molecular weight	$= 98.072$

Calcium phosphate $Ca_3(PO_4)_2$:

2 calciums	$= 2 \times 40.08$		$= 80.16$
2 phosphorus	$= 2 \times 30.974$		$= 61.948$
8 oxygens	$= 8 \times 15.999$		$= 127.992$
		molecular weight	$= 270.100$

Avogadro's Number

We have used the terms atomic weight and molecular weight to mean the mass relative to ^{12}Carbon (^{12}C) of one atom or one molecule. These terms are also used to mean a specific quantity of an element or compound of a number of grams equal to the atomic or molecular weight. Therefore, instead of saying the atomic weight of oxygen is 15.999, we say the gram-atomic weight of oxygen is 15.999 g. Similarly, the gram molecular weight of sulfuric acid (see above) is 98.072 g.

The number of carbon-12 atoms needed to add up to 12 g is 6.02×10^{23}. Similarly, the number of atoms of *any* element needed to add up to its atomic weight is 6.02×10^{23} atoms of that element. This is Avogadro's law and it states that:

> 6.02×10^{23} *atoms* are contained in 1 gram-atomic weight of an *element.*

and

> 6.02×10^{23} *molecules* are contained in 1 gram-molecular weight of a *compound.*

Therefore, since the *atomic weight* of any element contains the *same number of atoms,* but the atomic weights are different, then the *weight of the atoms must be different.*

FOR EXAMPLE:

Element	Gram-atomic weight	Number of atoms
Hydrogen	1.008	6.02×10^{23}
Carbon	12.00	6.02×10^{23}
Lead	207.2	6.02×10^{23}

A gram-molecular weight of a compound also contains 6.02×10^{23} molecules of the compound. We found the molecular weight of HCl to be 36.461, of H_2SO_4 to be 98.074, and that of $Ca_3(PO_4)_2$ to be 270.200. Each molecular weight therefore contains an Avogadro number of molecules.

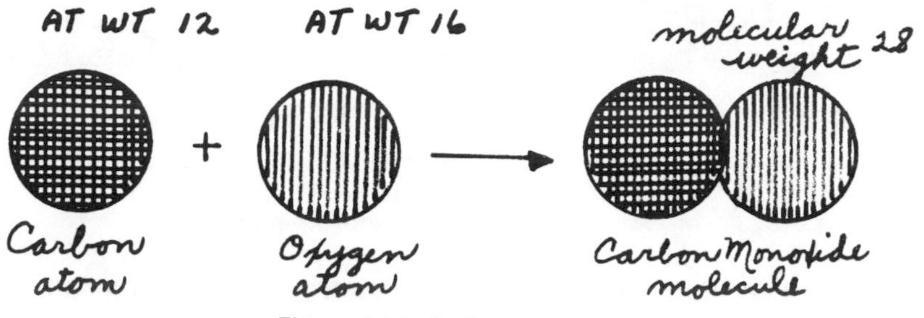

Figure 4-14. Carbon monoxide.

THE LAW OF DEFINITE COMPOSITION

This law states that when atoms of a given element combine with atoms of another element to form a particular molecule, every molecule contains the same elements combined in a definite and constant ratio by weight. This means that no matter how anyone prepares carbon monoxide (CO), every molecule of carbon monoxide will always contain 12 parts by weight of carbon and 16 parts by weight of oxygen (Fig. 4-14). This means that carbon monoxide, regardless of how it is prepared, found, or used, will always have the same composition: 1 atom of carbon and 1 atom of oxygen. This law of definite composition therefore enables us to identify particular substances and to use the chemical properties known about those substances to solve problems. The same compound *always* has the same formula, indicating the same number and kind of atoms. Carbon dioxide (Fig. 4-15) is a colorless, odorless and tasteless gas, but it is not poisonous like carbon monoxide. Carbon dioxide is found in soda pop, carbonated water, champagnes, beers, and in our breath. It is always composed of *1* atom of carbon for every *2* atoms of oxygen and is called "di" because it contains 2 atoms of oxygen. Therefore, the molecule contains 12 parts by weight of carbon and 2 × 16 or 32 parts by weight of oxygen.

Percentage Composition of a Molecule

Everyone knows that the formula for water is H_2O, but not everyone understands what H_2O actually means. You should know that the water molecule is composed of 2 parts by weight of hydrogen and 16 parts by weight of oxygen. The

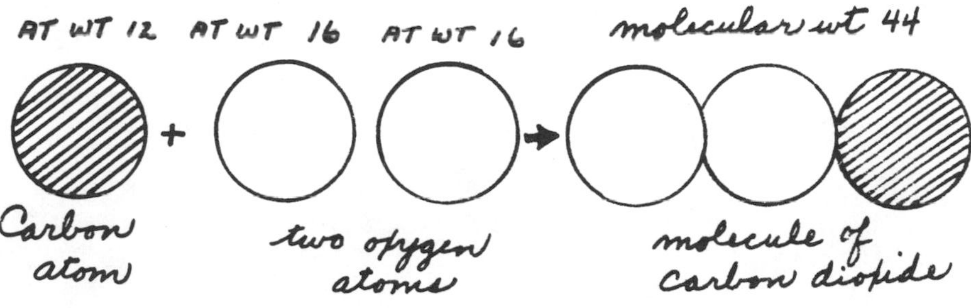

Figure 4-15. Carbon dioxide.

total weight of the molecule is therefore 18, and the percentage composition of each can be found by dividing the weight of the element by the total weight.

Hydrogen:	$2 \times 1 = 2$*	% H =	$2 \div 18 = 11\%$*
Oxygen:	$1 \times 16 = 16$	% O =	$16 \div 18 = 89\%$
Total	$= 18$	Total	$= 100\%$
			(*approximate)

If, however, 1 part of hydrogen (instead of 2 parts) combined with 16 parts of oxygen, then a poisonous substance called hydrogen peroxide would be formed. In this case only half the weight of hydrogen united with the same weight of oxygen.

SUMMARY

An atom is the smallest particle of an element that retains the chemical identity of that element. The atomic weight of elements is based on the carbon-12 atom. Elements are symbolized by abbrevaitions derived from English or Latin names. The atom is composed of a nucleus, which contains protons and neutrons, and electrons which orbit the nucleus in "electron clouds." There are seven energy levels of electrons, and they are filled in a definite manner. For the first 20 elements, the first shell can hold only 2 electrons, the second 8 electrons, and the third only 8 electrons. The atom is electrically neutral, and the number of protons (called the atomic number specific for each element) equals the number of electrons in orbit. The mass of an atom is calculated from the sum of the masses of the protons and neutrons, neglecting the very small mass of the electrons. Variation in the mass of an atom is due to the number of neutrons, and elements which have the same atomic number but different atomic masses are called isotopes. The smallest particle of a compound is the molecule, which has a formula listing the number and kinds of atoms composing it. The weight of all the atoms in the molecule is called the molecular weight. The molecule of a given compound always has the same formula. The percentage composition of the molecule can be calculated by dividing the weight of a component element by the total molecular weight.

EXERCISE

1. Write 1×10^{-6} as a positive exponent. What is its value as a fraction, a decimal?
2. What do the following symbols represent: Al, Cu, Pb, Na, K, H, Cl, Br, C, Ag, P, F, Ba, Ca, Ne, S, Zn, Sn, O, N, Mg, I.
3. Characterize the components of the atom as to the mass, electrical charge, and location.
4. Draw the electronic configurations for elements having the following atomic numbers, and identify them: 1, 4, 8, 12, 16, 20.
5. Look in the periodic table (Fig. 5-1) and write the atomic weights of the elements listed in question 4.
6. What contributes to the atomic mass of an element, and why are the atomic weights fractional numbers?
7. Name and draw a schematic representation of the three isotopes of hydrogen.
8. What does the formula of a compound tell you?
9. How many of each type of atom are in the compound $Ca_3(PO_4)_2$?
10. Calculate the following molecular weights: NaOH; $Ca(OH)_2$; $Fe_3(PO_4)_2$; Na_2CO_3.
11. How many atoms of carbon would be contained in 6 grams? How many molecules of NaOH would be contained in 20 grams?
12. What is the percentage of hydrogen and oxygen in water, and in hydrogen peroxide (H_2O_2)?

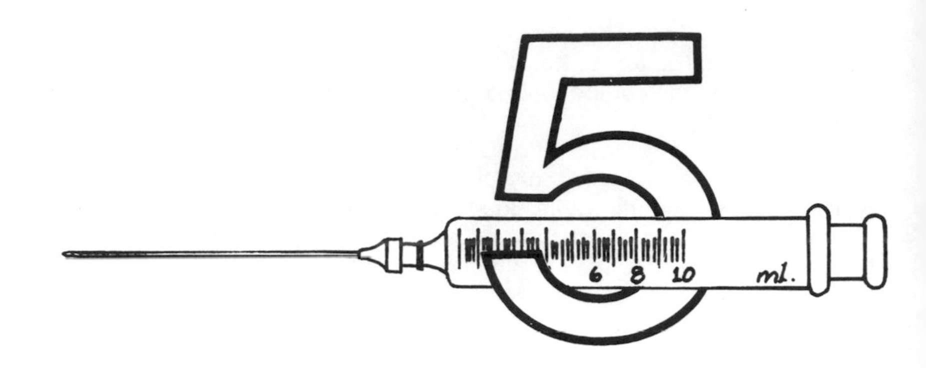

OBJECTIVES

When you have completed this chapter you will be able to:

1. Identify the scientist who created the periodic table.
2. Give the basis on which this scientist worked.
3. State the facts which are found in the box of an element in the periodic table.
4. Identify the Groups IA, IIA, VIIA, and VIIIA by their outer electron shell configuration.
5. Locate the metals, nonmetals, transition elements and rare gases in the periodic table.
6. Discuss in detail what a "family" is, and what causes a group of elements to act as a family.
7. Describe the general physical characteristics of the classes of elements.
8. List 11 important elements that health care personnel should be familiar with and give a short discussion of each.
9. Name 4 important trace elements.

THE PERIODIC TABLE

In the early part of the nineteenth century scientists were accumulating tremendous quantities of data and information about the elements and compounds. They saw that certain elements resembled each other in their chemical reactivity, but in order to use this great store of information, some sort of organization was necessary. A Russian chemist named Mendeleev recognized the recurrence of certain chemical properties of the elements, and he arranged them in a chart in order of increasing *atomic weights,* based on the weight of the carbon atom. He noted that elements with similar chemical properties recurred at regular intervals (periodic intervals), and placed those elements which had similar chemical and physical properties in a vertical column, just as days of the week are placed in a calendar. Later research eliminated some of the inconsistencies of his table when the elements were arranged in order of their *increasing atomic numbers,* and the rare gases of the atmosphere (helium, argon, neon, etc.) were found to fit satisfactorily into the table after their discovery (Fig. 5-1).

When Mendeleev first arranged the elements in a periodic table, there were a number of vacant spaces, and he claimed that those vacant spaces represented places for elements that were yet undiscovered. He was right, and he predicted the discovery of three new elements with amazing accuracy, and also predicted their properties. He was able to do this because he knew that the *properties of those elements would resemble the elements found in that family.*

THE STRUCTURE OF THE PERIODIC TABLE

If you will examine the periodic table you will see that every element occupies a box (Fig. 5-2) which contains the chemical symbol for that element. Also in the box are two numbers, a whole number which is the atomic number of the element (the number of protons in the nucleus) and the atomic weight. Calcium has an atomic number of 20, (20 protons), its symbol is Ca, and its atomic weight is 40.08. Calcium has an electron configuration of 2 electrons (first energy level), 8 electrons (second energy level), 8 electrons (third energy level),

Groups

Periods	IA	IAA	IIIA	IVA	VA	VIA	VIIA		VIIIA		IB	IIB	IIIB	IVB	VB	VIB	VIIB	VIIIA
1	1 **H** 1.008																1 H 1.008	2 **He** 4.003
2	3 **Li** 6.940	4 **Be** 9.013											5 **B** 10.82	6 **C** 12.001	7 **N** 14.008	8 **O** 16.00	9 **F** 19.00	10 **Ne** 20.183
3	11 **Na** 22.991	12 **Mg** 24.32											13 **Al** 26.98	14 **Si** 28.09	15 **P** 30.975	16 **S** 32.066	17 **Cl** 35.457	18 **Ar** 39.944
4	19 **K** 39.100	20 **Ca** 40.08	21 **Sc** 44.96	22 **Ti** 47.90	23 **V** 50.95	24 **Cr** 52.01	25 **Mn** 54.95	26 **Fe** 55.85	27 **Co** 58.94	28 **Ni** 58.71	29 **Cu** 63.54	30 **Zn** 65.38	31 **Ga** 69.72	32 **Ge** 72.60	33 **As** 74.91	34 **Se** 78.96	35 **Br** 79.916	36 **Kr** 83.80
5	37 **Rb** 85.48	38 **Sr** 87.63	39 **Y** 88.92	40 **Zr** 91.22	41 **Nb** 92.91	42 **Mo** 95.95	43 **Tc** 99	44 **Ru** 101.1	45 **Rh** 102.91	46 **Pd** 106.4	47 **Ag** 107.880	48 **Cd** 112.41	49 **In** 114.82	50 **Sn** 118.70	51 **Sb** 121.76	52 **Te** 127.61	53 **I** 126.91	54 **Xe** 131.30
6	55 **Cs** 132.91	56 **Ba** 137.36	57-71 **La-Lu** **Rare** **Earths**	72 **Hf** 178.50	73 **Ta** 180.95	74 **W** 183.86	75 **Re** 186.22	76 **Os** 190.2	77 **Ir** 192.2	78 **Pt** 195.09	79 **Au** 197.0	80 **Hg** 200.61	81 **Tl** 204.39	82 **Pb** 207.21	83 **Bi** 209.00	84 **Po** 210	85 **At** 210	86 **Rn** 222
7	87 **Fr** 223	88 **Ra** 226	89-96 **Actin-** **ides**	104 **Ku** 259	105 **Ha** 260													
Rare Earths			57 **La** 138.92	58 **Ce** 140.13	59 **Pr** 140.92	60 **Nd** 144.27	61 **Pm** 147	62 **Sm** 150.35	63 **Eu** 152.0	64 **Gd** 157.26	65 **Tb** 158.93	66 **Dy** 162.51	67 **Ho** 164.94	68 **Er** 167.27	69 **Tm** 168.94	70 **Yb** 173.04	71 **Lu** 174.99	
Actinides			89 **Ac** 227	90 **Th** 232.05	91 **Pa** 231	92 **U** 238.07	93 **Np** 237	94 **Pu** 242	95 **Am** 243	96 **Cm** 247	97 **Bk** 249	98 **Cf** 251	99 **Es** 254	100 **Fm** 253	101 **Md** 256	102 **No** 253	103 **Lr** 257	

Figure 5-1. The periodic table.

and 2 electrons (fourth energy level). One gram-atom of the element (or one gram-atomic weight) is the atomic weight of the element in grams.

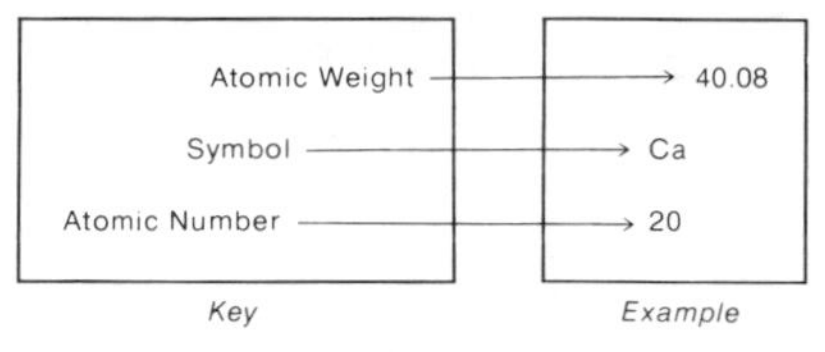

Figure 5-2. The calcium box in the periodic table.

The Group Numbers

The group numbers usually indicate the number of electrons in the outer shell, and therefore are useful in predicting chemical behavior of the element.

In Group IA: 1 electron is present in the outermost shell.
In Group IIA: 2 electrons are present in the outermost shell.
In Group VIIA: 7 electrons are present in the outermost shell.
In Group VIIIA: There are 8 electrons present in the outermost shell.

If you will remember these facts, you will find that the problem of writing chemical formulas and making chemical calculations will be simplified and that you will make fewer mistakes.

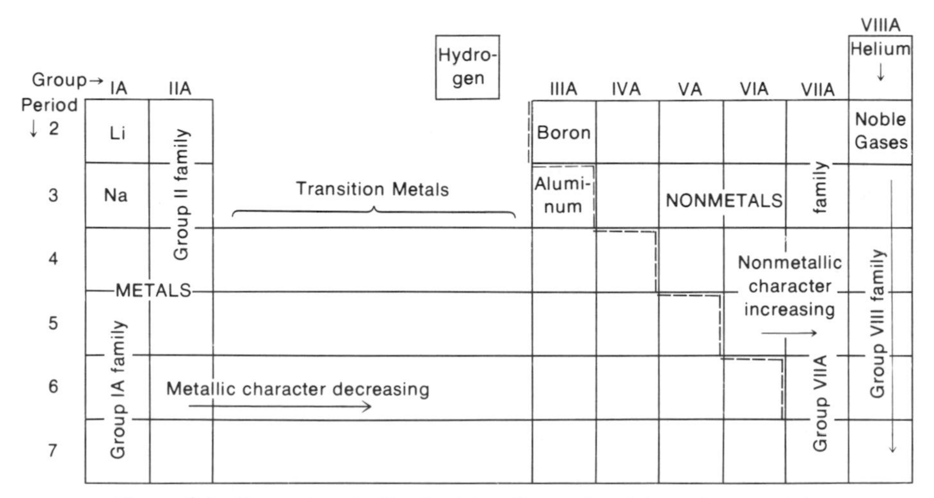

Figure 5-3. General periodic chart locations of metals and nonmetals.

Locations of the Metals, Nonmetals and Transition Elements

In the modern periodic table, metals are located to the left, nonmetals to the right, and the transition elements, metalloids (elements which can act as either metals or nonmetals), in between (Fig. 5-3).

FAMILIES OF ELEMENTS

Families of elements are those elements which behave in a similar manner, chemically and physically. The modern periodic table groups the elements in *vertical columns,* called groups or families, in which you find families of elements that are related in properties. This division of the elements into natural families is of great value to you, as you become familiar with the properties of the elements, enabling you to predict their behavior. For example, when you

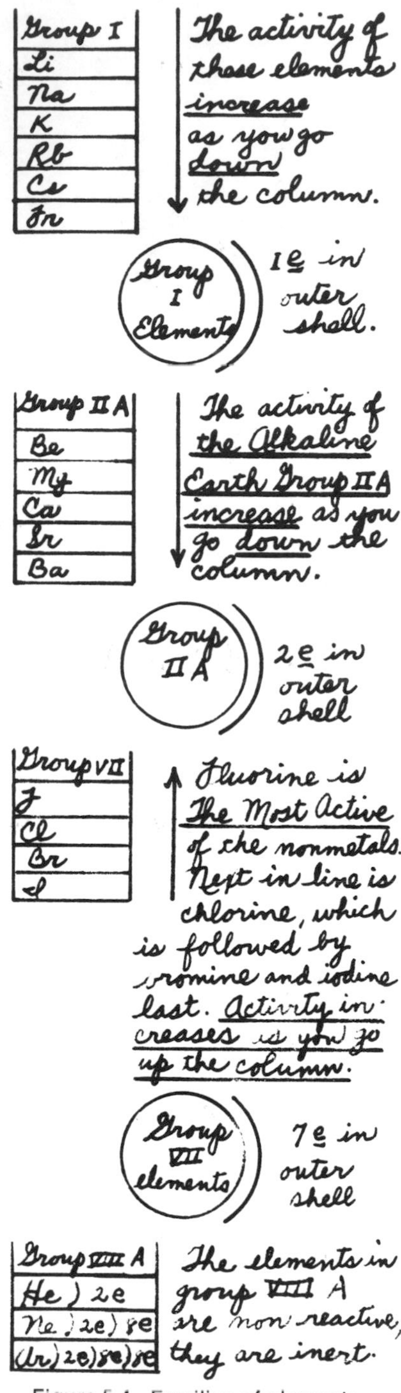

Figure 5-4. Families of elements.

study the compounds of the element sodium, Group IA, you will then know the characteristics of the other elements in this group, thus simplifying your study of chemistry. The similarity of the properties of the elements which places them in the periodic table by families or groups is a result of the *number of electrons in their outer shells.* There are differences in the activity of substances with the same number of outer shell electrons. You can relate the activity of elements in the same family or group to human beings, because some members of the same family are lazy, while others are energetic and active. The same holds true for the elements of the same group (Fig. 5-4). Potassium is more active than sodium, rubidium more than potassium, and francium is the most active element in the first group.

Periods

The horizontal rows of elements are called periods, each element having different properties because they belong to a different family or group (Fig. 5-5).

The Metals

The metals, located on the left side of the period table, are highly lustrous, have a silvery appearance, are soft with a low melting point, and have good thermal and electrical conductivity. Group IA (lithium, sodium, potassium, etc.) is called the *alkali metal group.* Group IIA (beryllium, magnesium, calcium, strontium, barium and radium) is called the *alkaline earth metal group,* or simply the alkaline earths.

The Nonmetals

The nonmetals are located on the right side of the periodic table. They have a wide range of properties; half are gases, one is a liquid, and the remainder are solids. They do not conduct electricity (except carbon) and they have a dull surface, not the shiny silvery one of metals. Group VIIA (fluorine, chlorine, bromine, iodine) is called the halogen or salt former group because many salt compounds contain these elements in their molecules.

The Transition Elements

In the middle of the periodic chart are those elements that can act as either acids or bases, and their action depends upon the environment to which they are subjected. They are also called amphoteric elements.

The Inert Gases

Group VIIIA is a family of gases which are also called the "rare" gases or the "noble" gases. They are inert, and do not react with other elements. Helium, neon, argon, krypton, xenon, and radon comprise this family. They are inert because of one important fact which you should remember: *Their outermost electron shell contains either 2 electrons or 8 electrons.*

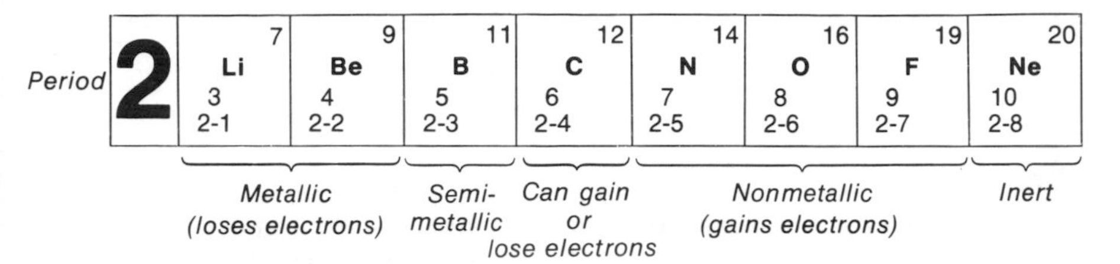

Figure 5-5. Arrangement of elements into periods.

IMPORTANT ELEMENTS

Of the more than 100 elements known at the present time only about 11 are of prime importance, and 6 or 7 of secondary importance to people in the health care profession. Since elements are the building blocks of the body, their chemical behavior affects the body. The percentage composition of chemical elements in the human body is shown below.

Chemical Elements in the Body

Oxygen	65.0%	Potassium	0.35%
Carbon	18.5%	Sulfur	0.25%
Hydrogen	9.5%	Chlorine	0.20%
Nitrogen	3.3%	Sodium	0.15%
Calcium	1.5%	Magnesium	0.05%
Phosphorus	1.0%	Trace elements	0.20%

From the standpoint of preparation for a career in the health sciences, certain elements are of utmost importance, and they are listed below with brief explanations for the reasons of their importance.

Oxygen

Oxygen is the most abundant element on the surface of the earth, constituting in its elemental form about 21 percent by weight of our atmosphere and 88.8 percent by weight of our water. It is a constituent of many important compounds, is involved in the combustion of fuels, is critical to our respiratory processes, and is necessary for metabolism of food.

Carbon

Carbon is probably the most important element for life because all plant and animal life is made up of compounds containing carbon. Carbon is essential for food, warmth, and clothing. Elemental carbon is found in many forms: charcoal, coke, coal, graphite, and diamonds. Many substances containing carbon in the form of its compounds are used as fuel: gasoline, natural gas, oil, wood, and kerosene. In the form of carbon dioxide, it is a product of body metabolism, and through photosynthesis carbon serves as a basic raw material in plants to complete the carbon cycle of nature.

Figure 5-6. Composition of atmosphere by weight.

Hydrogen

Hydrogen, the lightest of all elements, is a component of acids, water, and most organic substances including fats, proteins, and carbohydrates.

Nitrogen

Nitrogen is a gas which makes up about 79 percent of our atmosphere (Fig. 5-6) and it is indirectly essential to life. It is present in all living matter, animal and vegetable. As gaseous nitrogen goes through the nitrogen cycle, it is converted to usable compounds which are essential to life in all living cells. As amino acids, nitrogen compounds are the end point of digestion and, as proteins, are the building blocks of living tissues.

Sodium and Potassium

Compounds of these elements are soluble in water. They are intimately associated with the chemistry of the body fluids, and serve to regulate the electrolyte balance of the body and the normal functions of the body's organs.

Calcium

Calcium is an important component of the bones and teeth in animals and humans. It is found in milk products and is very important to the diet, especially to infants, who need it for building bones and teeth.

Iron

This element is extremely important from the health standpoint because it is a component of a protein substance called hemoglobin in the blood. Blood is vitally needed for the transport of oxygen and nutrients to the living cells and tissues of the body.

Chlorine

Chlorine in the form of the chloride ion (a negatively charged particle of chlorine) is vital for maintaining the proper electrolyte balance of fluids in the body. The element chlorine is a greenish-yellow gas, is very reactive chemically, and forms sodium chloride (ordinary table salt) when combined with sodium. Sodium chloride is an important component of all body fluids.

Sulfur

This is a yellowish solid and is used to make important medicines, such as the sulfa drugs. Certain important proteins contain sulfur.

Phosphorus

Phosphorus is essential to the formation of strong bones and teeth and is needed for the formation of brain and nervous tissue. It is a component of fertilizer serving to maintain the fertility of the soil.

TRACE ELEMENTS

Copper

Minute amounts of copper are believed to be necessary for the formation of hemoglobin, especially in the regeneration of hemoglobin after nutritional anemias in children. Copper occurs in certain enzymes, which are substances that cause chemical changes to take place in the body.

Manganese

Most nutritional authorities agree that manganese is an essential trace element, but they do not know the exact human requirements.

Cobalt

This is an essential element and it is required for the normal functioning of the pancreas. It is a component of vitamin B-12 and is deemed to be essential for hemoglobin formation.

Iodine

Iodine is found in the thyroid gland at a concentration more than 1,000 times that in the muscles and more than 10,000 times that in the blood. Iodine deficiency is related to the occurrence of goiter.

Other Trace Elements

Other trace elements necessary for mammalian life are zinc, molybdenum, and probably chromium, fluorine, selenium and vanadium.

SUMMARY

Elements have been arranged in order of their increasing atomic numbers in the periodic table, in a format which places elements with the same chemical characteristics in vertical columns called families. There are 8 main groups or families of elements, with the metals in the left of the table, the nonmetals in the right side, and the transition elements in the middle. Elements in a chemical family resemble each other because they have the same number of electrons in their outermost shell. The inert gases, having either 2 or 8 electrons in their outer energy shell, do not react. To health care personnel, there are 11 elements of prime importance and about 7 elements of secondary importance, along with some trace elements.

EXERCISE

1. Where are the metals located in the periodic table?
2. Where are the nonmetals located?
3. Where are the transition elements located?
4. What are the elements grouped in vertical columns called?
5. Why do elements in the same vertical columns resemble each other in their chemical behavior?
6. Which is a more active metal, sodium or potassium?
7. How many electrons are in the outer shell of Group IA; Group IIA, Group VIIA, Group VIIIA?
8. What is another name for the inert gases? Why do they not react with other elements?
9. What are the elements in the horizontal rows called?
10. Can elements in different periods have the same chemical behavior? Why not?
11. What information can you learn about an element by examining the periodic table?
12. Which element is present in the body in the highest percentage? Which element is second highest?
13. Name 7 of the 11 most important elements in the body, and give facts about them.

OBJECTIVES

When you have completed this chapter you will be able to:

1. State the reason why elements other than the inert gases react.
2. Draw the electron structure of compounds.
3. State the effect on the electrical charge of an atom when it loses or gains electrons.
4. Locate in the periodic table those elements which tend to lose electrons and those which tend to gain electrons.
5. Define electrovalent compounds.
6. Balance formulas of simple chemical compounds and draw their electronic structure.
7. Draw the electronic dot structure of atoms.
8. Define oxidation number and list the oxidation numbers of free elements and commonly encountered positive and negative ions.
9. Distinguish by two methods of nomenclature the elements that have two oxidation numbers.
10. Define a radical, write its formula, and give its oxidation number.
11. Describe the covalent bond and how it is formed.
12. Distinguish between compounds which have similar sounding names and state why the exact name is important.
13. Construct correct formulas for ionic compounds by illustrating the procedure to be followed.
14. Compare empirical and molecular formulas.
15. List five phenomena which accompany chemical reactions.
16. List and give an example for four types of chemical reaction.

CHEMICAL BONDING, IONS, CHEMICAL FORMULAS

17. Define oxidation-reduction reactions and give examples of oxidizing and reducing agents.
18. Balance chemical equations.

Elements react and combine with each other to form compounds, which are the substances that you will come in contact with. These compounds, as you already know, are composed of molecules which vary greatly in their size and complexity, from the simplest containing only two atoms, to the very complex protein and nucleic acid molecules containing tens of thousands of atoms of many different elements. The simple gases, such as hydrogen, oxygen, nitrogen, chlorine, etc., are *diatomic,* that is, they contain two atoms of the same element, but they are considered to be molecules (Fig. 6-1).

Each atom of an element is electrically neutral, because the number of protons (the positive charges) equals the number of electrons (the negative charges). Yet all elements, except the inert gases, react with other elements to form compounds which have entirely different physical and chemical characteristics. For example, gaseous oxygen reacts with gaseous hydrogen to form *liquid water.* They are no longer gases; you cannot physically separate the water back into gaseous oxygen and gaseous nitrogen, although you can do it with electricity (Fig. 6-2). The question arises then: Why are the inert gases nonreactive, while the other elements are reactive? That answer can be found in the electronic configuration of the elements.

You will remember that the outermost shells of the elements have different numbers of electrons, but that the inert gases have either 2 electrons (helium) or 8 electrons (all the other inert gases) in their outermost shell. *The 2 and the 8 configuration in the outer shell is an especially stable arrangement.* This is the reason why the inert gases do not react: they have a stable electron configuration. All of the other elements do react because their electron configuration is not stable, and upon reaction, a stable arrangement is achieved, a 2

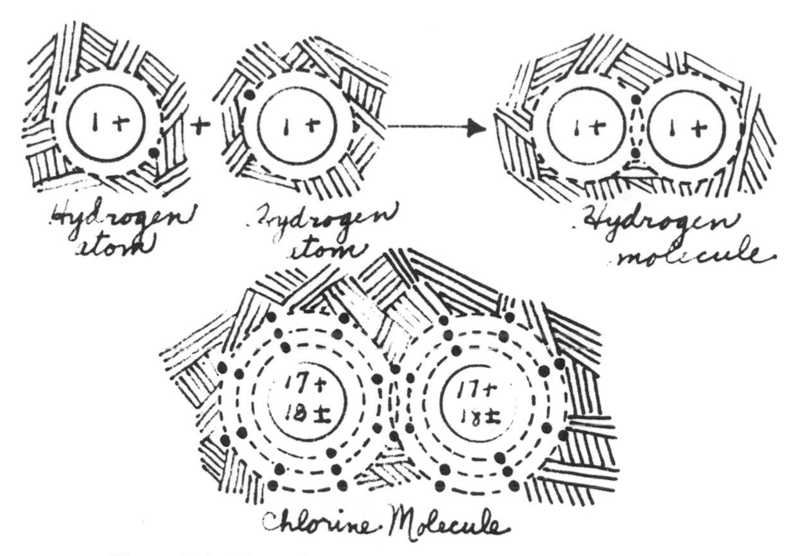

Figure 6-1. Diatomic molecules: hydrogen and chlorine.

or 8 electron outer shell. Therefore, we have the following situation. All of the elements except the inert gases have between 1 and 7 electrons in the outermost shell: the inert gases *always* have either 2 or 8 electrons in that shell. The inert gases do not react because their electronic configuration is stable. In order to achieve that configuration the other elements must:

1. Lose all of the electrons in the incomplete outer shell and thus end up with a complete lower number shell and achieve stability.

or

2. Borrow enough electrons from some other element to fill up their outer shell and thus end up with a complete outer shell and achieve stability.

For those elements having 1, 2 or 3 electrons in their outer shell, it is easier to lose those electrons rather than borrow more. In the case of sodium (at. no. 11, Group IA), whose electron configuration is 2-8-1, the sodium loses 1 electron and therefore drops to the 2-8 electron configuration which is standard (Fig. 6-3). Magnesium (at. no. 12, Group IIA) has a 2-8-2 electron configuration. It tends to lose 2 electrons, dropping to the 2-8 configuration, thereby achieving stability (Fig. 6-4).

For those elements having 5, 6, or 7 electrons, it is easier to borrow the needed electrons to achieve stability (an octet) than to lose electrons. Oxygen (at. no. 8, Group VIA) has an electron configuration of 2-6. It needs 2 electrons to achieve stability. Chlorine (at. no. 17, Group VIIA) has a configuration of 2-8-7. It requires only 1 electron to become 2-8-8 and achieve stability.

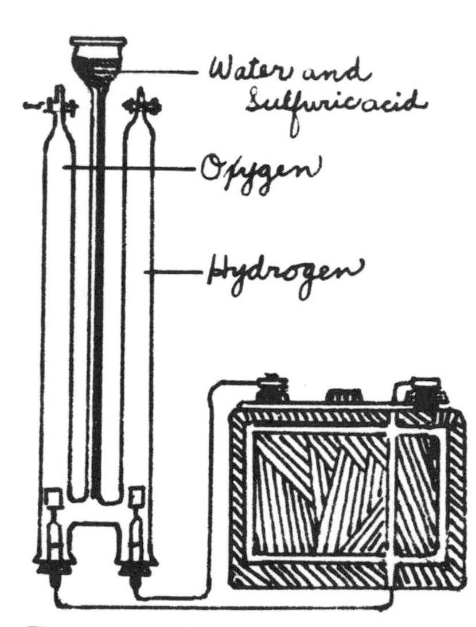

Figure 6-2. Electrical decomposition of water into the elements. When a current of electricity is sent through a mixture of water and sulfuric acid, the water is broken up into the elements hydrogen and oxygen.

EFFECT OF LOSS OR GAIN OF ELECTRONS

The atom of an element is electrically neutral, the number of protons equaling the number of electrons. If electrons are lost or gained, then the neutrality is destroyed, and particles result which are either positively or negatively charged;

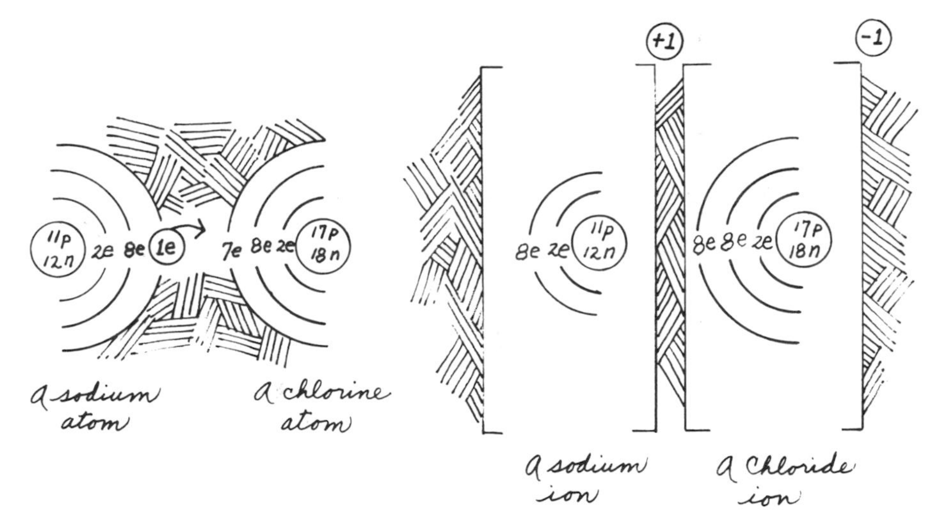

Figure 6-3. Sodium donates 1 electron.

these particles are called *ions.* If *electrons are lost,* the number of electrons (negative charges) is less than the number of positive charges, and therefore the particle is *positively charged.* This is called a *cation.* If *electrons are gained,* the number of electrons (negative charges) will be greater than the number of positive charges, and therefore the particle is *negatively charged.* This is called an *anion.*

NOTE: *Only electrons* can be transferred. *Protons are never* transferred. A particle becomes positively (+) charged due to *loss* of electrons, *never due to a gain of protons.*

When an atom becomes an ion, it loses the characteristics which it had as an atom, and acquires new chemical and physical characteristics. For example, chlorine is a yellow, poisonous, irritating gas, but, when it accepts an electron to become a chloride ion, it becomes harmless. As part of the compound NaCl, sodium chloride (ordinary table salt), the compound is a white crystalline solid, salty, and soluble in water.

Elements Tending to Lose Electrons

The metals, located on the left of the periodic table, tend to *lose* their electrons and thereby become positively charged cations. Thus sodium, as you just read, tends to form the sodium ion by losing one electron. Since it lost *1* negatively charged electron, the ion has a *plus one* charge, because now it has 11 positively charged protons in the nucleus and only 10 negatively charged electrons in orbit. Magnesium lost *2* electrons and the magnesium ion formed has a *plus two* charge because it now has 12 positively charged protons in the nucleus and only 10 negatively charged electrons. Therefore the *metals become positively charged* when they become ions by losing electrons.

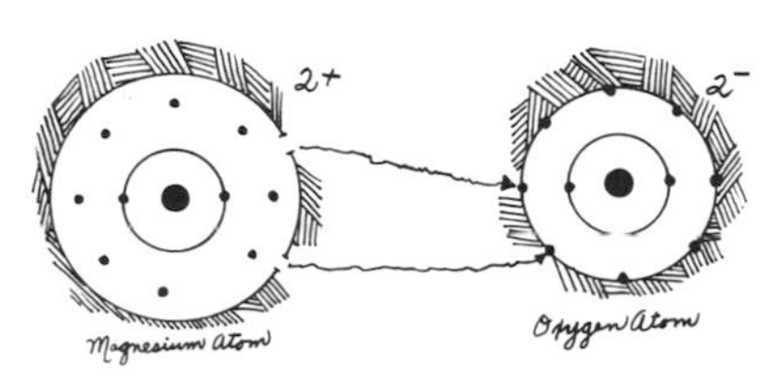

Figure 6-4. Magnesium donates 2 electrons.

Elements Tending to Gain Electrons

As you just read, *the nonmetals tend to gain electrons and thereby become negatively charged.* Chlorine gains 1 electron to form the chloride ion with one minus charge. Oxygen gains 2 electrons to complete its outer shell to form the oxide ion with a charge of 2^-.

Electrovalent Compounds (Ionic Compounds)

When a metal (losing electrons) and a nonmetal (gaining electrons) react chemically, the electrons are transferred, and an electrovalent compound is formed. It is called electrovalent because 2 oppositely charged ions attract each other forming the electrovalent chemical bond, also called an ionic bond. The number of electrons which are given up must always be equal in number to the electrons accepted, and conversely, for an atom to receive a certain number of electrons, that exact number must have been given up.

Naming Chemical Compounds

When a metal reacts with a nonmetal to form a compound, the compound is named by the following method:

1. The metal name is used first.
2. The nonmetal is used last, with the ending of the element modified to indicate that it is part of a *compound.*

Examples:

Fluorine	becomes	fluor*ide*
Chlorine	becomes	chlor*ide*
Bromine	becomes	iod*ide*
Oxygen	becomes	ox*ide*
Sulfur	becomes	sulf*ide*

Table 6-1 illustrates the naming of simple diatomic compounds.

Table 6-1. Names of simple diatomic compounds.

Metal reacting	*Nonmetal reacting*	*Compound formed*	*Name of compound*
Lithium	Chlorine	LiCl	Lithium chlor*ide*
Lithium	Bromine	LiBr	Lithium brom*ide*
Lithium	Iodine	LiI	Lithium iod*ide*
Sodium	Chlorine	NaCl	Sodium chlor*ide*
Potassium	Chlorine	KCl	Potassium chlor*ide*
Calcium	Oxygen	CaO	Calcium ox*ide*
Barium	Sulfur	BaS	Barium sulf*ide*

BALANCING ELECTRONS IN THE FORMULA

In the previous section, you noted that when 1 electron was given up, 1 was accepted in order to achieve stability. When 2 electrons were given up, as with magnesium, 2 electrons were accepted by the oxygen. However, in many cases,

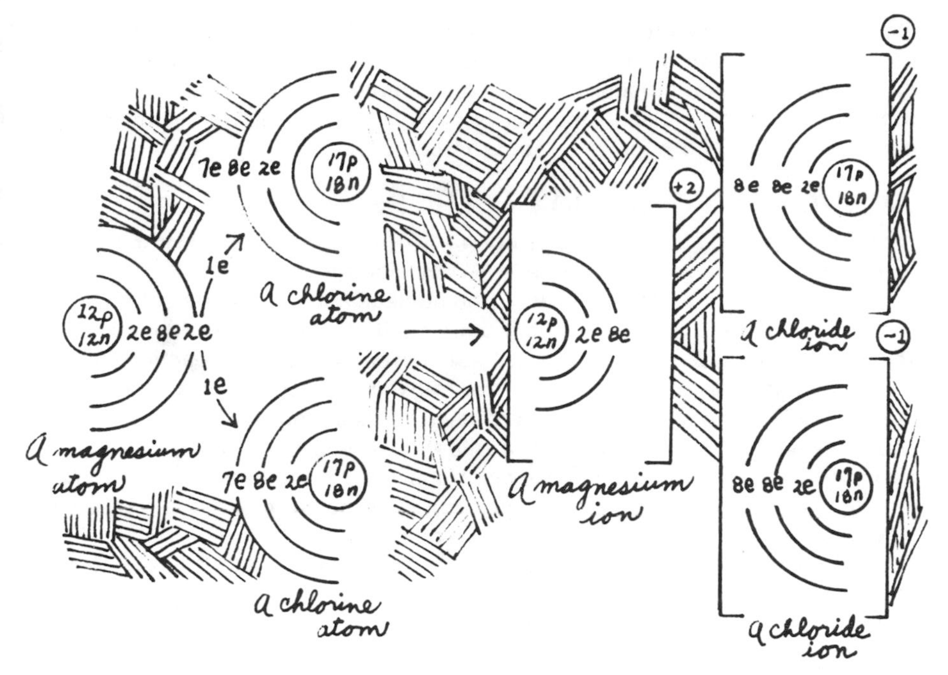

Figure 6-5. Balancing electrons.

there is an imbalance between the metal and the nonmetal electron capacity to give up or to accept electrons. In these cases, more than 1 atom of the metal or nonmetal must be used in order to achieve the balance of electrons transferred. The following example illustrates this point:

Magnesium (at. no. 12), electron configuration of 2-8-2, can achieve stability by *losing 2* electrons to become the magnesium ion, +2 charged, having a 2-8 configuration (Fig. 6-5).

Chlorine (at. no. 17), electron configuration of 2-8-7, can achieve stability by *accepting 1* electron to become the chloride ion, +1 charged, and having the 2-8-8 configuration.

The magnesium needs to give up *2* electrons, whereas the chlorine can accept only *1*. The magnesium therefore requires *2* chlorine atoms, each accepting 1 electron, in order to donate the 2 electrons in its outer shell. Table 6-2 will

Table 6-2. Names of simple polyatomic molecules.

Metal reacting	*Nonmetal reacting*	*Compound formed*	*Formula*
Calcium	Chlorine	Calcium chloride	$CaCl_2$
Calcium	Iodine	Calcium iodide	CaI_2
Sodium	Oxygen	Sodium oxide	Na_2O
Potassium	Sulfur	Potassium sulfide	K_2S

illustrate the balanced formulas of polyatomic compounds, and you should take special note to confirm that, in each case, the number of electrons donated by the metal equals the number of electrons accepted by the nonmetal. The electrical charge of *any* molecule must be equal to zero, the number of plus charges must be equal to the number of negative charges.

VALENCE AND OXIDATION NUMBERS

When changes in the electronic structure take place as a result of a chemical reaction, the changes can be represented by electron dot notation. The symbol of the atom represents all of the atom (the nucleus and the inner electron shells) and the (·), (x) and (o) represent the outer shell of electrons, which are termed valence electrons. When electrons are transferred in chemical reactions to form an electrovalent or ionic compound, the transfer can be shown as that seen in Figure 6-6. The different notations are used to designate their source, but you should remember that *all electrons are identical regardless of their source.* As far as we are concerned, only those electrons in the outer shell affect the chemistry of the atom, and we can therefore represent the atom on paper in a simple way, even though it is a two dimensional picture.

The oxidation number is also called the oxidation state, and it designates the electrical character of the atoms; whether they are negatively, positively, or neutrally charged. When electrons are donated by an atom and it becomes more positively charged, the atom is said to be in a *positive* oxidation state and is given a positive oxidation number. When the atom accepts electrons and becomes more negatively charged, it is said to be in a negative oxidation state and is given a negative oxidation number. The number of electrons accepted or donated by the atom determines the numerical value of the oxidation number. Table 6-3 gives the oxidation numbers of commonly encountered ions. Those elements in the free state, such as elemental sodium, oxygen, chlorine, silver, and hydrogen are assigned an oxidation number of *zero.* Therefore, when

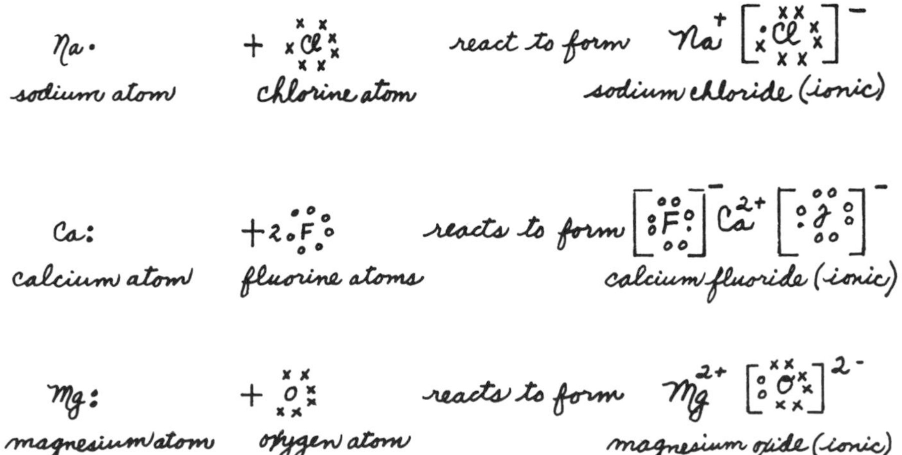

Na + Cl react to form $Na^{+}[Cl]^{-}$
sodium atom chlorine atom sodium chloride (ionic)

Ca + 2 F reacts to form $[F]^{-}Ca^{2+}[F]^{-}$
calcium atom fluorine atoms calcium fluoride (ionic)

Mg + O reacts to form $Mg^{2+}[O]^{2-}$
magnesium atom oxygen atom magnesium oxide (ionic)

Figure 6-6. Electron dot representations of chemical reactions.

Table 6-3. Oxidation numbers of commonly encountered ions.

Ion	Name	Ion	Name
Positive 1 (1+)		*Negative 1 (1−)*	
H^+	Hydrogen	Cl^-	Chloride
Na^+	Sodium	OH^-	Hydroxide
K^+	Potassium	NO_2^-	Nitrite
Ag^+	Silver	NO_3^-	Nitrate
Cu^+	Copper (1)	Br^-	Bromide
NH_4^+	Ammonium	I^-	Iodine
Hg^+	Mercury (1)	HCO_3^-	Bicarbonate
Positive 2 (2+)		*Negative 2 (2−)*	
Ca^{++}	Calcium	O^{--}	Oxide
Cu^{++}	Copper (11)	S^{--}	Sulfide
Ba^{++}	Barium	SO_3^{--}	Sulfite
Hg^{++}	Mercury (11)	SO_4^{--}	Sulfate
Pb^{++}	Lead (11)	CO_3^{--}	Carbonate
Zn^{++}	Zinc	HPO_4^{--}	Biphosphate
Mg^{++}	Magnesium		
Fe^{++}	Iron (11)		
Positive 3 (3+)		*Negative 3 (3−)*	
Al^{+++}	Aluminum	PO_4^{---}	Phosphate
Fe^{+++}	Iron (111)		

they lose electrons and become positively charged, they are in the positive oxidation state. When they accept electrons and become negatively charged, they are in the negative oxidation state. *The oxidation number only refers to one atom of an element.*

In any compound the algebraic sum of the positive oxidation numbers and the negative oxidation numbers of the atoms must always be equal to zero.

Examples:

Compound	Oxidation states	
NaCl	Na^{1+} Cl^{1-}	sum of $+1$ and $-1 = 0$
$CaCl_2$	Ca^{2+} 2 Cl^{1-}	sum of $+2$ and $2\times(-1) = 0$
BaO	Ba^{2+} O^{2-}	sum of $+2$ and $-2 = 0$
Na_2O	2 Na^{1+} O^{2-}	sum of $2 \times (+1)$ and $-2 = 0$

Multiple Oxidation Numbers

Some elements have more than one oxidation number, that is, they can exist in several oxidation states, and consequently the name of the compound must reflect the exact oxidation number to avoid confusion and possible mistakes. There are two methods used to name compounds where this situation exists. One uses the ending of the metal to indicate the oxidation state. For example iron can exist as Fe^{2+}, which is called fer*rous*, or Fe^{3+}, which is called fer*ric*. Compounds of iron and chlorine would be:

Fe^{2+} Cl_2^{1-} or $FeCl_2$ is fer*rous* chloride
Fe^{3+} Cl_3^{1-} or $FeCl_3$ is fer*ric* chloride

Similarly tin can exist as Sn^{2+}, which is called stan*nous*, or Sn^{4+}, which is called stann*ic*.

$Sn^{2+} Cl_2^{1-}$ or $SnCl_2$ is stan*nous* chloride
$Sn^{4+} Cl_4^{1-}$ or $SnCl_4$ is stann*ic* chloride

The *ous* ending designates the lower oxidation state, the *ic* ending, the higher oxidation state.

This system is satisfactory when there are only two oxidation states, but some elements have as many as five. The newer system in use today uses Roman numerals to indicate the oxidation state. The above examples would therefore be identified as:

$FeCl_2$ Iron (II) chloride
$FeCl_3$ Iron (III) chloride
$SnCl_2$ Tin (II) chloride
$SnCl_4$ Tin (IV) chloride

RADICALS

Groups of atoms can bond together and behave as individual atoms in chemical reactions. The individual character of the elements disappears and the group of atoms exhibits specific characteristics attributable to the radical, which can be either positively or negatively charged. The group of atoms has an oxidation number which depends upon the number and type of atom contained in the radical. They combine with oppositely charged ions to form electrically neutral molecules, following the laws of chemical bonding which apply to simple compounds.

These radicals have specific names that indicate the type and number of atoms comprising the radical. The chemistry of the body involves these radicals. Their names and formulas should be learned. The more common ones are listed in Table 6-4 along with their oxidation numbers and a sample compound. Some

Table 6-4. Common radicals, their oxidation numbers and sample compounds.

Radical	*Name*	*Oxidation number*	*Sample compound*	*Common name*
OH^{1-}	Hydroxide	1−	$NaOH$	Sodium hydroxide
NO_2^{1-}	Nitrite	1−	$NaNO_3$	Sodium Nitrite
NO_3^{1-}	Nitrate	1−	$NaNO_3$	Sodium Nitrate
SO_3^{2-}	Sulfite	2−	$NaSO_3$	Sodium sulfite
SO_4^{2-}	Sulfate	2−	$NaSO_4$	Sodium sulfate
CO_3^{2-}	Carbonate	2−	Na_2CO_3	Sodium carbonate
$C_2O_4^{2-}$	Oxalate	2−	$Na_2C_2O_4$	Sodium oxalate
PO_3^{3-}	Phosphite	3−	Na_3PO_3	Sodium phosphite
PO_4^{3-}	Phosphate	3−	Na_3PO_4	Sodium phosphate
NH_4^{1+}	Ammonium	1+	NH_4OH	Ammonium hydroxide
HCO_3^{1-}	Bicarbonate	1−	$NaHCO_3$	Sodium bicarbonate

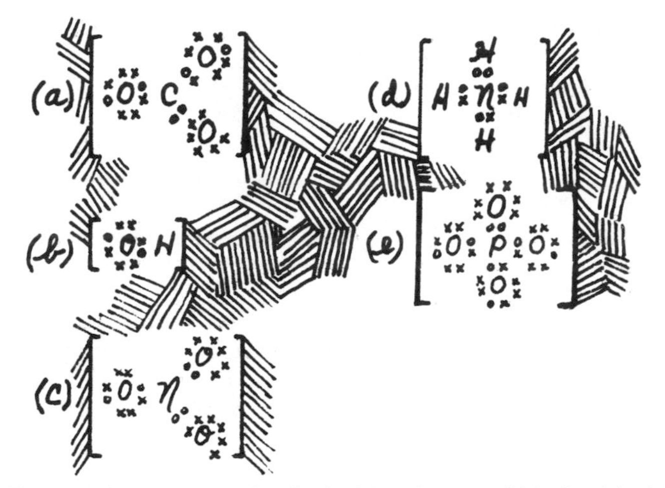

Figure 6-7. Electron dot structure of radicals. (a) carbonate, (b) hydroxide, (c) nitrate, (d) ammonium, and (e) phosphate.

electron dot structures are shown in Figure 6-7. The electron configuration of the entire sulfate radical is shown in Figure 6-8.

The simplest way to remember the oxidation numbers of the different radicals is to learn the formulas for the most commonly encountered acids, bases, and salts, keeping in mind that the:

Hydrogen ion always has an oxidation number of 1+
Hydroxyl ion always has an oxidation number of 1−
Ammonium ion always has an oxidation number of 1+
Sodium and potassium always have an oxidation number of 1+

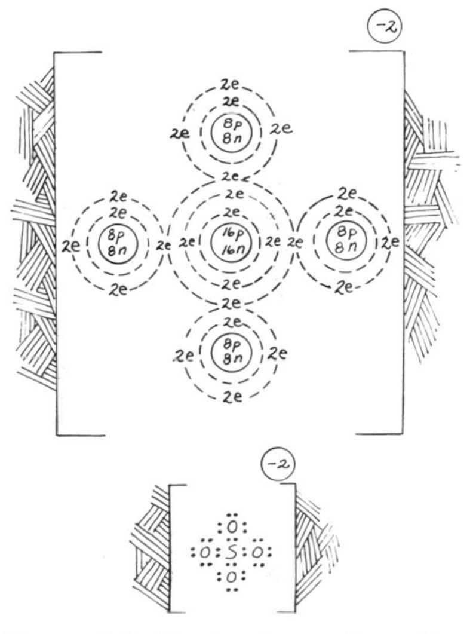

Figure 6-8. Electronic configuration of the sulfate radical.

If you remember the above facts, you can, in most cases, determine the value of the oxidation number of the radical. For example:

H_2SO_4 Sulfuric acid—2 hydrogens: $2 \times (+1) = +2$: SO_4 must be 2−
H_3PO_4 phosphoric acid—3 hydrogens: $3 \times (1+) = 3+$: PO_4 must be 3−
$NaNO_3$ sodium nitrate—1 sodium: $1 \times (1+) = 1+$; NO_3 must be 1−
$KMnO_4$ potassium permanganate—potassium 1+: MnO_4 must be 1−
HNO_3 nitric acid—1 hydrogen 1+; NO_3 must be 1−

THE COVALENT BOND

There are many compounds formed which are not ionic but consist of molecules whose atoms are tightly bound together. Chemical bonding in these cases is achieved by *sharing* (not lending and borrowing) electrons in order to achieve the stability of the inert gases, and it is called *covalent bonding or covalency.* The bonds which hold the atoms together are called covalent bonds or shared electron pair bonds.

When electrons are shared between atoms, each atom receives the full benefit

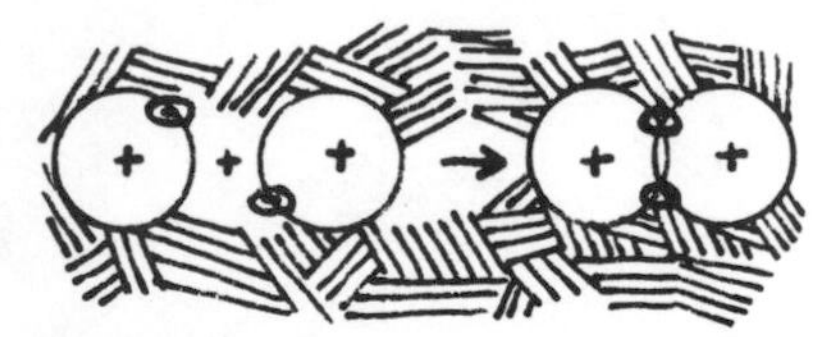

Figure 6-9. Covalent bonding of 2 hydrogen atoms to form the hydrogen gas molecule.

of the shared electrons, and their presence and number is counted in each atom's electron structure. The hydrogen molecule serves as an example of covalent bonding by the sharing of electrons between 2 hydrogen atoms (Fig. 6-9). A hydrogen atom has 1 proton and 1 electron; it needs 1 additional electron in its electron shell in order to reach the stable outer electron shell of the inert gas helium, which has 2 electrons in it. As a result, 2 hydrogen atoms share their electrons, each atom receiving the full benefit of the shared electrons. Each now has a complete outer energy shell of 2 electrons resembling the stable outer energy shell configuration of the inert gas helium. In the case of the hydrogen molecule, the 1 electron of the atom was shared, but sharing can take place as long as the need exists. Chlorine, (at. no. 17) with a 2-8-7 electron structure, needs only *1* electron to reach the stable 2-8-8 arrangement, because it has 7 electrons in its outer shell and needs to have 8. The 2 atoms of chlorine can share *1* of their outer 7 electrons between themselves and therefore reach the stable structure of argon.

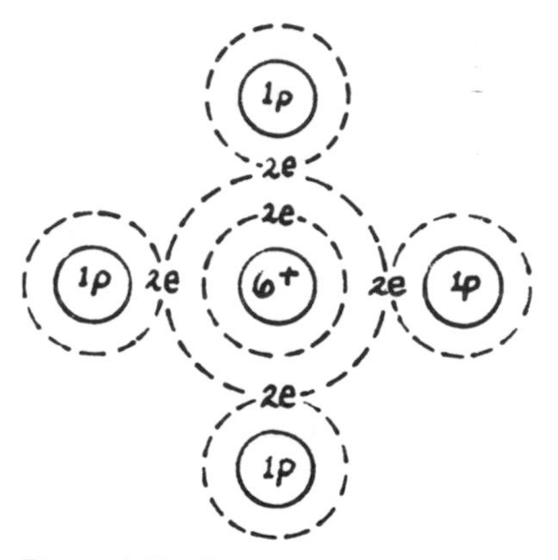

Figure 6-10. Covalent bonding: Methane CH_4.

Covalent Bonding between Unlike Atoms

Unlike atoms can share 1 or more of their outer shell electrons in order to reach the stability of the inert gases. This type of bonding is important to you because it is this type of covalent bond which results in the extremely large and complex molecules found in the body. Molecules are formed when the unlike atoms share 1 electron from each atom, as in the case of hydrogen chloride or methane (Fig. 6-10). In the case of methane, the carbon has 4 electrons in its outer shell. It shares 4 electrons, 1 with each hydrogen atom, each hydrogen in turn sharing its 1 electron with the carbon atom. Each reaches the stable state, carbon with 2-8, and hydrogen with 2 in its outer shell. Another type, where 1 atom shares its electrons with 2 other atoms is found in water (Fig. 6-11). The oxygen has 6 electrons and completes an octet of electrons (8) by sharing 2 of its electrons with *2* hydrogen atoms (1 with each hydrogen atom).

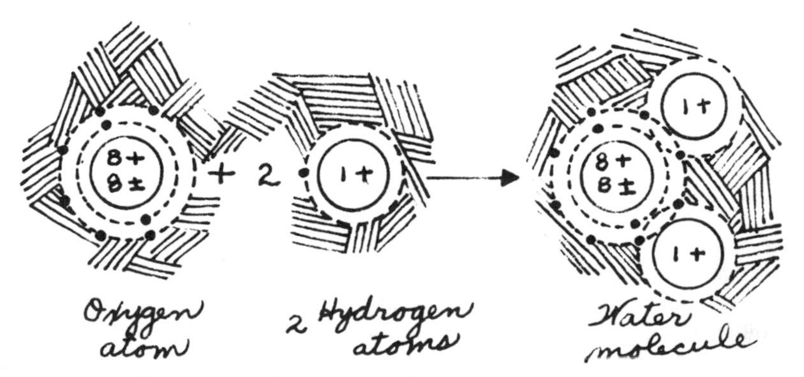

Figure 6-11. Covalent bonding in the water molecule.

WRITING CORRECT CHEMICAL FORMULAS

The importance of knowing, recognizing, and being able to write the correct formula for the wanted compound cannot be overemphasized. No one can afford to be careless in reading or writing the names of compounds or their formulas on medications or chemical test solutions. The difference in the oxidation number of the element determines the characteristics of the compound, and very serious damage can be caused by your being careless, inattentive, or just not knowing the difference.

For example: Sodium chlor*ide* is common table salt. It is essential to life, maintains the proper balance of fluid flow in the body, and stimulates the action of the heart muscle.

Sodium chlor*ate* decomposes when heated, giving off oxygen, forming explosive mixtures with other substances, acts as a purgative and is insoluble in water.

THESE ARE NOT THE SAME.

Mercury (1) chloride is calomel, a purgative, used medicinally, insoluble in water. It is also called mercur*ous* chloride.

Mercury (11) chloride is called corrosive sublimate. It is a poison, cannot be taken internally, and is used as an antiseptic and antisyphylitic. This is mecur*ic* chloride.

THESE ARE NOT THE SAME.

Other examples include carbon monoxide and carbon dioxide, amyl nitrite and amyl nitrate, sodium sulfite and sodium sulfate, benzine and benzene, and carbonic acid and carbolic acid.

Constructing Formulas for Ionic Compounds

In order to correctly write and read formulas, you must recognize the oxidation numbers of the atoms, ions and radicals, and always remember that the algebraic sum of the positive and negative charges must be *zero.* You already have learned the oxidation number of some of the elements and most of the radicals which you will come in contact with. The construction of a formula involves adjusting the ratio of the component parts of the molecule so that the molecule is electrically neutral.

PROCEDURE

1. Write down the component parts of the molecule.
   Example: To write sodium sulfate, write Na for sodium and SO_4 for sulfate.
2. Write the oxidation number of each molecule as a superscript.
   sodium would be Na^{1+}
   sulfate would be SO_4^{2-}
3. If the algebraic sum is equal to zero, the molecule is neutral and correct. If it is not, the number of each component must be adjusted to make the sum equal zero.
   The sum of oxidation numbers for sodium sulfate equals 1+ and $-2 = -1$
   The molecule, as written, is *not correct.*

4. "Crisscross" the oxidation numbers. This is the same thing as getting the lowest common denominator in arithmetic. The sum will now be zero.
   From Step 2: Na^{1+} SO_4^{2-}
   Crisscross: Na^{1+} SO_4^{2-}
   2 ↙ ↘ 1
   Check sum: $2 \times (1+) + 1 \times (2-) = 0$
   $+2$ and $-2 = 0$
   The formula is written correctly. Erase all oxidation numbers.

Additional examples:

Calcium chloride: Ca^{2+} Cl^{1-}; Ca^{2+} Cl^{1-}: $CaCl_2$
1 ↙ ↘ 2
$1 \times (2+) + 2 \times (1-) = 0$

Aluminum oxide: Al^{3+} O^{2-}; Al^{3+} O^{2-}: Al_2O_3
2 ↙ ↘ 3
$1 \times (3+) + 3 \times (1) = 0$

Sodium phosphate: Na^{1+} $(PO_4)^{3-}$; Na^{1+} $(PO_4)^{3-}$: Na_3PO_4
3 ↙ ↘ 1
$3 \times (1+) + 1 \times (3-) = 0$

Calcium phosphate: Ca^{2+} $(PO_4)^{3-}$; Ca^{2+} $(PO_4)^{3-}$: $Ca_3(PO_4)_2$
3 ↙ ↘ 2
$3 \times (2+) + 2 \times (3-) = 0$

NOTE: When you are calculating molecular weights or merely need to know how many atoms of each element are in the molecule, you must take special precautions in your calculations when the molecule is a radical.

Take for example the molecule $Al_2(SO_4)_3$:
There are *2* aluminums, Al_2
There are *3* sulfate radicals $(SO_4)_3$
But in each sulfate radical there is 1 sulfur atom and 4 oxygen atoms.
Therefore; since there are *3* radicals, there are *3 sulfur atoms* and *12 oxygen atoms* in the molecule. In the whole molecule there are:
2 aluminum atoms,
3 sulfur atoms, and
12 oxygen atoms.

Empirical Formulas

Empirical formulas are important to health care personnel because they provide the *ratio* of each type of atom in the molecule. They do not indicate the *actual* number of atoms of each element in the molecule, only the *relative* number. Some examples can be seen in Table 6-5.

Table 6-5. Molecular and empirical formulas.

Compound	Molecular formula	Empirical formula
Hydrogen peroxide	H_2O_2	HO
Mercury (1) chloride	Hg_2Cl_2	HgCl
Benzene	C_6H_6	CH

CHEMICAL REACTIONS

When chemical changes take place in the body or in our environment the compounds that react form compounds which have entirely different chemical and physical properties. These chemical changes are accompanied by the following phenomena:

1. Evolution or absorption of heat.
2. The formation of a precipitate.
3. The evolution of a gas.
4. A change in color of the solution.
5. The emission of an odor.
6. The development or loss of fluorescence in the solution.*
7. The emission of light (chemiluminescence).*

*Numbers 6 and 7 occur infrequently. Number 7 is illustrated by the firefly.

Chemical Equations

The statement that charcoal reacts with oxygen to form carbon dioxide is expressed by scientists in the form of an equation:

$$C + O_2 \longrightarrow CO_2$$

The carbon and the oxygen are the *reactants.* The arrow indicates that the reaction proceeded as shown, and the substance that was formed, the *product*, is on the right of the arrow. Chemical change can result in the conversion of complex substances into simple substances or the combination of simpler substances into more complex substances. Bonds between atoms are broken and new bonds are formed. Energy, in the form of heat, is always involved in chemical reactions. When it is given off the reaction is called *exothermic.* When heat is absorbed, the reaction is called *endothermic.* When other forms of energy are involved, including heat, the terms exergonic and endogonic are used.

We know, according to the law of the conservation of matter, that matter can be neither created nor destroyed in ordinary chemical reactions. No one kind of atom is ever changed (except in radioactivity) into another kind of atom. Therefore, in a chemical reaction, every atom involved in the reactants must be found in the product (Fig. 6-12). Of course it may now be part of a new compound having entirely different properties, but it must be accounted for in the product.

Figure 6-12. The law of conservation of matter. Every atom in the reactant must be accounted for in the product.

Types of Chemical Reactions

There are four basic types of ordinary chemical reactions.

SYNTHESIS

This is a reaction where more complex compounds are formed. Carbon dioxide from the air and water are converted by plants to a complex molecule, a sugar, in a process called photosynthesis, liberating oxygen.

$$6\ CO_2 + 6\ H_2O \longrightarrow C_6H_{12}O_6 + 6\ O_2$$

DECOMPOSITION

In this reaction more complex substances decompose to form simpler substances. The heating of a sugar decomposes it into pure carbon and water.

$$C_6H_{12}O_6 \xrightarrow[\text{heat}]{} 6\ C + 6\ H_2O$$

SINGLE DISPLACEMENT

In this case one element displaces another element in a compound and takes its place in the molecule. Zinc dropped into a solution of copper sulfate will replace the copper in the copper sulfate producing the element copper. The zinc has replaced the copper in the compound, forming a new compound, zinc sulfate, and liberating the element copper.

$$Zn + CuSO_4 \longrightarrow ZnSO_4 + Cu$$

DOUBLE DISPLACEMENT

Double displacement reactions take place when substances that are part of two different compounds displace each other. You might say that they exchange partners. When a solution of sodium chloride (NaCl), table salt, is added to a solution of silver nitrate, a white precipitate, a solid, appears. This white insoluble substance is silver chloride.

$$AgNO_3 + NaCl \longrightarrow AgCl + NaNO_3$$

OXIDATION-REDUCTION REACTIONS

Oxidation-reduction processes occur continuously in the body tissues. Carbohydrates, fats, and proteins are oxidized to provide energy for the body, and the body builds needed molecules through this reaction. The process takes place when atoms lose or gain electrons. When an atom *loses* electrons, it is said to be *oxidized,* and when it *gains electrons,* it is said to be *reduced.* In any event, all of the electrons must be accounted for, the number of electrons given up must be equal to the number of electrons accepted. Something can not be oxidized without something else being reduced.

Table 6-6. Oxidizing agents and reducing agents.

Name	Type	Formula	Application
Sodium hypochlorite Dakin's solution	Oxidizing	$NaOCl$	Wound treatment skin abrasions
Hydrogen peroxide	Oxidizing	H_2O_2	Wound treatment skin abrasions
Potassium permanganate	Oxidizing	$KMnO_4$	Antiseptic wound treatment
Chlorine water	Oxidizing	Cl_2/H_2O	Antiseptic
Tincture iodine	Oxidizing	I_2/alcohol	Antiseptic skin abrasions
Oxalic acid	Reducing	$(COOH)_2$	Removes rust spots and permanganate stains
Sodium thiosulfate	Reducing	$Na_2S_2O_3$	Removes silver and iodine stains

Example: In the example given under single displacement reaction, the elemental zinc displaces the copper. The oxidation number, you will recall, of free elements is always zero.

$$Zn^0 + Cu^{2+} (SO_4)^{2-} \longrightarrow Zn^{2+} (SO_4)^{2-} + Cu^0$$

The element zinc *lost* 2 electrons; it was oxidized, and became the zinc ion (Zn^{2+}). The copper ion (Cu^{2+}) *accepted* the 2 electrons; it was reduced, now electrically neutral, and became elemental copper, Cu^0.

Oxidizing agents are important to health care personnel because they kill bacteria and act as antiseptics. They also act as bleaching agents to remove stains, such as household bleaches. Reducing agents are used to disinfect rooms that have been occupied by patients who have contagious diseases, an example being formaldehyde. Some further examples can be seen in Table 6-6.

BALANCING CHEMICAL EQUATIONS

As you already know, matter cannot be created nor destroyed in ordinary chemical reactions. Therefore, every atom of every element must be accounted for in the reactants and the products. It is extremely important that each compound be written correctly and that the molecules be electrically neutral. Otherwise the equation will not represent the true facts because the formulas of the compounds are wrong to start with.

There is no hard and fast rule which will enable you to balance the chemical equations. However, you can observe certain key facts which will assist you in balancing any ordinary equation. They are:

1. Locate the part of the equation *which is most out of balance.*
2. Start at that point. Use whole numbers in *front* of *metals or molecules* which contain radicals in the reactants or products to make the number even on each side.

a. Formulas of compounds are fixed. You *cannot* alter any subscripts, but you can multiply the entire molecule by a whole number without affecting the formula of the molecule.
b. Gaseous elements, such as hydrogen, nitrogen, oxygen, and chlorine must always be indicated in the molecular state, e.g., H_2, N_2, O_2 and Cl_2.
c. The chemical equation is an expression of experimental facts. It cannot be altered or changed merely to accommodate your desire.

3. Leave the balancing of H_2O for the last operation. This will usually take care of the hydrogens and hydroxyls.

Concept of Balancing Equations

Sodium oxide reacts with hydrogen chloride to form water and sodium chloride. We would write this as a chemical equation as follows:

$$\underset{\text{Reactants}}{Na_2O + HCl} \longrightarrow \underset{\text{Products}}{H_2O + NaCl}$$

As this reaction is written, it violates the law of conservation of matter.

In the reactants there are: 2 atoms of sodium
1 atom of oxygen
1 atom of chlorine
1 atom of hydrogen

In the products there are: 1 atom of sodium
1 atom of oxygen
1 atom of chlorine
2 atoms of hydrogen

Obviously, the *reaction is not balanced,* because not all of the atoms in the reactants are accounted for in the products. In the reactants there are 2 atoms of *sodium,* but in the products, only *1* atom of sodium. The reverse is true for hydrogen: 2 hydrogens in the products and only 1 hydrogen in the reactants. Going back to the reaction $Na_2O + HCl \rightarrow H_2O + NaCl$, you can see 2 sodiums in the reactant Na_2O; you therefore need 2 sodiums in the product. However, NaCl has only 1 sodium. Therefore, according to rule 2 above, if you multiply the NaCl by the number 2, you will now have the required 2 sodiums in the product:

$$Na_2O + HCl \longrightarrow H_2O + 2\,NaCl$$

Now you have 2 sodiums on each side; the sodiums are balanced. But, in doing so, you now have *2* chlorines in the product, but only *1* chlorine in the reactant; they are out of balance. You need 2 chlorines in the reactants, so therefore, apply rule 2 above and multiply the HCl molecule by 2. This will account for the 2 chlorines in the reactants.

$$Na_2O + 2HCl \longrightarrow H_2O + 2NaCl$$

So far you have balanced the sodiums and the chlorines, and, in this equation, it so happens that the hydrogens and hydroxyls were automatically balanced.

In many reactions, it is necessary to multiply the water molecules by a whole number in order to account for them. The equation is balanced when the atoms in the reactants equal the atoms in the products.

Other Examples of Balancing Chemical Equations

$$NaOH + HCl \longrightarrow NaCl + \underset{(H\text{—}OH)}{H_2O}$$

Analysis:

In the reactants	*In the products*	*Need*	*What to do*
1 sodium	1 sodium	satisfied	nothing
1 chlorine	1 chlorine	satisfied	nothing
1 hydrogen	1 hydrogen	satisfied	nothing
1 hydroxyl	1 hydroxyl	satisfied	nothing

The equation is a balanced equation as written.

$$NaOH + H_2SO_4 \longrightarrow Na_2SO_4 + \overset{(H\text{—}OH)}{H_2O}$$

Analysis:

In the reactants	*In the products*	*Need*	*What to do*
1 sodium	2 sodiums	1 sodium	multiply NaOH by 2

The sodiums are balanced, but now there are 2 OH in the reactants and 1 OH in the products.

$$2\ NaOH + H_2SO_4 \longrightarrow Na_2SO_4 + H_2O$$

2 hydroxyls	1 hydroxyl	1 hydroxyl	multiply H_2O by 2

The reaction now looks like this:

$$2\ NaOH + H_2SO_4 \longrightarrow Na_2SO_4 + 2\ H_2O$$

The reaction is now balanced:
In the reactants: 2 Na; 2 OH; 2 H; 1 SO_4
In the products: 2 Na; 2 OH; 2 H; 1 SO_4

SUMMARY

Elements react and form compounds, and in doing so they achieve the stable outer shell of electrons of the inert gases, 2 or 8 electrons. They can react by lending and borrowing electrons (electrovalent bonding) to form compounds; the metals lending electrons, and the nonmetals borrowing electrons. The number of electrons lent must always equal the number borrowed. In lending or borrowing electrons, the electrical charge is changed: a positive oxidation number indicates the atom lost electrons, and a negative one indicates that it accepted electrons. In a molecule the algebraic sum of the oxidation numbers always equals zero. Radicals are groups of atoms bonded together acting as a single atom, having specific formulas, names, and oxidation numbers. Some compounds are formed by sharing electrons to achieve stability; they are covalent, and do not ionize. The names of chemical compounds must be specific, and in constructing formulas for ionic compounds, the sum of the positive and negative charges must equal zero. There are four types of chemical reactions, a chemical reaction being an expression of experimental fact. Reactions must be balanced, because every atom in the reactants must be accounted for in the products.

EXERCISE

1. What is the electrical charge on the atom of an element? Why?
2. Why do elements react with each other?
3. Draw the electronic representation of sodium reacting with chlorine to form sodium chloride.
4. If the atom loses electrons, how does it become charged? If it gains electrons?
5. Which elements tend to lose electrons? Which tend to accept electrons?
6. When is an electrovalent compound formed?
7. How are simple binary compounds named?
8. What is the electrical charge of any molecule?
9. Which of the following compounds have the correct formula

$CaBr$	KOH	HCl_2
$NaBr_2$	$Na(OH)_2$	H_2SO_4
HPO_4	$MgOH$	HNO_3
NaO	$MgSO_4$	Na_2SO_4

10. What are the oxidation numbers of the following *ions*?
    Na, K, Mg, Ca, Zn, Al, Ag, H, Cl, OH, NO_2, NO_3, PO_4, SO_4, SO_3, CO_3, HCO_3.
11. Write the electronic structure for the following compounds:
    HCl; H_2O; methane (CH_4); NaCl.
12. Match each compound with its name:

Sulfuric acid	$NaOH$
Hydrochloric acid	H_2SO_4
Sodium hydroxide	$Al_2(SO_4)_3$
Calcium hydroxide	$NaCl$
Phosphoric acid	CO
Aluminum sulfate	CO_2
Sodium chloride	HPO_4
Carbon dioxide	H_3PO_4
	HCl
	$Ca(OH)_2$

13. Match the types of chemical reactions with the following (numbers 1 through 4 have 2 answers):

a. Synthesis	1. $Zn + 2\,HCl \longrightarrow ZnCl_2 + H_2$
b. Decomposition	2. $2\,HgO \longrightarrow 2\,Hg + O_2$
c. Single displacement	3. $2\,Mg + O_2 \longrightarrow 2\,MgO$
d. Double displacement	4. $C + O_2 \longrightarrow CO_2$
e. Oxidation-reduction	5. $NaCl + AgNO_3 \longrightarrow AgCl + NaNO_3$

14. Balance the following:
    a. $H_3PO_4 + NaOH \longrightarrow Na_3PO_4 + H_2O$
    b. $Ca(OH)_2 + HCl \longrightarrow CaCl_2 + H_2O$
    c. $NaOH + HCl \longrightarrow NaCl + H_2O$
    d. $Al + H_2SO_4 \longrightarrow Al_2(SO_4)_3 + H_2$
15. What would be the empirical formula for a sugar that has a molecular formula of $C_6H_{12}O_6$?

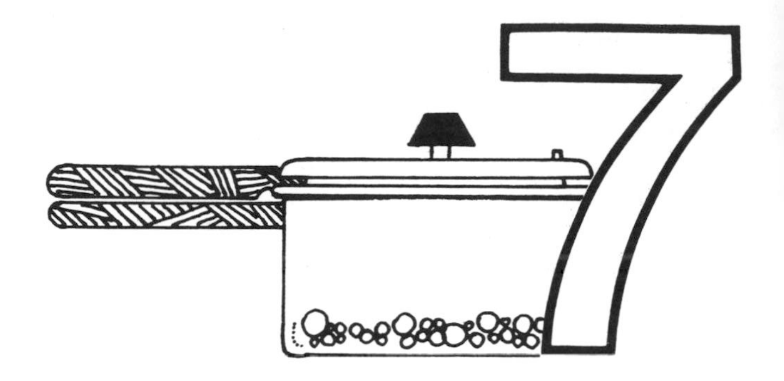

OBJECTIVES

When you have completed this chapter you will be able to:

1. State the physical properties of water.
2. Explain why living cells may die when they are frozen.
3. Distinguish between cohesion and adhesion of water molecules with other molecules.
4. Describe the phenomena of surface tension.
5. Explain the relationship between vapor pressure and temperature.
6. Compare the boiling points of water at subatmospheric, atmospheric, and above atmospheric pressures, and give the reason why this occurs.
7. Distinguish between the evaporation of solvent molecules in an open and a closed container.
8. Relate the chemical structure of water to hydrogen bonding.
9. Describe the solubilities of polar and nonpolar molecules.
10. Define the colligative properties of water and state the effects of dissolved solutes upon them.
11. Write the reactions of (where applicable) and state the following chemical reactions of water with:
    a. Metals
    b. Nonmetallic oxides
    c. Halogens
    d. Metal oxides
12. Distinguish between efflorescent, hygroscopic, and deliquescent compounds.
13. Name the elements whose salts cause water to be called "hard."
14. Distinguish between hard and soft water.
15. Characterize natural water, distilled water, and demineralized water.
16. Distinguish between temporary and permanent hard water.

WATER

Water, the most abundant and familiar chemical compound, is an absolute necessity for plant and animal life. Without water no life could exist. It is found as ice in the solid state, as water in the liquid state, and as vapor in the gaseous state. It can be considered as a "universal solvent" because more substances can be dissolved in it than any other solvent. In particular, health care personnel should be aware of how water relates to the fluid electrolyte balance in the body; electrolytes being substances which dissolve in water to yield ions that are capable of conducting electricity. Every activity in every cell in the body takes place in a watery environment, and therefore, knowledge of the physical and chemical properties of water is absolutely essential.

PHYSICAL PROPERTIES

Pure water is a colorless, odorless liquid which boils at 100°C and freezes at 0°C at 760 mm pressure. At 100°C and 760 mm pressure water changes to water vapor (steam) by the absorption of heat energy. Regardless of its physical state, the water molecule is unchanged. Changes in temperature or pressure simply convert water from one state to another. That is why liquid water can be obtained by melting pure ice and by cooling (condensing) pure steam.

Density

Normally water is considered to have a density of 1 g/cc, however, as water is warmed, the density will decrease because the volume increases. The extra heat energy tends to separate the water molecules; the same weight of water occupying a larger volume and consequently causing the density to decrease. For practical purposes, the density of water is considered to be 1 g/cc without significant error.

When water is cooled, its volume decreases until a temperature of 4°C is reached. At this point, further cooling causes an expansion in volume, and, at 0°C when water freezes, the water will have expanded almost 10 percent in

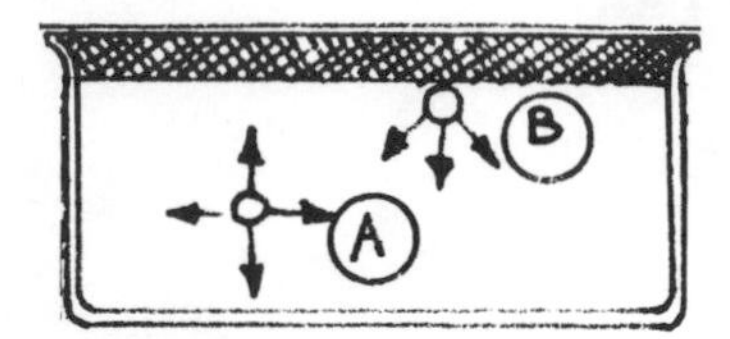

Figure 7-1. Surface tension of a liquid. A. Molecules in center are surrounded, equal attraction. B. Molecules at surface are unequally attracted in a downward direction.

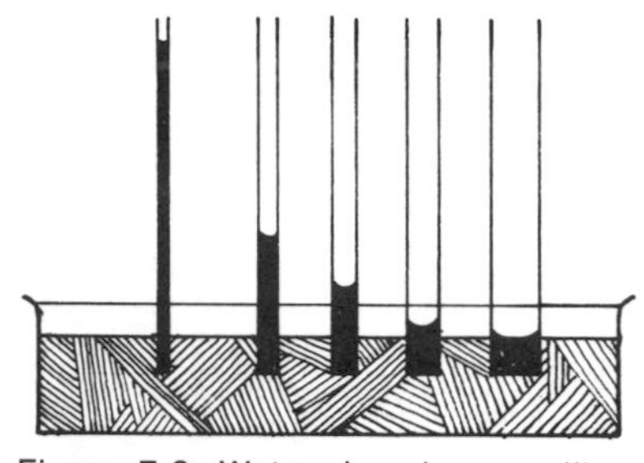

Figure 7-2. Water rises in a capillary tube due to the attractive forces between the water and glass molecules. The smaller the bore of the tube the higher the column of water will rise.

Figure 7-3. The lowest point of the curved upper surface (meniscus) should coincide with the calibration mark.

volume. This is the reason why water pipes and radiators burst when water contained in them freezes; the container must either expand or rupture. Individual body cells contain water (aqueous solutions), and when the water freezes and expands, the cell ruptures and dies.

Cohesion, Adhesion, Capillary Action

Molecules of liquids constantly attract each other; this is called *cohesion,* and the magnitude of the force of attraction is called the cohesive force. When different molecules attract each other this force is called *adhesion,* and the force of attraction is called the adhesive force.

When water drips out of a faucet, the drop of water assumes a spherical shape because the cohesive forces cause the molecules to cling together. Molecules in the center are completely surrounded by other molecules, and are therefore equally attracted from all sides. However, molecules which are near the surface of the liquid are attracted downward much more strongly, because there are so many more molecules below it than above it (Fig. 7-1). They therefore pack together to form a surface skin or film, and the force which is required to break that film is called *surface tension.* Surface tension causes water to draw into drops, instead of spreading out on greasy surfaces. This is the reason why pure water does not mix well with oils.

Surface tension of water can be reduced by soaps and detergents, thus allowing intimate mixing of the oils and water. The same action is effected by the bile salts in the aqueous intestinal fluids. These bile salts act as wetting agents to lower the surface tension of water and to facilitate the emulsification of fat by giving better penetration of the digestive enzymes.

Water molecules in a capillary tube (a very small bore tube) not only attract each other, but are attracted by the molecules of the glass. This attraction for the glass molecules causes the water to "creep" up the sides of the tube. Other water molecules follow and the liquid rises. This phenomena is very noticeable in a small bore tube but not very evident in a beaker or a large diameter tube (Fig. 7-2). However, the water will continue to rise until the weight of the water column counterbalances the attractive force pulling the liquid up. The smaller the bore of the tube, the higher the water will rise.

This attractive force between the water and glass causing the water to be pulled upward causes the water to assume a concave form. When reading aqueous volumes in burettes or pipettes, the lowest point of the curved upper surface of the liquid (the meniscus) is read as the measurement (Fig. 7-3). This property of a liquid "wetting" the tube surface is called capillarity (or capillary action) and is the principle used to take blood samples from blood drops of pricked fingers. Absorbent cotton, blotters, and wicks of candles and kerosene lamps work on this principle.

Mercury atoms have a cohesive force among themselves that is greater than their attractive force to glass molecules. The upper surface of a column of mercury is curved in a convex form. When reading barometers or manometers (pressure indicating instruments), you should always take the measurement at the highest point (Fig. 7-4).

Vapor Pressure of Water

Water molecules, as well as all liquids, are in a state of constant motion, the rapidity of the motion being dependant upon the amount of energy that they possess. Since heat is a form of energy, the greater the amount of heat, the greater the amount of energy, and the more rapid their motion. As heat energy is absorbed and the molecules vibrate faster, a greater percentage of them move fast enough to escape from the surface of the liquid. As the molecules escape from the liquid surface, they produce a pressure called "vapor pressure"; the higher the temperature, the higher the vapor pressure (Fig. 7-5).

When the vapor pressure of water is high enough to push the air out of the way, water boils (Fig. 7-6). The greater the air pressure, the greater the vapor pressure that the water molecules must have. This requires more energy. When the air pressure is less, as at higher altitudes, the vapor pressure needed to equal it is less, and this requires less energy. This is why water boils at lower temperatures at higher altitudes than at sea level; because at higher altitudes, the atmospheric pressure is less than 760 mm (Fig. 7-7).

Since the boiling point of water is the temperature at which its vapor pressure is equal to the pressure being exerted on it, steam sterilizers and pressure cookers using pressures higher than 760 mm can reach temperatures much higher than 100°C. Some spores of pathogenic microorganisms can survive temperatures of 100°C, and therefore higher temperatures require the use of pressure sterilizers. You may have thought that increasing the amount of heat of a container of boiling water increases the temperature, but this is not the case; the temperature remains the same. The increased temperature merely increases the *rate* of water evaporated per second.

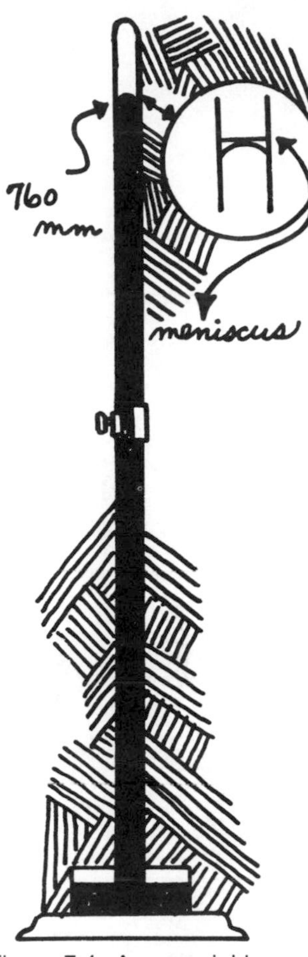

Figure 7-4. A mercurial barometer.

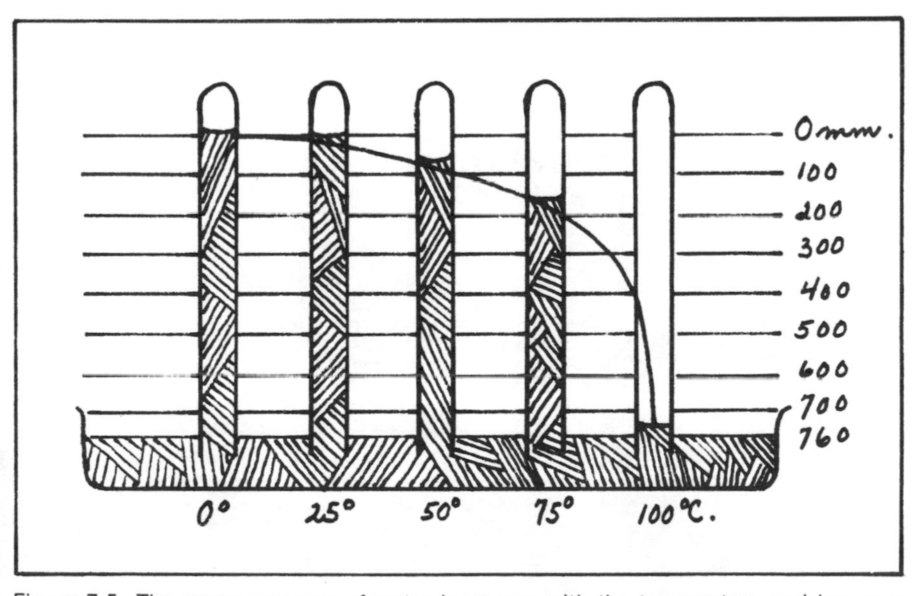

Figure 7-5. The vapor pressure of water increases with the temperature and becomes equal to 760 mm at the boiling point.

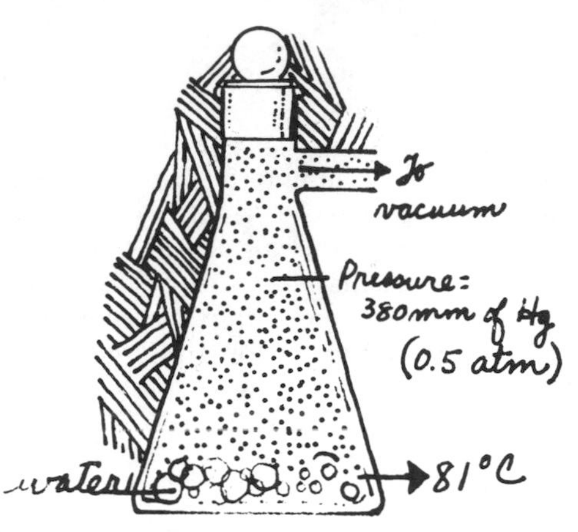

Figure 7-7. Reduced pressure simulating extremely high altitude, 380 mm or ½ atm. Water boils at 81°C.

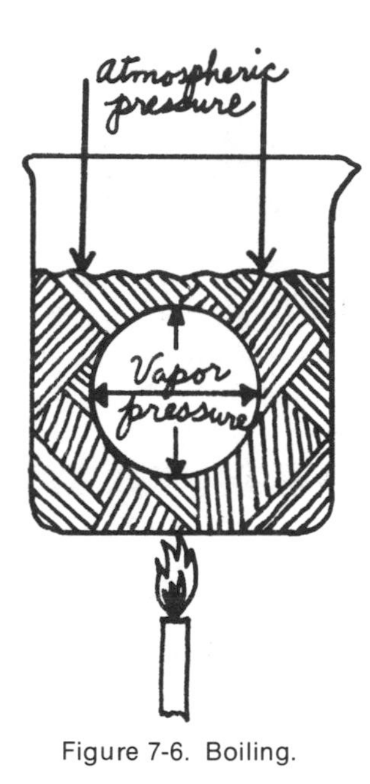

Figure 7-6. Boiling.

The pressure cooker and pressure steam sterilizer are sealed containers that prevent the escape of water vapor molecules above the boiling water. As more and more water molecules escape into the vapor state and their concentration increases, the pressure above the boiling water increases. As a result, the liquid water must acquire more energy in order to escape, and as they acquire more energy, the temperature rises (Fig. 7-8). These pressure apparatus merely cause the vapor pressure above the liquid to increase thus increasing the temperature of the boiling water.

If sterilization requires that materials be at 100°C for 10 minutes they cannot be sterilized at higher altitudes merely by boiling them in water in open containers. Pressure devices must be used to raise the temperature of the boiling water in order to sterilize adequately. Most proteins and bacteria are destroyed by heat, and sterilization of surgical and medical instruments and clothing requires high temperatures. The steam pressure sterilizer effectively destroys all bacteria, even though the exact temperature required varies from one species to another.

Figure 7-8. The pressure cooker (or steam sterilizer) operates at higher than 760 mm Hg, therefore the temperature of water is raised above 100°C.

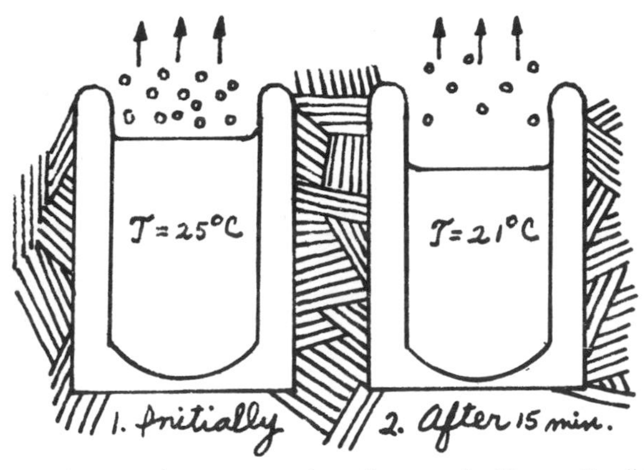

Figure 7-9. Evaporation requires energy. In a thermos bottle, as the liquid evaporates and removes heat energy from the liquid the temperature is lowered.

Evaporation

When perspiration on the skin surface evaporates it removes heat from the body, helping to maintain a normal heat balance. Sometimes a patient has a high temperature, and it is necessary to cool the patient by sponging him with cool water or alcohol. The heat from the body vaporizes the liquids, reducing the heat content of the body. This process is called evaporation (Fig. 7-9). When patients are treated with an alcohol sponge bath, care must be taken not to sponge too great an area at one time. The evaporation of the alcohol can chill the patient.

The rate at which liquids evaporate depends upon their vapor pressure. The higher the vapor pressure, the quicker the evaporation (Fig. 7-10). In fact some substances, such as ether or ethyl chloride, evaporate so rapidly that when they are sprayed on the surface of the skin, the area becomes frozen, and the nerve endings numbed enough to permit local surgery.

In closed containers evaporation occurs, but there is no loss of liquid. The rate of evaporation becomes equal to the rate of condensation, and a state of equilibrium is established: the number of molecules evaporating is equal to the number of molecules condensing (Fig. 7-11). Special care should be taken by

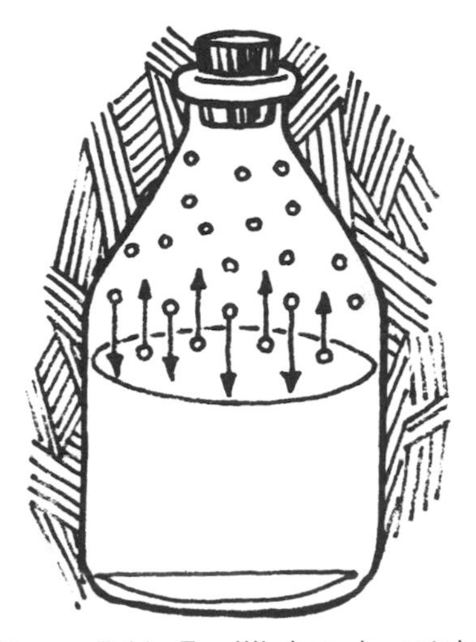

Figure 7-11. Equilibrium is established in a closed container. The rate at which the water molecules escape from the liquid is the same as the rate at which they return.

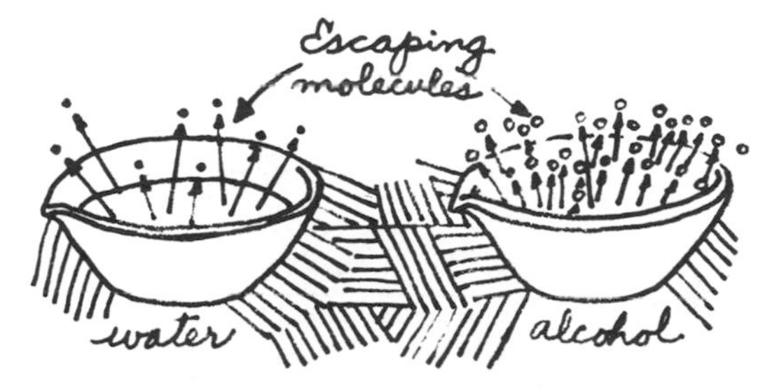

Figure 7-10. Alcohol has a higher vapor pressure than water; it evaporates at a quicker rate.

all health care personnel to securely close any containers which contain volatile liquids. Drug solutions, medications, test reagents and chemicals that contain liquids may evaporate, and the concentration of the active ingredients may change.

Hydrogen Bonding

Water possesses abnormal physical properties such as a high melting point, high boiling point, high heat of fusion, high heat of vaporization, and high surface tension. These properties can be attributed to abnormally high intermolecular attractions called *hydrogen bonds.* There is a very strong attraction between the nonbonding electrons of the oxygen atom of the water molecule (which is partially negatively charged) and a hydrogen atom of an adjacent water molecule (which is partially positively charged).

The water molecule is polar, although it is electrically neutral, because there is an uneven distribution of the positive and negative charges that are within the molecule itself (Fig. 7-12). As a result, one end of the molecule is negative and the other end positive. Therefore, the molecules tend to orient themselves in such a manner that the negative end of one molecule is next to the positive end of another molecule (Fig. 7-13). This takes place because of the laws of electricity; negative charged objects are attracted to positively charged objects, and positively charged objects are attracted to negatively charged objects.

Thus there is a bonding force between the water molecules, which is called *hydrogen bonding.* This very same phenomena is important to all living cells, and because the hydrogen bond is very weak compared to normal covalent bonds, the weakness of the hydrogen bond contributes to the relative instability of extremely complex molecules such as the enzymes, chromosomes, and protoplasm.

Since the water molecule does exhibit polarity, it is a good solvent for other polar molecules, such as methyl alcohol and ethyl alcohol (Fig. 7-14). It is, however, not a good solvent for nonpolar molecules, those substances in which

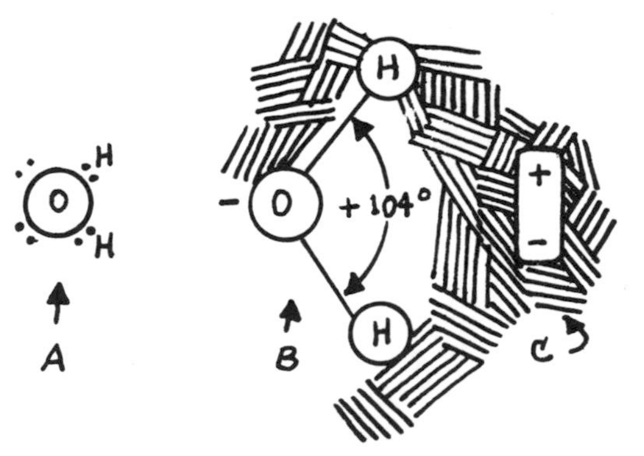

Figure 7-12. The unsymmetrical water molecule. A. The 2 hydrogen atoms form an angle of 104°, they are not evenly spaced around the oxygen atom. B. Although the molecule is neutral, the left side is negatively charged and the right side is positively charged, causing the molecule to have a polarity. C. The water molecule can be shown as a polar molecule, a dipole.

the centers of the positive and negative charges coincide with each other. The nonpolar solvents are oils, petroleum solvents, and greases. A good rule to remember is that "like dissolves like;" that polar solvents dissolve polar solvents, and nonpolar solvents dissolve nonpolar solvents (Fig. 7-15).

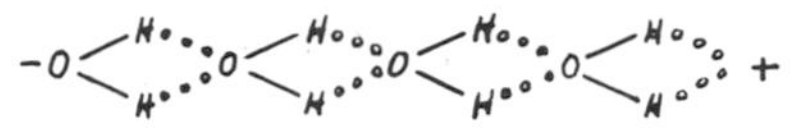

Figure 7-13. Hydrogen bonding in water. The chemical bonds are shown as solid lines and the hydrogen bonds are shown as dotted lines.

Effect of Dissolved Solutes on Properties of Water

When solutes are dissolved in water, yielding solutions containing cations, anions, radicals, or undissociated molecules, the physical properties of water change because the particles tend to break the normal structure of liquid water. Some of these changes, which depend solely upon the concentration of these particles (the number per unit volume) are characterized by the term *colligative* properties of water. They are: boiling point elevation, freezing point depression, vapor pressure lowering, and osmotic pressure.

Pure water boils at 100°C. Solutions of salts, sugar, or in fact any dissolved solute will raise the boiling point over 100°C. The greater the concentration of solute particles, the greater the elevation. Similarly, the freezing point of water is 0°C, but solutions will depress (lower) the freezing point below 0°C; again the greater the concentration of solute particles, the lower the freezing point (Fig. 7-16). In the winter, when salt is thrown onto frozen roads and sidewalks to melt the ice, it lowers the freezing point. The dissolved solutes endow the solution with the property called osmotic pressure, which will be covered in Chapter 9.

Pure water has a higher vapor pressure than water that contains dissolved solutes. The higher the solute concentration, the lower the vapor pressure, and conversely, the more dilute the solution, the higher its vapor pressure. Therefore pure water will evaporate more rapidly than aqueous solutions, and furthermore, solutions which contain substances that ionize will evaporate more slowly than solutions of equal concentration of nonelectrolytes. This means that when comparing the vapor pressures of identical concentrations of NaCl and sugar, the NaCl solution has a lower vapor pressure because each molecule yielded 2 ions, whereas the sugar molecule did not ionize (Fig. 7-17).

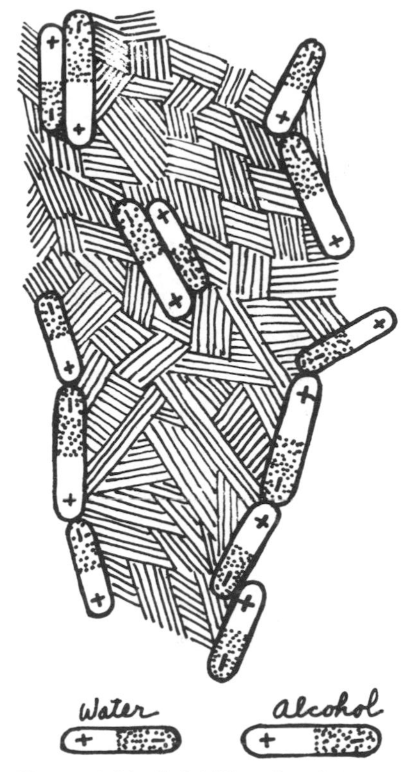

Figure 7-14. Solubility of polar compounds in each other. Polar compounds dissolve other polar compounds because their molecules can pair up with each other.

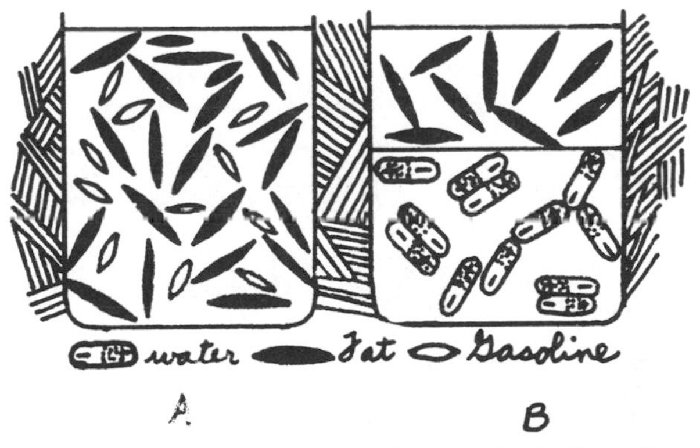

Figure 7-15. Gasoline and fat are nonpolar, water is polar. A. Nonpolar compounds will dissolve only in nonpolar compounds. B. Polar compounds tend to clump together and squeeze out the nonpolar compounds, making water and oils insoluble in each other.

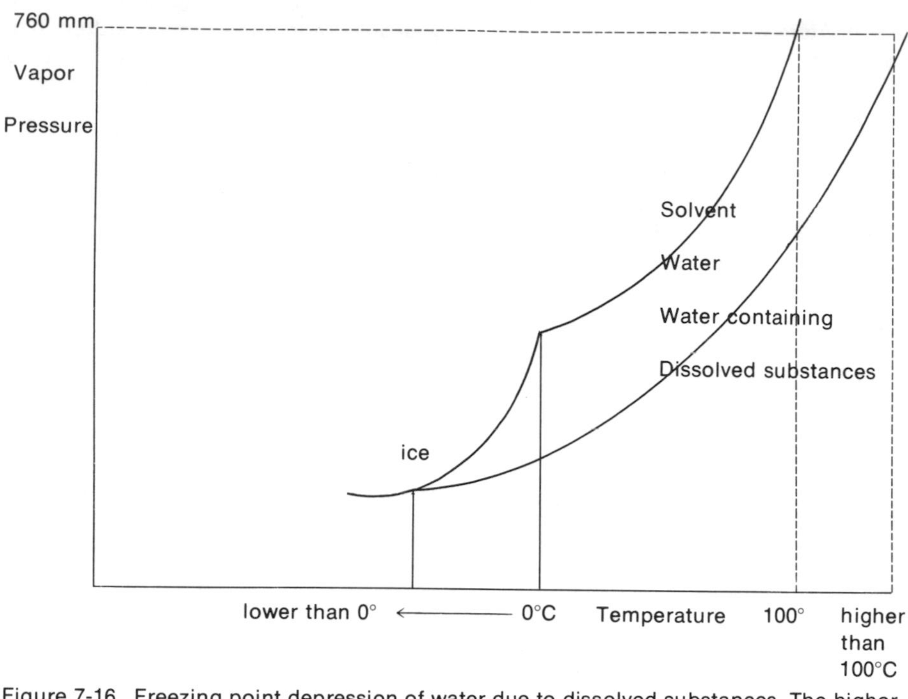

Figure 7-16. Freezing point depression of water due to dissolved substances. The higher the concentration of the dissolved substance, the lower the freezing point and the higher the boiling point of water.

Figure 7-17. Effect of solutes upon the vapor pressure of water in a vapor tight enclosure. A. At the start the concentration of solute particles is different because NaCl ionizes. Evaporation and condensation occur at different rates. B. At equilibrium a transfer of water has occurred from the more dilute to the more concentrated solution. The solutions attempt to attain identical vapor pressures.

CHEMICAL PROPERTIES

The water molecule is extremely stable; very little decomposition occurs even when water is heated to about 3000°F. However, water can be decomposed back into its separate elements by supplying water with sufficient energy in the form of direct current electricity to yield 2 volumes of hydrogen and 1 volume of oxygen (Fig. 7-18).

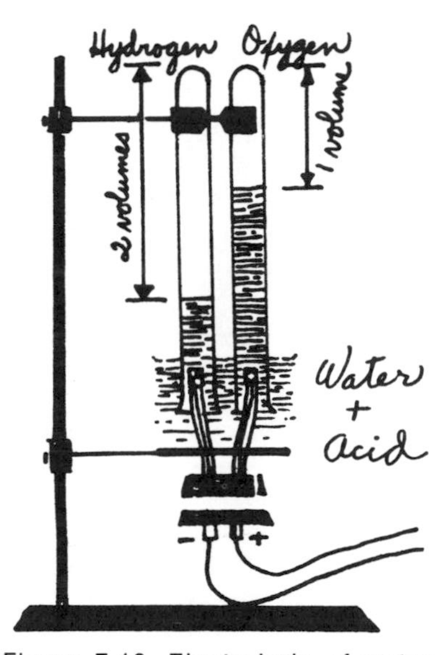

Figure 7-18. Electrolysis of water. Water can be decomposed into its components by the application of direct current.

Hydrates

One of the chemical reactions that water undergoes that is important to health care personnel is its ability to combine directly with other elements and compounds to form crystalline substances called "hydrates." These hydrates contain water in a definite proportion by weight, and the water contained in the hydrate is called water of crystallization. The reaction is reversible, which means that when the hydrate is heated, the hydrate can be decomposed into its original compound and the water is driven off. The original compound, without water, is called an *anhydrous compound.*

$$\underset{\text{water of crystallization}}{CuSO_4 \cdot 5H_2O} \xrightleftharpoons{\text{heat}} \underset{\text{anhydrous}}{CuSO_4} + 5\,H_2O$$

Some hydrates do not require heat to lose their water of hydration, but lose it merely by exposure to air. Washing soda, $Na_2CO_3 \cdot 10H_2O$ is an example. These compounds are called *efflorescent* compounds.

Other substances will actually absorb moisture from the air surrounding it, and are called *hygroscopic* compounds. Some even dissolve in the water that they absorb, and those are called *deliquescent.* Calcium chloride is extensively used as a drying agent in the laboratory and to maintain dry conditions in stored equipment. Many precision pieces of apparatus are sealed in plastic bagging containing calcium chloride which absorbs any traces of moisture in the air. By doing this, rusting (oxidation) of iron components is inhibited.

Plaster of Paris

Plaster of Paris is used extensively for making surgical casts. It is made from calcium sulfate

$$(CaSO_4)_2 \cdot H_2O + 3H_2O \longrightarrow 2CaSO_4 \cdot 2H_2O$$

which, when mixed with water, *sets* to a very hard mass called *gypsum.* Bandages are covered with the plaster of Paris, and they are either moistened or dipped in water, quickly wrung out and applied. The hard gypsum expands slightly as it forms, and care must be exercized not to cut off circulation of the affected part of the body by wrapping the bandage too tightly.

Reactions with Metals

Water reacts with metals to form bases (compounds which ionize in water to yield hydroxyl ions). For example, cold water plus sodium will react as:

$$2\ Na + 2\ H_2O \longrightarrow 2\ NaOH + H_2\uparrow$$

In the laboratory, metallic Na, K and Li are stored in oil to prevent them from reacting with the moisture in the air. The arrow by the H_2 indicates that hydrogen gas is given off. With the relatively inactive metals, steam is required, but the reaction still proceeds to form the bases.

$$Mg + 2\ H_2O \longrightarrow Mg(OH)_2 + H_2\uparrow$$

magnesium plus water yields magnesium hydroxide plus hydrogen

Metallic Oxides

Water reacts with soluble metallic oxides, such as Na_2O, K_2O, CaO, MgO, and BaO to yield bases:

$$Na_2O + H_2O \longrightarrow 2\ NaOH$$

Sodium hydroxide is lye, used in soap making, and it can cause severe chemical burns on contact with the skin.

$$CaO + H_2O \longrightarrow Ca(OH)_2$$

calcium oxide plus water yields calcium hydroxide

Calcium hydroxide is called lime water. It is used as whitewash, and can be added to the diet as an antacid.

Nonmetallic Oxides

With nonmetallic oxides, such as CO_2, SO_2, and NO_2, water reacts to form acids (substances which yield hydrogen ions (H^+) when dissolved in water.

$$CO_2 + H_2O \longrightarrow H_2CO_3$$

carbon dioxide plus water yields carbonic acid

Carbonic acid is found in beer, soda pop, and champagne. It is not a stable acid, and it tends to decompose:

$$H_2CO_3 \longrightarrow H_2O + CO_2$$

carbonic acid yields water and carbon dioxide

This is why carbonated beverages tend to go "flat" after being opened. Efficient closures can prevent the escape of the CO_2 gas. The CO_2 pressure, inside the container but above the liquid, eventually builds up to the point where the reaction stops. Equilibrium is established and the concentration of the carbonic acid remains steady.

Halogens

A halogen is an element of the chlorine group, which includes fluorine, chlorine, bromine, and iodine. Water reacts with chlorine gas, an active nonmetallic element, to form two substances, one of which decomposes to yield oxygen:

$$H_2O + Cl_2 \longrightarrow HCl + HOCl$$

water reacts with chlorine to yield hydrochloric acid and *hypochlorous acid*

The hypochlorous acid will further decompose:

$$2\ HOCL \longrightarrow 2\ HCl + O_2$$

hypochlorous acid yields hydrochloric acid and oxygen

Chlorine gas is added to city water supplies before the water is distributed in a process called chlorination. This kills most of the organisms in the water, making it safe to drink. However, viruses are resistant to chlorination. In certain parts of the country chlorine is added to the water supplies in such large amounts that it can be tasted. In swimming pools, where moderately excessive amounts are added, it has a stinging effect on the eyes and a detectable odor.

NATURAL WATER

Natural water is impure because it contains many gaseous, mineral and organic particles. In the health care field it is extremely important to know the degree of purity that is acceptable for specific procedures and purposes.

Hard Water

Water that picks up mineral matter, such as calcium, magnesium or iron ions, as it passes through the soil is called hard water. When these ions react with soap they produce a nonlathering quality, because they change the soap from a sodium salt, which is soluble to the Ca, Mg, or Fe salt, to a precipitate which is insoluble in water. These ions are responsible for the rings or "scum" around

Table 7-1. Impurities found in natural water.

Source	*Class impurities*	*Impurities*
Rain	Mineral Gaseous	Dirt, CO_2, N_2, O_2.
Tap	Minerals	Dissolved salts, minerals and suspended matter.
River and lake	Minerals Organic matter Gaseous Bacteria	Salts, clay, particles, bacteria, microorganisms, O_2, N_2, CO_2, pollutants from sewage and industry.

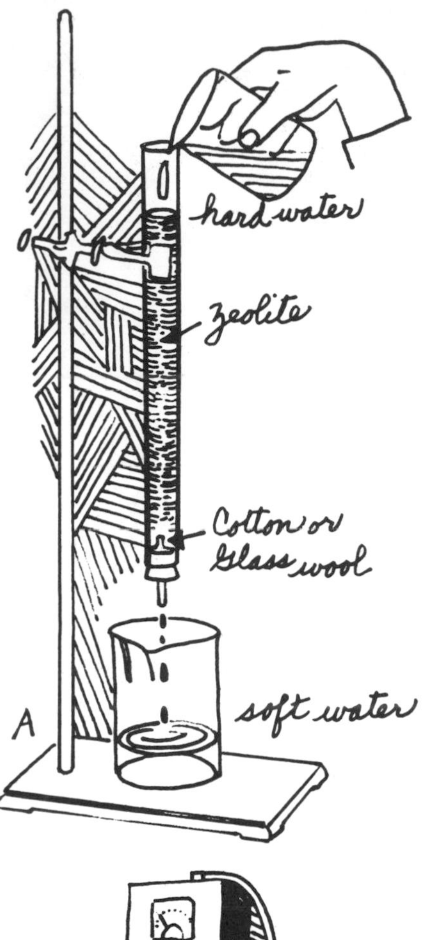

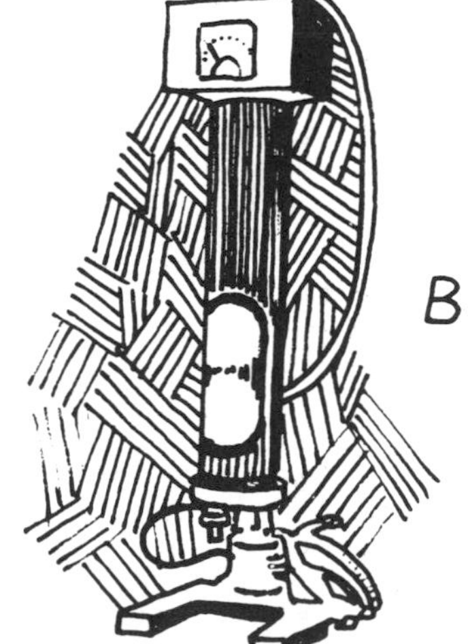

Figure 7-19. Water demineralizers. A. Laboratory constructed. B. Commercially available unit.

the bathtub and sink. Only when these ions are removed from the water can the water be effectively used to wash utensils and clothing.

Commercial equipment called water softeners are installed in homes, industrial, and health care facilities in areas that have a high mineral content in their water supplies. These softeners remove the troublesome ions, and therefore make the water "soft." In the laboratory, water containing these ions is passed through a demineralizer (Fig. 7-19), which removes the ions until the chemical resins inside are exhausted. The cartridge that contains the used resin must be periodically replaced to insure maximal demineralization. The time of replacement of the cartridge is usually indicated by a meter.

CAUTION: This water is called demineralized water. It is *not* distilled water. It may contain sodium ions and bacteria, and therefore is *not suitable* for parenteral (beneath the skin) injections.

DIFFERENCE BETWEEN TEMPORARY AND PERMANENT HARD WATER

Temporary hard water contains the bicarbonates of calcium and magnesium, $Ca(HCO_3)_2$ and $Mg(HCO_3)_2$. When temporary hard water is heated, the bicarbonates are changed into the carbonates $CaCO_3$ and $MgCO_3$, which precipitate from the solution. These carbonate precipitates are relatively insoluble, therefore the calcium and magnesium ions are removed from the solution and the water becomes soft. Permanent hard water contains the soluble salts of calcium and magnesium in the form of chlorides. When hard water is heated, no insoluble precipitate forms, and the ions are not removed from the solution. The water remains hard.

Soft Water

Soft water, as noted above, may contain the sodium ion (Na^+). It will allow soaps to lather because normal soap is a sodium salt itself and the sodium ions do not affect its lathering qualities.

Detergents

Synthetic detergents which do not form insoluble salts with calcium, magnesium or iron ions have been available for some time. They lather in hard water and do not leave the scum or film on utensils or clothing. They can be used in areas that have hard water without using demineralizers or water softeners.

Pure Water Requirements

Pure water, which is free from dissolved minerals, particulate suspended matter, and dangerous microorganisms, is needed in the delivery of health care. Certain procedures require only that the dissolved mineral ions be absent, and that requirement is met by the use of demineralizers yielding demineralized water. Other procedures require only the absence of microorganisms, some only the removal of suspended matter, while still other procedures require absolutely pure water. Whenever the procedure specifies specially treated or purified water, use only that water.

Water Purification

City water is normally collected and stored in large reservoirs and allowed to stand so that dirt, debris, and particulate matter will settle out. This process is called sedimentation. However, even after sedimentation some fine particles of dirt and debris remain, so the water is further purified. It is pumped to a water treatment plant where aluminum sulfate and lime are added. They react to form gelatinous aluminum hydroxide:

$$Al_2(SO_4)_3 + 3Ca(OH)_2 \longrightarrow 2Al(OH)_3 + 3\,CaSO_4$$

The gelatinous aluminum hydroxide coagulates (clumps together) and glues together most of the remaining fine particles of matter that did not settle out in the reservoir. The water is then filtered through several beds of sand, and finally treated with chlorine to kill any bacteria and make the water potable. Safety demands that the trained professional be aware of the dangers of contaminated water, because water that is unfit to drink can appear potable since it may have no unusual odor or taste.

Distilled Water

Distilled water, the process which involves the boiling of water and the cooling of the steam to condense it to yield liquid water, is the purest type of water (Fig. 7-20). Some water is even triple distilled in quartz vessels for maximum purity. All nonvolatile impurities remain behind, and the dissolved gases are expelled and not condensed along with the water.

Catastrophes, either natural or man made, may lead to a lack of normally treated safe drinking water. In such cases, raw water can be made safe for

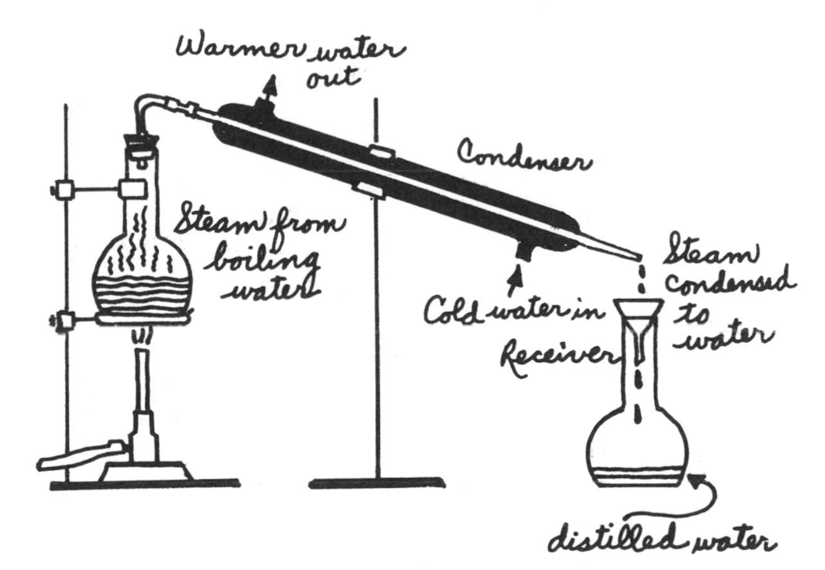

Figure 7-20. Distilled water preparation.

drinking purposes by boiling the water for at least 15 minutes at 100°C. This will kill pathogenic microorganisms, but it also makes the water taste "flat" because boiling expels the dissolved gases.
CAUTION: At extremely high altitudes, use pressure cookers to prepare drinking water, because the boiling temperature will be too low to kill microorganisms.

Nature's Purification of Water

If given sufficient time, the oxygen of the air will destroy bacteria in water. Contaminated water may not have had enough exposure to the air, so care must be exercised in drinking this type of water. If the water has a disagreeable taste or odor because of dissolved gases, the taste can be improved by merely pouring it back and forth between two containers. This repeated exposure to the air (*aeration*) may get rid of some of the objectionable odors in the water.

SUMMARY

Water has surface tension, a property which causes it to assume a spherical shape, and it also has an attraction for other molecules (capillary attraction). Its vapor pressure increases with the temperature, and it boils when that vapor pressure is equal to the pressure exerted on its surface. As it evaporates, it requires heat energy. Water is an asymmetric polar molecule, dissolving other polar molecules, but insoluble in nonpolar ones. Dissolved solutes reduce the vapor pressure of water by increasing its boiling point or reducing its freezing point, and the change is proportionate to the concentration of solute particles. Its important chemical properties include the formation of hydrates, reaction with metals, metallic oxides, nonmetallic oxides and halogens. Naturally occurring water contains impurities; metal ions can be removed with demineralizers, and pure water can be prepared by distillation.

EXERCISE

1. What is the effect of freezing upon cells?
2. Why does a drop of water assume a spherical shape?
3. What causes water to rise in a capillary tube?
4. What is vapor pressure? When will water (or any solvent) boil?
5. What is the effect of a pressure cooker (sterilizer) upon the boiling point of water? Why is it important?
6. Why do surfaces of evaporating liquids get colder?
7. Why is there no evaporation of a solvent from a closed container?
8. Why is water called a polar molecule, and why doesn't it dissolve oils?
9. What is the effect of dissolving a solute in water? On the boiling point? Freezing point? Vapor pressure?
10. What is a hydrate?
11. Write the reaction of (a) Na with water, (b) NaO with water, (c) CO_2 with water, and (d) Cl_2 with water.
12. What are the impurities found in natural water?
13. Which ions contribute to "hard" water? How are they removed?
14. Can demineralized water always be substituted for distilled water?
15. How can one make contaminated water potable?

OBJECTIVES

When you have completed this chapter you will be able to:

1. Define an acid and characterize the important component of every acid.
2. List the characteristics of acids.
3. Explain what is meant by the activity series.
4. Draw a diagram explaining why acids conduct electricity.
5. Write the chemical reaction of acids with:
   a. Metallic oxides
   b. Carbonates
   c. Bases
6. Discuss the properties of:
   a. HCl
   b. H_2SO_4
   c. HNO_3
   d. H_3BO_3
   e. $HOCl$
   f. Acetic acid
   g. Salicylic acid
   h. Ascorbic acid
   i. Lactic acid
   j. Amino acids
7. Define a base and characterize the important component of a base.
8. List the characteristics of bases.
9. Discuss the properties of:
   a. $NaOH$
   b. KOH
   c. $NaHCO_3$
   d. $Ca(OH)_2$
   e. NH_4OH
   f. $Mg(OH)_2$
10. Distinguish between strong and weak acids and bases.
11. Draw a diagram illustrating the difference between strong and weak acids and bases.
12. Give examples of what is meant by concentrated and dilute acids.

ACIDS, BASES AND SALTS

13. Calculate H^+ ion and OH^- ion concentrations.
14. Relate the hydrogen ion concentration to the pH.
15. Distinguish between pH and pOH.
16. Identify methods which are used to determine the pH.
17. Define and give examples of normal, acid, and basic salts.
18. Enumerate 4 functions of salts in the body.
19. Define, give the components of, and list three buffer systems in the body.
20. Show by chemical equations how a buffer system works to eliminate excess acids and bases.

One of the most common complaints that people always seem to have is an "acid stomach" or heartburn. As a viewer of television commercials, you are bombarded nightly with heartburn remedies, acid stomach relievers, antacids, and the like. The manufacturers of these products would not spend millions of dollars in advertising if people did not buy their products.

Even as a layman, you may have heard of various acids: citric acid in oranges, lemons, and grapefruits, phosphoric acid in soda pop (read the ingredients on the label), boric acid in eye wash, ascorbic acid (vitamin C), hydrochloric acid (secreted in the stomach), sulfuric acid (battery acid), and nicotinic acid (in cigarette smoke and tars). We can define an acid as a compound that will separate into ions, one of which must be a *hydrogen ion*, H^+. A substance that dissociates (separates) into a hydrogen ion and another ion, which can be a nonmetal (such as Cl, Br, and I) or a radical (such as phosphate, sulfate, nitrate, or carbonate) is another definition of an acid.

The hydrogen ion is actually only a proton, for it is the same particle that was formed when the hydrogen atom gave up one electron. The hydrogen ion (the proton) is a positively charged particle and it has an affinity for a neutral water

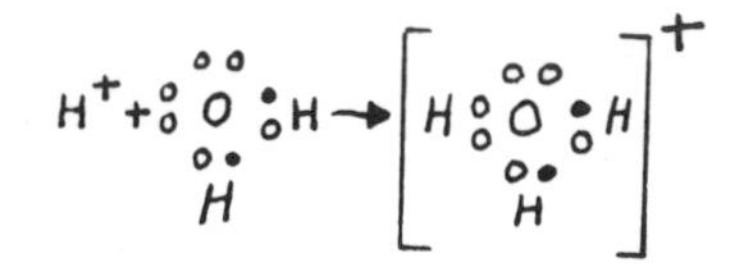

Figure 8-1. The hydrogen or hydronium ion.

molecule. When it attaches itself to a water molecule, it is called a hydronium ion, but the attachment is just strong enough to hold it onto the molecule. When the proton is needed in a chemical reaction, it is easily detached to react with another substance. Therefore, the particle can be called a hydrogen ion or a hydronium ion; both names are correct (Fig. 8-1).

CHARACTERISTICS OF ACIDS

Acids Have a Sour Taste

Citrus fruits, such as lemons, limes, oranges, and grapefruits contain citric acid. They have a sour, tart, and astringent taste. Similarly, vinegar contains acetic acid and it has a sour taste.

Acids Dissociate into Hydrogen Ions

Some acids dissociate yielding 1 hydrogen ion per molecule (HCl, HNO_3), while others yield 2 hydrogen ions (H_2SO_4, H_2CO_3), and some yield 3 hydrogen ions per molecule, H_3PO_4. In any case, whether the anion is a nonmetal or a radical, the compound *must* yield hydrogen ions in water to be called an acid.

Acids Change the Colors of Certain Dyes

When you squeeze a lemon into tea, the citric acid in the lemon changes the color of the tea.

Acids Are Corrosive

Spilled battery acid will destroy clothing and burn the skin. Excess HCl in the stomach causes a burning sensation.

Acids React with Certain Metals

Acids react with certain metals, but not with all of them, and the rate at which acids will react varies with the metal. Some metals are very reactive, and they react vigorously with acids. The various metals can be classified according to their activity (reactivity), with hydrogen being the reference (see Table 8-1).

1. Metals above hydrogen react with acids. The further away from hydrogen, the greater the acitivity; potassium is the most reactive, and lead the least reactive. Metals react with acids to form salts to liberate hydrogen gas:

$$Zn + H_2SO_4 \longrightarrow ZnSO_4 + H_2\uparrow$$
$$Fe + 2\,HCl \longrightarrow FeCl_2 + H_2\uparrow$$

2. Metals below hydrogen do not react with acids unless an oxidizing agent is present. Some acids, such as nitric acid, are oxidizing agents, and they will react with those metals.

Utensils, containers, and instruments which are made of those metals above hydrogen in Table 8-1 cannot be left in contact with or stored in acid solutions because the acids will react with the metals. Similarly, acids cannot be stored in these metals, but must be kept in glass or the newer plastic materials which are unaffected by acids.

Table 8-1. Activity series.

Property	Metals
Displace hydrogen from water. Vigorous reaction with acids.	K Na Ca
Displace hydrogen from steam and from acids.	Mg Al Zn Cr Fe
Displace hydrogen slowly from acids.	Ni Sn Pb H
Do not displace free hydrogen from acids. May react with acids.	Cu Sb Hg Ag Pt Au

Acids Are Electrolytes

Acids will conduct electricity because the aqueous solution contains the necessary ions to give up and accept electrons, the ions formed by the ionization of the acid (Fig. 8-2).

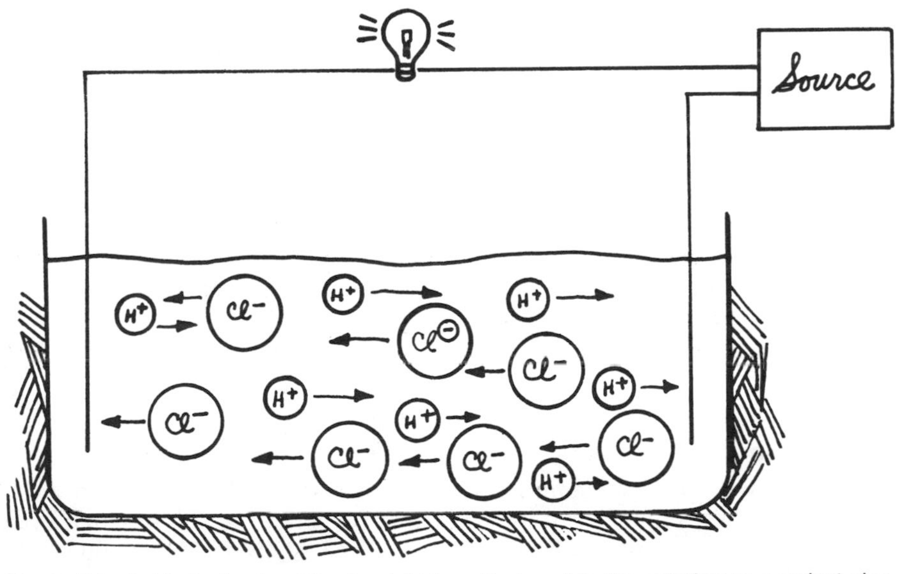

Figure 8-2. Acids ionize in water to yield ions that enable the solution to conduct electricity.

Acids Will Ionize

$$HCl \longrightarrow H^+ + Cl^{1-}$$
$$H_2SO_4 \longrightarrow 2\,H^+ + SO_4^{2-}$$
$$H_3PO_4 \longrightarrow 3\,H^+ + PO_4^{3-}$$

Actually, acids which contain more than 1 hydrogen ion ionize in successive steps. The ionization in the successive steps becomes progressively smaller, about 1/100,000 the degree of the previous step.

Ionization of H_2SO_4:

1. $H_2SO_4 \longrightarrow H^+ + HSO_4^{1-}$
2. $HSO_4^{1-} \longrightarrow H^+ + SO_4^{2-}$

Ionization of H_3PO_4:

1. $H_3PO_4 \longrightarrow H^+ + H_2PO_4^{1-}$
2. $H_2PO_4^{1-} \longrightarrow H^+ + HPO_4^{2-}$
3. $HPO_4^{2-} \longrightarrow H^+ + PO_4^{3-}$

Acids React with Metallic Oxides

Aqueous solutions of acids react with metallic oxides to form a salt and water.

$$ZnO + H_2SO_4 \longrightarrow ZnSO_4 + H_2O$$
$$CaO + 2\,HCl \longrightarrow CaCl_2 + H_2O$$

Acids React with Carbonates and Bicarbonates

Acids react with carbonates and bicarbonates to form carbonic acid, an unstable acid, which decomposes to yield carbon dioxide and water.

$$Na_2CO_3 + 2\,HCl \longrightarrow H_2CO_3 + 2\,NaCl$$
$$NaHCO_3 + HCl \longrightarrow H_2CO_3 + NaCl$$

The H_2CO_3 decomposes to yield CO_2 and H_2O

$$H_2CO_3 \longrightarrow H_2O + CO_2$$

When foam type fire extinguishers are inverted for use, the carbonate mixes with the acid. The carbon dioxide gas formed forces the water out through the nozzle which is directed at the base of the fire. As an antacid, sodium bicarbonate neutralizes excess HCl in the stomach to relieve "heartburn," as shown by the above equation.

Acids React with Bases

Acids react with bases in common chemical reactions to form salts and water. This reaction is very important to health care personnel because certain acids and bases are very corrosive to the tissues of the body, and their action can be neutralized by converting them into salts and water. First aid instructions for acid spills involve the use of $NaHCO_3$ to neutralize (render neither acid nor basic) the acid, after first flushing off the acid with excess water. The carbonic acid decomposes into CO_2 and H_2O.

$$HNO_3 + NaHCO_3 \longrightarrow NaNO_3 + H_2CO_3$$

PROPERTIES OF ACIDS

Some acids are strong because they ionize completely, while others are weak because they ionize only partially. Some are stable (H_2SO_4), and others are unstable (H_2CO_3) because they decompose easily. Some acids are poisonous while others are relatively harmless. The amount of the acid, *the concentration of the acid,* determines whether the acid is corrosive or not. Some properties of the common acids that you will encounter follow.

Hydrochloric Acid (HCl)

HCl is secreted in the stomach, at about one half of 1 percent concentration in the gastric juices, and it is essential for the digestion of food. Concentrated HCl (18 percent by weight) is extremely corrosive to tissue, animal fibers, and metals. In the stomach, an abnormally high concentration of HCl with the attending high concentration of hydrogen ions, is called hyperacidity. This hyperacidity can be temporarily corrected under a physician's care by taking antacids to reduce the H^+ ion concentration by neutralization. Older people sometimes have a condition called hypoacidity, too low a H^+ ion concentration, which can be remedied, under proper supervision, by drinking very dilute HCl solutions.

Sulfuric Acid (H_2SO_4)

This acid is found mainly in the laboratory as a chemical reagent, which is a substance that can be added to a solution or another substance to produce a chemical reaction. It is also used as a drying agent because it has such a strong affinity for water. It will actually dehydrate organic compounds, extracting the water and leaving a charcoal residue. The concentrated acid is extremely corrosive, *handle it with extreme care.* Flush off any spills on skin or clothing with excess water and follow first aid procedures. Extreme heat is evolved when it is mixed with water, as in making dilute sulfuric acid solutions, so always use caution and *always add acid to water, never add water to the acid.*

Nitric Acid (HNO_3)

Nitric acid is a yellowish-colored acid because it decomposes into the various colored nitrogen oxides and water. It is extremely corrosive to tissues, clothing, and metals. It is used as a reagent in the laboratory and should not be allowed to come in contact with your skin because of its corrosive action.

Boric Acid (H_3BO_3)

Boric acid is soluble to about 5 percent by weight in water. It was previously used as an eye wash in extremely weak concentrations, but recent data indicate that crystals exist in the solution form that can irritate the eye. It is used as a topical antiseptic for the mouth and nose. It should not be taken internally.

Citric Acid ($H_3—C_6H_5O_7$)

Citric acid, which is found in all citrus fruits, acts as a diuretic, a substance that increases the formation of urine. Its salts, such as magnesium citrate, act as purgatives, and sodium citrate is used as an anticoagulant for blood.

Hypochlorous Acid (HOCl)

Hypochlorous acid in the form of its salts, NaOCl and $Ca(OCl)_2$, is used to disinfect floors, walls, and bathrooms. The sodium salt, Dakin's solution, is relatively harmless to tissues, yet it has a very high bactericidal activity. Neither the acid nor the salts are stable, and they liberate oxygen upon decomposition. Most cities in the United States purify their drinking water by passing chlorine gas into the water. The chlorine gas reacts with the water as follows:

$$Cl_2 + H_2O \longrightarrow HCl + HClO$$

One part per million of HClO formed as shown above does a very effective job of killing bacteria. When excess Cl_2 is used, the water has a chlorine taste.

Acetic Acid ($H—C_2H_3O_2$)

Acetic acid in a dilute concentration (about 5 percent) is known as vinegar. In the concentrated form it is called glacial acetic acid, which is corrosive to tissues. It is very useful to remember that vinegar is a dilute acid and can be used in an emergency to neutralize lye, NaOH, or slaked lime, $Ca(OH)_2$, skin burns.

Salicylic Acid ($H—C_7H_5O_3$)

In a dilute alcoholic or cream solution salicylic acid is used to kill fungus on the skin, such as athlete's foot. A derivative of the acid, oil of wintergreen, is used for the treatment of muscular aches and rheumatism, and its sodium salt, aspirin, is universally used to reduce fever and to treat neuralgia.

Ascorbic Acid (Vitamin C)

Ascorbic acid, found in the citrus fruits, is used in the prevention and treatment of scurvy. It is a very essential vitamin.

Lactic Acid ($H—C_3H_5O_3$)

Lactic acid is formed when milk sours, and is found in sauerkraut and buttermilk. It inhibits the growth of putrefactive (decay causing) bacteria in the large intestine, and lactic acid milk is used to treat diarrhea in infants. Muscle fatigue can result from the accumulation of lactic acid due to excessive exercise.

Amino Acids

These acids are the building blocks of protein, the most important constituent of all living cells. They combine to form molecules which range in molecular weights from 50,000 to over 6,000,000. The protein in foods that you eat is broken down by the digestive process into individual amino acids; the body then takes those amino acids and synthesizes (combines them together) to make the specific proteins needed.

BASES

A substance that can counteract or neutralize an acid is called a base. For our purposes, a base is any substance that will neutralize the action of an acid, and a substance that will ionize in water to yield hydroxyl (OH^-) ions.

$$NaOH \longrightarrow Na^+ + OH^-$$
$$NH_4OH \longrightarrow NH_4^+ + OH^-$$
$$Mg(OH)_2 \longrightarrow Mg^{2+} + 2\,OH^-$$

The carbonates and bicarbonates are considered to be bases because they neutralize acids, even though they do not dissociate in water to give hydroxyl ions.

CHARACTERISTICS OF BASES

Bases Have a Greasy, Soapy Feeling

Watery solutions of bases feel soapy, and make the skin feel slippery.

Bases Have an Acrid Taste

Bases taste metallic; they have a bitter taste which can be detected in some toothpastes and soaps because of the alkali which is present.

Bases React with Acids

Bases react with acids (just as acids react with bases) to form a salt and water:

$$NaOH + HCl \longrightarrow NaCl + H_2O$$

Bases Change the Color of Certain Dyes

Many dyes and substances change color in the presence of bases. This principle is used in the laboratory to determine whether a solution is acid or basic. For example, phenolphthalein is a compound which is colorless in an acid solution and red in a basic solution. If phenolphthalein is present in an acid solution, the colorless solution will turn red when an excess of base is added to it.

Bases are Corrosive

Strong bases such as NaOH and KOH will burn the skin, changing the natural oils and fats of the skin to soaps. Their ability to dissolve and decompose fats and protein make bases useful as cleansing agents. Caution should be used when washing wool clothing with strongly alkaline soaps because wool is a protein.

Bases React with Salts

Soluble salts of calcium and magnesium react with bases to form the insoluble hydroxide (the arrow pointing down indicates a precipitate):

$$CaSO_4 + 2\,NaOH \longrightarrow Na_2SO_4 + Ca(OH)_2 \downarrow$$
$$MgCl_2 + 2NH_4OH \longrightarrow 2NH_4Cl + Mg(OH)_2 \downarrow$$

Bases React with Specific Metals

Strong bases, such as NaOH and KOH, react with active metals, Al and Zn, and liberate hydrogen just as the acids do.

$$Zn + 2\,KOH \longrightarrow Zn(OK)_2 + H_2\uparrow$$
$$2\,Al + 6\,NaOH \longrightarrow 2Al(ONa)_3 + 3\,H_2\uparrow$$

Strong bases, lye and caustic cleaning agents should not be used or stored in galvanized containers (coated with zinc) or aluminum utensils.

WEAK AND STRONG BASES

Sodium Hydroxide (NaOH)

Commonly called lye or caustic soda. Solid pellets of NaOH liberate large quantities of heat when they are dissolved in water. You have already learned that it is extremely corrosive to the skin, reacts with fats, oils, and greases to form soluble soaps, and that caution should be exercised when handling solid NaOH or solutions of it.

Potassium Hydroxide (KOH)

This base is as corrosive and reactive as NaOH. It is also called caustic potash, and in the laboratory it can usually be interchanged with NaOH as a reagent. When it is used to make soaps instead of NaOH, liquid soaps instead of solid soaps are formed.

Sodium Bicarbonate ($NaHCO_3$)

Sodium bicarbonate is also called sodium hydrogen carbonate or common baking soda. In water it yields a mildly basic solution, and is used to dissolve body mucous and secretions. As an antacid, it is an ingredient in commercial products, causing the product to effervesce (to bubble and foam) when it reacts with a weak acid, such as citric acid, to yield CO_2 and H_2O.

Calcium Hydroxide ($Ca(OH)_2$)

Calcium hydroxide is prepared by adding lime to water, producing a solution known as limewater. It is used in the treatment of hyperacidity of the stomach.

$$CaO + H_2O \longrightarrow Ca(OH)_2$$

It is only slightly soluble in water, is a weak base, and is used as a dietary supplement to build teeth and bones. As an antidote for acid poisoning, especially oxalic acid, it forms the insoluble calcium salt. It is often added to cow's milk to lower the acidity and impede rapid curdling.

Ammonium Hydroxide (NH_4OH)

Ammonium hydroxide is a weak base formed by the addition of NH_3 (ammonia) to water.

$$NH_3 + H_2O \longrightarrow NH_4OH$$

It slowly decomposes on standing to form water and liberate ammonia:

$$NH_4OH \longrightarrow H_2O + NH_3$$

Dilute solutions are known as spirits of ammonia and are used as antacids and to relieve rheumatic pains. Ammonia gas is a respiratory and heart stimulant.

Magnesium Hydroxide ($Mg(OH)_2$)

Magnesium hydroxide is known as milk of magnesia. It is a very weak base, used in toothpastes, mouthwashes, in the treatment of gastric hyperacidity, and as a laxative. It is sparingly soluble in water, and the suspension of the base gives the appearance of milk, therefore the common name, milk of magnesia.

STRENGTH OF ACIDS AND BASES

Certain terms are used to describe the activity of acids and bases. They are strong, weak, concentrated and dilute. You must not confuse them because they are not interchangeable.

Strong acids are strong because they are totally ionized in water. Weak acids are weak because they are only partially ionized in water. You can either have a high concentration (concentrated) or a low concentration (dilute) of an acid in water, depending upon *how much of the acid* is dissolved in the water. Hydrochloric acid is a strong acid. You can have a dilute solution, 1 percent, or a concentrated solution, 18 percent, of this strong acid. Acetic acid is a weak acid. You can have a dilute solution (vinegar, 5 percent) or a concentrated solution (99.5 percent, glacial acetic acid) of this weak acid.

The commonly encountered strong acids are HCl, H_2SO_4 and HNO_3, and they ionize completely in water:

$$HCl \longrightarrow H^+ + Cl^-$$
$$H_2SO_4 \longrightarrow 2\,H^+ + SO_4^{2-}$$
$$HNO_3 \longrightarrow H^+ + NO_3^-$$

Weak acids have little tendency to ionize and they yield few hydrogen ions as only a small fraction of the molecule dissociates. The commonly encountered weak acids are: acetic acid, carbonic acid, lactic acid, and boric acid. The strength of an acid can be determined by how much electricity they allow to pass in a conductivity apparatus.

Strong bases are strong because they are totally ionized in water. Examples of strong bases are NaOH and KOH. *Weak bases are weak because they are only partially ionized in water,* yielding few hydroxyl ions. Examples of weak bases are NH_4OH, Na_2CO_3 and $NaHCO_3$.

Advantages of Knowing the Strength of Acids and Bases

Weak acids or bases can be used where strong acids or bases would be too corrosive to tissue. Boric acid is a weak acid, and therefore can be used as a topical antiseptic. Topical use of a strong acid, such as HCl or H_2SO_4, would result in damage to the skin. When strong acids are spilled on the skin, weak bases are used to neutralize any acid remaining after first flooding with water to remove the acid. The use of a strong base instead of a weak base would cause further damage to the skin.

Concentration of Acids and Bases

The concentration of an acid or a base in water specifies *how much* of that substance is present in that particular solution. For example, concentrated HCl means that a *large amount* of HCl is dissolved in a small amount of water. There is a high concentration of H^+ ions and even a 5 percent HCl solution is corrosive. The gastric juices contain only about one half of 1 percent HCl, and therefore, the HCl in the stomach is tolerable and needed, because even though it is a *strong* acid, the concentration is extremely low. Acetic acid, on the other hand, is safe for internal use in a 5 percent concentration (this is vinegar). Only a small fraction of the acetic acid is ionized, yielding few hydrogen ions. The 5 percent acetic acid is safe but the 5 percent hydrochloric acid is corrosive. The concentrations of the two acids are the same, but the strength of the HCl is many times that of the acetic acid because of the total ionization of the HCl.

WATER, IONIZATION, AND ITS RELATIONSHIP TO ACIDITY AND BASICITY

The reason why acids neutralize bases and bases neutralize acids can be seen in the following statement: The water molecule dissociates only very minutely into H^+ and OH^- ions, only about 1 molecule in about 10,000,000.

$$H_2O \leftleftharpoons H^+ + OH^-$$

The large arrow indicates that when a H^+ ion comes in contact with an OH^- ion, the tendency is to form the neutral water molecule, thus the reaction tends to go to the left.

Whenever there is an acid solution with an excess of H^+ ions and a base containing an equal number of OH^- ions is added to that solution, they will combine to form neutral water. The acid qualities disappear because the H^+ ions become used up to form water molecules. The same thing happens when a base is neutralized. The basic solution has an excess of OH^- ions, which makes it basic. When an equal number of H^+ ions from an acid is added to the basic solution, the H^+ ions combine with the OH^- of the base to form the neutral molecule, water.

$$H^+ + O\overline{H} \rightleftharpoons H_2O$$

Pure water is neutral because when a molecule dissociates it yields 1 H^+ ion and 1 OH^- ion. There is always an equal number of each ion present. Therefore, pure water is neither acid nor basic, it is neutral because the number of H^+ ions equals the number of OH^- ions.

Scientists have developed the following mathematical relationship between the concentration of the H^+ and OH^- ions, which states: *The concentration of*

H^+ ions expressed in moles per liter multiplied by the concentration of OH^- ions expressed in moles per liter is always equal to 1×10^{-14}.

$$(H^+)\ (OH^-) = 1 \times 10^{-14}$$

In a neutral solution the number of H^+ ions must equal the number of OH^- ions and their concentrations must be equal. From your knowledge of scientific notation you can see that the only two equal values which would satisfy this equation would be 1×10^{-7} and 1×10^{-7}. This means that in a neutral solution, the concentration of hydrogen ions must be 1×10^{-7} moles/liter and that the concentration of hydroxyl ions must also be equal to 1×10^{-7} moles per liter. When these two values are substituted in the formula, you have an equality:

$$(1 \times 10^{-7})\ (1 \times 10^{-7}) = 1 \times 10^{-14}$$

NOTE: To multiply exponents, *add them.*
In an acid solution, the concentration of H^+ ions is *greater* than that of the OH^- ions, yet the *product* of the concentrations of the two ions must always equal 1×10^{-14}. In a basic solution, where the OH^- ion concentration is greater than that of the H^+ ion, the product of the two will still be equal to 1×10^{-14}.

Calculating Hydrogen and Hydroxyl Ion Concentrations

If you know the concentration of one of the ions in moles per liter, you can easily calculate the concentration of the other. The important fact to remember is that the product of the concentration equals 1×10^{-14}, and by subtracting the exponent of the known ion concentration, you will find the concentration of the other ion.
NOTE: It is important to remember that the *smaller* the exponent, the greater the value of the ion concentration that it represents, and the sum of the negative exponents must always equal -14.

Examples on calculating H^+ ion and OH^- ion concentrations:

1. If the H^+ ion concentration is 1×10^{-6} M/L, what is the OH^- concentration, acid or basic?

$$(-6) + ? = -14$$
$$? = -8$$

therefore the OH^- ion concentration is 1×10^{-8} M/L, and the solution is acid.

2. If the OH^- concentration is 1×10^{-9} M/L, what is the H^+ ion concentration, acid or basic?

$$(?) + (-9) = -14$$
$$? = -5$$

therefore the H^+ ion concentration is 1×10^{-5} M/L, and the solution is acid.

3. If the H^+ ion concentration is 1×10^{-10} M/L, what is the OH^- ion concentration, acid or basic?

$$(-10) + (?) = -14$$
$$? = -4$$

therefore the OH^- concentration is 1×10^{-4} M/L, and the solution is basic.

4. If the H^+ ion concentration = 1×10^{-7}, what is the OH^- ion concentration, acid, basic or neutral?

$$(-7) + (?) = -14$$
$$? = -7$$

therefore the OH^- ion concentration = 1×10^{-7}, and the solution is neutral.

pH AND THE HYDROGEN ION CONCENTRATION

One of the most important procedures used in the diagnosis and treatment of disease is an analysis of the acidity or basicity (alkalinity) of the various body fluids. Some fluids are normally acidic, such as the gastric juices or the urine, while the blood and intestinal fluids are normally basic. A very simple way to accurately express the acidity or basicity of fluids is known as the hydrogen ion scale, or more commonly referred to as the pH. The symbol pH indicates the concentration of the hydrogen ion in a simple way. It is a *whole* number and its *number value* is the same as the H^+ ion concentration when that concentration is written as some power of 10 in moles per liter.

Examples:

H^+ ion concentration	*pH*
1×10^{-5}	5
1×10^{-7}	7
1×10^{-10}	10

The procedure is:

1. Discard the 1
2. Discard the multiplication sign
3. Discard the 10
4. Discard the minus sign of the exponent
5. Use the whole number as a positive whole number for the value of the pH

When the H^+ ion concentration is:

a. from 1×10^{-1} to 1×10^{-7} the solution is acid.
b. 1×10^{-7}, the solution is neutral.
c. 1×10^{-7} to 1×10^{-14} the solution is basic.

NOTE: The expression 1×10^{-1} can be written as 10^{-1}; 1×10^{-7} can be written as 10^{-7}.

The pOH Relationship to pH

Just as the pH denotes the H^+ ion concentration, the term pOH denotes the OH^- ion concentration. And, just as you were able to calculate the OH^- ion concentration knowing the H^+ ion concentration by the formula $(H^+)(OH^-) = 1 \times 10^{-14}$, you can calculate the pOH by subtracting the numerical value of pH from 14.

$$pH + pOH = 14$$

If you know the pH, subtract it from 14 to get the pOH.
If you know the pOH, subtract it from 14 to get the pH.

Examples involving pOH:

1. The pH of a solution is 3, what is the pOH?

$$14 - 3 = 11$$
$$pOH = 11$$

The solution is acid. The H^+ ion concentration is greater than the OH^- ion concentration.

2. The pH of a solution is 8, what is the pOH?

$$14 - 8 = 6$$
$$pOH = 6$$

The solution is basic. The OH^- ion concentration is greater than the H^+ ion concentration.

Table 8-2. Relationship between acidic solutions,* basic solutions, pH and pOH.

	ACIDIC SOLUTIONS pH							*Pure water*	*BASIC SOLUTIONS pOH*						
	greater acidity			*medium*		*low*		*NEU-TRAL*	*low*		*medium*		*greatest basicity*		
H^+	10^0	10^{-1}	10^{-2}	10^{-3}	10^{-4}	10^{-5}	10^{-6}	10^{-7}	10^{-8}	10^{-9}	10^{-10}	10^{-11}	10^{-12}	10^{-13}	10^{-14}
OH^-	10^{-14}	10^{-13}	10^{-12}	10^{-11}	10^{-10}	10^{-9}	10^{-8}	10^{-7}	10^{-6}	10^{-5}	10^{-4}	10^{-3}	10^{-2}	10^{-1}	10^0
pH	0	1	2	3	4	5	6	7	8	9	10	11	12	13	14
pOH	14	13	12	11	10	9	8	7	6	5	4	3	2	1	0

*Concentrations are in moles per liter

The relationships that exist between the H^+ ion, OH^- ion concentrations and the pH and pOH can be seen in Table 8-2. Since the product of the H^+ ion and the OH^- ion concentration equals 1×10^{-14}, notice that the following relationship holds true.

When the H^+ ion concentration $= 1 \times 10^{-3}$,
the OH^- ion concentration $= 1 \times 10^{-11}$.
When the H^+ ion concentration $= 1 \times 10^{-12}$,
the OH^- ion concentration $= 1 \times 10^{-2}$

Also you can note that in every case the sum of the pH and pOH values equals 14.

This very same relationship is further explained in Figure 8-3. Note that in the pOH scale, values below 7 are basic, and those above 7 are acid. This is the reverse of the pH scale. The pH of some of the common substances that you will come in contact with is shown in Table 8-3.

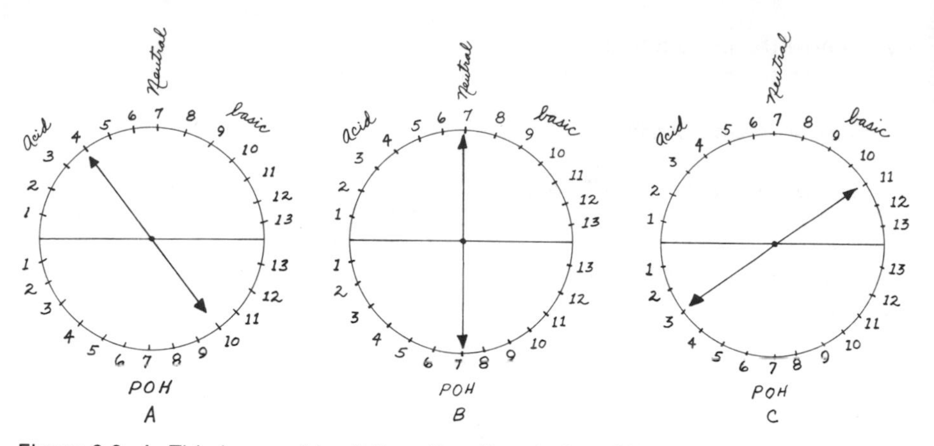

Figure 8-3. A. This is an acid solution; the pH = 4, the pOH = 10. B. This is a neutral solution; the pH = 7, the pOH = 7. C. This solution is basic; the pH = 11, the pOH = 3.

The pH Meter

An instrument called the pH meter is used to measure the pH or H^+ ion concentration. One model has two electrodes and a dial which is calibrated in pH units. Once the unit is standardized by calibrating it in a standard solution of known pH, the pH of unknown solutions can be easily determined by immersing the clean electrodes in that solution and reading the pH value directly on the dial.

Measurement of pH, Dyes and Indicators

Certain dyes and compounds are sensitive to the H^+ ion concentration in a solution. In acid solutions they have one color, in basic solutions, another color.

Table 8-3. pH of some common reagents, solutions and body fluids.

Hydrochloric acid (dilute)	1	Extremely strong acid
Sulfuric acid (dilute)	2	Very strong acid
Lemon juice	2	
Vinegar	3	Strong acid
Phosphoric acid	3	
Tomatoes	4	Acid
Black coffee	5	Very weak acid
Carbonated water	5	
Urine	6	Extremely weak acid
Milk	6.6	
Pure water	7	Neutral
Blood	7.35	
Sea Water	8.3	Extremely weak base
Sodium bicarbonate (dilute)	8.5	
Ammonium Hydroxide (dilute)	11	Base
Sodium carbonate	11.4	
Calcium Hydroxide (dilute)	12	Strong base
Sodium hydroxide	13	Very strong base

Some dyes show one color at one pH value, and another color at a different pH value. Some change over a broad range of pH units, while others change colors over a narrow range. The dye or indicator is usually used in a 1 percent concentration dissolved in an alcoholic solution. These dyes or indicators are complex organic compounds. Those individuals who are responsible for the maintenance to the proper pH of swimming pools add a few drops of the indicator to a container of the water and then compare the color with the acceptable standard. If the pH is too low (too acidic) they can then add lime to raise the pH to the proper level. If the pH is too high, they can then add acidic compounds to lower the pH to the proper value.

pH Indicator Papers

Laboratory test papers, pH indicator papers, are papers that are impregnated with combinations of indicators. They turn different colors at different pH values. One example is the test paper used to determine the acidity of urine.

SALTS

Salts are formed when acids neutralize bases; this is a continuous process in the human body. The presence and concentration of the various salts are extremely important and are measures of the health of the living body. Table 8-4 lists commonly encountered salts and their medical uses. The major cations of the body fluids are Na^+, K^+, Ca^{+2}, and Mg^{+2}; the major anions being HCO_3^-, Cl^-, HPO_4^{-2}, and SO_4^{-2}.

$$HCl + NaOH \longrightarrow NaCl + H_2O$$
$$H_2SO_4 + 2KOH \longrightarrow K_2SO_4 + 2\ H_2O$$

Normal Salts

A salt that does not have a replaceable hydrogen or hydroxyl group is called a *normal salt.* Examples of normal salts include NaCl, KCl, Na_2SO_4, Na_3PO_4, and $CaCl_2$.

Hydrogen Salts and Acid Salts

If the reacting acid has 2 or more replaceable hydrogens, and only part of the acidic hydrogen of the acid is neutralized, the salt is called an *acid salt.*

$$H_2SO_4 + NaOH \longrightarrow NaHSO_4 + H_2O$$

The sodium bisulfate still contains 1 active and replaceable hydrogen.

$$H_3PO_4 + NaOH \longrightarrow NaH_2PO_4 + H_2O$$

The sodium dihydrogen phosphate formed still contains 2 replaceable hydrogens. It can react with another NaOH to form sodium hydrogen phosphate:

$$NaH_2PO_4 + NaOH \longrightarrow Na_2HPO_4 + H_2O$$

Acid salts are extremely important to health care personnel because they are part of the chemical systems of the body that maintain the proper pH level of the body fluids. You will learn more about these systems, called buffers, very shortly.

Basic Salts, the Hydroxysalt

Basic salts are those in which only part of the hydroxyl groups of a base have been replaced. Examples of these salts include $Ba(OH)Cl$, $Bi(OH)_2Cl$, $Bi(OH)Cl_2$, $Mg(OH)Cl$, and $Ca(OH)Cl$. These salts retain the properties of a base as well as those of a salt. A typical reaction is:

$$Pb(OH)_2 + HNO_3 \longrightarrow Pb(OH)NO_3 + H_2O$$

The basic lead nitrate is formed because only 1 of the 2 hydroxyl groups was neutralized.

Table 8-4. Commonly encountered salts and their uses.

Salt	Use
$NaCl$	Composes 92% of the blood serum salts, affects irritability of the nerves, muscles and beating of the heart. Intravenous saline solution (0.85%) is used as an intravenous replacement fluid.
$NaHCO_3$	Antacid, baking soda.
$Na_2CO_3 \cdot 10H_2O$	Washing soda, alkaline baths.
$BaSO_4$	Opaque pigment for x-ray diagnosis. Although barium salts are poisonous, $BaSO_4$ is so insoluble that none dissolves; therefore it has no toxic effect on the body.
KI	Expectorant, it facilitates the expulsion of phlegm and mucous from the lungs and throat.
$MgSO_4$	Epsom salts, acts as a purgative.
$HgCl_2$	Antiseptic, germicide.
$AgNO_3$	Antiseptic, astringent, germicide.
$CaCO_3$	Antacid.
KNO_3	Diuretic (promotes formation of urine).
KBr	Sedative.
$(CaSO_4)_2 \cdot H_2O$	Plaster of Paris.
NaF	Applied to teeth to reduce caries.
$FeSO_4$	Treatment of anemia.
NaI	Treatment of thyroid, iodine deficiency.

FUNCTION OF SALTS IN THE BODY

Normal metabolism causes salts to be formed in the body. With the exception of the gastric juices and the urine, the body fluids are either neutral or alkaline. Although salts do not furnish energy to the body they are absolutely essential to life, acting in 4 major ways:

1. They maintain the proper pH of the body fluids.
2. They are essential in maintaining normal osmotic pressure of the body fluids.
3. They are needed for body structure, body tissues and protoplasm.
4. They are required for the maintenance of the elasticity and irritability of the nerves and muscles.

The Sodium Salts

The sodium salts act to stimulate the heart muscle, and a lack of sodium salts slows down the heart action, resulting in low blood pressure. The sodium ion concentration determines how much water is held in the tissues. Edema or dropsy are conditions caused by too high a sodium ion concentration. Excessive intake of NaCl can cause death because high concentrations interfere with the absorption and utilization of food. Excessive sweating can cause a depletion in the salt content of your body, since perspiration contains NaCl.

Calcium Salts

Calcium salts are essential for blood clotting and the contraction of the heart muscle. They are also necessary for firm skeletal structure and good teeth. Lack of calcium can cause poor teeth and rickets in children.

Iron Salts

Iron salts are needed in the formation of hemoglobin, a component of blood. Iron gives hemoglobin its red color. It is an absolute requirement for life. The lack of iron causes anemia, insufficient oxidation in the cells, and lowered vitality.

Potassium Salts

Potassium excess or deficiency can cause arrhythmias of the heart.

BUFFERS

The body fluids must stay within their normal pH limits for the body to function normally. The gastric juices, intestinal fluids, and blood must therefore have the ability to resist changes in pH. Acids can accumulate in the respiratory system causing respiratory acidosis due to hypoventilation. Acids can accumulate in the body as a result of nephritic acidosis (decreased excretion of acid metabolites) and can lead to metabolic acidosis. Metabolic alkalosis can also be caused by excessive vomiting (loss of gastric HCl) or antacid overdoses; respiratory alkalosis can result from fever and hyperventilation. Enzymes in the digestive processes have an optimum pH where they exhibit their greatest activity.

The body has various mechanisms for maintaining the normal pH for each particular fluid and metabolic system. They are called *buffer systems,* and they consist of a *pair of components, a weak acid and the sodium or potassium salt of that weak acid.* This pair serves to maintain an essentially constant pH, but the buffer system cannot take care of excessive amounts of acid or basic components that might appear as a result of pathologic respiration or metabolism.

A buffer system works in the following manner. If an excess of acid appears, the system pair neutralizes the acid returning the pH to the normal limits. Should excess basic compounds appear, the pair neutralizes the excess base, returning the pH to the normal limits (Fig. 8-4).

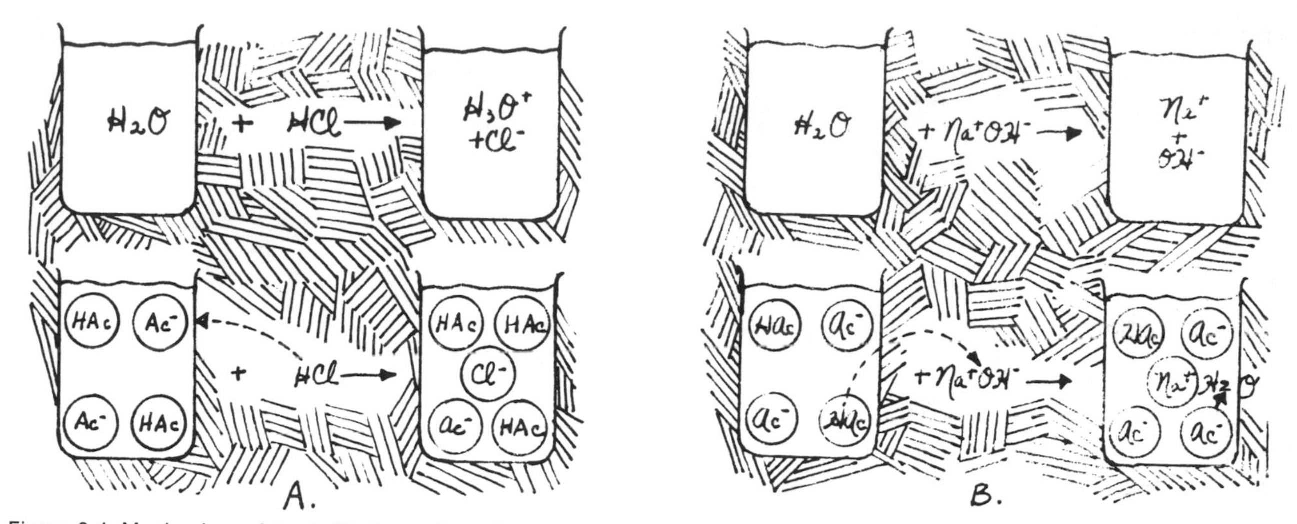

Figure 8-4. Mechanism of the buffering action of the acetic acid and acetate ion buffer. A. When acid appears the acetate ion ties up the hydrogen ion of the acid to form undissociated acetic acid. B. When base appears the acetic acid reacts with the OH^- ion of the base to form water molecules.

Carbonate Bicarbonate Buffer System

The carbonate bicarbonate system is an important buffer pair. Carbon dioxide in the blood reacts with water to form carbonic acid.

$$CO_2 - I_2O \longrightarrow H_2CO_3$$

Carbonic acid is a weak acid, dissociating very little, yielding few hydrogen ions in solution.

$$H_2CO_3 \rightleftharpoons H^+ + HCO_3^-$$

The large arrow pointing to the left indicates that the acid is only dissociated to a small extent, yielding few hydrogen ions plus the bicarbonate radical. Therefore the carbonic acid and the bicarbonate ion exist in solution.

When acid (H^+ ions) enters the system, the bicarbonate ion combines with the extra hydrogen ions to form the weakly dissociated carbonic acid. By removing the hydrogen ions, the bicarbonate ion removes the acid, forming carbonic acid. Thus the excess H^+ ions are tied up, the excess acid is buffered, and the system maintains its normal pH.

$$H^+ + HCO_3^- \longrightarrow H_2CO_3$$

If a base enters the system (extra OH^- ions), it is neutralized by the presence of the carbonic acid because acids react with bases. The OH^- ions that react with the carbonic acid are removed from the solution, the excess base is buffered, and the system maintains its normal pH. The body maintains its constant pH by keeping the ratio of the bicarbonate ion to the carbonic acid at 20:1.

The Monohydrogen Dihydrogen Phosphate Buffer

The phosphate buffer system consists of the HPO_4^{2-} and $H_2PO_4^{1-}$ pair. When acids appear, they are neutralized by the HPO_4^{2-} according to the reaction:

$$H^+ + HPO_4^{2-} \longrightarrow H_2PO_4^{1-}$$

The formation of the $H_2PO_4^-$ thus removes the hydrogen ions, and excesses of the $H_2PO_4^-$ are eliminated from the body by urine. When bases appear (OH^- ions), they react with $H_2PO_4^-$ of the buffer pair forming the monohydrogen phosphate:

$$OH^- + H_2PO_4^{1-} \longrightarrow HPO_4^{2-} + H_2O$$

Excessive quantities of the $H_2PO_4^-$ which are formed are eliminated by the urine.

SUMMARY

Acids are compounds that ionize yielding hydrogen (hydronium) ions in water. They react with metals to form salts, liberating hydrogen gas. The important common acids are HCl, H_2SO_4, HNO_3, H_3BO_4, HOCl, citric, acetic, and lactic acid. Bases are compounds that ionize in water to yield OH^- ions, and bases react with acids to form a salt and water. The important common bases are NaOH, KOH, $Ca(OH)_2$, $NaHCO_3$, Na_2CO_3, and $Mg(OH)_2$. Both strong acids and bases are totally ionized in water, while weak acids and bases are only partially ionized. Neutralization occurs between acids and bases because water is only minutely ionized into H^+ and OH^-. The concentration of H^+ in M/L multiplied by OH^- ion concentration in M/L is 1×10^{-14}. The concentration of H^+ ions can be expressed as pH; the concentration of OH^- as pOH. The sum of pH and pOH equals 14. A pH meter can be used to determine the H^+ ion concentration. The product of neutralization can yield normal, acid or basic salts. A combination of a weak acid and the salt of a weak acid forms a buffer, a system that serves to maintain the normal pH of body fluids.

EXERCISE

1. Name the characteristic ion of an acid. Give two names for it.
2. List five characteristics of acids.
3. What is meant by the activity series of metals? Which metal is most reactive with acids?
4. Complete the reactions: $Zn + H_2SO_4 \longrightarrow$
   $Fe + 2HCl \longrightarrow$
5. Write the ionization reactions of HCl, H_2SO_4 and H_3PO_4.
6. What is formed when an acid reacts with a base?
7. What is a remedy for hypoacidity?
8. Why is HOCl an effective bacteriocide?
9. Define a base and give three examples of bases.
10. List five characteristics of bases.
11. Complete the following reactions: $Zn + KOH \longrightarrow$
    $CaSO_4 + NaOH \longrightarrow$
12. Name five common bases and state what they are used for.
13. What is the difference between strong and weak acids? Strong and weak bases?
14. Write the mathematical formula that relates the H^+ ion and OH^- concentrations of water.
15. What is the OH^- ion concentration if the H^+ ion concentration is 1×10^{-4}? 1×10^{-7}? 1×10^{-11}? Are the solutions acid, basic or neutral?
16. What is meant by the term pH?
17. If the H^+ concentrations are: 1×10^{-2}, 1×10^{-7}, and 1×10^{-13}, what are the pH values? Are the solutions acid, basic or neutral?
18. If the pH is 4, what is the pOH?
19. If the pOH is 8, what is the pH? Is the solution acid or basic?
20. How can one determine the hydrogen ion concentration of a solution?
21. Which of the following is an acid salt, a neutral salt, and a basic salt: NaCl, $NaHCO_3$, Mg(OH)Cl.
22. What is a buffer? What are the components of a buffer system?
23. Explain how a buffer system eliminates excess acids and bases.

OBJECTIVES

When you have completed this chapter you will be able to:

1. List and give examples of different types of solutions.
2. Define a solution and the components of a solution.
3. Distinguish between a solution and a heterogeneous mixture.
4. State the characteristics of a solution.
5. Discuss the effect of temperature on the solubility of salts in water and determine the solubility of a particular salt at different temperatures.
6. Distinguish between dilute, concentrated and saturated solutions.
7. Discuss the effect of temperature and pressure on the solubility of gases in water.
8. Explain in detail osmosis and osmotic pressure, and how these phenomena affect the water balance of the body.
9. Describe the effect of the injection of isotonic, hypotonic, and hypertonic solutions on red blood cells.
10. Distinguish between true solutions, colloidal dispersions and suspensions.
11. Define dialysis and state the important difference between osmosis and dialysis.
12. Describe the artificial kidney machine and explain how it works.
13. Calculate and make solutions based on:
    a. Percent concentration
    b. Molar
    c. Osmolar
    d. Millimolar
    e. Milliosmolar
    f. Normal
14. Give the relationship between equivalents and milliequivalents, and between milliequivalents and milliosmols.

SOLUTIONS

In the health care field you will use liquids, solids, and gases in the delivery of medical care. In the majority of cases, one or more of the substances that you administer will be dissolved in another substance in the form of a solution.

A solution is a homogeneous mixture of two or more substances; the solution is completely uniform, with every part identical to every other part. These substances may be in the form of atoms, such as metallic gold dissolved in metallic copper to make 10 carat gold. Pure gold is 100 percent gold and called 24 carats; 10 carat gold is 10 parts gold and 14 parts copper. Substances making up a solution may consist of molecules such as sugar dissolved in water or soluble drugs dissolved in water and alcohol (Table 9-1). These substances may also be gases such as the oxygen dissolved in nitrogen in our atmosphere or special gas mixtures such as the oxygen dissolved in helium used in inhalation gas therapy. Furthermore, they may be ions dissolved among molecules, such as NaCl dissolved in water.

Table 9-1. Types of solutions.

Components		
Solute Phase	*Solution Phase*	*Examples*
Gas	Gas	All mixtures of gases, air
	Liquid	Carbonated beverages
	Solid	Hydrogen in palladium
Liquid	Gas	—
	Liquid	Alcoholic beverages (alcohol and water)
	Solid	*Amalgams,* mercury in gold
Solid	Gas	—
	Liquid	Sugar in water
	Solid	*Alloys,* brass, silver in lead

Other solutions can be composed of gas molecules dissolved in water, such as CO_2 dissolved in water, making carbonated water.

It is necessary first to understand the terminology and definitions of solution chemistry in order to fully understand the purpose and technique of using or administering medications in the form of solutions. Some of the important solutions used in the health care field include physiologic saline solution (commonly called Ringer's solution), boric acid solution, intravenous (I.V.) isotonic solutions, dextrose solution, and respiratory inhalation therapy gas solutions.

TERMINOLOGY OF SOLUTION CHEMISTRY

Solutions are *homogeneous* mixtures of two or more substances. Examples include salt water, homogenized milk, tea, atmospheric air, and alcoholic solutions of drugs.

A *solute* is a substance that is dissolved; it is present in the smaller amount, usually less than 50 percent. Examples include salt in salt water, all dissolved components of milk, oxygen in the air, and a drug in alcoholic solution.

A *solvent* is that substance that does the dissolving; it dissolves the solute and is present in the greater amount, over 50 percent. Examples include water in a salt water solution, water in tea, water in milk, nitrogen in air, and alcohol in the drug solution.

Difference Between a True Solution and a Heterogeneous Mixture

A mixture of iron dust and sulfur is not a solution because it is not homogeneous. There are distinct particles of iron and sulfur visible and detectable. It is a mixture because the iron and sulfur both retain their individual properties. They may be separated from each other by a magnet (which attracts the iron particles) or by a solvent called carbon disulfide that dissolves the sulfur particles but not the iron. The iron can be filtered from the solution, and the sulfur can be recovered intact by allowing the solvent to evaporate.

CHARACTERISTICS OF SOLUTIONS

The most commonly used solutions have solvents of either water or alcohol, or mixtures of the two. Solutions are clear, homogeneous, may be colored or colorless, will not settle out after long periods of standing (if the solvent is not allowed to evaporate), and will not hinder the passage of light through the solution. The amount of solute in a solution can be varied, thus changing the concentration, and by allowing the solvent to evaporate (loosely stoppered drug solution bottles) the concentration of the solute can be increased. Care should always be taken to securely close or stopper solution, medication, and reagent bottles to prevent evaporation of the solvent and a change in the concentration of the solution.

Separation of a Solution Into Its Components

Solutions can be separated into their individual components by using heat to take advantage of the difference in the vapor pressures of the substances and effect a separation. For example, a salt solution can be allowed to evaporate; the salt remains as a residue. If the same solution were put into a distillation apparatus, as is used industrially to make drinking water from sea water, the application of heat would vaporize the water from the salt, and the water would be collected as the distillate (see Fig. 7-20).

SOLUBILITY OF SOLUTES IN SOLVENTS

The quantity of solute that will dissolve in a definite amount of solvent is fixed at a particular temperature. Some substances are very soluble in water but insoluble in alcohol. Others are soluble in alcohol but insoluble in water. Some are insoluble in both alcohol and water but soluble in petroleum solvents and chemicals. In fact, no material is absolutely insoluble; even marble dissolves ever so slightly in water.

We define the solubility of a solute as the weight in grams of the solute that will dissolve in 100 grams of solvent *at a particular temperature.* From your everyday experiences you know that when a liquid solvent is heated more of the solid solute will dissolve. Hot water dissolves more sugar than cold water. Therefore, with the exception of certain calcium compounds and a few others, the higher the temperature of the solvent the more the solid solute dissolves (see Table 9-2). Knowledge of this will enable you to prepare solutions more quickly when you need them.

Table 9-2. Solubility of salts at various temperatures (g/100 g H_2O).

Temperature °C	*Potassium nitrate*	*Sodium chloride*	*Potassium alum*	*Calcium chromate*	*Potassium chloride*	*Sodium sulfate*	*Sodium nitrate*
0	13	35.7	4	13.0	28	4.8	73
10	21	35.8	10	12.0		9.0	80
20	31	36.0	15	10.4	34	19.5	85
30	45	36.3	23	9.4		40.9	92
40	64	36.6	31	8.5	40	48.8	98
50	86	37.0	49	7.3		46.7	104
60	111	37.3	67	6.0	46	45.3	
70	139	37.9	101	5.3		44.4	
80		38.4	135	4.4	51	43.7	133
90		39.1		3.8		43.1	
100	249	39.8		3.0	57	42.5	163

Dilute, Concentrated, and Saturated Solutions

Under normal conditions, the capability of the solvent (at a specific temperature) to dissolve a solid solute eventually stops. Regardless of what you do, the solvent *will not* dissolve any more solute. Solid crystals and particles will remain undissolved at the bottom of the liquid solution. At that point the solution is saturated. It will not dissolve any more solute *at that temperature.*

The following happens when you start dissolving a solute in a solvent. When you begin to dissolve solid solute in liquid solvent, you progress from a very dilute solution to a more and more *concentrated* solution and eventually reach the *saturation solution* point. There is no sharp line that separates the dilute from the concentrated solutions. A concentrated solution of boric acid in water is 5 g/100 ml, a 5 percent solution. A concentrated solution of NaCl may be 35 g/100 ml, a 35 percent solution, and a concentrated solution of sugar may be

49 g/100 ml, a 49 percent solution. Therefore whether a solution is dilute or concentrated depends upon the solute and the solvent, and how much of the solute is dissolved in a particular amount of the solvent at a definite temperature.

A substance such as silver chloride will only dissolve very slightly in water. Thus, you can have a saturated solution of silver chloride that is extremely dilute because of its low solubility. You cannot get a concentrated solution of silver chloride (or for that matter any other slightly soluble material) in water simply because of its low solubility.

Note in Figure 9-1 the solubilities of some inorganic compounds and how temperature affects the solubility. With some compounds, an increase in 100°C makes very little difference in the solubility; with sodium chloride there is a difference of only 4 grams for the 100°C temperature change. However, KNO_3 has a solubility of 12 g at 0°C and about 62 g at 40°C, almost five times as much for a 40°C change. Obviously, the effect of an increase in temperature on the solubility of the compound depends upon the particular compound. Certain calcium compounds are less soluble at higher temperatures than at lower temperatures.

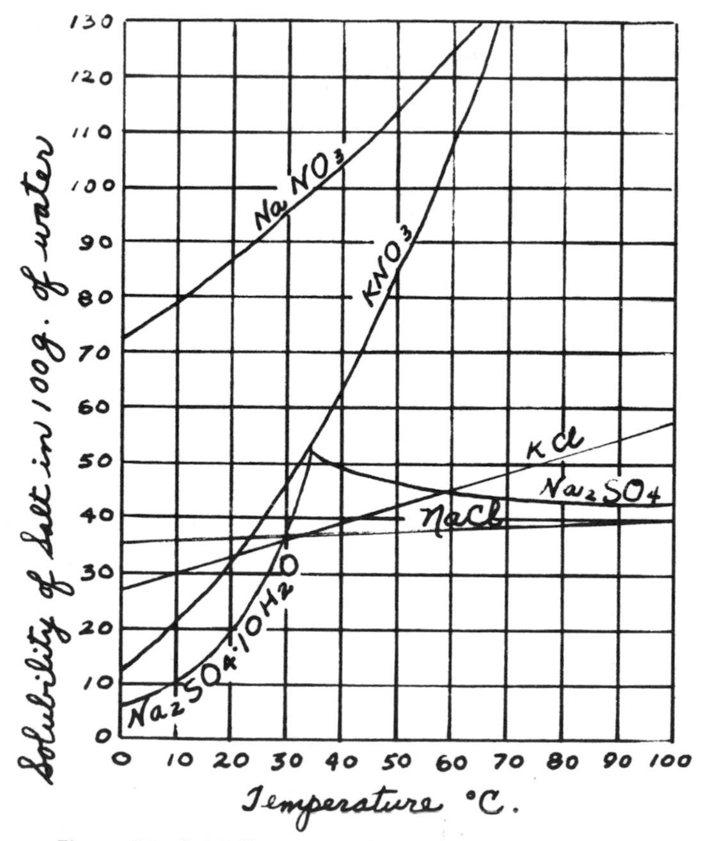

Figure 9-1. Solubility curves of some ionic solids in water.

Changing Dilute Solutions to Concentrated Solutions

Dilute solutions can be changed to more concentrated solutions by two methods: (1) adding more solute, or (2) evaporating or boiling off (if practical) some of the solvent.

Supersaturation

Normally, when a saturated solution at a high temperature is cooled, some of the dissolved solute will crystallize out to reach the saturation concentration at the lower temperature. Sometimes, however, you will find that the solute does not crystallize out at the lower temperature, but remains in solution. This is an unstable condition because the solution is supersaturated. The supersaturated solution, being unstable, will revert back to the saturated state if it is agitated, mixed, or allowed to stand. The excess solute will crystallize out.

Increasing the Rate of Dissolving a Solid Solute

The rate at which a solute dissolves in a solvent can be increased by using one or more of the following procedures:

1. *Stirring*: This increases the rate because fresh solvent (unsaturated solvent) is brought in contact with the undissolved solute.
2. *Heating*: Heating increases the rate at which the solute will dissolve.
3. *Powdering*: By grinding up the solid (pulverizing it) the surface area of the solute is greatly increased, exposing a greater surface area to the solvent.

Effect of Temperature on Gas Solubility in Liquids

Gases dissolved in liquids behave differently than in solids because as the temperature of the liquid solvent is increased, the solubility of the gas *decreases.* Conversely, as the temperature of the liquid is decreased, the solubility of the gas *increases.* Usually, a gas that does not react with its solvent can be driven out of solution completely by boiling the solution, as with carbon dioxide in water. However, many gases that do react with water, such as HCl, cannot be completely driven out of solution.

If the temperature of a water solution containing a gas is reduced to freezing, the solubility of the gas becomes almost zero. For this reason, bottles of carbonated beverages should *never* be frozen, because as the water freezes the dissolved gases are set free and the gas pressure on the bottle increases enormously. In fact, the tremendous increase in the gas pressure coupled with the smaller volume available to the confined gas because of the expansion of the water on freezing, can cause the bottle to explode with extreme violence and cause injury.

Effect of Pressure on Dissolved Gases

The solubility of a gas in a solvent depends on the pressure exerted in the liquid. This is known as Henry's law; which states that the greater the gas pressure, the more gas will dissolve in the liquid, and the less the gas pressure, the less gas will be dissolved. This is quite evident when you open a bottle of a carbonated beverage; there is an immediate loss of carbon dioxide because the pressure on the liquid is now reduced to that of the atmosphere. You also know that opened bottles of carbonated beverages will eventually go flat by gradually losing the CO_2.

From Table 9-3 you can see the effect of pressure on the solubility of nitrogen in water; by doubling the pressure of the water the weight of nitrogen dissolved is doubled. This phenomena is extremely important in inhalation therapy and in the treatment of divers. For example, a diver may die or suffer extremely painful injury as a result of the quick loss of pressure by quickly ascending from deep water. If he comes to the surface too quickly, the dissolved gases in the blood will bubble out (just as the CO_2 does in the opened bottle), because of the lowering of the pressure on them. These gas bubbles block the flow of the blood through the veins and arteries. This bubbling out of the dissolved gases is commonly known as the "bends," and the only possible way to treat this is to increase the pressure on the diver back to the original higher pressure experienced in deep water. By doing this (increasing the pressure on the diver) the bubbles of gas are dissolved back into the blood. Once that has been accomplished the pressure is slowly and gradually reduced to allow the dissolved gases to diffuse out of the blood, finally exposing him to atmospheric pressure. This is usually done by taking the diver back down to his original depth in the water or by placing him in a hyperbaric chamber.

Table 9-3. Solubility of nitrogen gas in water (g/100 g*) at 20°C.

0.001	at	380 mm	½ atmosphere
0.002		760 mm	1 atmosphere
0.004		1520 mm	2 atmosphere

*Weight rounded off to 3 places.

OSMOSIS

The importance of osmosis in the physiological processes of the body cannot be overemphasized. The principles of osmosis are important in the formation of urine, the maintenance of blood volume, the assimilation of nutrients, the distribution of water throughout the body tissues, and in the clinical treatment of edema, a condition where the body tissues contain an excessive amount of tissue fluid.

When a solution and a solvent are separated by a semipermeable membrane, this membrane permits the free movement of water molecules through it, but completely blocks the passage of dissolved particles, such as ions and molecules (Fig. 9-2). This process is called osmosis. When the semipermeable membrane separates two aqueous solutions that have different concentrations of ions and molecules, the water molecules flow through the membrane, the major flow of water being into the solution that has the higher concentration of the dissolved solute. The direction of the water flow is always from the dilute to the more concentrated. The force that causes this diffusion of water is called the osmotic pressure, and the water will flow through the membrane diluting the more concentrated solution until the osmotic pressure on both sides of the membrane are equal (Fig. 9-3). The water actually diffuses through the membrane in both directions simultaneously, but the rate of flow from the dilute or

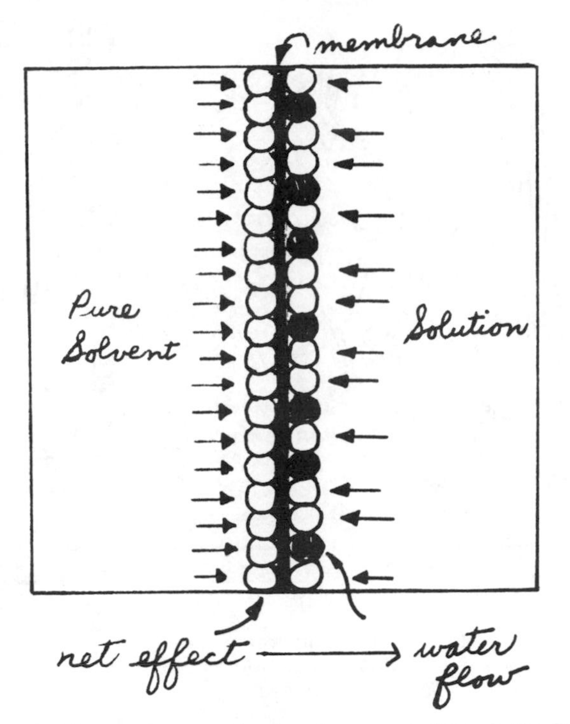

Figure 9-2. Osmosis. The semipermeable membrane permits only the flow of water molecules through it. On the pure solvent side (less concentrated) *all* of the membrane is in contact with water molecules. On the solution side (more concentrated) only *part* of the membrane is in contact with water molecules. There is a greater probability for the water molecule to pass from the more dilute to the more concentrated side.

pure solvent is greater than in the opposite direction. The tendency is for the concentrations inside and outside the membranes to equalize, but this cannot be completely attained. Always remember that the least concentrated solution has the most water.

Physiological Effects of Water Imbalance

The extracellular and intracellular body fluids contain electrolytes as cations and anions in definite concentrations (see Table 9-4). When their concentrations are varied from the normal, serious problems can result.

If an individual drinks excessive quantities of water, a serious condition called water intoxication can occur. The abnormally large quantities of water are absorbed in the blood stream and increase the blood volume and the blood pressure. The increased blood pressure forces water through the capillary walls into the intercellular (between cells) compartment, thus diluting the concentration of solute in that compartment. The intercellular compartment fluid, having been diluted, is now less concentrated than the intracellular (within the cell) compartment, and the process of osmotic flow begins. Water leaves the diluted intercellular compartment to flow into the now more concentrated (by com-

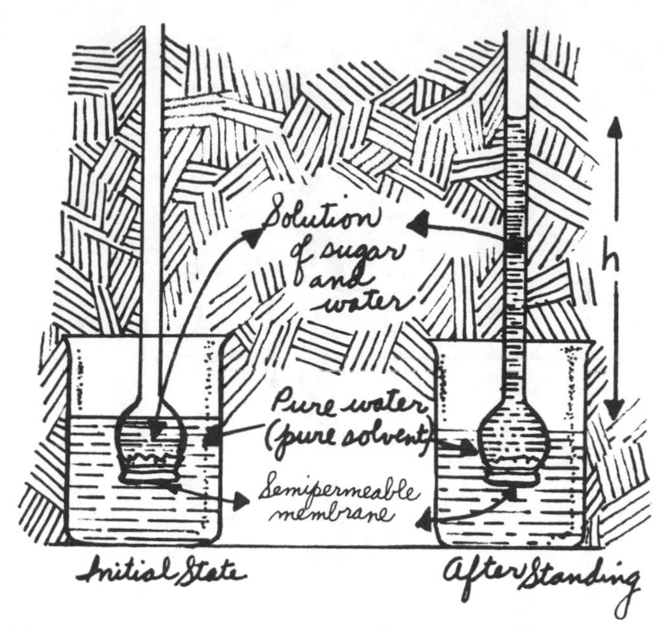

Figure 9-3. A laboratory experiment demonstrating osmosis. The osmotic pressure is the pressure that is necessary to maintain equilibrium conditions. The height to which the water rises is a measure of the osmotic pressure.

parison) intracellular compartment. When the water leaves the intercellular compartment to flow into the intracellular compartment, the intracellular compartment becomes excessively hydrated. This is called excessive cellular hydration.

Table 9-4. Electrolyte composition of body fluids.*

	Extracellular Fluid		
Electrolytes	*Plasma*	*Interstitial Fluid*	*Intracellular Fluid*
Cations:			
Sodium, Na^+	142	145	15
Potassium, K^+	5	4	157
Calcium, Ca^{++}	5	3	5
Magnesium, Mg^{++}	3	2	27
Total cations	155	154	204
Anions:			
Chloride, Cl^-	103	116	4
Bicarbonate$^-$, HCO_3^-	27	27	10
Sulfate $^=$, $SO_4^=$	1	2	18
Biphosphate$^=$, $HPO_4^=$	2	3	100
Proteins$^-$	16	1	72
Organic acid$^-$	6	5	
Total anions	155	154	204

*These values are expressed in milliequivalents per liter (mEq/L). Refer to the end of this chapter for a detailed explanation of this term.

On the other hand, when one drinks excessive quantities of salt water, the reverse osmotic flow takes place. The concentration of salts in the blood stream increases, causing a flow of water from the intercellular compartment to the plasma (the liquid part of the blood). This causes a higher concentration of salts to build up in the intercellular compartment which exceeds that of the intracellular compartment. As a result, water now flows from the intracellular compartment into the intercellular compartment and then into the blood stream. This increases the volume of blood, which in turn increases the blood pressure. The volume of urine discharged increases and therefore the cells become dehydrated, usually resulting in death.

Thus the body is able to maintain the proper balance of water in the various cellular compartments by the free movement of the water molecules through the cellular membranes. It is the relationship of the concentration of solute particles (such as NaCl) in the intercellular compartment that determines the direction of the water flow. Increases of intercellular salt concentrations cause water to flow into the cells; decreases in the intercellular salt concentrations cause water to flow out of the cells.

Isotonic, Hypotonic and Hypertonic Solutions

When two solutions have the same osmotic pressure, they are said to be *isotonic* to each other; this condition is extremely important when using intravenous saline solutions. I.V. saline solutions are needed when the body has lost a great deal of blood from injury or hemorrhage, in postoperative shock, when the patient is dehydrated, or in certain septic conditions. The I.V. fluid has the same concentration of salts as the blood, therefore the salt concentration of the blood is unchanged (Fig. 9-4). There is no osmotic flow of water and red

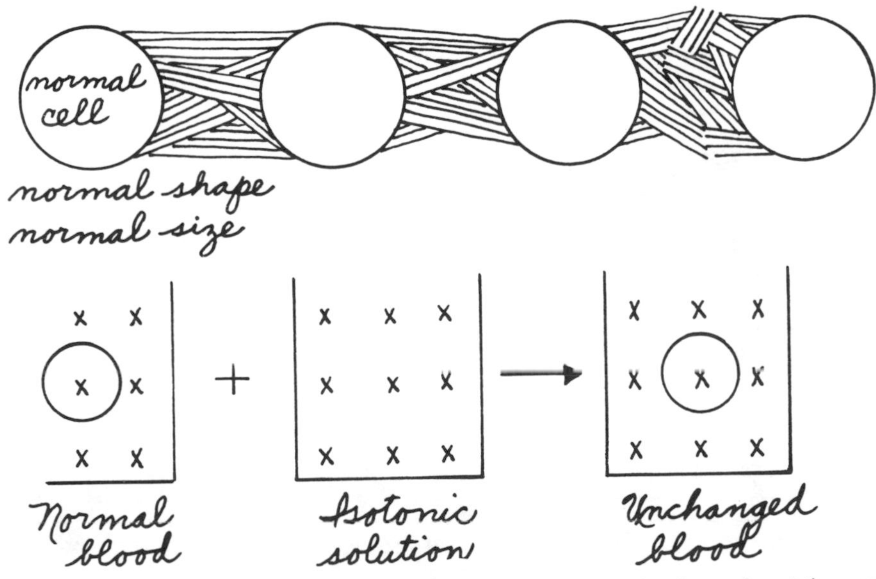

Figure 9-4. The red blood cell in an isotonic solution. The blood cell remains at the same size as there is no net change in the water content of the cell.

blood cells and tissues are not changed by osmosis. This fact must be considered when injecting drugs intravenously; they are dissolved in physiological saline solution so as not to adversely affect tissues or red blood cells.

Physiological I.V. saline solution is isotonic to the blood with respect to salinity and has a concentration of between 0.85 percent and 0.90 percent NaCl in sterile water. Ringer's solution contains 0.86 percent NaCl, 0.03 percent KCl and 0.033 percent $CaCl_2$ and is considered to contain the same proportions of salts as the blood.

HYPOTONIC SOLUTIONS

A hypotonic solution contains a lesser concentration of salts than the blood. If the blood cells are exposed to a hypotonic solution they will begin to swell and rupture as a result of osmosis (Fig. 9-5). This takes place because the water tends to pass through the membrane into the red blood cell where the salt concentration is high. Eventually the cell wall ruptures because the water attempted to dilute the more concentrated salt solution of the cell. This is called *hemolysis,* the blood is said to lake, and death can result.

HYPERTONIC SOLUTIONS

If an intravenous salt solution has a higher concentration of salts than the red blood cells, the process of osmosis will begin. The water within the red blood cell will flow out faster than it flows in. Consequently, the cell will shrivel and shrink (Fig. 9-6). This is called *crenation,* the cells are said to crenate, and in physiology this process is called plasmolysis.

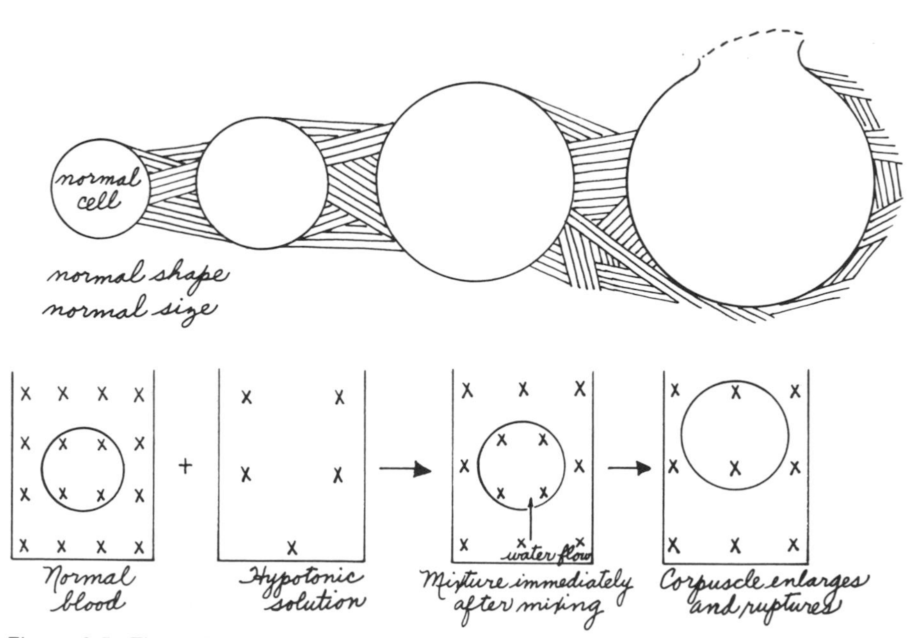

Figure 9-5. The red blood cell in a hypotonic solution. The osmotic flow of water into the cell causes the cell to swell and rupture.

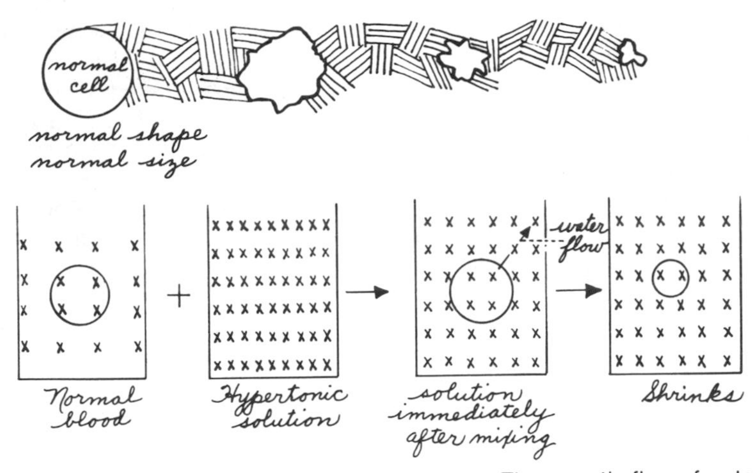

Figure 9-6. The red blood cell in a hypertonic solution. The osmotic flow of water out of the cell causes it to shrink, crenate.

The effects of osmosis hold true for the cleansing of open wounds. Physiological saline solutions should be used instead of pure water so that exposed tissue cells are not affected by osmosis.

SUSPENSIONS AND COLLOIDS

Extremely fine particles of magnesium hydroxide in water (milk of magnesia) or calcium hydroxide (lime water) will eventually separate and settle out. This is a suspension of particles in water. In a true solution, such as salt in water, the particles will not settle out. In between the true solution and the suspension is a type of dispersion in which the particles are too large for a true solution, yet are too small for a true suspension. This type of system normally never separates; it is called a *colloidal dispersion.* The particles may be extremely large molecules or aggregates of smaller molecules. Starch, plasma, and milk are examples of colloidal dispersions (see Table 9-5). When a light beam is passed through a colloidal dispersion, the particles actually scatter the light

Table 9-5. Types of colloidal systems.

Dispersed phase	*Dispersing phase*	*Example*
Liquid	Gas	Fog
Solid	Gas	Smoke, dust
Gas	Liquid	Foam, whipped cream
Liquid	Liquid	Emulsions, such as milk, mayonnaise
Solid	Liquid	Suspensions, such as paint, medicine
Gas	Solid	Solid foam, such as marshmallow, cake, bread
Liquid	Solid	Butter, oil-bearing shale
Solid	Solid	Many alloys, selenium in glass (tail light)

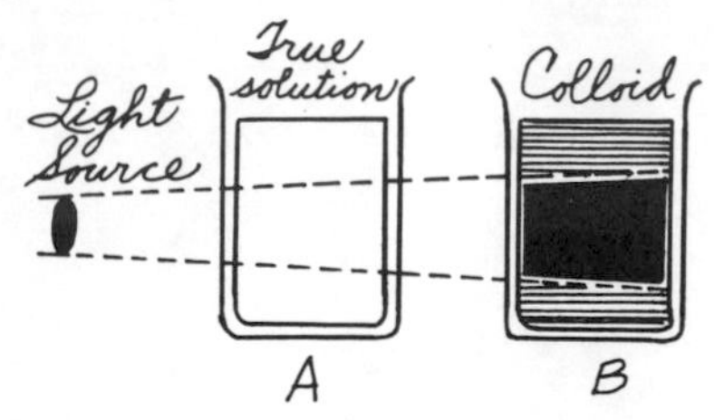

Figure 9-7. Comparison of the passage of light through a true solution and a colloidal solution. A. Light cannot be seen in a true solution. B. Light is scattered in a colloidal solution.

beam (the Tyndall effect) and the passage of light is easily seen. With the true solution, the light beam passes through unimpeded (Fig. 9-7).

In the digestive process pepsin in the gastric juices breaks down the very large protein molecules to form colloidal peptones. The enzyme catalysts in the body disintegrate ingested foods into colloidal particles ranging in size from 1 to 100 millimicrons, nanometers.

Colloids have an extremely large surface area, because they consist of so many extremely small particles. This tremendous surface area provides them with the property of adsorption, one of their most important properties. Colloidal particles have either negative or positive electrical charges (Fig. 9-8). They have little tendency to come together to form larger particles because like charges repel each other. A comparison of the properties of true solutions, suspensions, and colloidal dispersions can be seen in Table 9-6.

Table 9-6. Properties of solutions, colloidal dispersions, and suspensions

Property	*Solutions (molecular dispersions)*	*Colloidal dispersion*	*Suspension*
Homogeneity	Completely uniform	May or may not be uniform	Heterogeneous
Component separation	By physical means: by evaporation of solvent; *not by simple filtration.*	Partially separable by filtration; separable by membranes	Particles can be filtered from the liquid. Particles do not pass through membranes
Size of particles	Size of the molecule or the ion	Very large individual molecules or aggregates of small molecules.	Extremely large particles, can be seen with naked eye
	1/10,000,000 mm to 1/1,000,000 mm invisible	1/1,000,000 mm to 1/10,000 mm Can be seen with ultramicroscope	1/10,000 mm to 1 mm
Osmotic pressure	High	Low	None
Settling	Will not settle out	Particles can eventually settle out	Rapid settling
Tyndall effect	No, solution is transparent	Yes	Yes

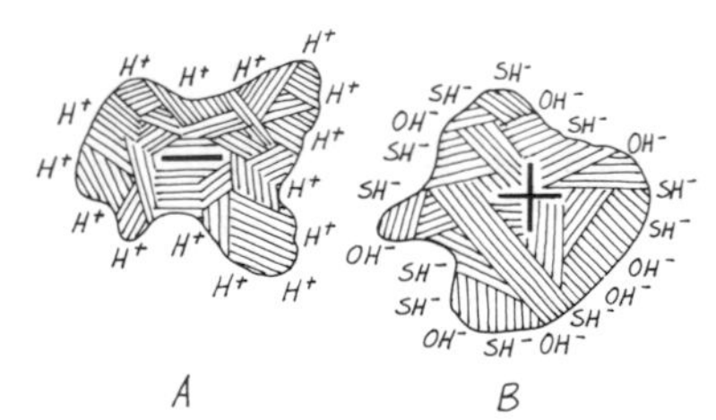

Figure 9-8. Adsorption of charged particles on colloidal aggregates. A. Adsorption of positively charged particles, the aggregate being negatively charged. B. The adsorption of negatively charged particles, the aggregate being positively charged.

DIALYSIS:

The membranes that *enclose* cells within living systems are not osmotic membranes, but instead membranes that are much more permeable, with larger openings. These membranes will not only permit the passage of the solvent water molecules, but also will allow IONS and SMALL molecules to pass through. This is necessary in order to permit the passage of nutrient molecules as well as waste products through the cell membrane, while at the same time preventing the passage of large molecules.

Membranes that *hold back* large molecules and particles of colloidal size but permit the passage of ions and small molecules are called *dialyzing membranes.* DIALYSIS is defined as the selective passage of small molecules and ions through membranes of this type. The only difference between dialysis and osmosis is that with osmosis, only the solvent is permitted to pass through the membrane. In dialysis both the solvent and certain small molecules and ions pass through the membrane.

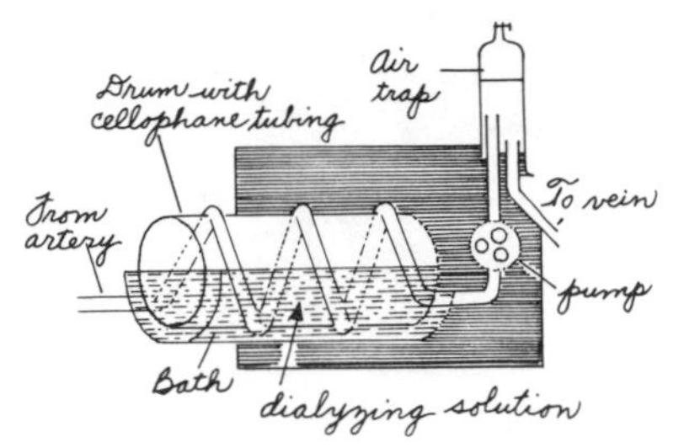

Figure 9-9. Artificial kidney machine.

The kidney is the filter of the body for waste products; its membranes allow the dissolved waste products to pass through, yet, at the same time, prevent the passage of very large protein molecules. The artificial kidney machine uses tubular coiled cellophane as the dialyzing membrane to purify the blood of patients having renal excretion problems (Fig. 9-9). This process is called hemodialysis. When the patient's blood is pumped through the coils, dialysis takes place and the soluble waste products of protein catabolism (breakdown of the proteins in the body) and toxic products pass through the dialyzing membrane into the dialyzing solution. Because that dialyzing solution is isotonic with the blood, no sodium ions diffuse. The composition of the dialyzing solution must be carefully controlled, because waste products begin to accumulate in it. If the composition is not controlled and the waste product concentration builds up too much, the blood returning to the patient will be as full of waste products and poisons as it was when it left the patient.

CONCENTRATIONS

We have learned that pure water does not conduct electricity and that water solutions containing electrolytes do conduct electricity. The fluids of the body contain electrolytes and conduct electricity, as evidenced by the many accidental electrocution deaths reported yearly. The concentrations of these dissolved solutes play an important part in the water balance of the body through osmosis and dialysis. It is of paramount importance that you become competent in definitions, terminology and calculations involved in solution concentrations, because the concentration of the solutes in the cellular compartments determines the water balance and direction of water flow.

Weight of Solute per Volume of Solvent

Weight-volume solutions are percent solutions, because the percent equals the number of grams of solute which are dissolved in 100 ml of solution. They are most frequently used in the health care field because once the weight of the solute has been measured out, it can be diluted in suitable volumetric measuring glassware. Because the density of water is 1 g/ml, weight-volume solutions can easily be prepared in volumetric flasks or graduated cylinders. It is much easier to prepare weight-volume solutions than weight-weight solutions, and in most cases, the small difference is within tolerance. For example, a 10 percent salt solution would have 10 g of salt/100 ml solution, a 1 percent salt solution would have 1 g of salt/100 ml solution, and a 25 percent NaOH solution would have 25 g NaOH/100 ml solution.

To prepare volumes of solution which are either more than or less than 100 ml, use the following procedure:

Step 1. Change the percent to a decimal by moving the decimal point 2 places *to the left.*

Step 2. Change the volume to milliliters (if other volumes such as liters or quarts are given).

Step 3. Multiply the decimal fraction by the volume in milliliters to get the weight of the solute, then dilute in a graduated cylinder or vessel to the desired volume.

Examples:

Prepare 3 liters of 0.9 percent I.V. saline solution.

Step 1. 0.9 percent = 0.009

Step 2. 3 liters = 3,000 ml

Step 3. Multiply: 3,000 × 0.009 = 27 g NaCl
Dilute to 3,000 ml.

Prepare 500 ml of a 5 percent dextrose solution.

Step 1. 5 percent = 0.05

Step 2. 500 ml

Step 3. Multiply: 500 × 0.05 = 25 g dextrose
Dilute to 500 ml.

Molar Solutions

A mole (mol) is the molecular weight (or formula weight) of a compound in grams. For example, one molecular weight of NaOH is 40 grams, one molecular weight of H_2SO_4 is 98 grams, and 1 molecular weight of HCl is 36.5 grams. When *one molecular weight of a compound* is dissolved in *1 liter of solution,* that solution is called a 1 molar (1 M) solution. If 2 molecular weights are dissolved, it is a 2 M solution, 5 molecular weights, a 5 M solution, and 1/10 molecular weight is a 0.1 M solution. Therefore, to prepare *1 liter* of a solution of a specified molarity, use the following procedure:

Step 1. Calculate the molecular (formula) weight.

Step 2. Multiply the molecular weight by the molarity desired to find the number of grams; then dissolve in water and dilute to 1 liter (see Fig. 9-10).

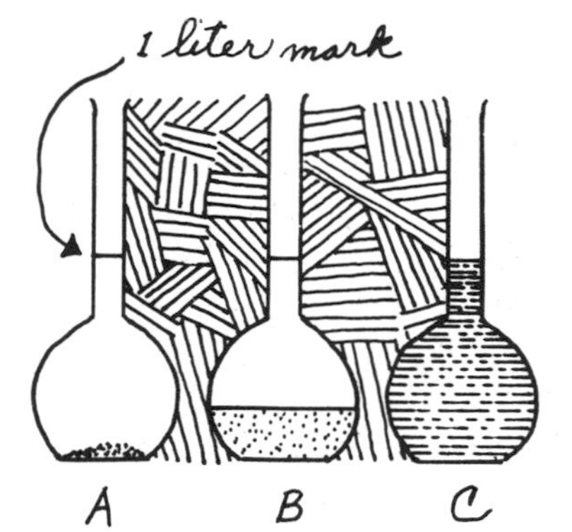

Figure 9-10. Preparing a solution of desired molarity. A. Place required weight of the solute in the flask. B. Dissolve the solute by adding a small amount of water. C. Add sufficient water to bring the volume of water up to the calibration mark.

Example:

Prepare 1 liter of a 2M NaOH solution.

Step 1. Molecular weight = 40 g

Step 2. Multiply: 40 × 2 = 80 g: dissolve 80 g of NaOH in water and dilute to 1 liter.

Prepare 1,000 ml of a 0.1 M H_2SO_4 solution.

Step 1. Molecular weight = 98

Step 2. Multiply: 98 × 0.1 = 9.8 g H_2SO_4
Dissolve in water and dilute to 1 liter.

Many times however, you will not want to make up 1 liter of a solution of specific molarity, but will want either less than or more than 1 liter. In these cases, you must add a Step 3 to the above procedure before dissolving in water to dilute. Steps 1 and 2 are identical to those above:

Step 1. Calculate the molecular weight.
Step 2. Write down the molarity desired.
Step 3. Change final volume to liters, then multiply Step 1 × Step 2 × Step 3 to find the number of grams needed which are to be dissolved in water to prepare the volume of the desired solution.

Example:
Prepare 1,500 ml of a 0.5 M H_2SO_4 solution.
Step 1. Molecular weight = 98 g
Step 2. Molarity = 0.5
Step 3. Volume wanted: 1,500 ml = 1.5 liter
Multiply: 98 × 0.5 × 1.5 = 73.5 g H_2SO_4
Dissolve in water, dilute to 1,500 ml (1.5 L).

Prepare 250 ml of 0.1 M NaOH solution.
Step 1. Molecular weight = 40 g
Step 2. Molarity = 0.1
Step 3. Volume wanted = 250 ml = 0.25 liter
Multiply: 40 × 0.1 × 0.25 = 1 g NaOH
Dissolve in water, dilute to 250 ml (0.25 L).

Osmolar

Solutions of nonelectrolytes (compounds which do not ionize) and *which have the same molarity,* yield the *same number of molecules or solute particles* in solution, as any other nonelectrolyte. On the other hand, solutions of compounds that ionize yield ions, not molecules, in solution, the number of ions depending upon the molecule.

Nonelectrolyte:
1 M sucrose solution = 1 M of sucrose molecules
$$C_{12}H_{22}O_{11} \longrightarrow C_{12}H_{22}O_{11}$$

Electrolytes: 1 M solution

Two ions: $NaCl \longrightarrow Na^{+} Cl^{-}$
1 mole ⟶ 1 mole + 1 mole = 2 moles ions

Three ions: $CaCl_2 \longrightarrow Ca^{2+} + Cl^{-} + Cl^{-}$
1 mole ⟶ 1 mole + 1 mole + 1 mole = 3 moles ions

Four ions: $AlCl_3 \longrightarrow Al^{3+} + Cl^{-} + Cl^{-} + Cl^{-}$
1 mole ⟶ 1 mole + 1 mole + 1 mole + 1 mole = 4 moles of ions

You can now see why equal volumes of identical molarity solutions can yield different molarities of solute particles in solution, depending upon whether the molecule ionizes and how many ions are formed from each molecule. Because the very important osmotic pressure depends upon the concentration of solute particles in the maintenance of the proper water balance of the body, we need a more specific term to describe concentrations of compounds. That term is called the "osmol," and with it, we have a common basis to describe the concentration of solute particles.

Since a nonelectrolyte does not ionize, *1 mol* of *any* nonelectrolyte will yield *1 osmol* (1 osm) of dissolved particles. Accordingly, 2 moles would yield 2 osm of dissolved particles, and 0.1 mole would yield 0.1 osm of dissolved particles. For the electrolyte, the osmolarity depends upon the number of ions formed by the ionized molecule:

1 mole NaCl yields 2 osmols of ions (1 Na^+ + 1 Cl^-) = 2 osm
1 mole $CaCl_2$ yields 3 osmols of ions (1 Ca^{2+} + 2 Cl^-) = 3 osm
1 mole $AlCl_3$ yields 4 osmols of ions (1 Al^{3+} + 3 Cl^-) = 4 osm

Fortunately most of the inorganic compounds of sodium, calcium, magnesium, and potassium, which play such important parts in the chemistry of the body are *totally dissociated* into their ions in water. This is the reason we can multiply the molarity of a solution by the number of ions which the molecule yielded, to obtain the osmolarity, as you have just seen.

We therefore define osmolar solutions, those having the same osmolarity, as solutions which have the same particle concentration, and the same osmotic activity. Steps for obtaining the osmolar concentration of a solution are as follows:

1. The *molarity* of the solution is specified or calculated as covered under the section Molar Solutions.
2. The ionization characteristics of the solute molecule are analyzed, and the number of ion particles obtained from 1 molecule are calculated.
3. The *molarity* value is multiplied by the number of ions to obtain the osmolar concentration.

Examples:

1. A solution that contains 1 mole of NaCl per liter (a 1 M solution) yields 2 ions, therefore it is a 2 osm/L solution. A solution of NaCl which is 2 M/L would be a 4 osm/L solution (2 × 2 = 4 osm).
2. A solution of glucose that contains 1 mol of glucose/liter yields only *1* solute particle (the molecule does not ionize), therefore it is a 1 osm/L solution. A solution of glucose which is 2 M/L would therefore be a 2 osm/L solution.
3. A solution that contains 1 mole of $MgCl_2$/L yields 3 ions, therefore it would be a 3 osm/L solution. A 3 M/L $MgCl_2$ solution would be a 9 osm/L solution.

Solutions of *different* solute particles that have the same *molarity* and ionize identically to yield the same number of ions, have the same *osmolarity* and therefore the same osmotic pressure.

0.1 M KCl has the same osmotic pressure as 0.1 M NaCl
1 M $CaCl_2$ has the same osmotic pressure as 1 M $MgCl_2$
1 M sucrose has the same osmotic pressure as 1 M glucose

The Milliosmol and the Millimol

In the chemical laboratory, the use of large beakers and flasks containing many liters of solutions is a common practice and the use of the terms molar and osmolar are practical. However, when working with the various body fluids, the health care professional is concerned with milliliters, because the liter is too large, and with millimol (mM) and milliosmol (mOsm), because molar and osmolar units are too large.

A milliliter is 1/1,000 L; a millimol is 1/1,000 mol
A milliosmol is 1/1,000 osmol

Therefore a solution which is 2 osm/L would also be 2 mOsm/ml; a 0.5 M/L solution would be 0.5 mM/ml. If you divide the numerator and the denominator by 1,000 you will see that the value of the fraction is unchanged:

$$\frac{2\text{ osm}}{1\text{ liter}} = \frac{\frac{2\text{ osm}}{1{,}000}}{\frac{1\text{ liter}}{1{,}000}} = \frac{2\text{ mOsm}}{\text{ml}} \quad \text{or 2 osm/ml}$$

Normality, Equivalents, and Milliequivalents

The equivalent is a measure of the chemical activity of a substance, the reactive ability of acids, bases and salts, and the concentration is expressed in *normality* (N). The equivalent weight of a substance is the weight in grams, the gram-equivalent weight, that will react with 1 g of hydrogen or 8 g of oxygen. This means that different elements are unequal in weight but equal in combining power (see Fig. 9-11). The equivalent weight of a substance can be found by dividing the gram-molecular weight by the positive valence (see Table 9-7).

Dissolving 58 g NaCl, 55 g $CaCl_2$ or 40 g NaOH in 1 liter of solution will yield a solution that contains *1 gram-equivalent weight in a liter of solution* and *that is a 1 normal solution.*

When 2 equivalent weights of a substance are dissolved in 1 liter of solution, the solution is said to be 2 normal. If 5 equivalent weights are dissolved in a liter, it is a 5 N solution; 0.1 equivalent weight, a 0.1 N solution (see Fig. 9-12). *An equivalent weight of a substance is equal in reactive or combining power*

Table 9-7. Equivalent weights of substances.

Substance	*Mol. Wt.**	*Positive Valence*	*Equivalent Wt.*
NaCl	58	1	58
$CaCl_2(OH)_2$	110	2	55
NaOH	40	1	40
$Mg(OH)_2$	58	2	29
$Al(OH)_3$	78	3	26
H_2SO_4	98	2	49
HCl	36	1	36

*Molecular weight in round numbers.

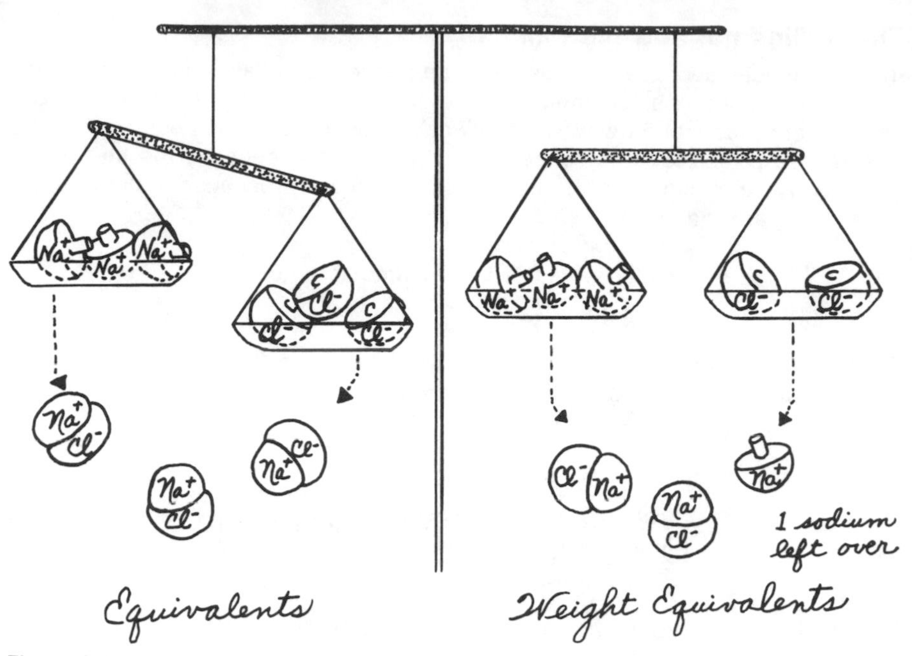

Figure 9-11. Equivalents of different elements are unequal in weight but equal in combining power. Weight equivalents of different elements are equal in weight but unequal in combining power.

Figure 9-12. Preparation of a solution of known normality. A. Put the number of grams of solute into the flask. B. Add some water to the flask and shake until all solute dissolves. C. Dilute with water to the calibration mark.

with an equivalent weight of another substance. Equal volumes of acid solutions of the same normality are equivalent to each other in reactive ability.

$$1\ N\ HCl = 1\ N\ H_2SO_4 = 1\ N\ H_3PO_4$$

Equal volumes of base solutions of the same normality are equivalent to each other in reactive ability:

$$2\ N\ NaOH = 2\ N\ Mg(OH)_2 = 2\ N\ Ca(OH)_2 = 2\ N\ Al(OH)_3$$

CONVERSION OF MOLARITY TO NORMALITY

Because the equivalent weight of a substance involves the consideration of the positive valence, the following procedure can be used to convert solutions of known molarity into solutions of known normality:

Step 1. Take the molarity.

Step 2. Multiply by the positive valence to get the normality.

PROBLEM

Change the following 2 M solution to normal solutions: NaOH, HCl, H_2SO_4, $Ca(OH)_2$, H_3PO_4, $Al(OH)_3$.

$$2\ M\ NaOH = 2 \times 1\ (Na^+ = +1) = 2\ N$$
$$2\ M\ HCl = 2 \times 1\ (H^+ = +1) = 2\ N$$

$$2\ M\ H_2SO_4 = 2 \times 2\ (2H^+ = +2) = 4\ N$$
$$2\ M\ Ca(OH)_2 = 2 \times 2\ (Ca^{+2} = +2) = 4\ N$$
$$2\ M\ H_3PO_4 = 2 \times 3\ (3H^+ = +3) = 6\ N$$
$$2\ M\ Al(OH)_3 = 2 \times 3\ (Al^{3+} = +3) = 6\ N$$

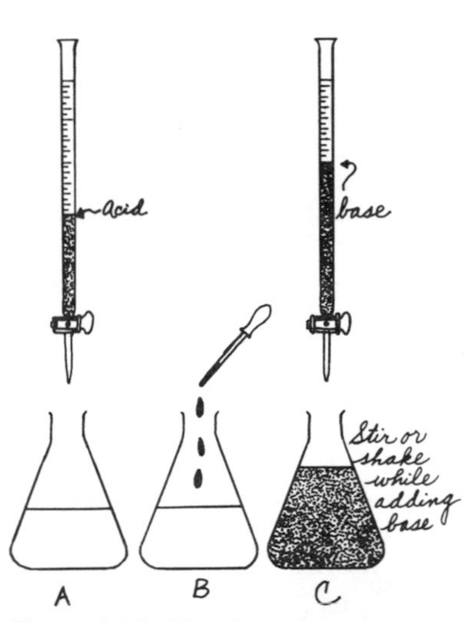

Figure 9-13. Titration of an acid with a base. A. Measure volume of acid. B. Add several drops of indicator. C. Add base until indicator changes color.

CONVERSION OF NORMALITY TO MOLARITY

Use the reverse procedure to change normality to molarity; divide the normality by the positive valence to get the molarity.

Example:

$3\ N\ HCl = 3\ M\ HCl$ (divide by 1)
$6\ N\ H_2SO_4 = 3\ M\ H_2SO_4$ (divide by 2)
$9\ N\ H_3PO_4 = 3\ M\ H_3PO_4$ (divide by 3)

CALCULATIONS INVOLVING NORMALITY

Since solutions of the same normality have the same reactive ability, this principle is used in chemical analytical procedures to find the normality of unknown solutions. In the determination of the normality of an unknown base solution, a definite volume of base is titrated. Titration is a process whereby the strength (normality) of a solution is determined by measuring the exact amount of another solution of known normality that is required to neutralize that volume (Fig. 9-13). The exact volume is observed by indicators or electronic instruments such as the pH meter.

In titrations of acid and bases:

$$\text{Volume(acid)} \times \text{Normality(acid)} = \text{Volume(base)} \times \text{Normality(base)}$$
$$V_a \times N_a = V_b \times N_b$$

NOTE: Always use identical units of volume.

Example:
How many ml of 2N HCl are required to neutralize 50 ml of 1 N HCl?

Substitute in formula: volume HCl $\times 2 = 50 \times 1$
$$2\ V_a = 50$$
$$V_a = 25 \text{ ml of 2N HCl}$$

How many ml of 1 N H_2SO_4 are required to neutralize 200 ml of 0.5 N NaOH?

$$\text{Volume } H_2SO_4 \times 1 = 200 \times 0.5$$
$$1\ V_a = 100$$
$$V_a = 100 \text{ ml of 0.5 N NaOH}$$

How many ml of 3 N H_3PO_4 are required to neutralize 25 ml of 2 N KOH?

$$\text{Volume } H_3PO_4 \times 3 = 25 \times 2$$
$$3\ V_a = 50$$
$$V_a = 16.67 \text{ ml of 2 N KOH}$$

Table 9-8. Relationship of equivalent weights to milliequivalent weights.

Substance	*Mol. Wt.** *grams*	*Equiv. Wt.* *grams*	*Milliequivalent Wt.* *milligrams*
NaOH	40	40	30
H_2SO_4	98	49	49
$CaCl_2$	110	55	55
$Al(OH)_3$	78	26	26
Na^+	23	23	23
Mg^{2+}	24	12	12
Cl^-	35	35	35
Ca^{2+}	40	20	20

*Weights expressed in whole numbers.

MILLIEQUIVALENTS

In the chemical laboratory and in large scale pharmaceutical preparations, quantities such as liters are commonly used. However, in the analysis of the body fluids, body secretions, and glandular and hormone procedures, we are working with extremely minute quantities, milligrams and micrograms. We could express their concentration in equivalents per liter (normality), but we would always be using extremely small values. Calculations would be cumbersome leading to the possibility of making mistakes. Therefore, the term milliequivalents, mEq, was developed.

A milliequivalent is equal to 1/1,000 of an equivalent.

If the volume is expressed in milliliters instead of liters, and milliequivalents are used instead of equivalents, then the identical ratio exists:

$$\frac{\text{Milliequivalents}}{\text{milliliters}}, \text{(mEq/ml)} = \frac{\text{Equivalents}}{\text{liter}}, \text{Eq/L}$$

This is to say that the *number value of the normality,* equivalents/liter is identical with the number value for milliequivalents/milliliter. Both expressions express the same ratio, *the milliequivalent weight in milligrams* is equal to the *equivalent weight in grams* (Table 9-8). The term mEq is a common measure for all plasma electrolytes, as can be seen in Table 9-9.

Table 9-9. Blood plasma, electrolyte composition.

Cations		*Anions*	
Na^+	142 mEq	Cl^{-1}	102 mEq
K^+	4 mEq	HCO_3^-	26 mEq
Mg^{2+}	2 mEq	protein	17 mEq
Ca^{2+}	5 mEq	HPO_4^{2-}	2
		balance	6
	153 mEq total		153 mEq total

RELATIONSHIP OF THE MILLIEQUIVALENT AND THE MILLIOSMOL

The milliequivalent is based upon the positive charge of the cation, and therefore it is a measure of the chemical activity of solute particles. The milliosmol is based upon the number of solute particles present which governs its osmotic activity in solution. Therefore, they do have a relationship which can be calculated if you know the positive charge of the cation.

For *univalent cations* (Na^+, K^+, NH_4^+):
1 mEq of the cation = 1 mOsm

For *nonelectrolytes* (glucose, sucrose):
1 mEq of the nonelectrolyte = 1 osm

For *bivalent cations* (Ca^{+2}, Mg^{2+}, Ba^{2+}):
Divide the mEq by the positive valence
1 mEq of the cation = ½ mOsm
2 mEq of the cation = 1 mOsm

For *trivalent cations* (Al^{3+}):
1 mEq of the cation = ⅓ mOsm
3 mEq of the cation = 1 mOsm

Therefore, intravenous NaCl solution that contains 308 mOsm/L contains 154 mOsm Na^+ and 154 osm/L Cl^-, or 154 mEq Na^+ and 154 mEq Cl^-.

SUMMARY

True solutions are composed of two or more substances, are homogeneous, will not settle out, and allow light to pass through unimpeded. They may be dilute, concentrated or saturated. Salts have a definite solubility at various temperatures. Gases dissolve less at higher temperatures, and an increase in pressure increases their solubility in water. Osmosis enables the body to maintain the proper water balance and dialysis enables the body to assimilate nutrients and dispose of waste products. The concentration of salts in the various cellular compartments plays an important part in the fluid balance, and injections of isotonic solutions do not disturb that balance. Colloids are a dispersion of particles that do not settle out, are electrically charged, and impede the passage of light. Concentrations of solutes can be expressed as percent weight/volume, molar, osmolar, millimolar, milliosmolar, and normal. Components of plasma are usually expressed in milliequivalent/liter (mEq/L) or milliosmols per liter (mOsm/L).

EXERCISE

1. Define a solute, solvent and a solution.
2. What is the difference between a true solution and a heterogeneous mixture? Can a solution be separated into its components?
3. How is the solubility of a salt affected by temperature of the solution? Do all salts have the same solubility in water?
4. When is a solution said to be saturated?
5. Can you have a saturated solution that is dilute? A saturated solution that is concentrated?
6. What is the solubility of $NaNO_3$ at 30°C?
7. How can you increase the rate of solution of a solute?
8. Why is it dangerous to freeze a carbonated beverage bottle?
9. Describe two procedures to treat a person suffering from the bends. What do these procedures accomplish?
10. Describe how osmotic flow causes water intoxication with an excess water intake? Dehydration with an excess salt water intake?
11. What are isotonic, hypotonic and hypertonic solutions? What is their effect on red blood cells in the blood stream?
12. Compare the size of particles in true solutions, colloidal dispersions, and suspensions.
13. What is the difference between osmosis and dialysis?
14. Why does the dialysis bath in the kidney machine have to be analyzed while in use?
15. How would you prepare the following solutions:
    (a) 500 ml of a 5 percent dextrose solution
    (b) 1,000 ml of a 0.85 percent NaCl solution
    (c) 1 liter of a 2 M NaOH solution
    (d) 0.1 liter of a 0.1 M H_2SO_4 solution
    (e) 1 liter of a 1 Osm NaCl solution
    (f) 1 liter of a 1 Osm glucose ($C_6H_{12}O_6$) solution
    (g) 0.5 liter of a 0.1 N NaOH solution
    (h) 100 ml of a 0.1 N H_2SO_4 solution
    (i) 50 ml of a 3 N H_3PO_4 solution
16. Which of the following are true?
    (a) 1 M NaOH = 1 N NaOH
    (b) 1 M H_2SO_4 = 2 N H_2SO_4
    (c) 1 M NaCl = 1 Osm NaCl
    (d) 1 M H_3PO_4 = 3 N H_3PO_4
    (e) 1 M $CaCl_2$ = 1 Osm $CaCl_2$
    (f) 1 M/L NaOH = 1 mM/ml NaOH
    (g) 1 N/L NaCl = 1 mEq/ml NaCl

OBJECTIVES

When you have completed this chapter you will be able to:

1. Distinguish between the terms force and pressure and calculate values of force, pressure and area.
2. Describe the spatial relationship between the particles of solids, liquids, and gases, and explain how pressure affects each.
3. Explain the Brownian movement in terms of the kinetic theory of matter.
4. Enumerate the physical factors which are used to describe a gas.
5. State the formulas for Boyle's law, Charles' law, Gay-Lussac's law, and the general gas law.
6. Make calculations using the gas laws, and explain these laws using practical illustrations.
7. Calculate the partial pressures of component gases of a mixture using Dalton's law.
8. Discuss how the gas laws apply to:
   (a) Inspiration (breathing in)
   (b) Expiration (breathing out)
   (c) External respiration
   (d) Internal respiration
9. Draw the components of a portable oxygen system and discuss the function of:
   (a) Tank
   (b) Regulator
   (c) Flowmeter
   (d) Humidifier
   (e) Mask

GASES, PRESSURE, AND INHALATION THERAPY

10. Define humidity therapy and state how the moisture content of dry gases can be increased by two methods.
11. Relate the two characteristics that determine liquid pressure with flow rates of liquids.
12. State Pascal's law and illustrate with two examples how that law can affect patients.
13. Describe the siphon and give two examples where that principle is used in the treatment of patients.
14. Relate Bernouilli's law to pressure exerted by flowing liquids and show its use in the water aspirator, the Bunsen burner, and the bulb atomizer.

Both liquids and gases, collectively called fluids, have the following properties in common: they have mass (weight) and they exhibit the phenomenon called pressure. A good understanding of pressure is an absolute necessity for understanding the functions of the human body. Body cavities and organs are affected by pressure both in health and disease. Normal breathing depends partially on the difference between the interpleural (between the pleural cavities) and the intrapleural (within the pleural cavities) pressure. Treatments such as enemas and irrigations are affected by pressure. Changes in the blood pressure and the cerebrospinal fluid pressure may be signs of great significance to health care personnel. Inhalation therapy in all of its complex applications depends upon pressure and pressure differences. Numerous other applications and examples will be detailed for you in this chapter.

The terms pressure and force mean entirely different things, and it is important that you understand what they mean and how to distinguish between them. Force is the *total* amount of energy or strength that is exerted against a surface or an object. For example, a force of 10 pounds or a force of 500 grams acts upon a surface. *The force term only contains the number value and one of*

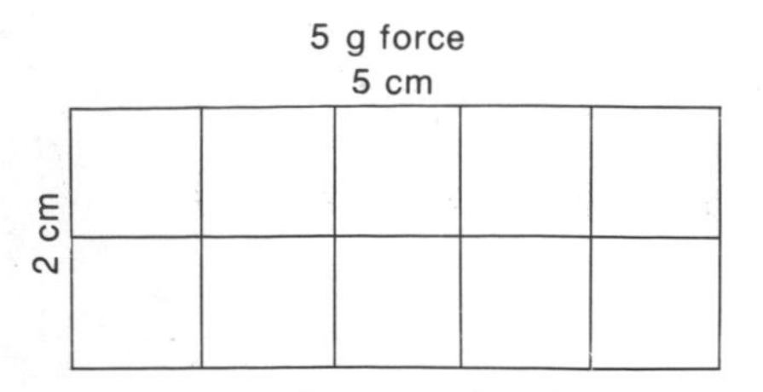

Figure 10-1. Relationship between pressure and force. Each and every square centimeter of area has a pressure of 5 g/cm² being applied to it. Since the total area is 10 cm², the force equals 50 g over the entire area.

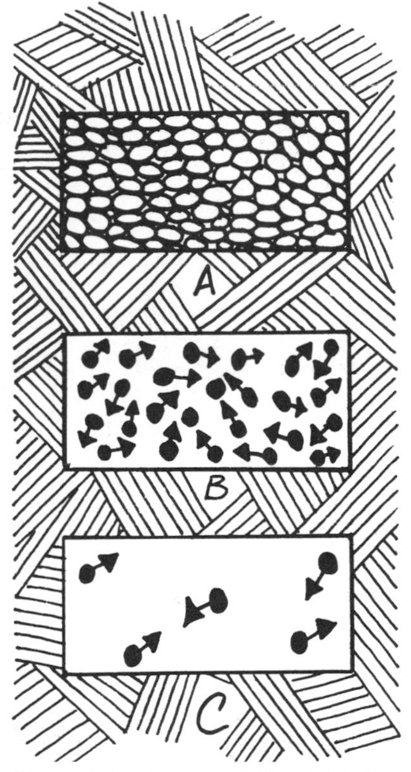

Figure 10-2. Space relationships between solids, liquids and gases. A. Solids: particles are held in fixed position by the attractive forces between them. B. Liquids: particles are close to each other as in solids, but are able to slide over each other. C. Gases: a great deal of space is between the particles, therefore gases can be compressed.

those units. Pressure is the force that is exerted against one unit area, such as a square centimeter (cm²), square inch (in²), square meter (m²), or square foot (ft²). For example, the atmospheric pressure is 14.7 lb/in², the gas pressure is 50 g/cm². *Therefore, the pressure term has the unit of force, such as the pound or the gram, and also the unit of area, the cm² or the ft².* The force and the pressure are related by the formula (Fig. 10-1):

$$\text{Pressure} = \frac{\text{Force}}{\text{Area}}$$

Example: A force of 50 g is applied to an area of 10 cm². What is the pressure?

Substituting in formula: $\text{Pressure} = \frac{50\text{ g}}{10\text{ cm}^2}$

$\text{Pressure} = 5\text{ g/cm}^2$

A pressure of 5 g/cm² is applied to a diaphragm. What is the force applied to it, if the area = 20 cm²?

Substituting in formula: $5\text{ g/cm}^2 = \frac{\text{Force}}{20\text{ cm}^2}$

or: $\text{Force} = 5\text{ g/}\cancel{\text{cm}^2} \times 20\ \cancel{\text{cm}^2}$

which equals: $\text{Force} = 100\text{ g}$

What would be the surface area if a force of 200 g were to give a pressure of 2 g/cm² on it?

Substituting in formula: $2\text{ g/cm}^2 = \frac{200\text{ g}}{\text{area}}$

or: $2\text{ g/cm}^2 \times \text{area} = 200\text{ g}$

then: $\text{area} = \frac{200\ \cancel{\text{g}}}{2\ \cancel{\text{g}}\text{/cm}^2}$

$\text{area} = 100\text{ cm}^2$

Health care personnel will find that a general knowledge of gases is valuable, especially their behavior during temperature, volume and pressure changes. Oxygen therapy, artificial respiration, and the treatment of surgical pneumothorax (a collection of air or gas in the pleural cavity) depend upon the gas laws, as do many normal physiological processes such as internal and external respiration.

THE STRUCTURE OF GASES

The molecules of a gas are widely separated from each other; there is a great deal of space between them and the molecules have very little attraction for each other. Because of this space between the molecules, gases can be compressed (contained in smaller volumes). Air is a gas; it can be compressed, as you know when the service station attendant fills your tire with air. Even though the gauge may read 28 lb, what it really means is that the air is at a pressure of 28 lb/in^2. Just as gases can be compressed, they can expand indefinitely, as you learned in an earlier chapter. Solids and liquids, on the other hand, have little or no space between the molecules, because the molecules practically touch each other (see Fig. 10-2). This fact is very important, because neither solids nor liquids can be compressed.

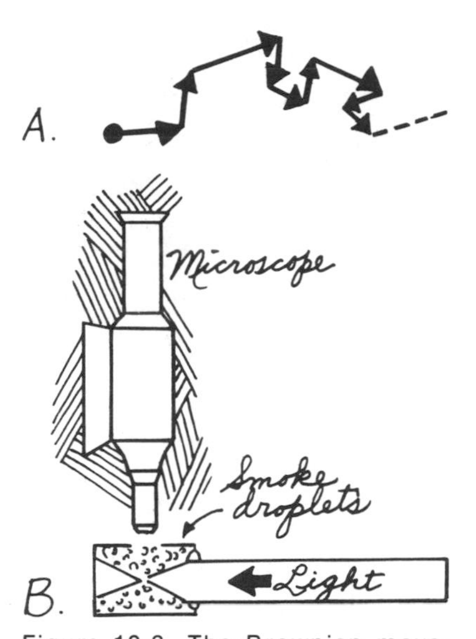

Figure 10-3. The Brownian movement. A. Motion of a molecule caused by collisions with other molecules. B. Cigarette smoke particles can be seen in motion under a microscope.

THE KINETIC THEORY OF MATTER

You already know that all matter is composed of atoms or molecules, and the kinetic theory relates to the movement of those particles. Many of your everyday observations suggest particle motion: the odor of a drop of perfume can be detected at a distance after a short time, a drop of ink in a container of water will eventually color the entire water (even without stirring), and a teaspoon of sugar dropped into a glass of water will eventually dissolve and the sweetness can be detected throughout the water.

In solid substances, the particles are held in fixed positions by the attractive forces between them, but they also have a slight random motion. When heat energy is supplied and their motion increases they gain sufficient energy to break away from each other. The particles then overcome the forces of attraction between them, and the solid is said to melt. In the liquid state, the molecules are free to move about, sliding over one another, but they still remain in contact. As more heat energy is added, they move with greater speeds, their vapor pressure continues to increase, and eventually the vapor pressure equals atmospheric pressure. At that point the liquid is said to boil.

Brownian Movement

Although we cannot observe molecules with the most powerful electron microscope, we can observe their behavior by considering large masses of molecules. The Scottish botanist, Robert Brown, observed that tiny specks of pollen in water were constantly in motion, the smaller the specks, the more violent the motion. This irregular motion is caused by the chance bombardment of these tiny particles by the water molecules, which move at high velocities. This random motion is called the "Brownian movement" (Fig. 10-3). Cigarette smoke particles under a microscope, and dust particles in a dark room illuminated by a flashlight show this motion.

Diffusion

Diffusion of gases in gases, solids in liquids, and liquids in liquids can be explained by the random motion of molecules from places of higher concentration to places of lower concentration (Fig. 10-4). This is called diffusion, and it takes place within the human body in many processes. Oxygen and carbon dioxide diffuse continuously in the process of internal respiration in the alveoli (air cells of the lungs) and in the passage of waste products from the tissues. A teaspoon of sugar will eventually dissolve and diffuse throughout a container of water, producing a completely homogeneous solution without stirring (Fig. 10-5).

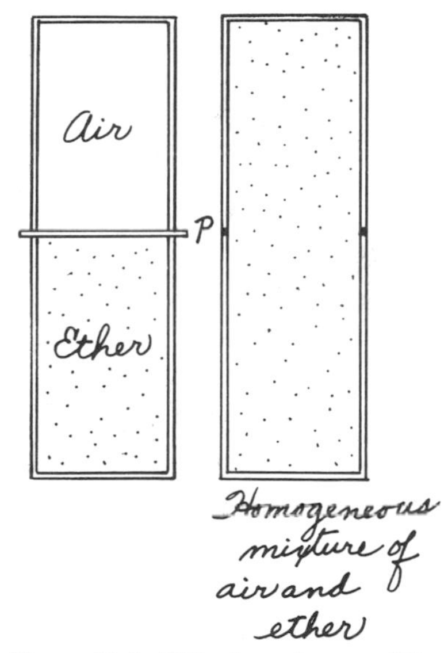

Figure 10-4. Diffusion of gases. When the dividing plate "P" is removed the two gases quickly diffuse to form a uniform mixture, going from their areas of high concentration to lower (or zero) concentration.

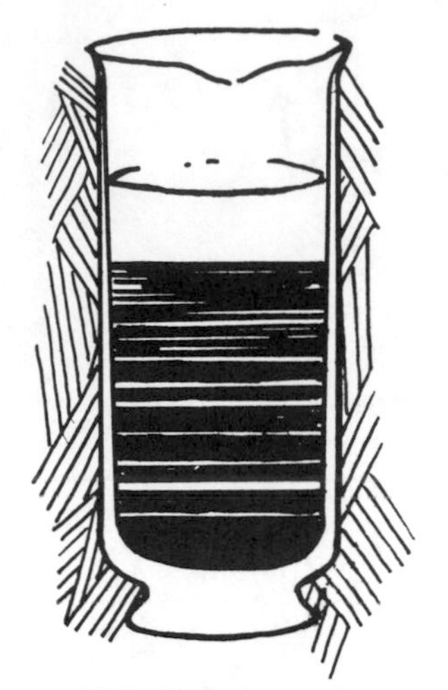

Figure 10-5. Diffusion of sugar in water. The sugar (represented by the darkened area) eventually diffuses throughout the water to produce a homogenous solution.

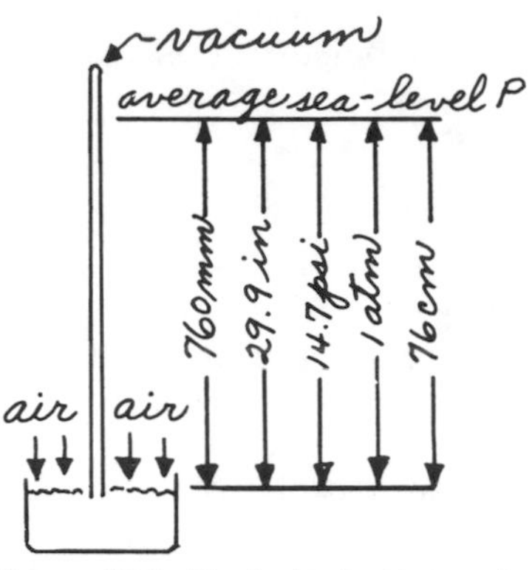

Figure 10-6. Equivalent values of atmospheric pressure. These terms express the same values: 1 atm = 760 mm Hg = 76 cm Hg = 29.9 in Hg = 14 lb/in².

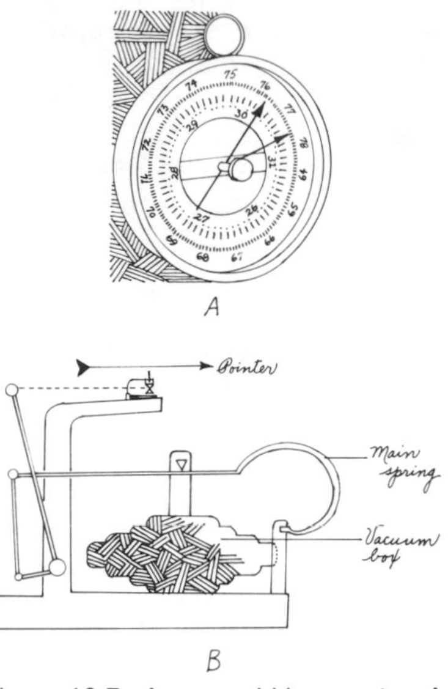

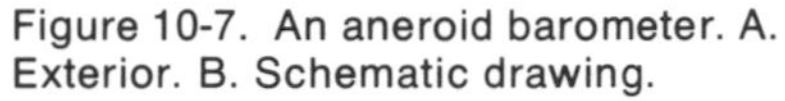

Figure 10-7. An aneroid barometer. A. Exterior. B. Schematic drawing.

Atmospheric Pressure

Atmospheric pressure is measured by a mercury barometer, an inverted (sealed at one end) glass tube filled with mercury. The height of the mercury column is a measure of the atmospheric pressure, which can be stated in a number of terms, all being equivalent in value to each other (Fig. 10-6).

The mercury barometer is large and cumbersome and is subject to breakage. A more convenient type is called the aneroid barometer (Fig. 10-7). It is durable, compact, and more easily read. This barometer translates the atmospheric pressure on a thin diaphragm to a dial by means of levers and gears. Most aneroid barometers have an indicator pointer in addition, which can be moved at will. This permits us to follow atmospheric pressure changes and is useful for predicting the weather.

Vapor Pressure of Compounds

We already know that water has a certain vapor pressure and that the pressure increases with the temperature. The vapor pressure of other liquids may be lesser or greater, depending upon the attraction of the molecules for each other (Fig. 10-8). When heat energy is absorbed by molecules they move faster and they have a higher vapor pressure. This enables them to escape more quickly from the liquid.

CHARACTERISTICS THAT GOVERN THE BEHAVIOR OF GASES

There are three factors that are used to describe gases: pressure (P), volume (V), and temperature (T). When all three factors are specified, the particular gas is completely described, because identical quantities of a gas confined under

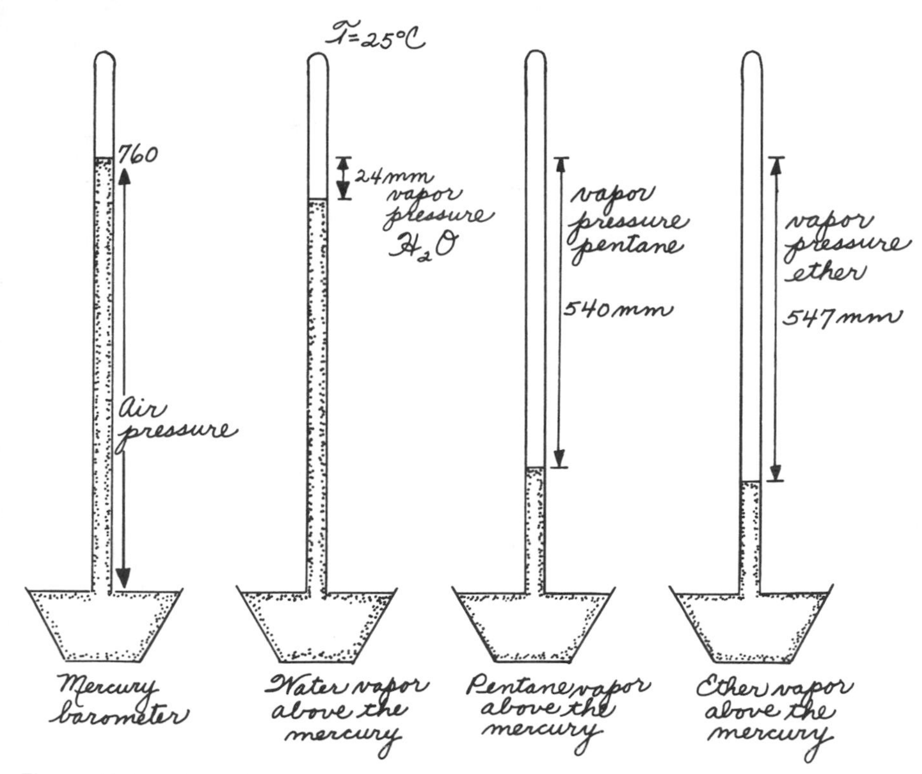

Figure 10-8. Effect of the vapor pressure of compounds on a column of mercury (a barometer).

the same three conditions will always behave in the same way. Gas volumes can be increased or decreased, gas pressures can be increased or decreased, and gas temperatures can be increased or decreased. When any two of these variables are changed the third one will automatically have a fixed value. When using the gas laws to make calculations, the initial and final pressures can be expressed in any unit, but they must be in the same unit. The same restriction holds for the initial and final volumes: they can be in any unit, but they must be expressed in the same unit. The temperature, however, must be expressed in °Kelvin, °K (or °Absolute, which is the same scale). Temperature in °K = °C + 273, therefore convert °F (if given) to °C, and then add 273 to obtain °K.

Boyle's Law

Boyle's law states: when the temperature of a gas is kept constant, the volume of a gas varies inversely with the pressure. This means that when the pressure is increased, the volume decreases: when the pressure is decreased, that the volume is increased. When one term goes up in value, the other term goes down. The formula is: *(temperature is constant, unchanged)*:

$$\text{Pressure}_{\text{original}} \times \text{Volume}_{\text{original}} = \text{Pressure}_{\text{final}} \times \text{Volume}_{\text{final}}$$

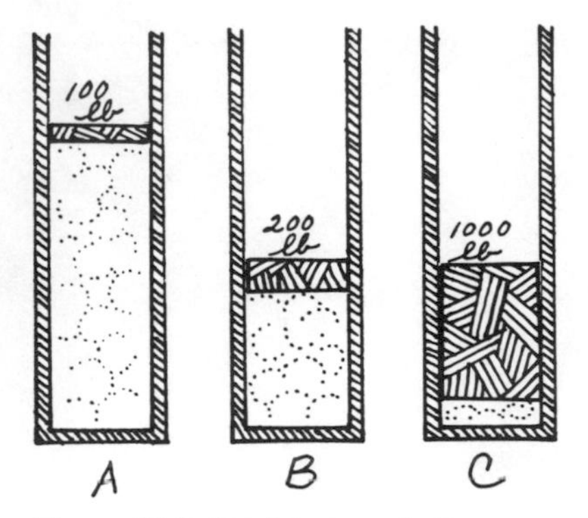

Figure 10-9. Boyle's law. A. Pressure applied = 100 lb/in². B. Pressure is doubled and volume is halved. C. Pressure is increased to ten times the original pressure and volume is 1/10 original volume.

NOTE: The pressures can be expressed in any units, but both the initial and final pressures must be expressed in the *same units.* The volumes can be expressed in any units, but both the initial and final volumes must be expressed in the *same units.*

Therefore, if a gas is under 100 lb/in² pressure and the pressure is increased to 200 lb/in² (doubled), the volume will be halved. If that same gas at 100 lb/in² were subjected to 1,000 lb/in² (ten times the pressure) the volume would be reduced to 1/10 the original volume (Fig. 10-9).

Example: A cylinder contains 750 ml of a gas at 1 atm pressure. The pressure is decreased to 0.25 atm. What is the new volume, if there is no change in temperature?

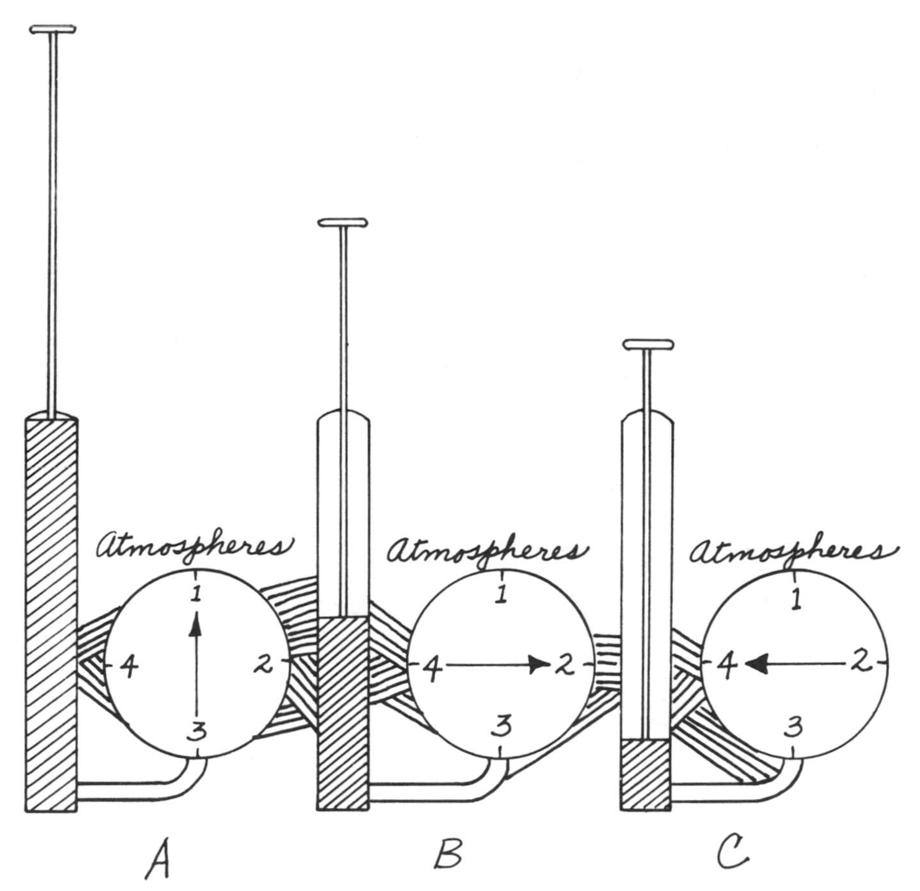

Figure 10-10. Explanation of Boyle's law with a bicycle pump. A. Piston is completely withdrawn, the contained air is at 1 atm pressure. B. The pressure is doubled to 2 atm and the volume is halved. C. The pressure is quadrupled to 4 atm and the volume is reduced to ¼ that of the original.

Substitute in formula:

$$(1 \text{ atm})(750 \text{ ml}) = (0.25 \text{ atm})(V_f \text{ ml})$$
$$0.25\, V_f = 750 \text{ ml}$$
$$V_f = 3{,}000 \text{ ml (final volume)}$$

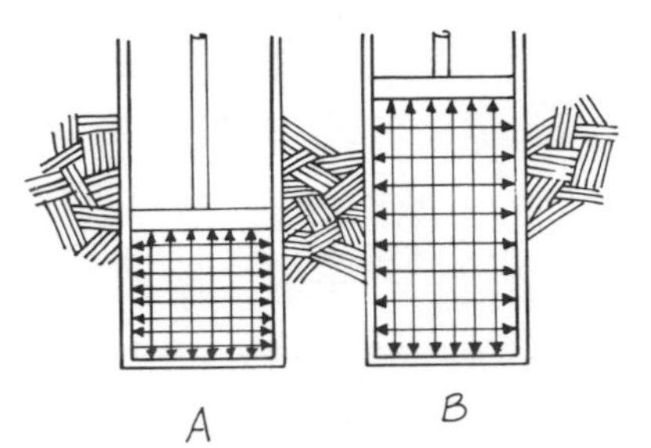

Figure 10-11. Kinetic theory explanation of Boyle's law. Expansion of the gas from original volume A to new volume B means that the molecules must travel further between the successive impacts on the walls of the container. The pressure is reduced because their impacts are spread over a larger surface.

An ordinary bicycle pump that has the outlet of the hose sealed can be used to demonstrate Boyle's law. The cylinder can contain a certain volume of air when the piston is withdrawn as far as possible. As you press the piston down, the volume of air decreases. When the piston is pushed down as far as possible, the pressure being applied is the greatest, and the volume of air is compressed to its smallest volume (see Fig. 10-10).

Boyle's law explains why the pressure rises in a given gas system as the volume is decreased. When the volume is *decreased,* the same number of molecules of gas strike a smaller area, each molecule is involved in more collisions per second, and therefore, the pressure rises (Fig. 10-11). This law partly explains the mechanism used in breathing, artificial respiration, the vasculator, and the compressed gas gauges.

When the muscles of respiration are affected (as in poliomyelitis or other neurological conditions) so that a patient's breathing is impaired or stops, the respirator is used to maintain life. Respirators may be large and fixed, such as the tank respirator (the iron lung, Fig. 10-12), or portable, only covering the anterior part of the patient's chest. A movable piston is withdrawn, thus increasing the volume of air surrounding the chest of the patient, reducing the external pressure on the patient's chest. Air rushes into the patient's lungs causing an inhalation of air. The second part of the cycle is compression of the bellows, which decreases the volume of air on the patient's chest, increasing the pressure and causing exhalation.

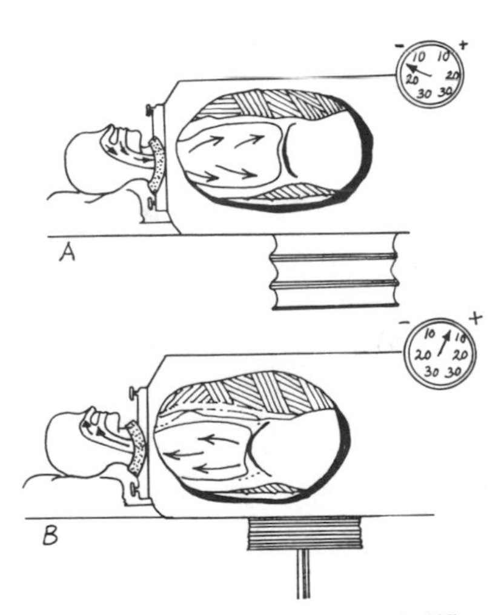

Figure 10-12. The iron lung. A. When the bellows are expanded the pressure on the chest is reduced. Atmospheric pressure forces air into lungs. B. When the bellows are compressed the pressure on the chest is increased above atmospheric pressure and air is forced out of the lungs.

Charles' Law

Charles' law relates how the volume of a gas can change with the *absolute temperature,* if the pressure of the gas is not changed (Fig. 10-13). When the pressure on a gas remains the same (constant), if the temperature rises, the volume increases; and if the temperature falls, the volume decreases. The formula for calculating volumes and temperatures according to Charles' law is:

$$\frac{\text{Volume (initial)}}{\text{Volume (final)}} = \frac{\text{Temperature (initial)}}{\text{Temperature (final)}}$$

When containers, syringes, or any sealed object is subject to sterilization by high heat, the effects of Charles' law must be taken into consideration, because any contained gas will expand under the influence of heat. Closures on tightly sealed containers must be loosened and you should be aware that plastic objects may deform due to the expanded gas.

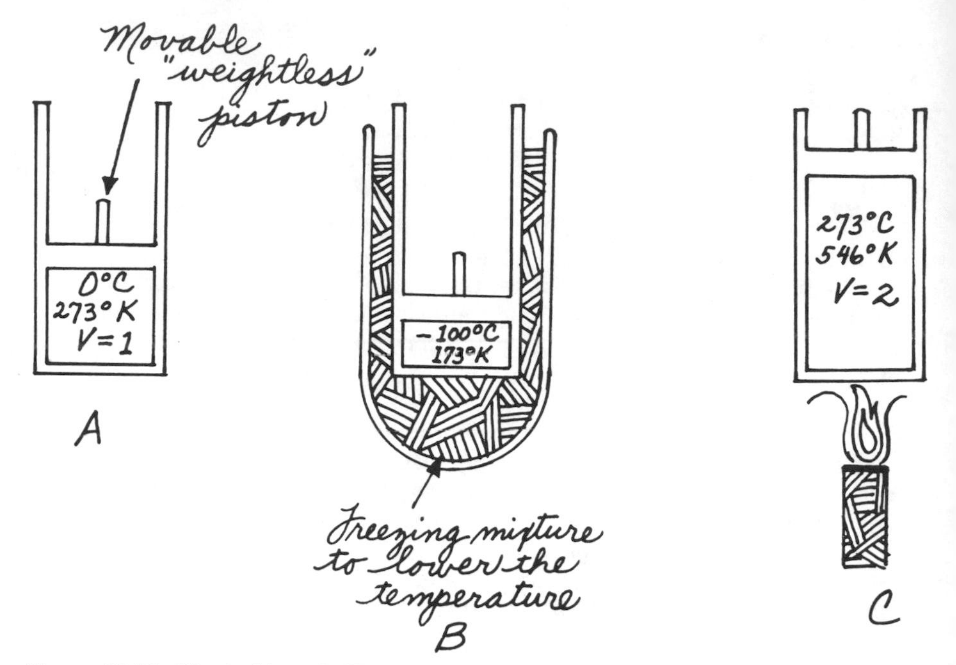

Figure 10-13. Charles' law. A. The original volume (V = 1). B. Cooling the gas reduces its volume to less than 1. C. Heating the gas to 273°C doubles the volume (V = 2).

Gay-Lussac's Law

Gay-Lussac's law relates how the pressure on a gas system changes with changes in absolute temperature if the *volume* of the gas is not changed. This law applies to compressed gas cylinders, aerosol spray cans, or any container of gases that cannot expand or contract. Therefore, when the *volume of a gas does not change,* the pressure will go up when the temperature goes up, and the pressure will go down when the temperature goes down. The formula is:

$$\frac{\text{Pressure}_{\text{initial}}}{\text{Pressure}_{\text{final}}} = \frac{\text{Temperature}_{\text{initial}}}{\text{Temperature}_{\text{final}}}$$

The automobile tire "blows out" partly due to Gay Lussac's law, because of the increase in pressure due to the temperature rise of the air in the tire. You probably have taken a wire coat hanger and, in order to break it apart, bent the wire back and forth many times until the wire weakens and breaks. The wire became very hot as a result of the bending, and the same thing happens with the automobile tire. As the car moves and the tire rolls, it continually flexes and bends. The temperature rises, the pressure increases, and the heat generated by the flexing weakens the tire body. Eventually, the pressure exceeds the capability of the tire, and the tire blows out.

The effect of increased temperatures on gases is the reason why compressed gas cylinders or containers, such as oxygen, carbon dioxide, nitrogen, or hydrogen, should never be stored in direct sunlight, near hot radiators, or in hot storage areas. The caution label on aerosol spray cans (hair spray, deodorants, disinfectants, or insecticides) often read:

> CAUTION: Warning. The contents of this can are under pressure. Do not use near open flame. Do not set on stove or radiator or keep where the temperature will exceed 120°F, as the container may explode. Do not throw into fire or incinerator.

Even though no spray comes out of a can when you depress the button because the contents have been used up, the can is not empty. It still contains the propellant gas, at atmospheric pressure, plus possibly volatile chemicals and flammable solids.

The General Gas Law

All three laws (Boyle's, Charles' and Gay-Lussac's) can be combined to give a general gas law, one formula which will solve any gas problem, regardless of how many of the variables change or remain constant. Gases can be heated, cooled, compressed, expanded, and subjected to increased or decreased pressures in any combinations. All that has to be done to solve the problem is to substitute the proper values in their correct places in the formula and then solve for the unknown value. The formula is:

$$\frac{\text{pressure}_{\text{initial}} \times \text{volume}_{\text{initial}}}{\text{temperature}_{\text{initial}}} = \frac{\text{pressure}_{\text{final}} \times \text{volume}_{\text{final}}}{\text{temperature}_{\text{final}}}$$

NOTE: Pressures must be in the same units and volumes must be in the same units. Temperatures must be in °K (°A).

INHALATION THERAPY AND THE GAS LAWS

The inhalation therapist must be able to apply the three gas laws, because the mechanism of breathing, the treatment of patients with compressed gases, the functioning of the Drinker respirator, artificial respiration, and resuscitation depend upon those laws. About 50 percent of the confinement period of patients who are sick in bed for over 7 days is due to respiratory conditions. The two basic purposes of respiration in the body are to supply oxygen to the cells and to eliminate the carbon dioxide. The respiratory processes in the body can be divided into three categories: ventilation, external respiraton and internal respiration.

Ventilation

Ventilation is another name for breathing and consists of inspiration (breathing in) and expiration (breathing out). Breathing is an example of Boyle's law. The size of the thoracic cavity (the pleural cavity, the chest, containing the heart, lungs, and bronchi) is changed by using the diaphragm (the musculomembra-

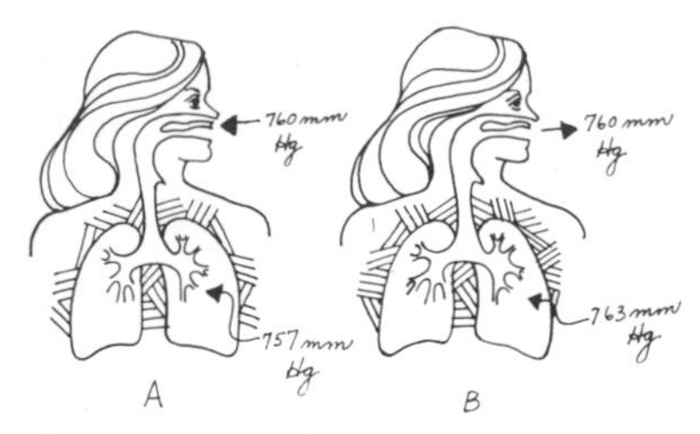

Figure 10-14. Pressure differentials and the breathing cycle. A. Inspiration. B. Exhalation.

nous wall separating the abdomen from the thoracic cavity). As the diaphragm contracts, the size of the cage (the framework enclosing the chest cavity) expands, increasing the volume of the lung. This causes inspiration. The lung has a negative pressure of 3 mm Hg, which is 3 mm Hg lower than atmospheric pressure (Fig. 10-14). The higher atmospheric pressure forces air into the lungs, which is at lower pressure. When the diaphragm relaxes during expiration, the size of the cage decreases, compressing the lungs and creating a positive 3 mm Hg pressure in the lungs. This higher pressure (higher than atmospheric pressure) is the positive pressure that forces the air out of the lungs. Normal adults inhale and exhale about 500 cc of air during each breath, about 150 cc remaining in the bronchial tubes and about 350 cc actually reaching the air (alveolar) sacs of the lungs.

LAW OF PARTIAL PRESSURES

In the respiratory process we breathe a mixture of many gases, each component gas of the mixture behaving as if it were there alone. Its pressure depends upon how many molecules of the gas are present without regard for any other gas that may be there. This is the law of partial pressures, as formulated by John Dalton. It states that since each gas exerts its own pressure independently of the other gases, the total pressure of the gas mixture is equal to the sum of the partial pressure of each gas.

The major components of the atmosphere and their concentrations are 20.96 percent oxygen, 79 percent nitrogen and 0.04 percent carbon dioxide. Other gases have been included with the nitrogen, therefore the atmospheric pressure equals the partial pressure of oxygen plus the partial pressure of nitrogen plus the partial pressure of carbon dioxide. The partial pressure of each gas is equal to the concentration times the total pressure of the gas mixture, and is designated by the symbol P_{O_2}, P_{CO_2}, and P_{N_2}. If the atmospheric pressure equals 760 mm Hg, then the partial pressures of oxygen, nitrogen and carbon dioxide can be calculated:

$$
\begin{array}{lcll}
P_{O_2} & = & .2096 \times 760 = & 159.296 \text{ mm Hg} \\
P_{CO_2} & = & 0.0004 \times 760 = & 0.304 \text{ mm Hg} \\
P_{N_2} & = & .79 \times 760 = & \underline{600.40 \text{ mm Hg}} \\
 & & & 760.000 \text{ mm Hg}
\end{array}
$$

External Respiration

The interchange of gases takes place in the lungs, where the blood loses carbon dioxide and water vapor and gains oxygen. The P_{CO_2} of the inspired air is about 0.3 mm Hg, while the alveolar air is rich in carbon dioxide, the P_{CO_2} equalling 40 mm Hg. On the other hand, the inspired air is rich in oxygen, P_{O_2} equalling 160 mm (approximately) and the alveolar air is poor in oxygen, P_{O_2} equalling 100 mm Hg. It is the difference in the partial pressures of the gases that enables the lungs to interchange the gases, because the gases move from points of higher pressure to points of lower pressure. The movement of the oxygen is in the opposite direction of the carbon dioxide, and the nitrogen gas, being inert, acts as a carrier and a diluent for both oxygen and carbon dioxide (Fig. 10-15).

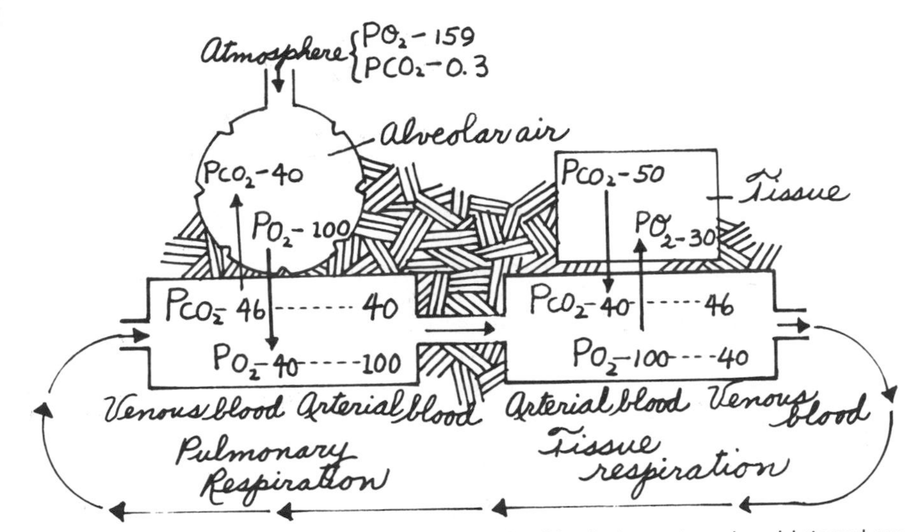

Figure 10-15. Diffusion of oxygen and carbon dioxide during external and internal respiration. The direction of the gas flow is determined by the difference in partial pressure.

Internal Respiration

Oxygen from the blood, oxyhemoglobin, diffuses through the tissue fluid into the cell because of the difference in the partial pressures of the oxygen in the two areas. The carbon dioxide diffuses from the cell into the blood forming carbaminohemoglobin, which returns to the lungs. The pressure gradients favor the chemical changes that take place in the blood between the carbon dioxide and the oxygen, thus enabling the oxygen to flow from the blood to the tissue and the carbon dioxide to flow from the tissue to the blood. Since a continuous supply of oxygen is absolutely essential to life, any factor that reduces that supply endangers cell life.

Oxygen Therapy

The purpose of oxygen therapy is to increase the oxygen tension, P_{O_2}, of the blood plasma to normal conditions that meet the tissue cell needs. When there is a decrease in the oxyhemoglobin, a condition called cyanosis develops. Patients who are anemic have a low hemoglobin level, and therefore may lack normal levels of oxygen.

Carbon monoxide gas is toxic. It poisons the blood because hemoglobin will preferentially combine with carbon monoxide rather than with oxygen, thus reducing the amount of hemoglobin available to combine with the oxygen. Carbon monoxide poisoning is indicated by a cherry red color of the skin, whereas cyanosis, oxygen need, is characterized by a bluish skin color.

OXYGEN ADMINISTRATION

When moderate to high concentrations of oxygen are desired, face masks, tents or hood-like devices may be used because of the factors of comfort, isolation, and air conditioning (Fig. 10-16). These devices do pose some danger because

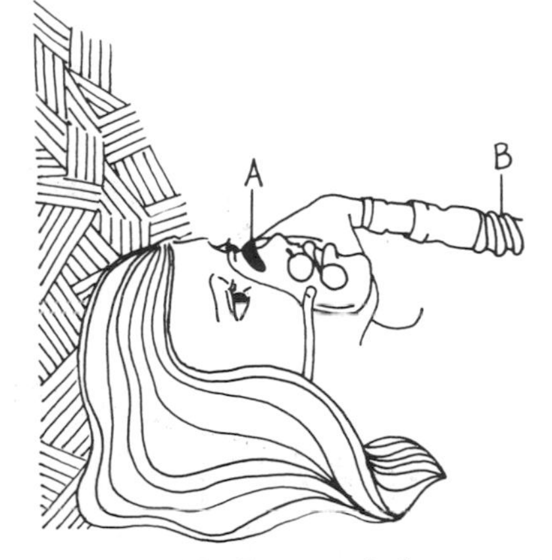

Figure 10-16. Face mask for oxygen therapy. A. Malleable clamp to seal mask to nose. B. Gas inlet tube.

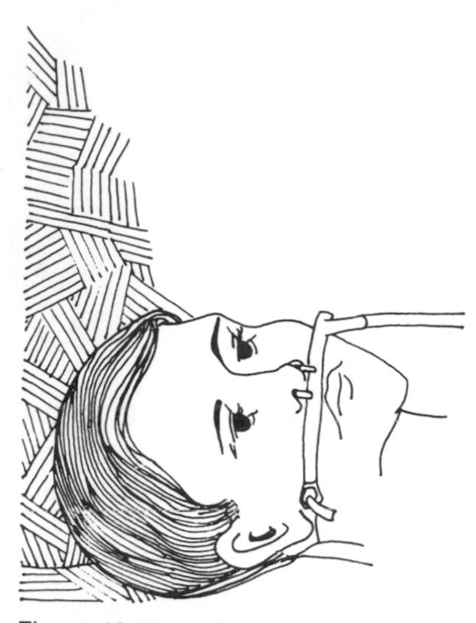

Figure 10-17. Nasal cannula used for oxygen therapy.

they require a high oxygen flow and they can serve as a source of bacterial growth and contamination.

When moderate to low concentrations of oxygen are to be delivered, the nasal cannula (a tube whose two tips are inserted in the nostrils) is a safe and economic method (Fig. 10-17). The danger of carbon dioxide retention, even at low flow rates, is minimal.

Oxygen is available from portable tanks (Fig. 10-18) or from outlets near the patient, but it is dry and can irritate the membranes of the trachea, bronchi and alveoli. Therefore it must be humidified (relating to the moisture content) and the high pressure of the compressed oxygen must be reduced to atmospheric pressure prior to administration. Finally, the flow of the oxygen in liters per minute must be precisely controlled according to the treatment procedure prescribed. By using the method that meets the needs of the patient, oxygen therapy will enable the hemoglobin to become saturated with oxygen. If successful, carbon monoxide poisoning can be eliminated and symptoms of hypoxia (a lack of oxygen) will disappear.

OXYGEN INTOXICATION

Too much oxygen in the blood means that the extra oxygen is actually dissolved in the watery fluid of the blood because the hemoglobin is saturated. This dissolved oxygen at extremely high concentrations causes changes in the metabolic rates of the tissue cells, especially affecting the nervous system. This can result in convulsions, and also can disrupt the hemoglobin-oxygen buffer system. Premature infants are highly susceptible to blindness if they are exposed to higher oxygen concentrations than they are conditioned for, the oxygen destroying the capillary network supplying the cells of the retina.

CAUTIONS WHEN WORKING WITH OXYGEN

As the concentration of oxygen increases, the danger of fire increases. Substances will ignite and burn at lower temperatures when the oxygen concentration is increased. Always be safety conscious, avoiding all possible sources of ignition, never smoking around oxygen, and never using oil or grease on oxygen valves.

Humidity Therapy

Pulmonary water balance is important to the patient because of the irritations that are caused by dry gases. The inhalation therapist can correct any deficiency in the moisture content of a gas by taking into consideration how much water vapor is needed and how much water vapor is delivered to the airways of the patient. Two terms are used to describe the moisture content of gases: *Relative humidity* is the amount of water vapor in a gas compared to the amount of water vapor that is necessary to saturate the gas at the same temperature (Fig. 10-19). *Absolute humidity* is the weight of water contained in a cubic meter of air.

Healthy adults may lose about 40 ml of water daily from their respiratory tract because the expirated air from the alveoli is saturated with water, 100 percent. Those who suffer from high temperatures and poor fluid balance may lose as

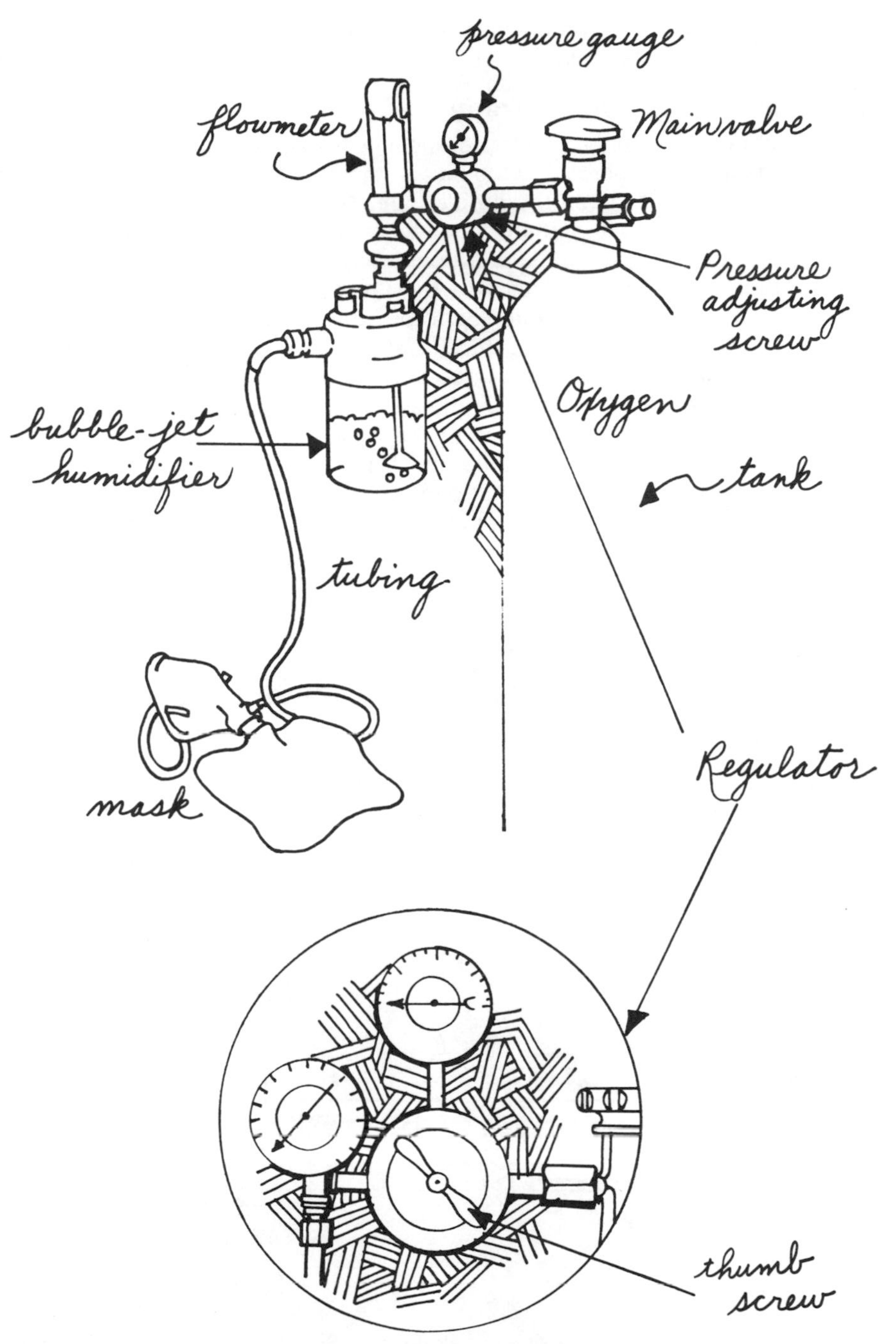

Figure 10-18. Portable oxygen system.

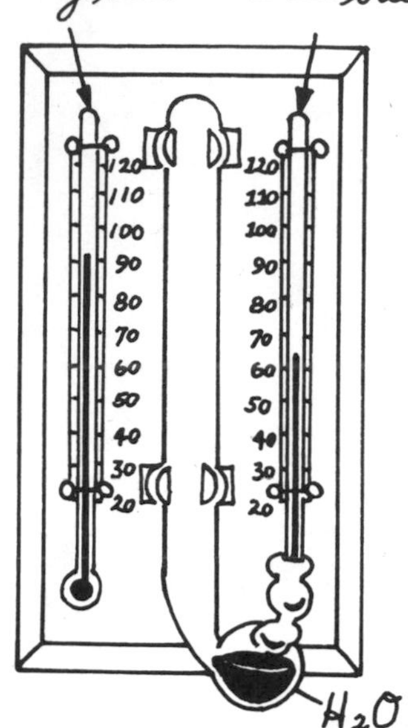

Figure 10-19. Relative humidity instrument (a psychometer). The dry bulb thermometer measures the air temperature. The wet bulb thermometer registers a lower temperature due to the evaporation of moisture. Knowing these two temperatures the relative humidity can be found in tables.

much as 200 ml daily. Therefore gases must be humidified to correct any humidity defect in the respiratory tract.

Aerosols and Aerosol Therapy

Aerosols are suspensions of matter (particulates) suspended in air (gases). The important factor in an aerosol is the size of the particles, because the amount of medication depends upon the droplet size. The number of droplets formed, the amount of particulate matter contained in the droplet, and the percent of droplets that are deposited in the different areas depend upon the droplet size.

Gas Bubblers are designed so that the dry gas passes through the water reservoir in the form of small bubbles to provide a large number of gas-liquid surfaces (see Fig. 10-18). By doing this, the evaporation of the water is increased because the small gas bubble is surrounded by a large quantity of water. The size of the bubbles is determined by the size of the openings in the tube or by a porous diffusion head at the end of the tube. This method, however, has certain limitations, because the amount of gas that can be humidified is restricted by the amount of water that can be evaporated (absorbed by the gas bubbles). Furthermore, as the water evaporates, the temperature of the water is lowered, thus cooling the gas.

Increasing the Moisture Content of Gases

Nebulization is a process that creates a mist with or without medication by using an arrangement that breaks up large droplets and passes only those that are light enough to float (Fig. 10-20). Saturation of gases by nebulization is very effective.

Aerosol therapy and humidifying inspired gases are needed to: obtain a decongestant effect, obtain a bronchodilator effect, thin secretions, soothe inflamed mucous membranes, fight systemic dehydration, and deliver medications to affected areas.

Large droplets will not pass beyond the upper airways, and must be smaller than 3 micrometers (formerly micron, a micron being about 1/25,000 in) to reach the alveoli. If they are too small, over one half are carried out in the expired air because of their lightness.

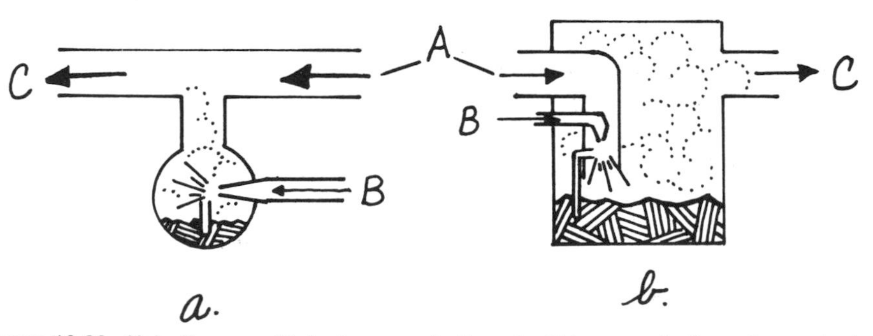

Figure 10-20. Nebulizers. a. Mainstream nebulizer. b. Sidearm nebulizer. Aerosol mists are delivered from exits (C), the therapeutic gas enters at A, the nebulizer power gas enters at B, and the outlet is at C.

Hyperbaric Oxygen Chambers

Patients breathing air at 760 mm Hg with the oxygen content at a partial pressure of about 160 mm Hg, receive oxygen in their blood stream at about 100 mm Hg pressure. Hyperbaric chambers are used to increase the amount of dissolved gases in the blood stream because they operate at pressures higher than atmospheric. A patient breathing pure (and humidified) oxygen, 100 percent, at 760 mm Hg would have a pressure of about 700 mm Hg in the arterial blood stream.

The hyperbaric chamber enables the therapist to increase the oxygen in the blood over 700 mm Hg by increasing the environmental pressure. Therefore, if a patient is subjected to a pressure of 1,520 mm Hg (2 atm) of pure oxygen, the partial pressure of the oxygen in the blood would be 1,400 mm Hg. Theoretically under an environmental pressure of 3 atmospheres, the patient's blood would have oxygen at a pressure of 2,100 mm Hg.

There are problems in the use of hyperbaric chambers that not only involve the gas laws, but safety in the handling of oxygen. Both patients and personnel are subject to accidents resulting from a too rapid decompression, as in the case of diver's "bends," and oxygen toxicity. Factors that must be considered in the environmental control systems are:

1. Gas content and the percentage of the gas
2. Pressure of the gas
3. Temperature of the environment and the gas
4. Humidification

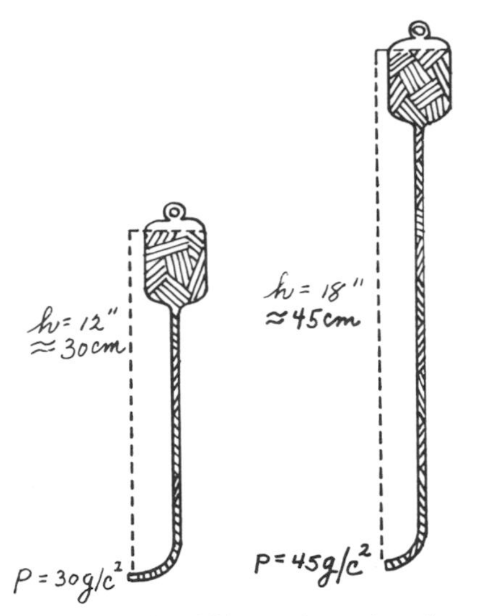

Figure 10-21. Effect of varying the height of a column of liquid on pressure. Increasing the height of the container increases the pressure.

LIQUIDS, PRESSURE AND FLOW RATE

When liquids are used in treatments such as enemas, blood transfusions, intravenous administration, bladder irrigation, or tidal drainage, the pressure exerted by the liquid (which governs the flow rate) must be carefully controlled so the full benefit of the treatment will be achieved without causing a problem for the patient. The rate of flow of liquids depends upon the height of the liquid and the density (Fig. 10-21). Increasing the height of the liquid column by lifting the container will increase the rate of flow of the liquid; changing the shape of the container has no effect (Fig. 10-22) because the height of the liquid is unchanged. At any depth below the surface of the liquid, the pressure exerted by the liquid is the same in all directions (Fig. 10-23).

Flow Rates of Liquids

Even though the pressure of different liquids may be the same, their rate of flow will vary due to the following factors:

1. Length of the tubing: Any tubing offers resistance to the flow of liquid, the longer the tube the greater the resistance.
2. Type of tubing: The surface condition of the inside of the tubing affects the rate of flow. Rough and irregular surfaces cause turbulent flow, and this condition is found in rubber tubing. On the other hand, plastic tubing offers much less resistance to flow.
3. Diameter of the bore of tubing: The greater the diameter of the tubing, the greater the liquid flow rate.

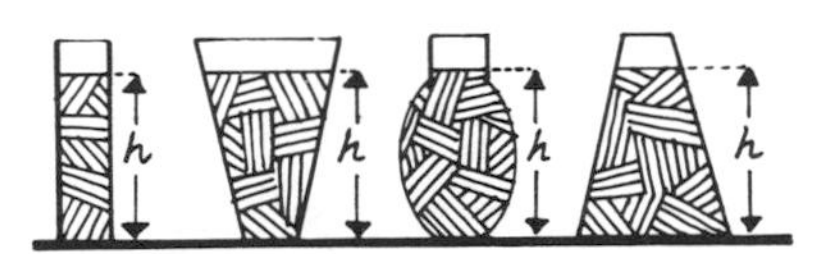

Figure 10-22. Effect of shape upon pressure. The pressure exerted by a liquid is independent of the shape of the container; it depends solely on the depth (h).

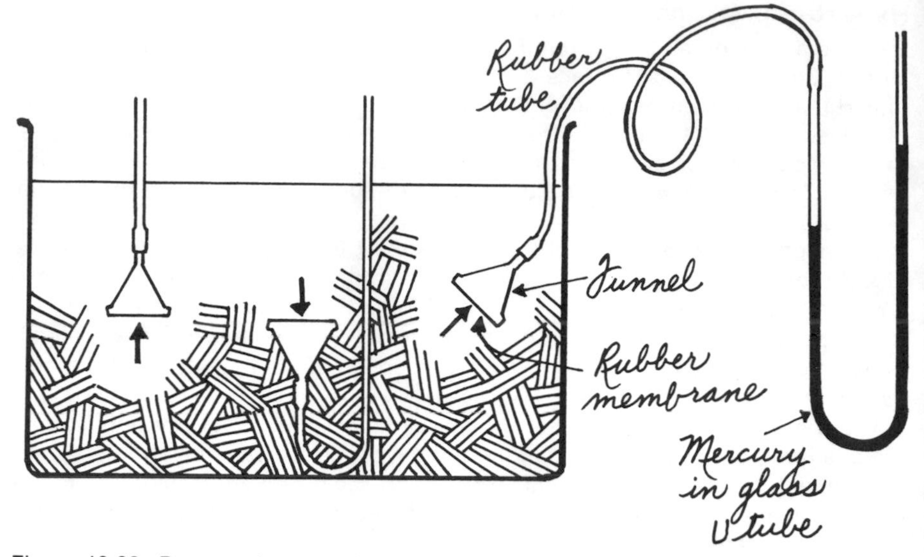

Figure 10-23. Pressure beneath the surface of a liquid. The pressure is the same in all directions, depending only on the depth to which the object is submerged.

4. The viscosity of the liquid: Some liquids are more viscous than others, and they will flow through identical tubings with a slower rate. Isotonic I.V. saline solutions have a faster rate of flow than blood.

The Siphon

The siphon is a useful device that can transfer liquids as long as the outlet end of the tube is lower than the surface of the liquid being transferred. When the tube is completely filled with liquid with *no air bubbles,* atmospheric pressure forces the liquid up the short arm of the tube (Fig. 10-24). The liquid will flow out of the long arm of the tube under the influence of gravity, which is the identical principle used in gastric lavage (Fig. 10-25). In the irrigation of the bladder, the Munro tidal drainage apparatus will periodically fill the bladder with a solution, and then empty the bladder by means of the siphon action (Fig. 10-26).

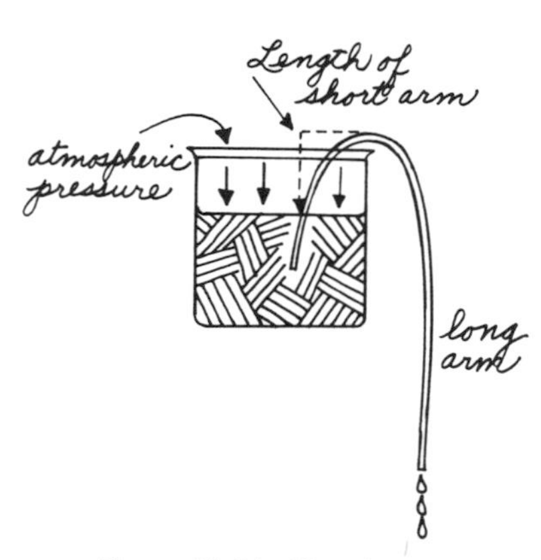

Figure 10-24. The siphon.

Pascal's Law

Pascal's law states that when pressure is applied to a confined liquid, the applied pressure is transmitted undiminished to all parts of the liquid. This law has many clinical applications because fluids in the body are confined, and when pressure is applied at one part it is exerted to all other parts of that confined liquid.

The fetus of a pregnant woman is surrounded by the confined amniotic fluid, and should an unusually large pressure be applied to the abdominal wall, that

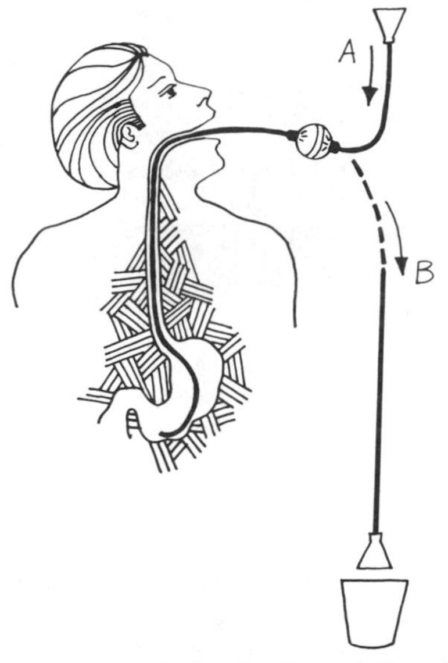

Figure 10-25. Application of the siphon principle to gastric lavage. A. The stomach is filled with liquid. B. The funnel and tube are turned downward to form the long arm of the siphon.

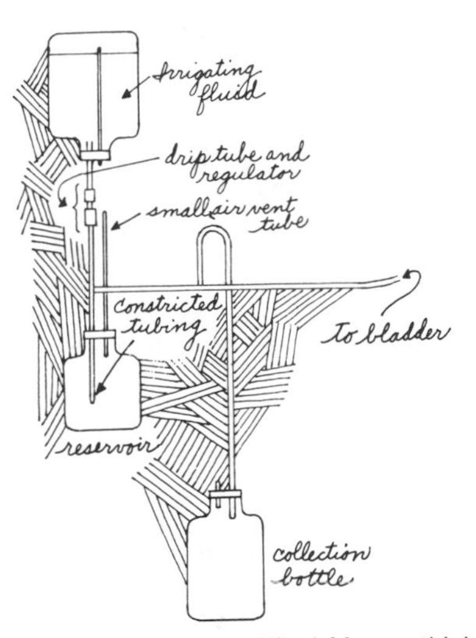

Figure 10-26. A modified Munro tidal drainage apparatus.

pressure would be transmitted to the fetus, possibly causing fetal damage (Fig. 10-27). Another vital organ, the eye, contains confined fluid, and a severe blow can cause damage to the optic nerve due to the transmitted pressure. Increased pressure on the cerebrospinal fluids may cause paralysis.

In the hospital, Pascal's law is used in the treatment of patients who are confined to the bed for long periods of time. They tend to develop decubitus ulcers (bed sores) on parts of their body that receive the largest pressure from regular mattresses. Water or air mattresses are substituted because every part of the body that is in contact with the mattress will receive the same pressure.

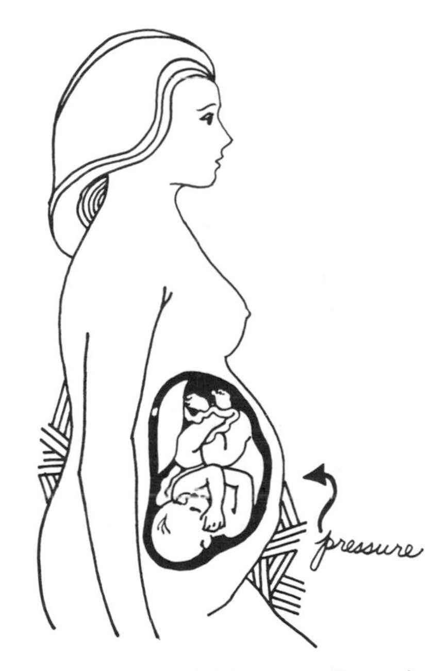

Figure 10-27. Pascal's law. Excessive pressure on the abdomen is transmitted through the amniotic fluid and may be harmful to the fetus.

Bernoulli's Law

Bernoulli's law relates to the pressure on the sides of the tubing that liquids flow through. It states that the greater the rate of flow of the fluids the less the pressure exerted on the sides. You will find applications of this law in the water aspirator (Fig. 10-28), which enables you to get a source of reduced pressure for vacuum needs. Another important application can be found in the bulb atomizer (Fig. 10-29). Air is forced through a constriction causing reduced pressure. The atmospheric pressure forces the liquid up the tube where it mixes with the air to form a spray. Even the common laboratory Bunsen burner uses this law when it is functioning (Fig. 10-30).

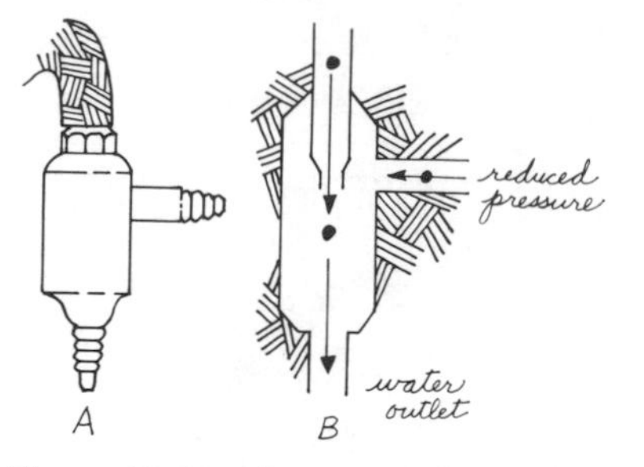

Figure 10-28. Water aspirator. A. External view. B. Diagrammatic view.

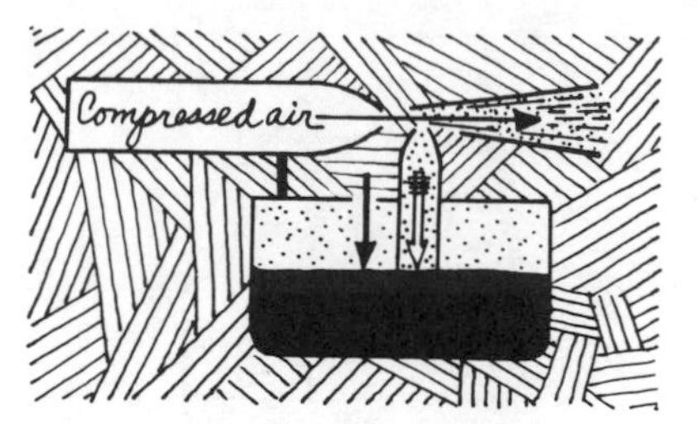

Figure 10-29. Bernouilli's law and the bulb atomizer.

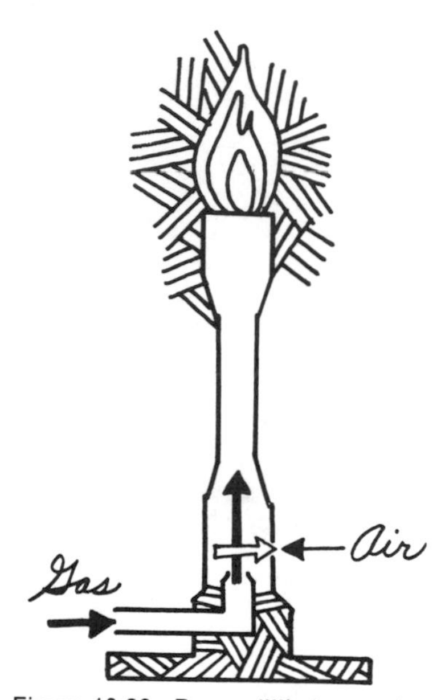

Figure 10-30. Bernouilli's law and the Bunsen burner. The high velocity of the gas through the small opening causes a reduction in air pressure inside. The higher outside atmospheric pressure forces the air to mix with the gas.

SUMMARY

Force is the total amount of energy exerted against a surface, whereas pressure is the force exerted against one unit area. They are related by the formula: $P = F/A$. Pressure of gases can be explained by the kinetic theory of matter which states that gas molecules strike the surfaces of the container with high velocities, and that the greater their velocity (which varies directly as to the absolute temperature), the greater the pressure. All gases can be described by pressure, volume, and temperature, and these factors are related by Boyle's, Charles', and Gay-Lussac's laws. The practice of inhalation therapy depends upon the use of gas laws and chemical principles. In a mixture of gases, each gas behaves as if it were there alone, its partial pressure depending upon its concentration, and the total pressure equals the sum of the individual partial pressures of the components. Through humidity therapy, increasing the moisture content of gases, the inhalation therapist can correct deficiencies in the moisture content of gases with gas bubblers and nebulizers. The pressure exerted by a column of liquid depends solely on the density and the height of the column, independent of the shape of the container. Pascal's law states that the pressure applied to a confined liquid is transmitted undiminished throughout the liquid. Bernouilli's law relates the side (lateral) pressure of a flowing liquid to its velocity; the greater the velocity, the smaller the lateral pressure exerted.

EXERCISE

1. The cobalt-60 unit weighing 6,000 lb and having a base 2 ft by 3 ft is to be installed on a floor that can support 500 lb/ft^2. Is it safe to install the unit?
2. An adult has a surface area of about 3,000 in^2. If the atmospheric pressure is 15 lb/in^2, what is the total force being exerted on the body?
3. Label each statement true or false:
   a. Molecules in a solid are at rest; they are not in motion.
   b. Molecules in a liquid have a greater intermolecular attraction for each other than do the molecules in a solid.
   c. Liquids can be compressed because of the space between the molecules.
   d. Gases cannot be compressed.
   e. Temperature is a measure of the kinetic energy of gas molecules.
   f. Diffusion is caused by the random motion of gas molecules from a point of lower to a point of higher pressure.
   g. It is the impact of the gas molecules upon the walls of a container that generates gas pressure.
4. Give four equivalent values for atmospheric pressure expressed in different units.
5. When is the vapor pressure of all liquids equal to atmospheric pressure?
6. How many liters of oxygen could you get from a cylinder of compressed oxygen with a volume of 3,000 cc and a pressure of 15 atmospheres, if that gas were expanded at atmospheric pressure?
7. An adult inhales 500 ml of air at 20°C. When he exhales the air it has been heated by the body to 98.4°F. Is this correct?
8. The storage room for the compressed gases is kept at 0°C. Fire breaks out and the remote temperature gauge shows that the temperature is now 273°C. If the compressed gases are stored at 2,000 lb/in^2 in cylinders that have a safe pressure only to 3,500 lb/in^2, could the cylinders possibly explode or remain intact?
9. Five hundred milliliters of gas at 20°C and at 760 mm Hg pressure are compressed to a pressure of 5 atm and a temperature of 212°F. What is the new volume?
10. Describe the three categories of respiration and diagram the pressure gradients of oxygen and carbon dioxide in the atmosphere, alveoli, blood and tissues.
11. How are gases that are stored at high pressures used for inhalation therapy?
12. Give two methods for increasing the moisture content of gases, and give the advantages and disadvantages of each.
13. The difference in the water level height is 60 cm and the density of the I.V. fluid is 1.01 g/cm^3. What pressure is exerted by the column of liquid?
14. Explain how opening a window can draw gases out of a room using Bernouilli's principle.

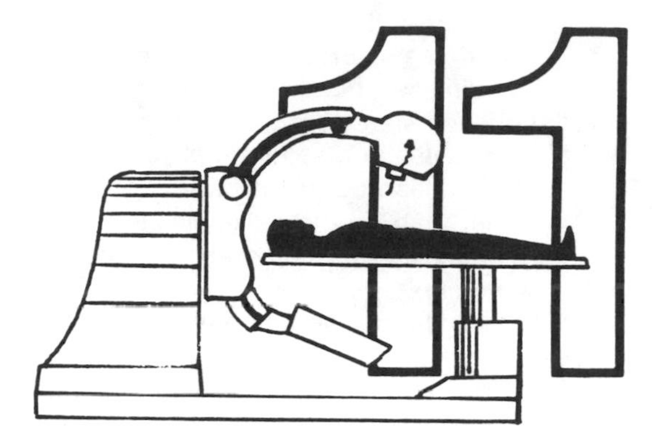

OBJECTIVES

When you have completed this chapter you will be able to:

1. Define the three radiations: alpha, beta, and gamma as to their electrical charges, composition, and mass.
2. State the effect of the emission of an alpha, beta and gamma radiation on the atomic number and the atomic mass.
3. Identify the location of the components of the atom and state the reason why isotopes exist.
4. Distinguish those atoms by atomic weight that may decompose and emit radiations.
5. Explain how beta radiations can be emitted from the nucleus of an atom by the decomposition of a neutron.
6. Diagram the decay of $^{238}_{92}U$ to $^{234}_{90}Th$ and to $^{234}_{91}Pa$ by the emission of an alpha particle followed by a beta ray.
7. Clarify the difference between a chemical reaction and a nuclear reaction.
8. Diagram the process of nuclear fission of U-235.
9. Distinguish between nuclear fission and nuclear fusion by giving examples of each.
10. List the instruments used to detect radiation and explain the principle used by each.
11. State the units used in radiology and where they are applied.
12. Discuss the penetrating power of radiations.
13. Enumerate the physiological effects of radiations and the most vulnerable and susceptible points in the body.
14. Calculate the maximum permissible dose of radiation.
15. Detail how one is exposed everyday to radiations.

NUCLEAR CHEMISTRY AND ITS APPLICATIONS

16. Define LD50/30 days.
17. Write the inverse square law and show (by calculations) the effect of distance upon radiation exposure.
18. List the protections that health care personnel can use to protect themselves against radiation exposure.
19. Make half-life calculations and state the relevance of the half-life when using radioisotopes.
20. Give some of the medical applications of radiation.
21. Name four commonly used radioisotopes used in treatment procedures and discuss how they are used.

BACKGROUND

Radioactivity is defined as the spontaneous decomposition of the *nucleus* of an atom. The net result is *transmutation,* the transformation of an element into a different element. When this happens, the *nucleus* emits certain high speed particles and/or electromagnetic waves. The emission of the particles causes the atomic number to increase or decrease, and may also change the atomic mass. A short background of the history of the development of natural radioactivity will be of value in assisting you to understand the basic concepts.

In 1895 William Roentgen discovered x-rays, very short wavelength electromagnetic waves, which were obtained when the rays of a cathode tube impacted upon other substances (Fig. 11-1). A cathode tube is similar to your television picture tube, wherein cathode rays impinge upon a fluorescent screen to produce the television picture. He called them x-rays, as they were unknown then, and found that they had the ability to penetrate opaque substances to yield an image on photographic plates.

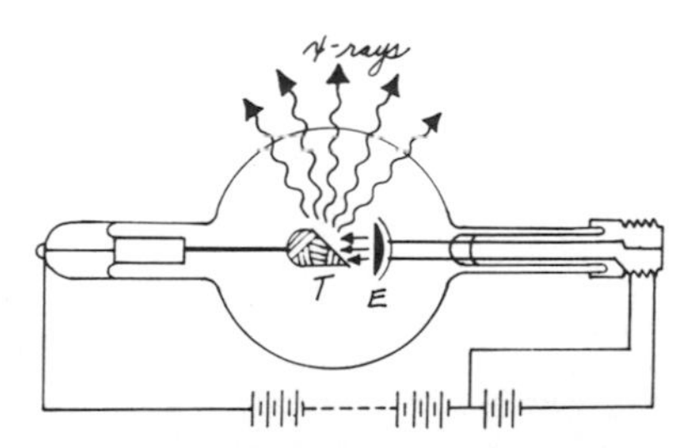

Figure 11-1. Diagram of an x-ray tube. The x-ray tube is an evacuated tube. When high speed electrons (E) impact the target (T) x-rays are emitted.

At about the same time a Frenchman, Henri Bequerel, noticed that uranium samples glowed in the dark, and they also had the ability to affect photographic plates. He termed this *glow radiation,* and he was the first to witness radioactivity, in which an unstable nucleus decomposed and emitted radioactive rays. Later the famous Madame Curie of France discovered that both radium and another element, which she named polonium, were present in pitchblende, and that these elements decomposed giving off radiations.

The discovery of the emission of radioactive waves by the decomposition of elements posed two important questions that were to be answered. First, what were these rays and of what were they composed? Secondly, what happened to an element after it had given off these rays?

Continued research by scientists led to the discovery of three different types of radiation. When the radiations were subjected to an electrical field charge, one was attracted to the negative pole, one was attracted to the positive pole, and one was not affected at all (Fig. 11-2). From your knowledge of electricity, you may know that like charges repel and unlike charges attract. Therefore, the one that was attracted to the positive pole must have been negatively charged, the one attracted to the negative pole must have been positively charged, and that the one that was not attracted to either pole must have been neutral. These radiations are called beta, alpha, and gamma radiations, and they cannot be detected by the senses of the human body, as can the radiations of light and heat.

Alpha particles consist of the helium ion, which is the nucleus of the helium atom stripped of its two orbiting electrons. It therefore has 2 units of positive charge and a mass of 4 atomic units (amu) with the symbol $^{4}_{2}He$. *Beta particles* are electrons and have no appreciable mass and have a unit negative charge, represented by the symbol $^{0}_{-1}e$. The third radiation is called the *gamma ray,*

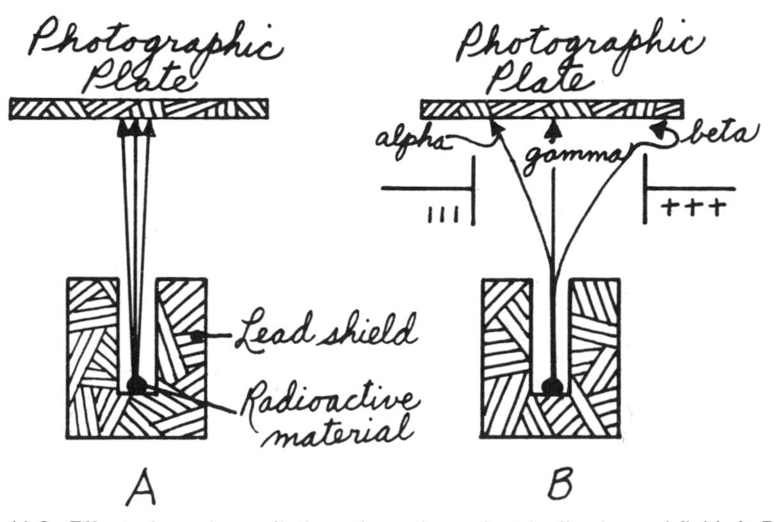

Figure 11-2. Effect of passing radiations through an electrically charged field. A. Path of alpha, beta, and gamma radiations without an electrically charged field. B. Path of these radiations when passed between a pair of electrically charged plates.

which is not a particle of material substance, but consists of electromagnetic radiations of higher frequency (shorter wavelength) than x-rays.

Since the emergence of the era of nuclear energy and atomic power, everyone is aware of the fantastic power that is locked in the incredibly small atom which was set free by the work of many scientists. Today you are exposed through every information medium to the many technical terms which are requisite to nuclear science: terms such as radioactive isotopes, atomic piles, chain reactions, radiation dangers and isotopes, nuclear fission, nuclear fusion, cobalt-60, strontium-90, mutations caused by radiations, uranium-235, tracer elements, irradiation, Geiger counters, and many more. Figure 11-3 shows the international symbol for radioactivity. One of the blessings of the atomic era is the use of artificially created radioactive isotopes to diagnose and treat patients. Knowledge of nuclear theory and radioactivity in its varied uses and applications is a must today for the professional in the health care field.

To understand radioactivity, you must first understand the language of the physicists and build an understanding of the basic concepts involved in the nature of the atoms that compose matter. Earlier in your studies you found out that: (1) elements are composed of atoms that have a nucleus made up of protons and neutrons with electrons orbiting the nucleus, and (2) elements may have different atomic masses due to differences in the number of neutrons contained in the nucleus (isotopes).

Figure 11-3. International sign of radioactivity.

INSTABILITY OF ISOTOPES

Some of these isotopes are unstable and emit radiations. The isotope U-235 is the unstable isotope of U-238 and is the isotope of uranium needed to make the atomic bomb. The nucleus of U-235 contains 3 less neutrons than the nucleus of the stable U-238, both isotopes containing the same number of protons (atomic number), which is 92 for uranium.

When unstable isotopes decompose, they give off radiations and energy, and this is the basis of atomic energy and radiology. Of the more than 100 elements that have been identified, about 81 are stable, 7 are naturally radioactive (such as radium, thorium, and polonium), and the rest are artificially radioactive. There are a large number of stable and radioactive isotopes of these elements.

Everything, whether it is matter or energy, must be accounted for. The same considerations hold true for nuclear science. When particles are emitted, something is lost and the residual substance changes in some way. Probably the easiest way to understand radiation is to examine the decomposition of isotopes.

It is the instability of the nucleus, that particular combination of protons and neutrons for an element, that leads to spontaneous disintegration. In fact, elements having atomic weights greater than 206 tend to decay or disintegrate at various rates forming elements that tend to decay or disintegrate at various rates forming other elements that do not decay. The important difference between a chemical reaction and a nuclear reaction is that in a chemical reaction, the atoms are rearranged by combination, decomposition, and displacement reactions, whereas in a nuclear reaction, the atoms are actually converted or changed into atoms of different elements or isotopes of the element.

THE ORIGIN OF BETA RAYS

So far, you have learned that alpha, beta, and gamma radiations are emitted from the *nucleus of the atom*, yet in an earlier chapter you found that the nucleus of the atom consists solely of protons and neutrons. So, if beta rays (electrons) are emitted from the nucleus *which does not contain electrons,* where did they come from? The answer lies in the neutron, a neutrally charged particle, which has about the same mass as a positively charged proton. If a proton were to combine with an electron, a neutrally charged particle would result. The mass of the neutron would be approximately the same as that of the proton because of the negligible mass of the electron. When a particle neutron decomposes and throws off an electron (beta particle), a proton is left. There is no change in the atomic mass because the mass of the electron emitted was negligible. However, 1 more proton is now contained in the nucleus and the atomic number of the element increases by 1. *When a beta particle is emitted from a radioactve substance, the atomic number of the resulting substance is one number higher.*

Examples of Radioactive Decay

Radon gives up an alpha particle to form polonium:

$$^{222}_{86}Rn \longrightarrow {}^{218}_{84}Po + {}^{4}_{2}He + \text{energy}$$

Since the laws of the conservation of mass and energy must be obeyed, you will notice that for a nuclear equation to be balanced, the sum of the mass numbers (superscripts) and the sum of the atomic numbers (subscripts) must be the same on each side of the equation.

Mass (reactants) = 222	mass products = 218 + 4 = 222
Atomic numbers = 86 (reactants)	Atomic numbers = 84 + 2 = 86 (products)

Uranium to thorium:

$^{238}_{92}U \longrightarrow {}^{234}_{90}Th$	$+ {}^{4}_{2}He + \text{energy}$
Mass (reactants) = 238	Mass (products) = 234 + 4 = 238
Atomic numbers = 92 (reactants)	Atomic numbers = 90 + 2 = 92 (products)

Thorium to proactinium:

$^{234}_{90}Th \longrightarrow {}^{234}_{91}Pa$	$+ {}^{0}_{-1}e + \text{energy}$
Mass (reactants) = 234	Mass (products) = 234
Atomic number = 90 (reactants)	Atomic number = 91 − 1 = 90 (products)

The emission of the beta particle (electron) from the neutron decomposition formed a new proton in the nucleus. Therefore, the atomic number increased by 1, from 90 to 91, forming proactinium-91.

Table 11-1. Decay of uranium.

Element	Atomic Weight	Atomic Number	Symbol	Half-Life	Particle Emission	Effect on At Wt.	Effect on At number
Uranium	238	92	$^{238}_{92}U$	4.6 billion years	alpha	−4	−2
Thorium	234	90	$^{234}_{90}Th$	24.1 days	beta gamma	0	+1
Proactinium	234	91	$^{234}_{91}Pa$	1.14 min	beta gamma	0	+1
Uranium	234	92	$^{234}_{92}U$	270,000 yrs	alpha	−4	−2
Thorium	230	90	$^{230}_{90}Th$	83,000 yrs	alpha gamma	−4	−2
Radium	226	88	$^{226}_{88}Ra$	1,590 yrs	alpha gamma	−4	−2
Radon	222	86	$^{222}_{86}Rn$	3.82 days	alpha	−4	−2
Polonium	218	84	$^{218}_{84}Po$	3.05 min	alpha	−4	−2
Lead	214	82	$^{214}_{82}Pb$	26.8	beta	0	+1
Bismuth	214	83	$^{214}_{83}Bi$	19.7 min	beta gamma	0	+1
Polonium	214	84	$^{214}_{84}Po$	0.0015 sec	alpha	−4	−2
Lead	210	82	$^{210}_{82}Pb$	22 yrs	beta gamma	0	+1
Bismuth	210	83	$^{210}_{83}Bi$	5 days	beta	0	+ 1
Polonium	210	84	$^{210}_{84}Po$	140 days	alpha	−4	−2
Lead	206	82	$^{206}_{82}Pb$				

Table 11-1 shows how radioactive uranium decomposes through radiation into other elements and eventually ends up as lead. All radioactive substances with an atomic number over 82 decompose and finally form lead, atomic number 82. None exist forever. You can see in Table 11-1 that as a result of the three successive disintegrations of $^{238}_{92}U$ you again have the uranium nucleus. Notice, however, that while the *atomic number* is the same, (92), (and it must be the same to be the same element) the *mass number* is now 234 instead of the original $^{238}_{92}U$ and the $^{234}_{92}U$ are isotopes, atoms of the same element that have different atomic masses. Thus, the heavy radioactive elements, such as uranium, transmute and decay into other elements, eventually terminating as an isotope of lead.

ARTIFICIAL RADIOACTIVITY

As a result of nuclear research, there are two kinds of radioactivity; natural and artificial. We know that certain naturally occurring substances such as radium, radon, thorium and uranium are radioactive when found in nature. These are obviously *naturally radioactive*. However, by bombarding the nuclei of ordinary (nonradioactive) elements with high speed subatomic particles, such as protons and neutrons, scientists can transmute them into elements which may or may not be radioactive. The ones formed that emit radioactivity are called *artificial radioactive elements.*

Atomic Energy

In a radioactive disintegration *energy is liberated.* In the famous Einstein equation ($E = mc^2$) the E represents energy, the m represents the mass which is to be converted into atomic energy and c represents the velocity of light (3×10^{10} cm/sec). When 1 gram of matter is converted into energy according to this equation, it is equivalent to about 20 billion kilocalories (which is equivalent to the heat energy obtained by burning 3,000 tons of coal).

The kinetic and potential energies should not be confused with the nuclear energy. They are separate and distinct. For example, equal masses of carbon and plutonium would have the same amounts of kinetic and potential energies, but the plutonium would also possess nuclear energy sufficient to destroy large cities. In the transformations involving potential and kinetic energies, the electron orbits are altered as we have seen in ordinary chemical reactions. The transformations that involve nuclear energy alter the proton and neutron structure of the nucleus.

Nuclear Fission

In 1938 a new type of nuclear disintegration was discovered in Germany. Scientists found that when the nucleus of U-235 was bombarded with neutrons, it split into approximately two equal parts, with the liberation of more neutrons and a much larger amount of energy than was found in ordinary radioactivity (Fig. 11-4). This behavior was called nuclear fission (or atomic fission). Biology students will recall that the amoeba reproduces itself by biological fission.

Nuclear Energy

Certain radioactive isotopes can be unleashed as bombs or controlled with nuclear reactors or atomic furnaces. The fissionable material is the nuclear fuel, which may be U-235. Nuclei split and emit at least 1 neutron for every fission. Those neutrons then bombard other U-235 nuclei causing them to split, each emitting 1 or more neutrons. Thus you have what is called a "chain reaction."

Depending upon the conditions of the reaction, you can obtain barium-56 and krypton-36, or zirconium-40 and tellurium-52 or other pairs. In any event, the sum of the atomic numbers of the products must be 92, the atomic number of uranium.

$$^{235}_{92}U + ^{1}_{0}n \longrightarrow \underset{\text{(unstable)}}{[^{236}_{92}U]} \longrightarrow ^{144}_{56}Ba + ^{90}_{36}Kr + \text{energy} + 2^{1}_{0}n$$

$$^{235}_{92}U + ^{1}_{0}n \longrightarrow \underset{\text{(unstable)}}{[^{236}_{92}U]} \longrightarrow ^{97}_{40}Zr + ^{137}_{52}Te + \text{energy} + 2^{1}_{0}n$$

Uranium-235, upon capturing a neutron, becomes U-236, which becomes unstable. It immediately divides or explodes into other elements while releasing large amounts of energy.

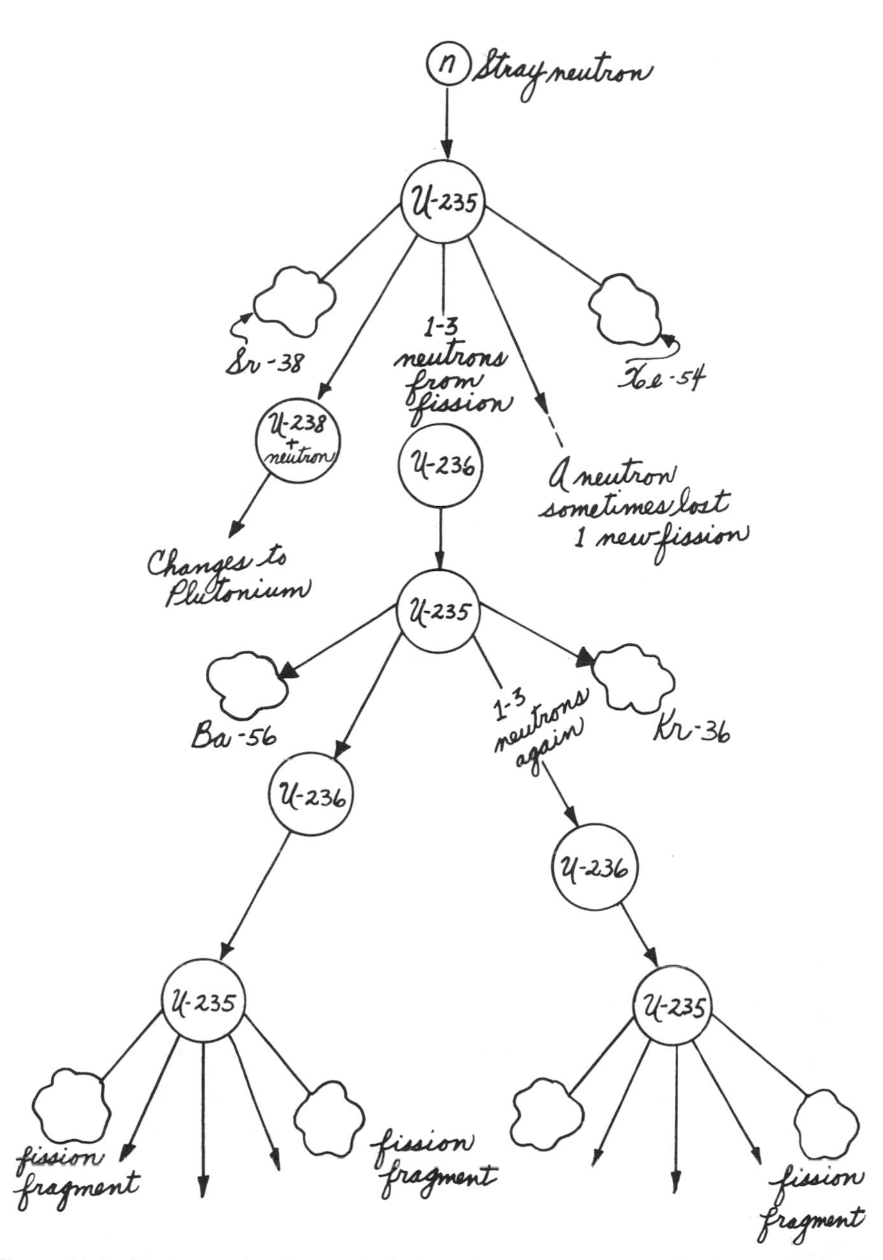

Figure 11-4. Chain reaction in atomic fission. A neutron starts the fission process when unstable U-236 is formed, which may split into Sr and Xe, liberating more neutrons. One neutron may strike U-235 and form plutonium, but others may form unstable U-236 which could split to form other fission products, Ba and Kr. The chain reaction is almost instantaneous. All U-235 in a bomb can be considered to fission instantaneously, thus producing large quantities of energy.

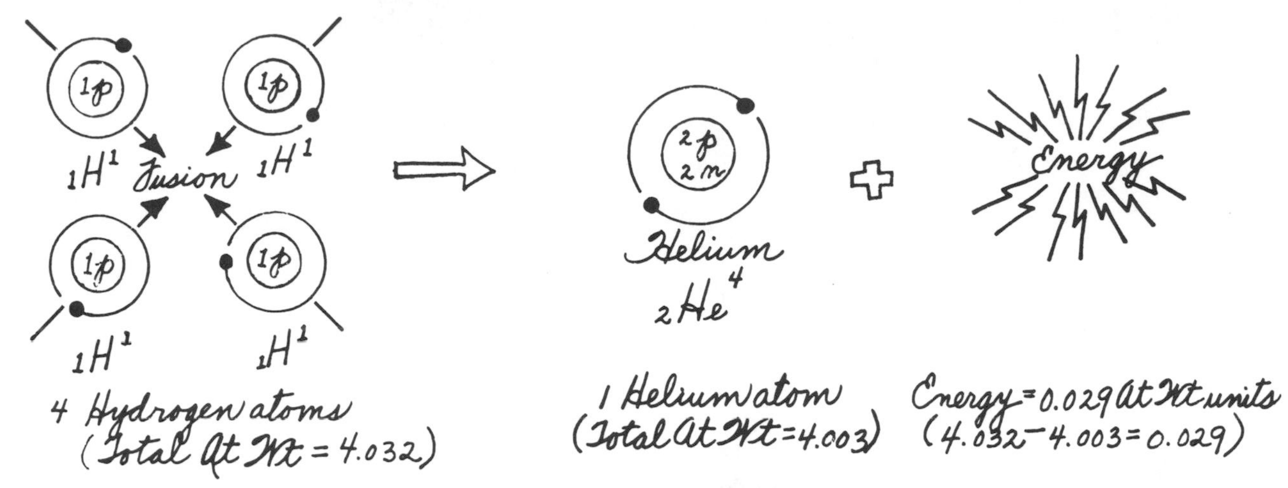

Figure 11-5. Nuclear fusion of 4 hydrogen atoms to form a helium atom, liberating energy in the process.

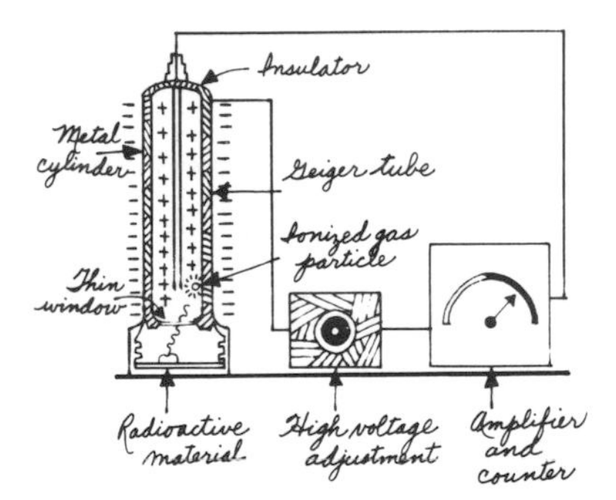

Figure 11-6. The Geiger-Müller counter. The radiation ionizes the gas in the tube causing the flow of electric current that is amplified and measured.

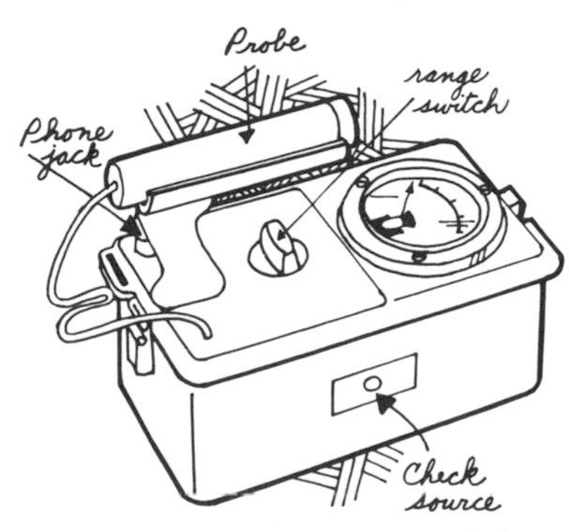

Figure 11-7. The Geiger-Müller counter.

In order to control the chain reaction, absorbers or control rods made of boron or cadmium are adjusted by insertion or removal into the reactor, to capture and absorb emitted neutrons from fission. If a sufficient number of neutrons are absorbed, the chain reaction stops. This nuclear fission was responsible for the earliest large scale utilization of nuclear energy. The explosion of the bomb over Hiroshima was the beginning of the so called "atomic age."

Nuclear Fusion

In nuclear fission, energy is created when large nuclei split to form several smaller nuclei. In *nuclear fusion* even more energy is created when lighter nuclei, such as the isotopes of hydrogen, are converted into heavier nuclei, helium (Fig. 11-5). Exceedingly high temperatures of millions of degrees Celsius are necessary for such reactions to occur. The large amounts of energy released by *nuclear fission* are used by scientists as the basis for the energy requirements for *nuclear fusion,* which is the principle of the hydrogen bomb.

DETECTION AND MEASUREMENT OF RADIATION

Radiation can be detected and measured by various devices such as the well known Geiger-Müller counter, the scintillation counter, and the film badge.

Geiger-Müller Counter

The Geiger-Müller counter is the best known device for detecting radiation. It easily detects the presence of a single ionizing radiation. Because ions are formed, electricity will flow and a current is set up that can be measured either by "clicks" or by a needle reading meter. The greater the amount of radiation, the greater the number of "clicks" or the greater the meter reading of current (Fig. 11-6).

In the hospital this device is used for diagnostic purposes to discover the location and amount of radioactive material in a patient being treated with radioactive isotopes. Many radioactive isotopes are now available, and technicians and physicians have at their disposal many new techniques for using them in the diagnosis and treatment of disease. The radioactive isotopes can be readily located or followed in their passage through the human body with the Geiger-Müller counter (Fig. 11-7).

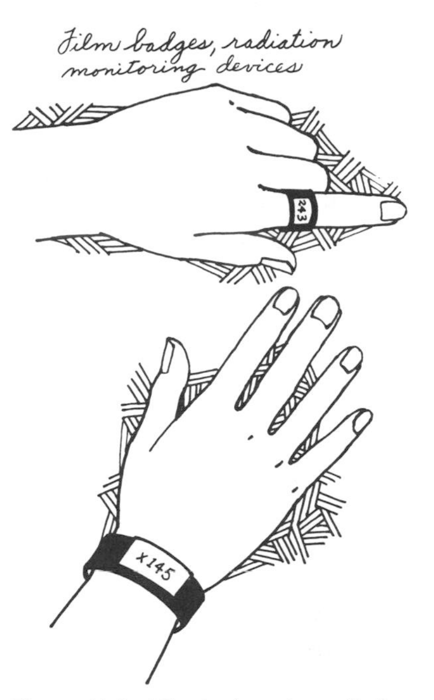

Figure 11-8. Film badges for radiation monitoring.

The Scintillation Counter

Certain compounds, such as zinc sulfide, have the property of glowing when struck by alpha particles. Other compounds, such as anthracene, glow when struck by beta particles. The scintillation counter measures the amount of radiation by electrical magnification of these scintillations (the flashes of light caused by impaction of the radiations).

Film Badges

Anyone in the health care field who may be exposed to radiation should wear a film badge, which is a simple device for measuring radiation. It consists essentially of a metal holder about the size of a pack of book matches containing a strip of unexposed photographic film bearing a code or identification number (see Fig. 11-8). It is worn on the part of the body nearest the expected source of radiation. The radiation is indicated and measured by the darkening of the x-ray film contained inside the badge when the film is developed. The greater the amount of darkening of the film, the greater the exposure to radiation.

Thermoluminescent Dosimeters

Newer types of personnel monitoring devices are the thermoluminescent dosimeters (TLD) that emit visible light when they are heated after exposure to radiation. They have several advantages over film monitoring devices in that they are more sensitive (measuring exposures of 5 millirems), more accurate, and they are unaffected by humidity. The light that is emitted upon heating is measured accurately by a photomultiplier tube device, the visible light emitted being proportional to the radiation exposure.

Monitoring Dosimeters

Monitoring dosimeters enable health care personnel to read the cumulative exposure to radiation directly on the scale at any time (Fig. 11-9). They are very convenient because the fiber indicator points to the milliroentgen exposure on the scale. They are called thimble type chambers, and should be periodically calibrated by the manufacturer.

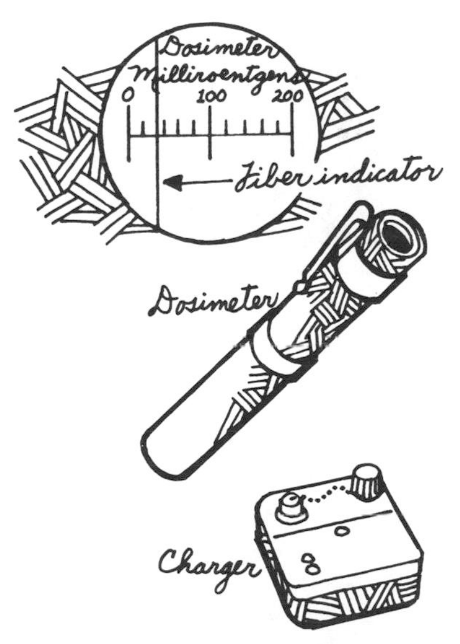

Figure 11-9. Monitoring dosimeter.

UNITS OF RADIATION

A number of units have been defined for the purpose of measuring the *intensity, the energy, the dose, and the biological effectiveness of any given radiation* (Fig. 11-10).

The CURIE (Ci) is the rate of disintegration of radium together with its radioactive decay products present in a 1 gram sample. More convenient units include the *millicurie* (1/1000 Ci). This unit is a measure of the *activity* and not the quantity of material.

The ROENTGEN is a unit that measures the exposure of x-ray or gamma radiation, and is limited to the effect of x-rays or gamma rays in the air.

The RAD (*r*adiation *a*bsorbed *d*ose) is the unit used to measure the *absorbed*

Figure 11-10. Relationship of radiation units.

dose. It is a very small amount of energy, but because of its ionizing effect on body tissue, it has been estimated that a total body dose of about 600 rads of gamma radiation would be lethal for most people.

The REM (*R*oentgen *e*quivalent for *man*) of any given radiation is that quantity that causes, when absorbed by man, an effect equivalent to the absorption of one roentgen.

The RBE (*r*elative *b*iological *e*ffectiveness) of a radiation is defined as the *ratio* of the absorbed dose delivered by gamma rays of cobalt-60 to the absorbed dose delivered by a particular radiation in question when both are compared in producing the same biological effect. The RBE of a radiation signifies its ability to produce a specified effect in a given tissue.

PENETRATING POWER OF THE RADIATIONS

Alpha particles have a velocity of 1/10 the speed of light, have a high ionizing power, but can be stopped by cardboard, paper, or skin.

Beta particles have a velocity 9 times the speed of alpha particles, are about 100 times as penetrating, but have a lower ionizing power than alpha particles, and are stopped by a sheet of aluminum. They can penetrate several millimeters into human tissue.

Gamma rays have shorter wavelengths than x-rays and have a velocity equal to the speed of light (3×10^{10} cm/sec), have an extremely high penetrating power (can penetrate several feet of concrete), and can easily penetrate the human body.

X-rays are ionizing radiations that can penetrate approximately 15 cm into tissue in diagnostic tests, the depth regulated by the voltage. They do not pass through bones and teeth as easily as through tissue, and therefore are useful tools for the dentist and physician.

Energy Content of the Radiations

Gamma radiations have the greatest penetrating power, but contain the least energy; the beta radiations are next in penetrating power, but their energy content is greater. Alpha radiations have the least penetrating power, but the greatest energy content.

PHYSIOLOGIC EFFECTS OF RADIATIONS

The three radiations and x-rays are all ionizing radiations; they convert molecules of matter through which they pass into ions and complex radicals. These radiations are dangerous to human tissues and cells because the high energy fragments and radicals may recombine to form different molecules. If this happens to an enzyme or a nucleic acid, the effects can be extremely serious. Ionization resulting from the radiation of tissues not only disrupts the chemical processes within the cell, but may also alter the genes, causing the cell to grow abnormally. This is because new molecules that are foreign to the cell may form. The result is *radiation sickness.*

Radiation Sickness

The seriousness of radiation sickness depends upon the type of radiation and the amount of exposure. The principle symptoms are nausea, vomiting, diarrhea, internal hemorrhage, and a feeling of weakness. Radiation damage includes the following: (1) severe burns and loss of hair, (2) destructive effect on bone marrow and reduction of red blood cells, causing anemia, (3) damage to genes and chromosomes, producing mutations, (4) decreased level of white blood cells and (5) damage to enzymes, interfering with cell metabolism.

The first evidence of danger involved with radioactive substances occurred in the uranium mines of Bohemia before the discovery of radioactivity. Many of those miners died of cancer. In the 1920s workers in a New Jersey watch factory pointed their brushes by moistening them with their lips to paint radium watch dials. Dental examinations of the workers revealed that their jaw bones were disintegrating, and those that died showed unmistakable signs of radiation poisoning because their bodies had the soft luminescent glow of radium. Many scientists who worked with radium and its compounds also began to develop skin sores that were difficult to heal. Further investigation showed that the radioactive rays could destroy bacteria and other organisms much more rapidly than they would injure healthy tissue.

The best available evidence seems to indicate that the primary site of radiation damage is in the nucleus of a cell. The damage occurs impartially in all cells exposed, but its effect is more evident when cells undergo division, and even more pronounced in rapidly dividing cells, such as cancer cells. The symptoms of overexposure will appear first among the cells that most frequently undergo division. Bone marrow cells fall into this category, and as a result of overexposure, bone marrow is no longer able to produce sufficient blood cells, and a decrease in the white cell count is observed. The white blood cells are the body's first line of defense against infection, therefore resistance to infection is reduced as a result of overexposure. Many of the cases of fatal irradiation from overexposure to the rays of radioactive substances or x-rays show a decrease in blood cells in the following sequence: white cells, platelets, and then red blood cells, with related susceptibility to infection, hemorrhage, and finally anemia.

A more sensitive tissue is the genetic material contained within the egg cell or sperm. Since the protein material in these cells contains the entire genetic code for reproduction of the species any change caused by radiation, however small, causes an irreversible change in the reproductive pattern (a mutation). Some common chromosomal mutations are shown in Figure 11-11. Almost all mutations are lethal and the remainder are undesirable, so it is absolutely imperative to avoid the irradiation of genetic material whenever possible. When a pregnant woman is exposed to radiation it frequently causes damage to the fetus, particularly during the period of formation of the major organs (from the second to the sixth week of fetal life). Some of the resulting abnormalities include: malformations of the head, hands, feet, or genitals, bone defects, cataracts, blindness, mongolism, and other types of subnormal mentalities.

Because a living body is a highly complex organism composed of many interdependent units that are incapable of survival by themselves but are dependent upon other units for oxygen, nutrients, and coordination, these units are similar to the links in a chain. Any serious impairment of one unit has an immediate effect on the other units in the complicated network, so the effect of radiation damage is multiplied and the overall activity of the entire organism is seriously affected.

With these considerations in mind, it is the personal concern of every health

Normal

Figure 11-11. Normal and radiation damaged chromosomes.

care employee to protect themselves, their coworkers and their patients. Successful radiotherapy requires a high degree of knowledge, skill, competence, and experience. Serious if not fatal injuries may result from x-rays and radioactive materials in the hands of careless or inexperienced personnel.

Effect of a Single Exposure to Radiation

Scientists now can predict the effect of a single exposure to radiation on the human body based upon their accumulated evidence. This could happen from an accident occurring while handling radioactive substances, should a nuclear reactor rupture, or from exposure to the radiations of a nuclear bomb.

MAXIMUM PERMISSIBLE DOSE

The National Council on Radiation Protection has established the maximum permissible dose (MPD) for the general population and for personnel engaged in radiology. They have set the permissible level for total-body radiation for individuals subject to long term daily radiations according to the following formula:

MPD = 5 (N-18) rems
where MPD = maximum permissible dose
5 = 5 rems
N = person's age in years

The total-body radiation must be of sufficient strength and penetrating power to significantly affect certain important parts of the body, such as the genitals, head, blood forming organs, and upper and lower portions of the trunk to exceed the MPD. However, the MPD for the general population is reduced to one tenth of that set for those actively working with radiations.

Everyday Exposure to Radiations

COSMIC RAYS. We are constantly being bombarded by cosmic rays (ions travelling at high speeds such as H^+ and He^{+2}) from outer space that penetrate the earth's atmosphere.

RADIOACTIVE FALLOUT to date has increased the exposure of the general population by less than 1 percent. However, there are areas where the concentration of fallout has been greater, and the consequences much more serious. Two isotopes produced by nuclear fallout, Sr-90 and Cs-137, have been the source of much concern. Strontium-90 is similar to calcium in its chemical activity (they belong to the same group in the periodic chart), and it is absorbed by grass and soil, ingested by cows, and their milk contains strontium-90 which then becomes concentrated in the bones of growing children. Strontium-90 has a half-life of about 50 years, which means that a child who absorbs Sr-90 from milk will keep most of the Sr-90 in his body for his entire lifetime. Cesium-137 is similar to sodium (they belong to the same group in the periodic chart), and it can replace sodium, which is a constituent of all human body fluids. Cesium-137 can cause considerable damage, although not as much as strontium-90, because its half-life is 140 days. This means that half of the cesium-137 in the human body decomposes in that time.

Possibly the greatest danger from fallout is the presence of dangerous iso-

Table 11-2. Thirty day lethal dose equivalent.

Species	LD_{50} *(roentgens)*
Dog	315
Pig	375
Man	450
Monkey	600
Rat	675
Yeast	10,000
Bacteria	100,000
Viruses	1,000,000

topes in the air we breathe. Carbon-14 reaches us constantly in the air. Alpha particles are harmless when they strike the outside of the body because they cannot penetrate more than 0.002 inch; they are absorbed by the dead outer layers of skin and do not reach the living cells. However, when they are inhaled they may pass through 5 cell layers and can severely damage them. Beta particles can travel several millimeters through tissue. The amount of C-14 that we normally receive is tolerable. Dangerous levels would only result from a nuclear explosion.

LD 50/30 DAYS. This is the *30 day medium lethal dose equivalent,* which is a measure of radiation toxicity. It is the dose (in *rems* or millirems) that will kill 50 percent of the exposed individuals within 30 days. Table 11-2 lists the LD50's for various species.

THE EFFECT OF DISTANCE UPON RADIATION EXPOSURE

The closer one is to a radiator, a source of radiant energy, the more heat one receives. Similarly, the closer you are to an electric light bulb, the more light you are exposed to. The *change in the exposure* that you get depends upon the change in the distance between you and source of radiations, and it obeys the *inverse square law* (Fig. 11-12). This law states that the exposure to radia-

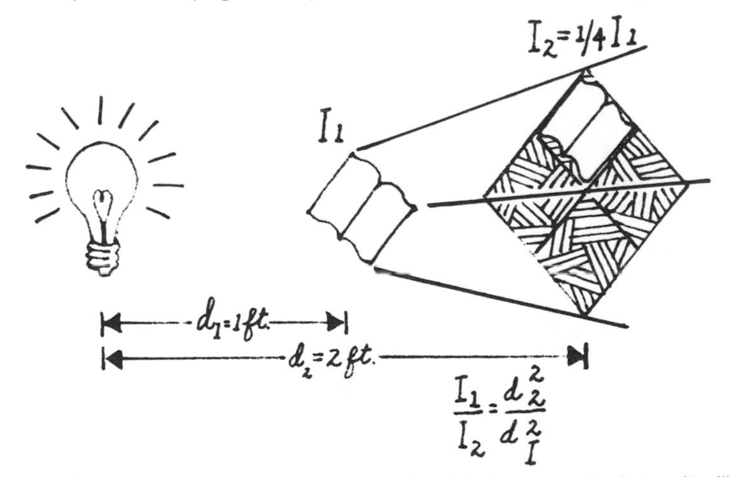

Figure 11-12. The inverse square law. The relationship between the intensity (I) and the distance (d) from a radiation source is described by this law.

Table 11-3. Effect of distance on radiation exposure.

Distance from the radiation source in cm	*Relative intensity*
1	1
5	1/25
10	1/100
20	1/400
50	1/2,500
100	1/10,000
200	1/40,000
500	1/250,000

tions caused by changing the distance is inversely proportional to the square of the distance (see Table 11-3).

Naturally, if you put a greater distance between you and the radiating source you will receive less radiation. If you get closer to a radiating source, you will receive more radiation. However, the effect of cutting the distance in half does not mean that you get double the radiation, you get the square of the double, or four times the radiation. The following procedure will enable you to calculate the change in the exposure based upon the final and original distances involved.

Procedure to Calculate Change in Intensity of Exposure

1. Write the final distance over the original distance
$$\frac{\text{Final distance}}{\text{Original distance}}$$
2. Reduce the fraction obtained
3. Invert the fraction
4. Square the fraction to get the answer.

Example: You are 15 feet from a source of radiation and you move closer until you are 3 feet away.

A. Do you receive more or less radiation?
B. How much more or less do you receive?

A. More, you are closer.
B. 1. Final distance = 3 ft, Original distance = 15 ft
$$\frac{\text{Final}}{\text{Original}} = \frac{3\text{ ft}}{15\text{ ft}}$$
2. Reduce: $\frac{3}{15} = \frac{1}{5}$
3. Invert the fraction: $\frac{1}{5}$ becomes $\frac{5}{1}$
4. Square: $\frac{5^2}{1} = \frac{25}{1}$ = 25 times the radiation.

IMPORTANT: *Always put as much distance between you and the radiation source as possible.* Use long handled tongs to handle "hot" (radioactive) substances.

PROTECTION AGAINST RADIATIONS

Shielding

Shielding materials such as lead or concrete and protective clothing should always be used to minimize exposure when you are working with x-rays, radioactive isotopes, and all sources of radiation.

Time of Exposure

The amount of radiation received is directly proportional with the time of exposure. The greater the time, the more the exposure. You should make sure that you are only exposed for the minimum time. Radiation injury may be caused by repeated and continued exposure to low levels of radiation, because the radiation is cumulative (accumulates by successive exposures). The time of exposure can be monitored and determined by film badges, geiger counters, dosimeters, proportional counters, and ionization chambers.

IMPORTANT: Keep an accurate accounting of your exposure to radiations. Ultraspeed films reduce the exposure time for x-ray procedures.

It is impossible for anyone to avoid exposure to radiations completely, but you should take every possible precaution to avoid unnecessary exposure and always treat radiation with respect.

HALF-LIFE

An extremely important property of all radioactive elements and isotopes is that the rate at which they decay or disintegrate is *absolutely constant,* and is unique for each element and isotope. This means that regardless of what you could possibly do to the radioactive substance (heating, cooling, freezing, fusing, fluxing, powdering, chemically reacting, beating, or melting) you *cannot alter* the rate of the decay of the substance. It is constant and unchanging, and this is a major difference from an ordinary chemical reaction.

In order to measure this rate of disintegration or decay, scientists have measured *how long it takes for one half of the substance to decay.* This time is called the *half-life period. This half-life period is therefore the time that is required for one half of the substance to decompose.* One half of the amount will decay in that time, then half of what is left will decay in the same length of time, and so on. Physicists count the number of alpha or beta particles that are emitted in one second. These half-life periods range from a small fraction of a second ($^{214}_{84}Po$: 1.6×10^{-4} sec.) to almost 14 billion years ($^{232}_{90}Th$: 1.39×10^{10} years). Here are two methods to calculate how much radioactive substance remains after certain periods of time.

PROBLEM

Radioactive iodine has a half-life of 8.0 days. If a patient is given a dose of 24 micrograms (24 mg), how much I-131 will be left in the patient's body after 24 days?

Method 1: The half-life is 8.0 days. That means that in 8 days, ½ of the substance will decompose, leaving ½, 12 micrograms (½ of 24). At the end of 16 days (the second half-life period) ½ of the 12 micrograms decompose, leaving

6 micrograms. And, at the end of 24 days (the third half-life period) ½ of the 6 micrograms decompose, leaving 3 micrograms remaining in the patient. Make a time chart: half-life = 8 days:

Time (in days)	0	8	16	24	
Amount left	24 mg	12 mg	6 mg	3 mg	(answer)

Method 2: Use the mathematical formula: $Q = A(\frac{1}{2})^n$. Where Q is the amount remaining, A is the starting amount, and n is *the number of half-life periods.* Divide the total time period (in this case, 24 days) by the half-life period ($24 \div 8 = 3$) to get "n", the number of time periods. Then substitute directly in the formula:

$$Q = 24\ \mu g\ (\tfrac{1}{2})^3 = 24\ \mu g\ (\tfrac{1}{2})(\tfrac{1}{2})(\tfrac{1}{2}) = 24 \mu g\ (\tfrac{1}{8})$$
$$Q = 3 \mu g \text{ (answer)}$$

Biological Half-Life

Radioactive elements and their compounds are utilized by the body exactly as their nonradioactive counterparts. Since the body constantly eliminates waste substances through normal biological processes, some of the radioactive substances taken into the body will be eliminated along with the nonradioactive ones. Substances such as strontium-90 pose serious problems because a long period of time is required to eliminate one half of it from the body through natural processes. It is said to have a long biologic half-life: the time that is required for half of the amount of radioactive substance taken into the body to be eliminated through natural processes. Substances that are eliminated quickly are said to have a short biologic half-life.

APPLICATIONS OF RADIOACTIVE ISOTOPES

The first "practical" use of nuclear energy was the atomic bomb of World War II, followed by the test explosion of the first hydrogen bomb a few years later. Since that time, great progress has been made in the use of nuclear energy for nonmilitary purposes.

Medical Applications of Radiation

You have learned that radiation (including x-rays and radioactive isotopes) has been used for purposes of diagnosis, therapy and research. X-rays are used by the dentist and the physician. The x-rays are in many respects similar to ordinary light rays, except that x-rays have a much shorter wave length (higher frequency). X-rays are invisible, yet they cause many substances to fluoresce or glow and become luminous. Because of their very short wave lengths they are able to pass through many substances, including human flesh, which is opaque to ordinary light (Fig. 11-13). Denser objects such as bones, teeth, and metals largely stop this radiation, and in this way produce shadows on photographic film. Thus an x-ray photograph can be made and a bone fracture can be studied, cavities located in teeth, or pieces of metal located in the body.

WHAT AN X-RAY NEGATIVE CAN REVEAL

Figure 11-14 shows a number of x-ray negatives taken by a dentist, and we can now see what conditions exist that were not detectable by a visual examination

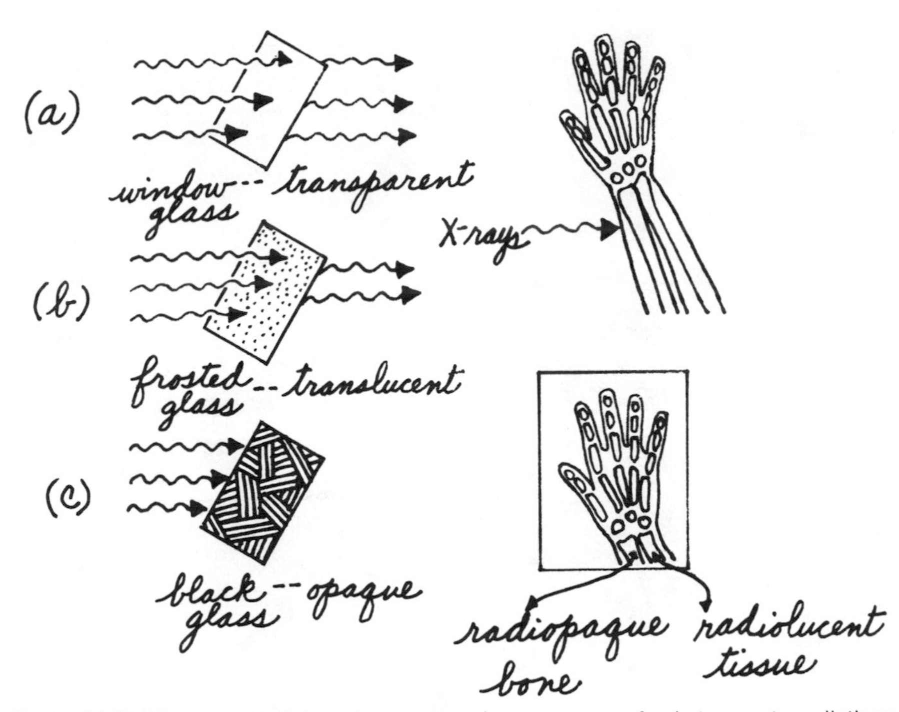

Figure 11-13. Transparency, transluscency and opaqueness of substances to radiations. Radiations are absorbed in three degrees: (a) transmission (no absorption), (b) attenuated (partially absorbed), and (c) complete absorption.

of the teeth. The x-ray negative therefore enables the professional to diagnose certain conditions.

Barium sulfate is opaque to x-rays, and when a water suspension of this salt is administered by mouth (upper gastrointestinal) or by rectum (barium enema, lower gastrointestinal), it coats the inner surfaces of stomach or intestinal tract (see Fig. 11-15). This procedure makes it possible to obtain x-ray photographs of these organs for diagnostic purposes. The representative values of a radiation dose in millirads of a typical x-ray examination is given in Table 11-4.

Table 11-4. Representative values of radiation dose in millirads.

x-ray examination of	*Skin exposure*	*Gonadal exposure*
Skull (lateral)	1,500	Less than 1
Chest	30	Less than 5
Spine (lumbar)	Around 3,500	Around 900
Abdomen	Around 600	Around 200
Pelvimetry (to fetus)	1,000	1,000
Dental (full mouth)	500	Less than 5

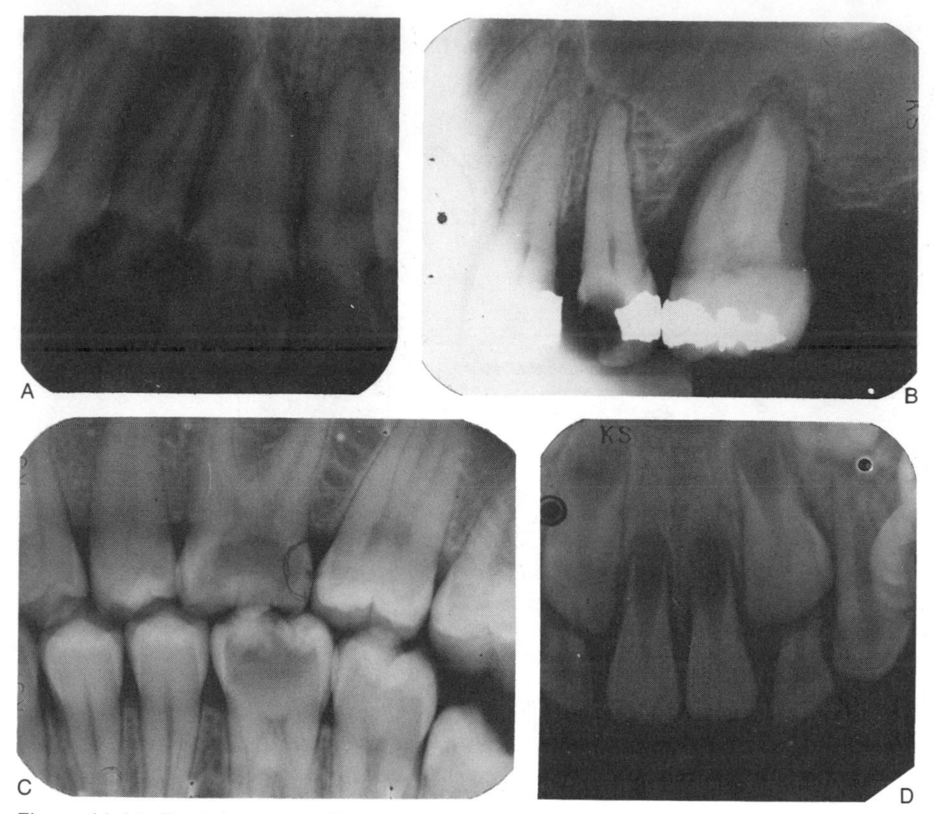

Figure 11-14. Dental x-rays. A. Rampart decay and abscess. B. Severe bone loss. C. Large area of decay. D. Tooth development.

Therapeutic Applications of Radiation

Radium therapy was originally used to treat cancerous tissues and other malignant growths by placing a radium salt in a hollow platinum needle that was placed into or next to the cancerous growth. The alpha and gamma rays from the radium (and its decomposition product, radon gas) acted as the source of ionizing radiations that had a destructive action upon the cancer cells. Unfortunately, they also destroy normal tissue, and for this reason their action must be strictly confined to the diseased structure. Since radium has a half-life of 1,620 years, a radium tube inserted in a patient's body must soon be removed to prevent undue damage to normal tissues because its radioactivity is continuous for thousands of years. Cobalt-60 (half-life of 5.3 years) has recently been utilized as a substitute for radium because it does not have this undesirable long term effect.

Recent developments include a newer form of x-ray therapy involving the use of oxygen at pressures of 3 atmospheres, called hybaroxic cancer radiation. This treatment makes use of the fact that cancer cells are almost 3 times as sensitive to x-rays at this pressure as they are at normal pressure, while normal

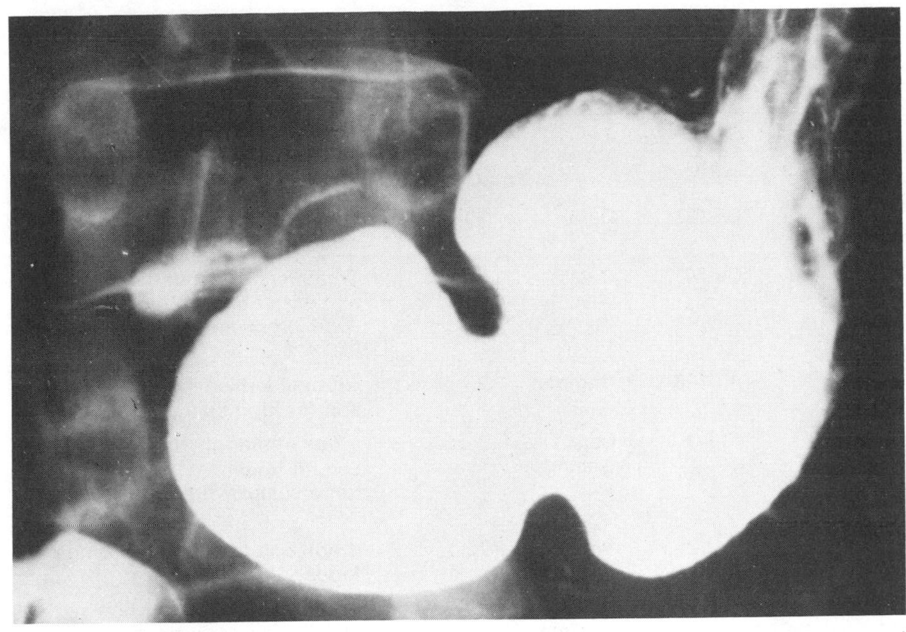

Figure 11-15. X-ray of the upper G. I. tract. The barium sulfate is opaque to x-rays and the stomach is clearly outlined.

cells are not affected to the same extent. Another recent development is the use of a beam of neutrons rather than x-rays. Neutrons, unlike x-rays, are absorbed by the atomic nuclei, particularly hydrogen, which have a high concentration in cancerous tissue. Consequently, this process of N-radiography (neutron radiography) is very useful in bone cancer studies.

The chemical and biological effects of the radioactive elements and isotopes are the same as their stable nonradiating counterparts. The human body cannot distinguish between an element and its radioactive isotope, and they are treated by the body as though they were the same. However, the radioactive substances can be detected in the body in minute quantities by the Geiger-Müller counter. A summary of the radioactive isotopes used in medicine can be seen in Table 11-5.

COBALT-60

Cobalt-60 has replaced radium in the treatment of cancer. It is cheaper; $100 worth of Co-60 is equivalent to $20,000 worth of radium. Another reason is its short half-life (5.2 years) as compared to radium (1,620 years). Cobalt-60 is used to treat many types of cancer, and many hospitals use a large machine to supply the strong gamma radiations to the cancer site (Fig. 11-16).

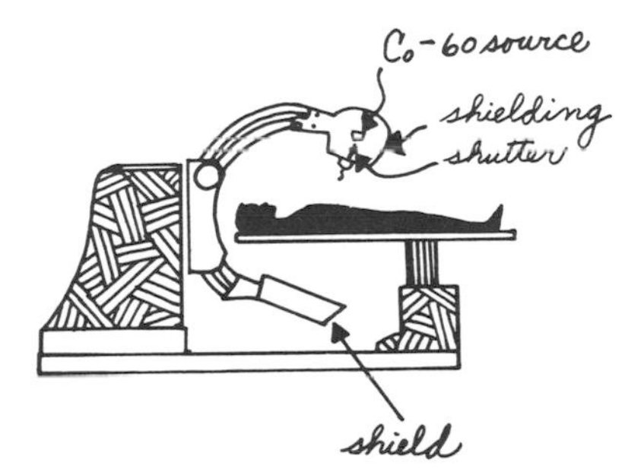

Figure 11-16. Rotational teletherapy using Co-60 gamma rays.

IODINE-131

Iodine is concentrated in the thyroid gland of the body, which absorbs iodine and concentrates it to become thyroxin. Normally acting thyroid glands quickly remove iodine from the bloodstream. If a malfunctioning thyroid is suspected,

Table 11-5. Summary of radioactive isotopes used in medicine.

Element	*Isotope*	*Rays Emitted*	*Half-life*	*Summary*
Carbon	C-14	Beta	5760 yrs	Study metabolism of carbohydrates, fats, proteins, mechanism of photosynthesis
Cesium	Cs-137	Beta	30 yrs	Occurs in fallout. Body fluid research
Cobalt	Co-60	Beta Gamma	5.2 yrs	Treatment of cancer, cheaper and easier to use than Ra-226 (Co-60) has much shorter half-life
Gold	Au-198	Gamma	2.7 days	Chronic leukemia treatment, and for lung cancer
Iodine	I-131	Beta Gamma	8 days	To determine thyroid condition and for treatment, locating brain tumors, study function of kidneys
Irridium	Ir-192	Beta Gamma	75 days	Irradiation of and processing of foods
Iron	Fe-59	Beta Gamma	45 days	Determine where and how red blood cells are formed and destroyed, diagnosis of anemia
Phosphorus	P-32	Beta	14.3 days	Locating brain tumors, studying formation of bones and teeth, treatment of overproduction red blood cells, treatment of skin cancer, locating breast cancer, determining blood volume
Radium	Ra-226	Alpha Gamma	1620 yrs	Used to treat cancer
Sodium	Na-24	Beta Gamma	15 hrs	Locating blood clots, study rate of circulation of blood
Strontium	Sr-90	Beta	28 yrs	Occurs in fallout, used to study healing process in bone, possible treatment of bone cancer and malignant growths in the eye.
Uranium	U-235	Alpha	7.1×10^8 yrs	Source of radioactive isotopes
	U-239	Beta Gamma	24 min	Source of radioactive isotopes

the patient is given a drink of water containing a small amount of I-131 in the form of sodium iodide. A Geiger-Müller counter is placed on the thyroid to indicate whether the thyroid is functioning properly, absorbing iodine at the normal rate. If less than the normal amount of iodine is absorbed, the patient may have a hypothyroid condition, and if the rate is greater than normal, the patient may have a hyperthyroid condition. The I-131 is useful for treating cancer of the thyroid gland. It is added to the diet so that the patient will receive a continuous radiation treatment. The short half-life of I-131 (8 days) makes it possible to control radiation dosages since less than 10 percent of the original activity remains after 30 days.

METASTASIS is the breaking off of malignant cancer cells, which are then transported by the blood and lymph to other parts of the body where they set up a colony. If metastasis has occurred in the thyroid the patient is given radioactive NaI^{131} in a water solution (Fig. 11-17). The cancer colonies will still absorb the radioactive iodine and are destroyed by its gamma rays. Unless all malignancy is removed, the cancerous condition is not cured.

Figure 11-17. Radioactive iodine (I-131) for diagnosing and treating thyroid gland disorders.

In cases of hyperthyroidism, treatment with larger amounts of I-131 will effectively reduce the size of the thyroid gland by irradiation of the tissues involved in the disorder. Also, in chronic congestive heart disease where a healthy thyroid has too much activity for a damaged heart, I-131 inactivates part of the thyroid and decreases the amount of thyroxin secreted. It also seems to relieve some patients from the severe pains of angina pectoris.

I-131 has been used for determining the circulation time of the blood and for vascular flow measurements, determining the total blood volume, locating brain tumors (Fig. 11-18) and to study kidney function.

PHOSPHORUS-32

Phosphorus-32 has been used to detect and localize tumors in the brain, because brain tumors absorb phosphorus more readily than normal brain cells. It is an invaluable tool for diagnostic study of the formation of bones and teeth and for therapeutic treatment, because it affects both red and white cells in the blood. However, it must be administered with caution, since patients treated with P-32 have a higher than normal incidence of leukemia.

Blotting paper soaked with a solution of P-32 and applied over a skin cancer (held in place with adhesive tape) has been effective in many cases. Because it has no gamma radiations and a short half-life (14 days), it is less irritating to the tissue than other radioactive substances. In chronic leukemia, where the white blood cells become too numerous, P-32 is used as a radiation agent and also as a diagnostic tool to obtain more accurate estimations of red blood cells than would be obtained by a routine count.

SODIUM-24

This isotope can be added to an isotonic NaCl solution and injected into the blood stream for the purpose of detecting faulty circulation. If a Geiger-Müller counter gives a higher count for one foot than for the other, it reveals a poor circulation in the foot with the lower count. It can be used to locate a blood clot, because if a clot is present activity drops at the location of the clot, with a high count on one side of the clot and a low count on the other side. It may often tell whether amputation of a limb is necessary because of poor blood circulation. Radioactive Na-24 may be used intravenously, because in 15 hours, one half of the atoms will have become ordinary magnesium, a normal constituent of the blood, because a beta particle is emitted.

$$^{24}_{11}\text{Na} \longrightarrow ^{24}_{12}\text{Mg} + ^{0}_{-1}\text{e}$$

Figure 11-18. Diagnostic use of I-131.

RADIOACTIVE IRON

Radioactive iron, Fe-59, is rapidly absorbed in conditions of anemia and can be used to determine where and how red blood cells are formed and destroyed, and how to diagnose and study anemia.

RADIOACTIVE GOLD

Radioactive gold, Au-198, can be used to treat chronic leukemia and is implanted (after encapsulation) in the pleural cavity in the treatment of lung cancer. Gamma radiations of Au-198 destroy cancerous cells, and because of the short 3 day half-life period, quickly decays to negligible amounts. It is a beta emitter.

RADIOACTIVE CARBON

Radioactive carbon, C-14, is used to study the mechanism of photosynthesis and the metabolism of carbohydrates, fats, and proteins. Labelled isotopes are administered by injection or feeding, and later parts of the animal or the waste products are analyzed for the isotope. Carbon-14 is also used to date fossils, because by knowing the proportion of radioactive C-14 found in dead organic matter scientists can determine the age of the fossils.

SUMMARY

Radioactivity is the spontaneous decomposition of atoms with the emission of alpha, beta, and gamma rays. Radioactive elements decompose into other elements and isotopes due to the instability of the neutron-proton ratio in the nucleus; their transformation can be followed by the radiation emitted. The radiations can be detected and measured by several instruments, and different units of radiation are used to measure the activity, intensity, exposure and absorption of the radiations, all of which have different penetrating powers. Radiation exposure may cause desirable or undesirable physiological effects depending upon the type and exposure. Some parts of the body are especially susceptible to radiations, for example, exposure during pregnancy can result in fetal deformities and abnormalities. The effect of distance upon radiation exposure can be calculated by the inverse square law, and exposure can be minimized by using precautions. Each radioactive element and isotope has a definite half-life period, the time required for one half of the material to decompose. Radioactive isotopes are used in the treatment of certain diseases and in diagnostic procedures.

EXERCISE

1. Describe the composition of an alpha particle, a beta particle and a gamma ray.
2. How is the mass of an element affected by the emission of each radiation?
3. How is the atomic number affected by the emission of each radiation?
4. List in descending order the penetrating power of the radiations.
5. Explain how a beta particle can be emitted from a nucleus that does not contain electrons.
6. Complete the following equations:

   a. $^{31}_{15}P + ^{1}_{0}n \longrightarrow ^{31}_{14}Si + ?$

   b. $^{9}_{14}Be + ? \longrightarrow ^{8}_{3}Li + ^{1}_{0}n$

   c. $^{24}_{11}Na \longrightarrow ^{24}_{12}Mg + ?$

7. What are x-rays and how are they produced?
8. Why is barium sulfate used in x-ray procedures?
9. How do nuclear reactions differ from ordinary chemical reactions?
10. Draw a diagram to illustrate the nuclear fission of U-235.
11. Why is impossible to find an environment on earth that is free from radiation?
12. Why is strontium-90 dangerous to mankind?
13. List three ways to reduce exposure to radiations.
14. Which parts of the body are extremely susceptible to radiations and explain why.
15. Why is continued exposure to small amounts of radiation dangerous?
16. Describe the methods used to detect and measure radiation exposure.
17. What radioactive isotope is used to:
    a. Replace radium-226 in the treatment of cancer?
    b. Locate a blood clot?
    c. Treat overproduction of white blood cells?
    d. Treat cancer of the thyroid and malfunctioning thyroid glands?
    e. Locate brain tumors?
    f. Study the circulation of the blood?
    g. Determine how and where red blood cells are formed and destroyed?
    h. Study the formation of teeth and bones?
18. If the half-life of an isotope is 4 hours, and you prepared 100 mg of it, how much would be left after 24 hours?
19. A 16 mg sample of I-131 was administered to a patient. How many milligrams are left after about 1 month, 32 days?
20. A technician is standing 4 feet from an x-ray unit and is receiving radiations because of improper shielding. If the technician now comes closer and is 1 foot away from the unit, how much more radiation is he receiving?

OBJECTIVES

When you have completed this chapter you will be able to:

1. State the differences between organic and inorganic compounds.
2. Detail the importance of the carbon atom to health and daily living.
3. Show three ways that a carbon atom can achieve a stable configuration.
4. Define covalent bonds and illustrate their tetragonal angles in the carbon atom.
5. Write the molecular, structural, ball and stick, and condensed formulas for the first five alkanes.
6. Define the term saturated with respect to hydrocarbons.
7. List the physical states of the simple alkanes and what these compounds are used for.
8. Distinguish the difference between an alkane and an alkyl radical by writing the formulas of the simple alkyl radicals.
9. Identify isomers.
10. Utilize the IUPAC system of nomenclature for naming isomers.
11. Describe an alkene and what characteristic group it contains.
12. Relate *cis-trans* isomerism to the double bond.
13. Name compounds containing double bonds by the IUPAC system.
14. Give the chemical reactions of the alkenes by writing the reactions for:
    a. Hydrogenation
    b. Oxidation
    c. Addition of Cl_2, HCl, and H_2O
    d. Polymerization
15. Define an alkyne and state its characteristic group.

THE CARBON COMPOUNDS

16. Compare the chemical reactivities of the alkynes and the alkenes.
17. Name and draw the simple cyclohydrocarbons.
18. List the benefits of using cyclopropane as an anesthetic.

Until 1828 chemists believed that the chemical compounds produced in living organisms possessed a certain life characteristic that differentiated them from compounds found in nonliving substances. These compounds were called *organic compounds* in order to distinguish them from those *nonliving or inorganic compounds.* However, in that year a German chemist named Wohler prepared urea, a waste product excreted in the urine of living organisms, by heating ammonium cyanate. This was the first time that an organic substance had been synthesized or prepared from an inorganic substance.

Since Wohler's time, in less than 175 years, over 1 million organic chemical compounds have been synthesized, isolated, and identified. Yet during the past 2,000 years only about sixty thousand inorganic compounds have been made. The significance of this fact is that the chemistry of living organisms is based upon the carbon atom. All living organisms are composed of molecules that are structured primarily on the carbon atom with molecular weights ranging from about one hundred to over five million (Table 12-1). The molecules of the proteins, fats, carbohydrates, lipids, hormones, enzymes, genes, and chromosomes are extremely large and very complex, and they have chains and rings of carbon atoms as their basic structure. Therefore, a knowledge of the carbon atom and how it behaves is absolutely necessary to your understanding the chemistry of the body and of the matter in your environment (Table 12-2).

THE CARBON ATOM

The carbon atom has an atomic number of 6, which means that it has 6 protons in the nucleus and 6 electrons in orbit (Fig. 12-1). Since elements tend to achieve stability by attaining the inert gas configuration, the carbon atom will

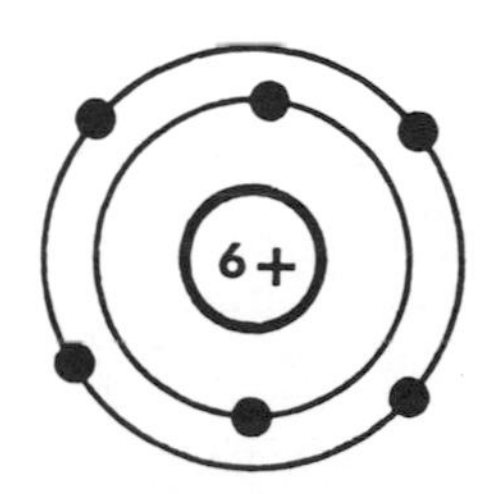

Figure 12-1. Electron structure of the carbon atom.

Table 12-1. Difference between organic compounds and inorganic compounds.

Organic Compounds	*Inorganic Compounds*
Are composed of few elements: always C, usually H, O, N, sometimes S, P, Cl, Br, I	Are composed of various combinations of 92 elements
Large and complex molecules	Relatively small number of atoms in the molecule
Nearly all burn	Very few will burn
Low melting points, well below 350°C, some char and decompose	Extremely high melting points, usually above 350°C, vaporizing only at high temperatures
Usually possess color and odor, soluble in organic solvents, usually insoluble in water	Usually odorless and colorless, and soluble in water to various percentages
Covalently bonded, few ionize	Ionic bonding, usually ionize in water solutions
Many isomers	Few isomers

tend to *share* its 4 outer electrons with other atoms. In doing so, it attains the stable outer shell of 8 electrons, and this process of sharing electrons is called covalent bonding.

THE COVALENT BOND

Organic compounds are formed by covalent bonding, the sharing of electrons between the atoms, each atom receiving the full benefit of the shared electrons. Covalent compounds do not ionize, and most of the compounds that compose living matter are covalently bonded compounds.

Table 12-2. The importance of the carbon atom.

To Health	Composition and metabolism of foods, digestion, biochemical processes of the body, drugs, vitamines, hormones.
To Daily Living	
Heat	Fuel oil, natural gas.
Light	Fuels to generate electricity.
Foodstuffs	Flavors, colorants, sweeteners, preservatives, containers and wrappings.
Transportation	Gasoline, diesel fuel, oil, lubricants, components of motor cars, plastics, rubber
Clothing	Synthetic fibers, plastics, synthetic textiles, cloth, and outerwear.
Personal	Detergents, soaps, colorants for cosmetics, perfumes and chemical specialty odors, paints plastics, insecticides, coatings, inks, explosives, photographic film and accessories, drugs, discardable items.

The carbon atom will covalently bond with many other elements including other carbon atoms to achieve stability, as will hydrogen atoms. The hydrogen atom will share its 1 electron with 1 electron of a carbon atom, therefore attaining the stable configuration of the helium atom, which has 2 electrons in its outer shell. The carbon atom can reach the stable configuration by any one or combination of the following ways:

1. Sharing 1, 2 or 3 electrons with the same number of electrons of another carbon atom.
2. Sharing 1, 2 or 3 electrons with the electrons of another element.
3. Sharing electrons in any combination of the above, as long as the carbon atom and the other atom complete their outer shells.

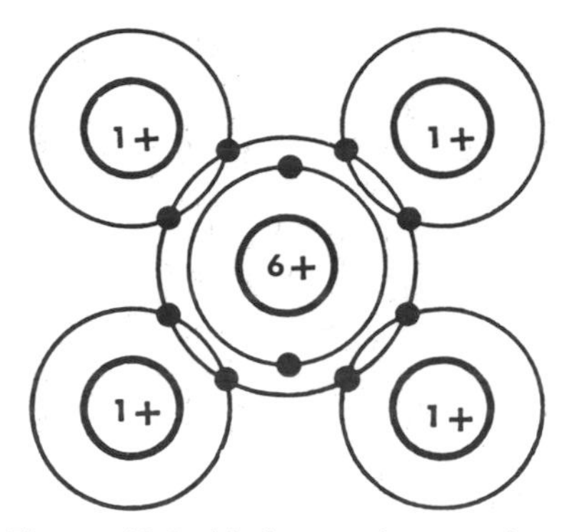

Figure 12-2. Methane; the covalent linkage between the carbon and hydrogen atoms.

Methane

Methane is the simplest organic hydrocarbon, and it contains 1 carbon atom with 4 hydrogen atoms attached to it by covalent bonding (Fig. 12-2). Methane can also be written with straight lines to represent the bond sites (a pair of shared electrons) which is called a single bond (Fig. 12-3).

This method of writing formulas is cumbersome as it requires additional space on paper. It can be condensed and written as CH_4.

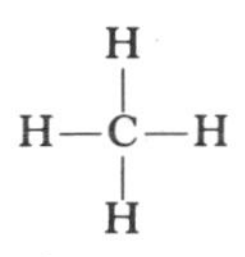

Figure 12-3. Structural formula of methane.

Ethane

The next compound would have 2 carbon atoms bonded together, each sharing 1 electron between themselves, and each having 5 electrons in their outer shells. Each carbon atom can acquire the 3 needed electrons by sharing electrons with 3 hydrogen atoms. This compound is called ethane, and its structure can be written as shown in Figure 12-4. It can therefore be conveniently written as CH_3-CH_3, which indicates that 3 hydrogen atoms are attached to each of the 2 carbon atoms bonded together.

We do not find hydrocarbons, as such, in the living body. However, many of the compounds found in the body are composed of hydrocarbon-like molecules such as fats and oils. The body, which is essentially aqueous, has developed mechanisms to handle, treat, oxidize, and reduce these water insoluble oils and fats. Your knowledge of the hydrocarbons themselves and how they behave will enable you to understand bodily processes.

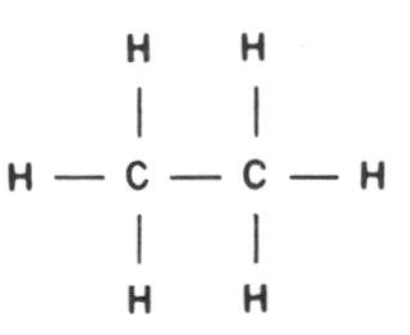

Figure 12-4. Structural formula of ethane.

Propane

When 3 carbon atoms share electrons and join together in a chain, each shares one of its neighboring carbon atom's electrons. The end carbon atoms form 3 covalent bonds with hydrogen atoms and 1 covalent bond with the middle carbon atom, while the middle carbon atom forms 2 covalent bonds with the end carbon atoms and 2 covalent bonds with 2 hydrogen atoms. The complete molecule contains 3 carbon atoms and 8 hydrogen atoms and is called propane. It can be written as C_3H_8, or more explicitly as CH_3—CH_2—CH_3 (Fig. 12-5).

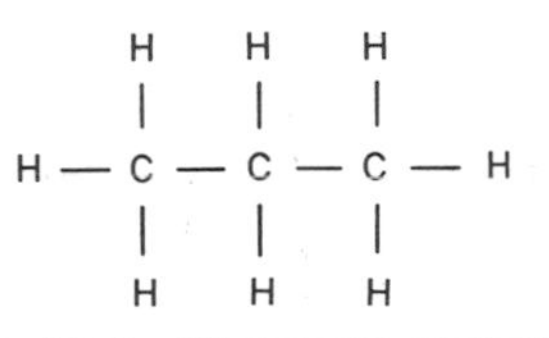

Figure 12-5. Structural formula of propane.

The Family of Hydrocarbons

As more and more carbon atoms join together in a straight chain sharing their electrons between themselves and completing their outer electron shells by sharing electrons with hydrogen atoms, a whole series of compounds known

Table 12-3. Physical state and uses of the saturated hydrocarbons. The physical state of the alkanes changes as the number of carbon atoms in the molecules increase.

Gases C_1—C_4	Heating gases, cigarette lighter fuel.
Volatile liquids C_5—C_6	Petroleum ether & naphtha used as cleaners and industrial solvents.
Gasoline C_6—C_{10}	Liquid mixture of hydrocarbons and their isomers.
Kerosene C_{10}—C_{14}	Oily liquid, used for lighting and heating.
Lubricating — mineral oils C_{14}—C_{20}	Heavy oils used for lubrication and engine fuels, as intestinal lubricants. Nondigestible.
Petroleum jelly, waxes, paraffins C_{22}—C_{30}	Jellies and waxes depending upon the molecular weight. Mixtures of the higher alkane series and their isomers. Used for candles, lubrication, sealing jars of preserved foods, and candles.

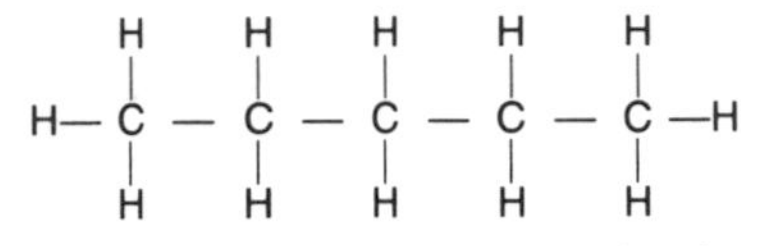

Figure 12-6. Structural formula of n-pentane.

as the alkanes are formed. The common ones that everyone knows about are methane (marsh gas), propane (heating fuel), butane (gas cigarette lighter fuel) and octane (gasoline). The name of the compound indicates how many carbon atoms are in the molecule (see Table 12-3).

The Single Covalent Carbon-to-Carbon Bond

We have already learned that the carbon atom can form chains of carbon atoms by covalent linkage, and we have written the structural formulas and also the more convenient molecular formulas in straight line representations. The structure of pentane, 5 carbon atoms, is shown in Figure 12-6. However, the 4 bonds of the carbon atom are positioned in tetragonal angles and therefore the carbon chains are not straight lines (see Fig. 12-7). Furthermore, the carbon atoms can rotate about that bond and can therefore assume other spatial positions as seen in Figure 12-8.

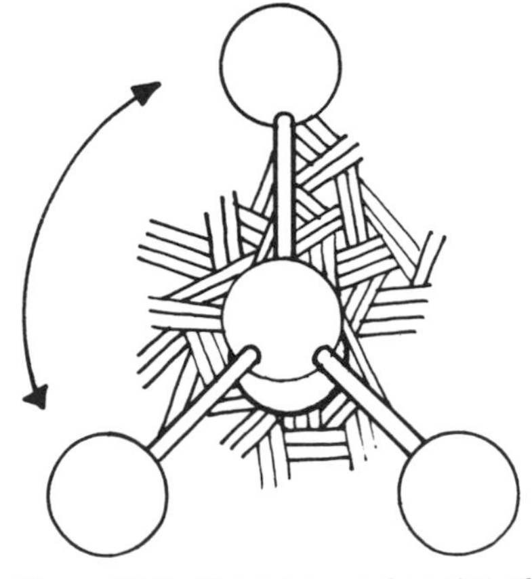

Figure 12-7. The tetragonal angles of the bonds of the carbon atom.

SATURATION IN THE ALKANES

Since the alkanes are composed solely of carbon atoms and hydrogen atoms and contain only single bonds, they are said to be saturated with respect to hydrogen. There is no further tendency to share any more electrons with additional hydrogen atoms, because all of the carbon atoms have already shared their electrons to achieve stability.

Names and Formulas of the Common Alkanes

The names of the common alkanes, their carbon and hydrogen atom content, and their formulas are shown in Table 12-4, and ways to write the condensed structural formulas shown in Figure 12-9.

The Alkyl Groups, Alkyl Radicals

The alkanes are complete and whole molecules, each atom having achieved a stable electronic configuration. However, should 1 hydrogen atom be removed from the molecule, you would have an incomplete molecule, because all of the

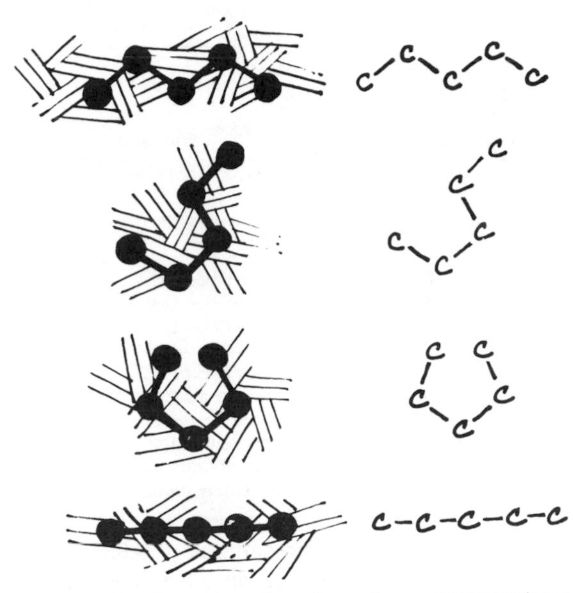

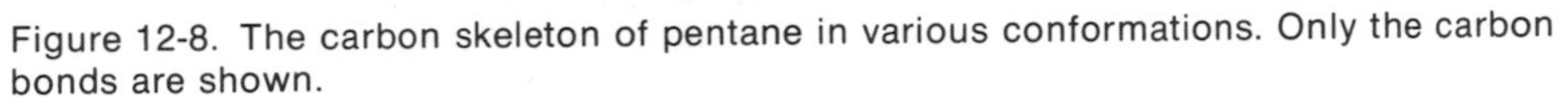

Figure 12-8. The carbon skeleton of pentane in various conformations. Only the carbon bonds are shown.

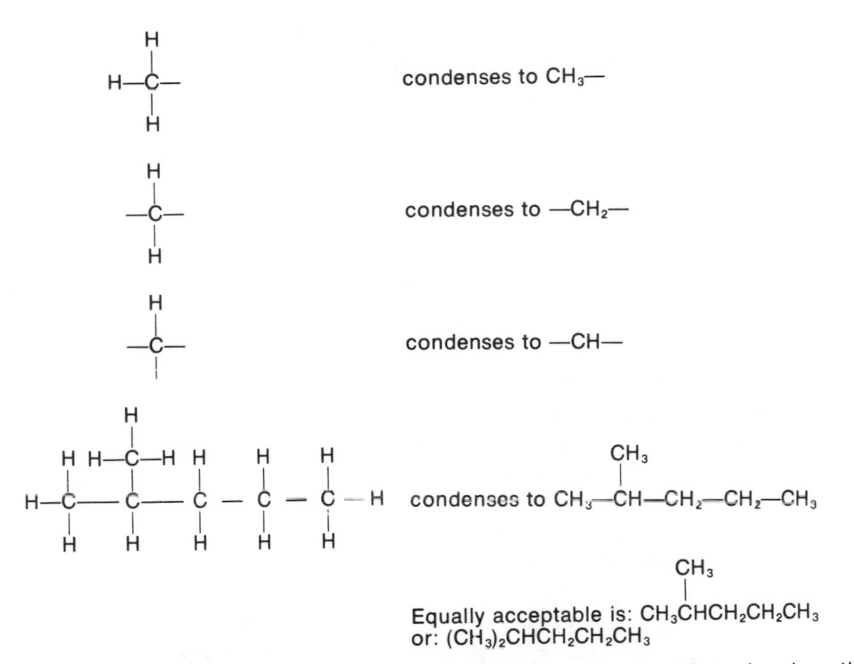

Figure 12-9. How to write a condensed structural formula. Structural formulas show the atom-to-atom sequence but many details are not shown. Structural formulas are condensed according to certain rules and conventions. Notice that the valence of the carbon atom is always 4 and that each connecting bond represents one valence.

Table 12-4. The Alkanes.

Common name		Number of carbon atoms	Number of hydrogen atoms*	Formula‡
Methane	CH_4	1	4	CH_4
Ethane	C_2H_6	2	6	$CH_3—CH_3$
Propane	C_3H_8	3	8	$CH_3—CH_2—CH_3$
Butane	C_4H_{10}	4	10	$CH_3—CH_2—CH_2—CH_3$
Pentane	C_5H_{12}	5	12	$CH_3—CH_2—CH_2—CH_2—CH_3$
Hexane	C_6H_{14}	6	14	$CH_3—CH_2—CH_2—CH_2—CH_2—CH_3$
Heptane	C_7H_{16}	7	16	$CH_3—CH_2—CH_2—CH_2—CH_2—CH_2—CH_3$
Octane	C_8H_{18}	8	18	$CH_3—CH_2—CH_2—CH_2—CH_2—CH_2—CH_2—CH_3$
Nonane	C_9H_{20}	9	20	$CH_3—CH_2—CH_2—CH_2—CH_2—CH_2—CH_2—CH_2—CH_3$
Decane	$C_{10}H_{22}$	10	22	$CH_3—CH_2—CH_2—CH_2—CH_2—CH_2—CH_2—CH_2—CH_2—CH_3$

*The number of hydrogen atoms for the carbon atoms in the molecule can be found by the general formula for the alkanes: C_nH_{2n+2} where n = the number of carbon atoms.
‡Each end carbon atom has 3 hydrogen atoms; each middle carbon atom has 2 hydrogen atoms.

outer shells of every atom would not be satisfied: 1 electron would be lacking. An incomplete molecule like this is called an alkyl group or alkyl radical.

These alkyl groups are extremely important, and their names and formulas follow the names and formulas of their parent alkane compounds exactly. The nomenclature of the radicals is done as follows:

1. Remove the *ane* ending from the alkane: this leaves a stem.
   Examples: methane becomes meth
   ethane becomes eth
   propane becomes prop
   butane becomes but
2. Add *yl* to the stem.
   Example: meth + yl becomes methyl
   eth + yl becomes ethyl
   prop + yl becomes propyl

You should remember that the groups are not complete molecules; they cannot exist by themselves. They must be completed by sharing that electron with another carbon atom or the atom of another element to achieve a stable configuration. The radical or groups are not important in themselves, but are important because they are the basis and the reason why there are thousands of different possible compounds when they do share that electron with another atom.

If a hydrogen atom were added back to the alkyl group, the original alkane would again be formed:

$$\text{ethyl} + \text{hydrogen} \longrightarrow \text{ethane}$$
$$CH_3—CH_2— + H \longrightarrow CH_3—CH_3$$

The names, derivations, carbon and hydrogen atom composition and formulas of the common alkyl groups are shown in Table 12-5.

Table 12-5. Stems, names and formulas of the alkyl groups.

Base alkane	Stem	Name of the group	Carbon atoms	Hydrogen atoms	Formula of group
Methane	Meth	methyl	1	3	$CH_3—$
Ethane	Eth	ethyl	2	5	$CH_3—CH_2—$
Propane	Prop	propyl	3	7	$CH_3—CH_2—CH_2—$
Butane	But	butyl	4	9	$CH_3—CH_2—CH_2—CH_2—$
Pentane	Pent	amyl*	5	11	$CH_3—CH_2—CH_2—CH_2—CH_2—$
Hexane	Hex	hexyl	6	13	$CH_3—CH_2—CH_2—CH_2—CH_2—CH_2—$
Heptane	Hept	heptyl	7	15	$Ch_3—CH_2—CH_2—CH_2—CH_2—CH_2—CH_2—$
Octane	Oct	octyl	8	17	$CH_3—CH_2—CH_2—CH_2—CH_2—CH_2—CH_2—CH_2—$
Nonane	Non	nonyl	9	19	$CH_3—CH_2—CH_2—CH_2—CH_2—CH_2—CH_2—CH_2—CH_2—$
Decane	Dec	decyl	10	21	$CH_3—CH_2—CH_2—CH_2—CH_2—CH_2—CH_2—CH_2—CH_2—CH_2—$

*Amyl is preferred to pentyl because of common usage

Modern internal combustion engines in cars and airplanes require fuel with increased amounts of power from a given weight of fuel (Fig. 12-10). A compound of lead with 4 ethyl radicals attached to it, lead tetraethyl, increases the power obtainable from gasoline by making the gasoline burn more completely without "knocking." This is the ethyl gasoline available at the gas pump, gasoline to which certain amounts of the tetraethyl lead is added.

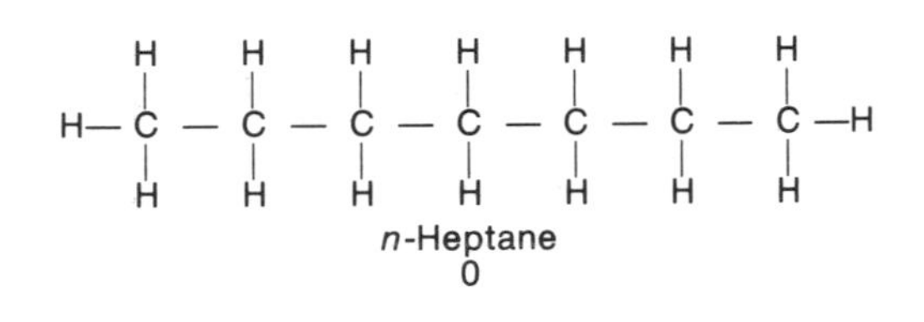

n-Heptane
0

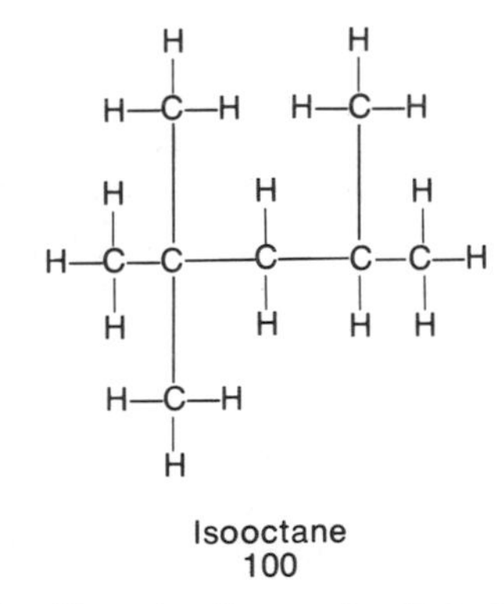

Isooctane
100

Figure 12-10. Gasoline and antiknock ratings. n-heptane as a fuel causes a "knocking." Isooctane is the standard and is rated at 100. The octane number or rating of gasoline is a measure of the performance of the gasoline as compared to isooctane. Various isomers of octane are produced that have different antiknock and power qualities.

Isomers

Isomers are compounds that have the same number of atoms and the same type of atoms, but a different arrangement of the atoms (a different molecular structure). Because the atoms have a different arrangement with respect to another molecule they form different compounds and therefore have different properties, which may be physical, chemical or both.

For example, butane is an alkane having 4 carbon atoms and 10 hydrogen atoms. The carbon atoms are linked together in a straight chain. On the other hand, isobutane also has 4 carbon atoms and 10 hydrogen atoms, but the carbon atoms are linked together in a branched chain. They are isomers, each having the same number and type of atoms in the molecule, but the arrangement of the atoms differs in the 2 molecules. They are isomers of each other and they have different physical properties.

Butane is written: $CH_3—CH_2—CH_2—CH_3$
Isobutane is written: $CH_3—CH(CH_3)—CH_3$

You will recall in the discussion on covalency that the carbon atom was written first and then any atoms that were attached to that carbon atom were written. Therefore, you can see that the methyl group ($CH_3—$) is written following the second carbon atom, and is attached to that second carbon atom by a covalent bond.

As the length of the carbon chain increases, the number of possible arrangements of carbon atom increases tremendously, and the number of possible isomers increases. For example, with a molecule that contains 10 carbon atoms and 22 hydrogen atoms (decane) there are 75 different possible arrangements yielding 75 different compounds, each having different physical properties.

It should be obvious now, that we cannot merely use the term *iso* to indicate specific isomers of decane. We must have a positive method to identify the location of any alkyl group attached to the chain, and to locate groups with respect to each other in the molecule. Therefore, a system of nomenclature was devised to detail the exact spatial location of any group or element in the carbon chain, and it was called the IUPAC system (International Union for Pure and Applied Chemistry).

In this system, the carbon atoms in a chain are sequentially numbered starting with 1 for the end carbon atom, using the *longest straight chain in the molecule* as the basis for determining the root name of the molecule. The root name of the compound is based on the number of carbon atoms in the straight chain.

FOR EXAMPLE: if there are 4 carbon atoms, it would be a butane, 5 carbon atoms, a pentane. If it were 6, 7, 8, 9 or 10 carbon atoms, it would be a hexane, heptane, octane, nonane or decane, respectively.

The basic rule for naming a compound is:

1. Use the longest straight chain as the base
2. Position the substituting group on the carbon chain by prefixing the group with the number of the carbon atom to which it is attached.

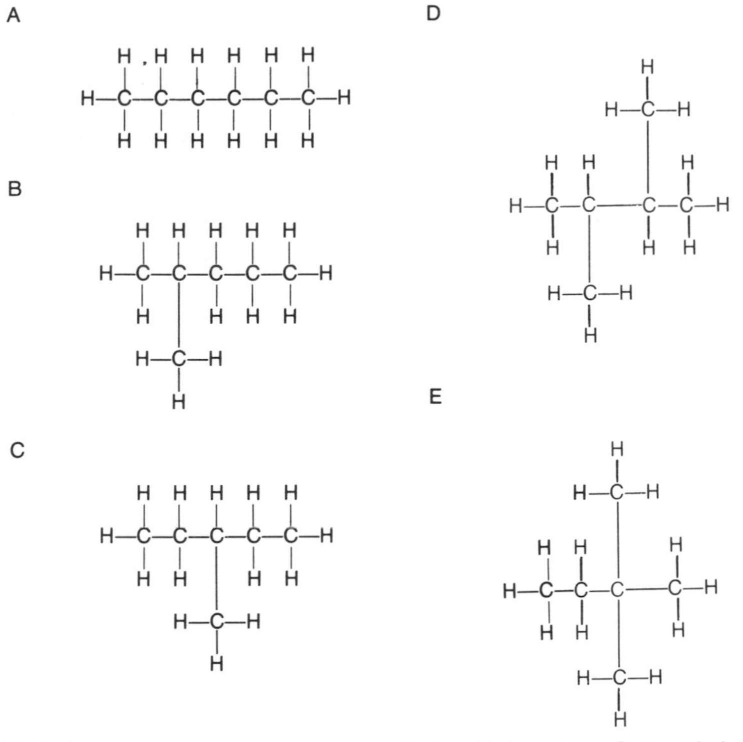

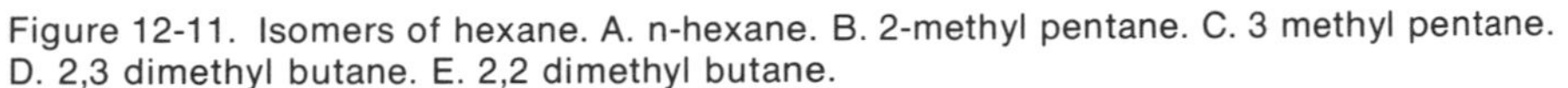

Figure 12-11. Isomers of hexane. A. n-hexane. B. 2-methyl pentane. C. 3 methyl pentane. D. 2,3 dimethyl butane. E. 2,2 dimethyl butane.

Hexane (Fig. 12-11) contains 6 carbon atoms which can be arranged in five different ways to form 5 isomeric hydrocarbons. It is the ability of carbon atoms to link up in an almost limitless number of variations that make the creation of different molecules possible.

Physical Characteristics of the Alkanes

The alkanes are nonpolar molecules and do not dissolve in water. As you already know they are physically gases, oils and waxes, depending upon their molecular weight and are extremely soluble in each other, petroleum solvents and in the chlorinated dry cleaning solvents.

Chemical Reactivity of the Alkanes

The alkanes are saturated with hydrogen atoms. They are relatively inert and do not tend to react with other compounds or elements except under vigorous conditions. One exception, however, is their strong tendency to react with oxygen. They will oxidize or burn; methane ignites almost spontaneously. When they react with oxygen they form carbon dioxide and water.

The Unsaturated Hydrocarbons, Alkenes, Olefins

Earlier in this chapter under covalency you found out that the carbon atom would share more than 1 electron with other carbon atoms. When it shares 2 of its electrons with another carbon atom, which in turn shares 2 of its electrons with it, a new series of hydrocarbons are formed. These are called the *alkenes* or olefins, and because 4 electrons are shared between the 2 carbon atoms, a double bond is formed. When they shared only 2 electrons between them, the linkage was a single bond. Since each carbon atom now has the equivalent of 6 electrons in its outer shell, it needs only 2 more electrons.

The alkenes are named after their alkane counterparts and the simplest compound has 2 carbon atoms. It is called ethene. The next larger molecule would be propene, and following that would be butene.

Cis-Trans Isomerism

The introduction of a double bond allows for the possibility of isomers because of its location in the molecule. All single bonds are identical, but a double bond is different than a single bond. The compounds that have the same arrangement of atoms but a different position for the double bond are different compounds and have different physical and chemical properties, therefore they are isomers.

In the ethane molecule there is a carbon-to-carbon single bond, $CH_3—CH_3$, and the carbon atoms are relatively free to rotate around the axis of the bond. There is no restriction in the rotation. However, in the alkenes, where there is a carbon-to-carbon double bond, the presence of this double bond restricts the free rotation of the carbon atoms. As a result of this restriction, geometric isomers can exist (Fig. 12-12), that differ from each other according to the location of substituents on the same or opposite side of the double bond. They are named *cis* and *trans* according to their spatial relationship to each other, followed by the IUPAC name.

No geometrical isomerism can exist if 2 of the substituents bound to 1 of the carbon atoms are identical. A geometric isomer can only exist if a proper number of different substituents are attached to the 2 double bonded carbon atoms.

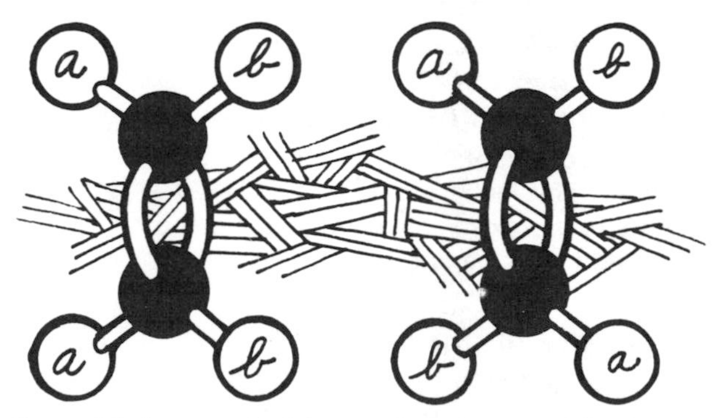

Figure 12-12. Geometric isomers. In this model the isomers differ from each other only in the position of the atoms "a" and "b" that are attached to the lower carbon atom. These positions are reversed. The upper carbon atoms are identical but the lower ones are mirror images of each other. The carbon atoms are prevented from rotating by the presence of the double bond.

Nomenclature with the Double Bond

The ending ene denotes a double bond and its location is designated by the IUPAC system by a number at the beginning of the compound (Table 12-6). If there are two double bonds, the ending is diene and two numbers at the beginning of the name would identify their locations (Table 12-7).

Table 12-6. Names, formulas, and carbon-hydrogen composition of alkenes.

Name	*Carbon atoms*	*Hydrogen atoms*	*Formula‡*	*Structure*
Ethene	2	4	C_2H_4	$CH_2{=}CH_2$
Propene	3	6	C_3H_6	$CH_2{=}CH{-}CH_3$
1-Butene	4	8	C_4H_8	$CH_2{=}CH{-}CH_2{-}CH_3$
2-Butene*	4	8	C_4H_8	$CH_3{-}CH{=}CH{-}CH_3$
1-Pentene	5	10	C_5H_{10}	$CH_2{=}CH{-}CH_2{-}CH_2{-}CH_3$
2-Pentene*	5	10	C_5H_{10}	$CH_3{-}CH{=}CH{-}CH_2{-}CH_3$

*2-butene is an isomer of 1-butene and 2-pentene is an isomer of 1-pentene due to the location of the double bond.
‡Notice that the general formula for the alkene is C_nH_{2n} (where n = the number of carbon atoms).

As the number of carbon atoms in the chain is increased, the number of possible locations for double bonds increases the number of possible isomers. The number of isomers is also increased if alkyl groups are substituted for hydrogen atoms.

Table 12-7. Nomenclature of the dienes.

1,3 butadiene	$CH_2{=}CH{-}CH{=}CH_2$
1,2 butadiene	$CH_2{=}C{=}CH{-}CH_3$

Chemical Reactivity of the Alkenes

The alkenes are very reactive compounds compared to their alkane counterparts because of the presence of the double bond. The carbon atoms connected by the double bond are very susceptible to reactions with elements and compounds. The alkenes are said to be unsaturated because atoms of hydrogen and other elements can add to the carbon atoms linked by the double bond, making it a single bond.

ADDITION OF HYDROGEN TO ALKENES

Alkenes will accept hydrogen atoms to become saturated alkanes, 1 hydrogen molecule (H_2) adding to each double bond:

$$\text{Ethene} + H_2 \longrightarrow \text{ethane}$$
$$CH_2{=}CH_2 + H_2 \longrightarrow CH_3{-}CH_3$$

$$\text{1-butene} + H_2 \longrightarrow \text{butane}$$
$$CH_2{=}CH{-}CH_2{-}CH_3 + H_2 \longrightarrow CH_3{-}CH_2{-}CH_2{-}CH_3$$

A diene, an alkene having 2 double bonds, can accept 1 hydrogen molecule to become an ene and then the ene formed can accept another hydrogen molecule to become the parent alkane. This is selective hydrogenation, and this type of reaction can be found in certain biochemical processes in the body:

$$\text{1-3 butadiene} + H_2 \longrightarrow \text{1-butene}$$
$$CH_2{=}CH{-}CH{=}CH_2 + H_2 \longrightarrow CH_2{=}CH{-}CH_2{-}CH_3$$

$$\text{1-butene} + H_2 \longrightarrow \text{butane}$$
$$CH_2{=}CH{-}CH_2{-}CH_3 + H_2 \longrightarrow CH_3{-}CH_2{-}CH_2{-}CH_3$$

Some of the cooking oils are unsaturated, but hydrogen can be added to the molecules to give a waxy solid cooking fat. You will find the term hydrogenated vegetable oil on some of the commercially sold cooking fats.

OXIDATION OF THE ALKENES

The double bond of the alkene, as you already know, is reactive, and will react with the oxygen in the air. In fact, some of the spoilage of foods, the rancidity of butter, and the biochemical conversion of many compounds in the body to make other compounds are based upon the oxidation the double bond.

Addition Reactions of Alkenes

Generally under the proper reaction conditions elements such as bromine (Br_2), chlorine (Cl_2), compounds such as HCl and H_2O and other molecules will add to the double bond to form a saturated molecule (the double bond becoming a single bond) with the reactants linking to the carbon atoms. This is a modification of the addition of the hydrogen molecule reaction.

Addition of Cl_2:
$$CH_2{=}CH_2 + Cl_2 \longrightarrow CH_2Cl{-}CH_2Cl$$

Addition of HCl:
$$CH_2{=}CH_2 + HCl \longrightarrow CH_3{-}CH_2Cl$$

Addition of H_2O (H-OH):
$$CH_2{=}CH_2 + HOH \longrightarrow CH_3{-}CH_2OH$$

Alkenes can be identified from the alkanes by the addition reaction of bromine, which is a reddish colored element and very reactive towards alkenes. If bromine is added to separate containers, one containing an alkane and the other an alkene, the one containing the alkene will decolorize as the bromine reacts. The alkane will remain colored:

$$\text{Alkane} + Br_2 \longrightarrow \text{no reaction}$$
$$\text{Alkene} + Br_2 \longrightarrow \text{decolorization of the } Br_2$$
$$CH_2{=}CH_2 + Br_2 \longrightarrow CH_2Br{-}CH_2Br$$

Polymerization of the Alkenes

Modern plastics have names such as polyethylene and polypropylene. They are extremely large molecules (polymers) that are made up from recurring small molecules (monomers). The poly plastics, such as polyethylene, are made by subjecting alkenes to extreme conditions of temperature and pressure in the presence of catalysts. Under these conditions, the double bond of the alkene opens up and the open bond connects the monomer molecules together to form gigantic molecular chains that are the plastics with molecular weights that range in the hundreds of thousands.

$$\begin{array}{cccc} \text{Ethene +} & \text{ethene +} & \text{ethene +} & \text{ethene} \\ CH_2{=}CH_2 & CH_2{=}CH_2 & CH_2{=}CH_2 & CH_2{=}CH_2 \end{array}$$

Reactive form: This is an unstable form of the molecule:

$$\{-CH_2-CH_2-\}\ \{-CH_2-CH_2-\}\ \{-CH_2-CH_2-\}\ \{-CH_2-CH_2-\}$$

The single bonds of each molecule join their neighboring molecule to form the long molecular chain:

$$-CH_2-CH_2-CH_2-CH_2-CH_2-CH_2-CH_2-CH_2-\ldots.$$

THE ALKYNES

The carbon atom will share 3 of its electrons with another carbon atom, each then having the equivalent of 7 electrons in its outer shell, needing only 1 more electron. It can get this by sharing 1 electron with a hydrogen atom to form a new series of compound called alkynes (Table 12-8).

The Triple Bond

Since each carbon atom shares 3 electrons (there are now 3 pairs) we have a triple bond which is represented by "≡." The names of the alkynes follow their alkane counterparts, however the endings of the names are yne instead of ane. The location of the triple bond follows the same system as used in the alkenes, the position of the triple bond being indicated by a number that precedes the name.

Compounds containing triple bonds are not easily digested by the body enzymes, and the structure of the oral birth control pill is protected by a triple bond.

Table 12-8. Nomenclature of the alkynes.

Propyne	$CH{\equiv}C{=}CH_3$
1-butyne	$CH{\equiv}C-CH_2-CH_3$
2-butyne	$CH_3-C{\equiv}C-CH_3$
1-pentyne	$CH{\equiv}C-CH_2-CH_2-CH_3$
2-pentyne	$CH_3-C{\equiv}C-CH_2-CH_3$

The simplest alkyne is ethyne, which has the formula, $CH{\equiv}CH$. It is more commonly known as acetylene, a welding gas, which reacts vigorously with air (oxygen) to give intense heat.

Chemical Reactivity of the Alkynes

The alkynes undergo reactions with the same compounds as the alkenes, but they will form different compounds because of the triple bond. They add 1 molecule of hydrogen to form an alkene:

$$CH{\equiv}CH + H_2 \longrightarrow CH_2{=}CH_2 \text{ (ethene)}$$

Addition of another molecule of hydrogen to ethene will form ethane, as you have already seen under the chemical reactivity of the alkenes.

There is, however, one reaction of the alkynes that differentiates them from their alkene and alkane counterparts. This is their reaction with certain metals, because of the extreme reactivity that the triple bond gives the carbon atom to which it is attached. Silver and mercury derivatives of acetylene:

$$Ag{-}C{\equiv}C{-}Ag \quad \text{and} \quad Hg{-}C{\equiv}C{-}Hg$$

These derivatives are very explosive substances. They are so sensitive to shock that the slightest vibrations will cause them to explode. They are called the acetylides and should be handled only by experienced personnel.

THE CYCLIC HYDROCARBONS

Thousands upon thousands of possible carbon-to-carbon configurations (straight chain, branched chains, double bonds, and triple bonds) demonstrate the reason why over one million organic carbon compounds have been made to date. However, another variation in the possible molecular structure increases the possibilities many times, and that variation is the formation of carbon rings or cyclic compounds.

A chain of 3 carbon atoms can form a ring structure or a cyclic compound (Fig. 12-13), the simplest cyclic hydrocarbon being cyclopropane. This compound is saturated because it contains only single bonds, but cyclic hydrocarbons are reactive as compared to straight chain saturated hydrocarbons. The cyclic ring structure is found in many very important compounds in the body and also present in many potent drugs.

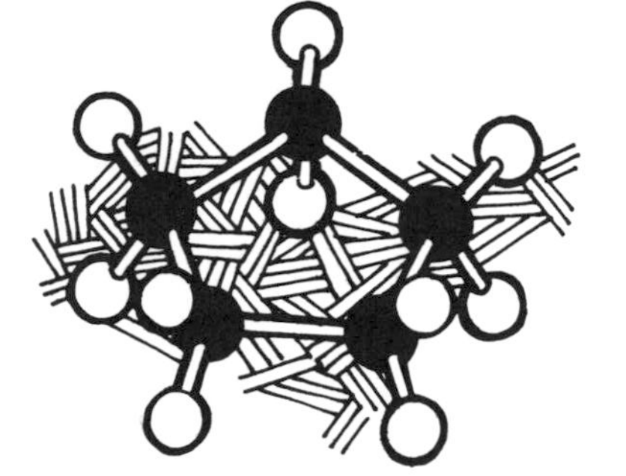

Figure 12-13. Cyclopentane. Ball and stick model of cyclopentane shows 5 carbon atoms linked in a closed ring, each having 2 hydrogen atoms attached to each carbon atom.

Nomenclature of the Cyclic Hydrocarbons

The cyclohydrocarbons are named by counting the number of carbon atoms in the ring, and using the alkane nomenclature prefixed by cyclo. Thus we can have cyclobutane, cyclopentane, and cyclohexane compounds with the location of the groups or elements designated by the number of the carbon atom (Fig. 12-14).

Chemical Reactivity of the Cycloalkanes

They are extremely reactive, the smaller ring structures being more reactive than the larger ones because the bonds are under greater strain. Under proper

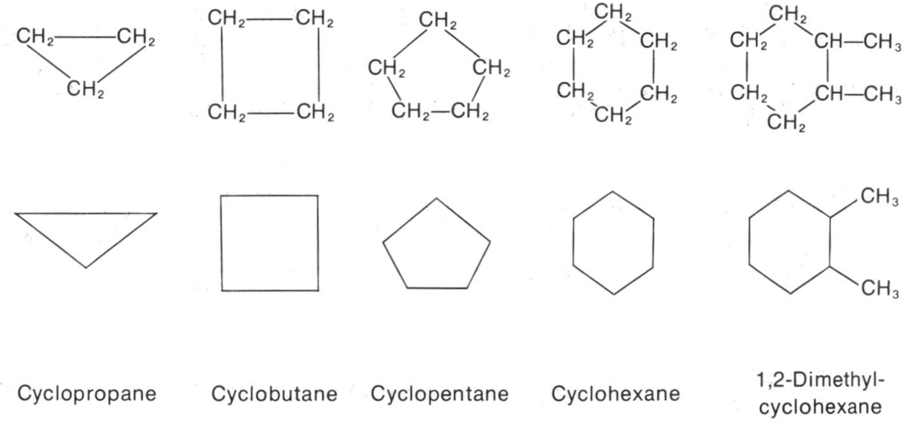

Figure 12-14. Cycloalkanes.

conditions, the ring will break open and substances can react with the active form.

The most important compound is cyclopropane, which is used in both minor and major surgery as an anesthetic because it induces anesthesia very quickly and it has a short recovery time. When compared to nitrous oxide, cyclopropane effects muscular relaxation much more quickly. However it is dangerous to use because it is extremely explosive when mixed with air or oxygen.

SUMMARY

Carbon atoms can covalently bond to each other to form straight chains or closed rings to form very large and complex molecules. When it shares its other valence bonds solely with hydrogen, hydrocarbons are formed. If these compounds contain only single covalent bonds, they are saturated compounds. When they contain double or triple bonds, they are said to be unsaturated compounds with respect to hydrogen. The removal of one hydrogen atom from a complete hydrocarbon molecule leaves an organic radical which cannot exist by itself, but leads to the formation of many different compounds when they bond with other elements. Isomers are compounds that have the same number and kind of atoms, but the atoms are arranged differently, forming different compounds. Alkenes contain double bonds, and the double bond can lead to *cis-trans* isomerism. The alkenes are chemically reactive, and can add H_2, Cl_2, HCl or H_2O, and can be oxidized and polymerized. The presence of a triple bond yields an alkyne, which is very reactive. When carbon atoms form rings or closed chains, cyclohydrocarbons are formed, cyclopropane being used as an anesthetic.

EXERCISE

1. Why are there so many more organic compounds than inorganic compounds?
2. List the differences between organic and inorganic compounds.
3. What is covalency and how does the carbon atom achieve stability?
4. Write the molecular, structural and condensed formulas for the first five aliphatic hydrocarbons.
5. Why is it important to know the chemistry of the hydrocarbons?
6. What is meant by saturation in a hydrocarbon molecule?
7. What is an alkyl group and why is it important?
8. Write the condensed formulas for the first five alkyl groups.
9. What is an isomer?
10. Write the structural and condensed formulas for butane and isobutane.
11. Name the following compounds (use the IUPAC system):
    a. $CH_3—CH_2—CH_2—CH_3$
    b. $CH_3—CH(CH_3)—CH_2—CH_3$
    c. $CH_3—CH_2—C(CH_3)—CH(CH_3)—CH_3$
    d. $CH_3—CH_2—C(CH_3)_2—CH_2—CH_3$
12. What chemical reaction are the alkanes most susceptible to?
13. What is an alkene and what characteristic bond does it contain?
14. Name the following alkenes:
    a. $CH_3—CH{=}CH—CH_3$
    b. $CH_2{=}CH—CH{=}CH_2$
    c. $CH_3—CH_2—CH{=}CH—CH_2—CH_3$
    d. $CH_2{=}CH—CH{=}CH—CH_3$
15. Define *cis-trans* isomerism and state why it exists.
16. Complete the following reactions (assume reaction conditions):
    a. $CH_2{=}CH_2 + HCl \longrightarrow$
    b. $CH_2{=}CH_2 + Br_2 \longrightarrow$
    c. $CH_2{=}CH_2 + H_2O \longrightarrow$
    d. $CH_2{=}CH_2 + H_2 \longrightarrow$
17. Describe how the alkenes polymerize.
18. What is the characteristic bond in an alkyne, and what is the importance of that bond in certain medications?
19. What compounds do the alkynes resemble in their chemical reactivity?
20. Draw the structural formulas for the first five cyclohydrocarbons.
21. What is the importance of cyclopropane to health care personnel?

OBJECTIVES

When you have finished this chapter you will be able to:

1. Write the common names, the IUPAC names and the formulas for the alkyl alcohols.
2. Give the uses for methyl alcohol, ethyl alcohol, and propyl alcohol in the health care field.
3. Characterize the effect of various amounts of ethyl alcohol in the body.
4. State why denatured ethyl alcohol cannot be taken internally.
5. Identify polyhydric alcohols and what they are used for.
6. Write the reactions and products obtained in:
   a. Oxidation of primary, secondary and tertiary alcohols
   b. Dehydration of alcohols
   c. Reaction of alcohols with sulfuric acid
7. Explain molecular isomers and be able to identify them.
8. Describe the functional group of ethers, and be able to name alkyl ethers.
9. Discuss how diethyl ether is used as an anesthetic and how it can be made.
10. Draw the formula for the functional group of aldehydes and name the commonly encountered ones.
11. Show how the Benedict, Fehling, and Tollen tests show the presence of the aldehyde group, and its importance in blood sugar analysis.
12. Give uses for formaldehyde, acetaldehyde, and urotropine.
13. Draw the ketone functional group and name the common alkyl ketones.
14. Show by reactions how ketones are formed and the importance of acetone diagnosis in diabetes mellitus.
15. Describe the carboxyl group and given common usage names and IUPAC names for commonly encountered acids.

THE OXYGEN DERIVATIVES

16. Explain by equations how organic acids are neutralized.
17. Name polycarboxylic acids and write the formulas for those that are important to health care personnel.
18. Write the formula for an alpha and beta hydroxy acid and an alpha and beta keto acid.
19. Describe the reaction between an acid and an alcohol to give an ester.
20. List where esters are found in nature and what they can be hydrolyzed to yield.
21. Name any ester, knowing the name of the alcohol and the acid.
22. Distinguish between hydrolysis and saponification.
23. Describe the epoxide structure and name an important epoxide.

FUNCTIONAL GROUPS

When oxygen atoms are substituted for carbon or hydrogen atoms or are inserted into a hydrocarbon molecule we obtain a series of compounds called derivatives. These derivatives differ from the parent hydrocarbon molecules in their physical and chemical properties because of the atoms of oxygen now in the molecule, which gives rise to functional groups. These functional groups, or radicals, are a combination of atoms that have special properties characteristic of that particular grouping. A different grouping of atoms results in a different functional group with characteristics partially or totally unlike other functional groups. These functional groups cannot exist by themselves, but must be attached to another incomplete grouping, such as an alkyl radical, or another functional group to be stable. The physical characteristics of the complete molecule (the compound formed by this union) depend upon the number of carbon atoms, just as the characteristics of the hydrocarbon varied as the number of carbon atoms.

THE ALCOHOLS

An arrangement of atoms in a functional group such as a $-\overset{|}{\underset{|}{C}}-OH$ group, or simply an $-OH$ group attached to a carbon atom, is an alcohol; the $-OH$ group being called a hydroxyl group. This $-OH$ group is bound by covalent linkage; it does not ionize. This covalent $-OH$ group can be attached to any alkyl radical to give the corresponding alcohol. If the complete molecule contains 1 $-OH$ group, it is called a monohydric alcohol, those with 2 $-OH$ groups are dihydric alcohols, and those with 3 $-OH$ groups are trihydric alcohols.

Naming of Alcohols

The older and more commonly used method of naming the simple alcohols involved the coupling of the name of the alkyl radical plus the word alcohol. Thus the methyl radical, CH_3-, plus the $-OH$ group is methyl alcohol:

$$CH_3-OH$$

ethyl radical, CH_3-CH_2-, plus the $-OH$ group is ethyl alcohol:

$$CH_3-CH_2-OH$$

propyl alcohol, $CH_3-CH_2-CH_2-$, plus the $-OH$ group is propyl alcohol:

$$CH_3-CH_2-CH_2-OH$$

isopropyl radical, $(CH_3)_2CH-$, plus the $-OH$ group is isopropyl alcohol:

$$(CH_3)_2CH-OH$$

This method of naming alcohols was satisfactory for the simpler alkyl radicals, but as the number of carbon atoms increased and the location of the $-OH$ group varied, complications arose. The older method incorporated nomenclature that identified the kind of carbon atom to which the $-OH$ group was

Table 13-1. Primary alcohols.

Methyl alcohol	CH_3-OH
Ethyl alcohol	$CH_3-\underset{*}{C}H_2-OH$
Propyl alcohol	$CH_3-CH_2-\underset{*}{C}H_2-OH$
Butyl alcohol	$CH_3-CH_2-CH_2-\underset{*}{C}H_2-OH$
Hexyl alcohol	$CH_3-CH_2-CH_2-CH_2-CH_2-\underset{*}{C}H_2-OH$

*The asterisk indicates that the carbon atom to which the $-OH$ group is attached is bonded to only 1 other carbon atom.

Table 13-2. Secondary alcohols.

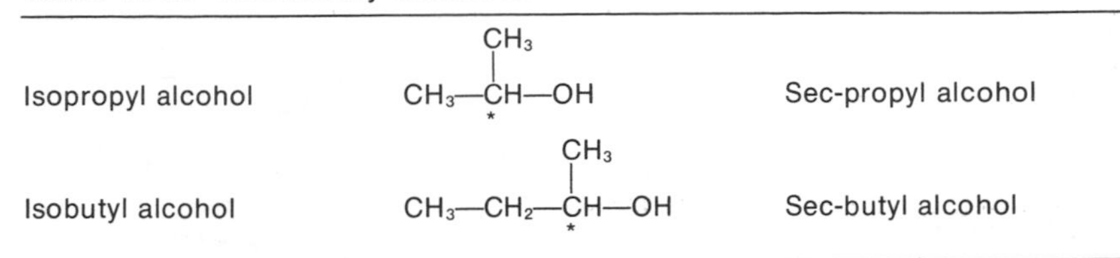

Isopropyl alcohol	$CH_3—\overset{CH_3}{\underset{*}{CH}}—OH$	Sec-propyl alcohol
Isobutyl alcohol	$CH_3—CH_2—\overset{CH_3}{\underset{*}{CH}}—OH$	Sec-butyl alcohol

*The asterisk indicates that the carbon atom to which the —OH group is attached is bonded to 2 other carbon atoms.

attached, and the terms primary, secondary and tertiary were added to the names of the alcohols.

Primary alcohols are alcohols in which the —OH group is attached to a carbon atom, which in turn is covalently bound to only 1 carbon atom (Table 13-1). Secondary alcohols are alcohols in which the —OH group is attached to a carbon atom, which is in turn covalently bound to *2* carbon atoms, abbreviated "sec-" (Table 13-2). Tertiary alcohols are alcohols in which the —OH group is attached to a carbon atom, which is in turn covalently bound to *3* carbon atoms, abbreviated "tert-" (Table 13-3).

Table 13-3. Tertiary alcohols.

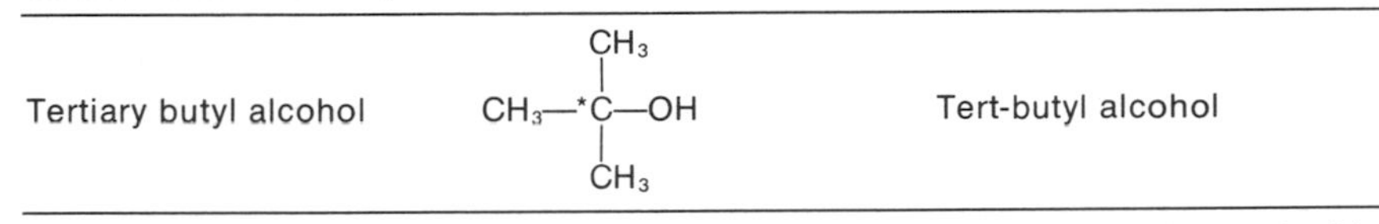

Tertiary butyl alcohol	$CH_3—{}^{*}\overset{CH_3}{\underset{CH_3}{C}}—OH$	Tert-butyl alcohol

*The asterisk indicates that the carbon atom to which the —OH group is attached is bonded to 3 other carbon atoms.

However, the problem with this method of naming alcohols is the specific identification of the various isomers, because of the branched chains and the location of the hydroxyl group that cannot be specified in the name. As a result, the IUPAC method was developed, which specifically detailed the structure of the molecule and eliminated the ambiguous nomenclature and structure of the molecules.

The IUPAC System of Naming Alcohols

1. The name of the alkyl radical plus the ending ol is the name of the alcohol.
2. The location of the —OH group is designated by the number of the carbon atom before the name, counting from the —OH group.
3. If there are 2 —OH groups in the same molecule, the *complete name of the alkane* is used instead of the alkyl radical, and the ending diol added to the name of the alkane. The locations of the 2 hydroxyl groups are designated, as above, by numbers preceding the name. With 3 —OH groups in the molecule, the name of the alkane is used and the ending triol added to the name of the alkane. The locations of the hydroxyl groups are designated by numbers of the carbon atoms as usual (Table 13-4).

Table 13-4. IUPAC nomenclature of alcohols.

Common name	*Formula*	IUPAC *name**
Methyl alcohol	CH_3—OH	Methanol
Ethyl alcohol	CH_3—CH_2—OH	Ethanol
n-propyl alcohol	CH_3—CH_2—CH_2—OH	1-propanol
Isopropyl alcohol	CH_3—CH(—CH_3)—OH	2-propanol
n-butyl alcohol	CH_3—CH_2—CH_2—CH_2—OH	1-butanol
Isobutyl alcohol	CH_3—CH(CH_3)—CH_2OH	2-methyl-1-propanol
Sec-butyl alcohol	CH_3—CH(OH)—CH_2—CH_3	2-butanol
Tert-butyl alcohol	CH_3—C(CH_3)(CH_3)—OH	2-methyl-2-propanol
Ethylene glycol	CH_2(OH)—CH_2(OH)	1,2, ethanediol
Glycerol	CH_2(OH)—CH(OH)—CH_2(OH)	1,2,3 propanetriol

*Using longest straight chain as base name.

Methyl Alcohol

Methyl alcohol is commonly called wood alcohol. It is poisonous and cannot be made nonpoisonous. Any substance that contains methyl alcohol, even in small amounts, is poisonous and cannot be taken internally. To prevent mistakes and avoid any possible misunderstanding it is always called by its full name, methyl alcohol (methanol or wood alcohol), and *never* by the name alcohol.

It is poisonous to living cells, because it oxidizes to formaldehyde, which is used to embalm corpses and to preserve laboratory specimens. Methyl alcohol affects the optic nerve causing blindness. During prohibition people drank alcoholic beverages that contained some methyl alcohol and became temporarily blinded, hence the expression "blind drunk."

Ethyl Alcohol

Ethanol (ethyl alcohol) is made by the fermentation of carbohydrates, grains, fruits, and sugar containing compounds. In the presence of enzymes from yeast, carbohydrates or sugars undergo a chemical change to form ethyl alcohol and carbon dioxide.

$$C_6H_{12}O_6 \xrightarrow[\text{(zymase)}]{\text{enzyme}} 2\ C_2H_5OH + 2\ CO_2$$

It can also be made from ethene (ethylene) by a series of reactions that can be illustrated by the simplified reaction with water in the presence of a catalyst:

$$CH_2{=}CH_2 + H{-}OH \xrightarrow{\text{catalyst}} CH_3{-}CH_2OH$$

Ethyl alcohol is a clear, colorless, volatile liquid that has a characteristic odor. Ordinary ethyl alcohol, 95 percent laboratory ethyl alcohol, is an excellent solvent for many substances and is used in the preparation of medicines (tinctures), flavoring extracts, and as a disinfectant. When it is mixed with water to form a 50 to 70 percent solution and used as a disinfectant it destroys organisms because it coagulates protoplasm. In the hospital, it is used as a solvent for medications and as a rubbing compound to cleanse the skin and to lower a patient's temperature.

When it is taken internally, it is absorbed without digestion, releasing 7,000 calories for each gram. It is therefore a readily available source of energy and can be used to overcome shock or collapse. When used in an alcoholic beverage, it gives a "lift" to the individual; however, when large quantities are consumed it tends to slow metabolic processes and to depress the central nervous system. This results in lack of coordination, mental confusion, drowsiness, lowering of the normal inhibitions, and finally stupor. It is for these reasons that people who consume too much alcohol from any source should not drive automobiles. The individual may feel relaxed but does not realize that his sense of judgment, sense of timing, and muscular coordination have been seriously impaired. Addiction to alcohol, alcoholism, is a serious illness affecting millions of people.

Denatured ethyl alcohol is undrinkable because the government requires the addition of methyl alcohol, or other nonremovable poisonous substance, be added to the ethyl alcohol to make it unfit for internal human consumption. This has been done to require those who wish to drink alcoholic beverages to purchase the highly taxed ones from the liquor store, and not from the hardware or drugstore selling untaxed denatured ethyl alcohol. All denatured alcoholic preparations such as bay rum, hair tonic, external medical preparations, household alcohol-containing cleaning products, and the various cosmetic preparations are poisonous. Commonly believed myths and superstitions such as passing denatured alcohol over charcoal, distilling it, and even saying incantations over it cannot make it drinkable. Drinking substances that contain methyl alcohol can cause blindness, paralysis, and death.

Isopropyl Alcohol

Isopropyl alcohol (sec-propyl alcohol) is commonly substituted for ethyl alcohol as a rubbing compound and as an astringent, an agent that checks secretions of mucous membranes and contracts and hardens tissues, limiting secretions of glands. It is a toxic compound and cannot be taken internally. Therefore, it should never be consumed as a beverage substitute for ethyl alcohol.

Polyhydric Alcohols

The two more important polyhydric simple alcohols are ethylene glycol (1,2, ethanediol) and glycerol (glycerin, 1,2,3, propanetriol). Polyhydric alcohols have the —OH groups attached to different carbon atoms, because when 2 —OH groups are attached to the same carbon atom, the compound is unstable and will decompose.

Ethylene Glycol (1,2,-ethanediol)

$$\begin{array}{cc} CH_2 — & CH_2 \\ | & | \\ OH & OH \end{array}$$

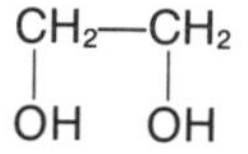

Ethylene glycol, with 2 —OH groups, is completely soluble in water. It is extremely nonvolatile with a very high boiling point. It is used as an antifreeze, a solvent in the paint and varnish industry, and in skin preparations because of its tendency to moisten the skin. Even though it is a derivative of ethyl alcohol, it is extremely poisonous and cannot be taken internally.

Glycerol (1,2,3-trihydroxy propane)

$$CH_2(OH)—CH(OH)—CH_2OH \quad \text{or} \quad \begin{array}{l} CH_2—OH \\ | \\ CH—OH \\ | \\ CH_2—OH \end{array}$$

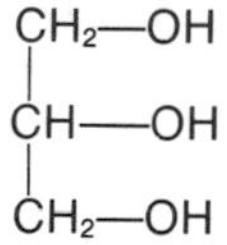

Glycerol (glycerine) is a colorless, syrupy liquid with a sweetish taste. It is nontoxic, hygroscopic (takes up water) and is used in skin preparations for softening the skin, in cosmetics and in food preparations. Glycerol is a part of the molecule of the fats of living organisms.

Glyceryl trinitrate is made by reacting glycerol with nitric acid in the presence of sulfuric acid. It is a raw material used in the manufacture of dynamite and trinitroglycerine (glycerol trinitrate), which is a heart stimulant commonly called nitroglycerine (Fig. 13-1).

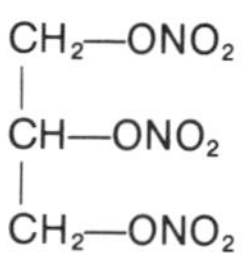

$$\begin{array}{l} CH_2—ONO_2 \\ | \\ CH—ONO_2 \\ | \\ CH_2—ONO_2 \end{array}$$

Figure 13-1. Glyceryl trinitrate (nitroglycerin).

Chemical Reactions

The products of the oxidation of alkyl alcohols vary as to whether the alcohol is primary, secondary or tertiary (Fig. 13-2).

Primary alcohols in the presence of an oxidizing agent are oxidized first to an aldehyde, which has the functional group $—\overset{\displaystyle O}{\overset{\|}{C}}—H$ (the condensed formula is written —CHO and not —COH, which represents the functional group of an alcohol). The secondary stage of the oxidation (Fig. 13-3) gives a carboxylic

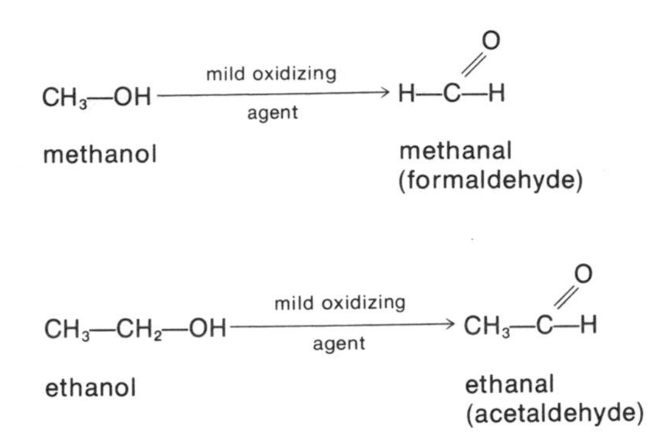

Figure 13-2. Oxidation of a primary alcohol: first stage.

acid, which has the functional group —C(=O)—OH (the condensed formula is written —COOH, which represents the functional group of the carboxylic acid). The third and final stage of the oxidation of a primary alcohol is the oxidation of the acid to CO_2 and H_2O (Fig. 13-4).

Secondary alcohols in the presence of an oxidizing agent are oxidized to ketone (Fig. 13-5) and have the functional group —C(=O)— (the condensed formula is written —CO—, which represents the functional group of the ketone). Carbon atoms that have a hydroxyl group attached to them are more easily oxidized than other carbon atoms that have only hydrogen atoms attached to them. Therefore, a dihydroxy molecule is formed first (2 hydroxyl groups on the same carbon atom). This structure is unstable, splitting out water and forming a ketone. This is one of the biochemical reactions of the body by which ketones (keto bodies) are formed. Ketones resist simple oxidation. However, drastic oxidation procedures break up the molecule into CO_2 and H_2O.

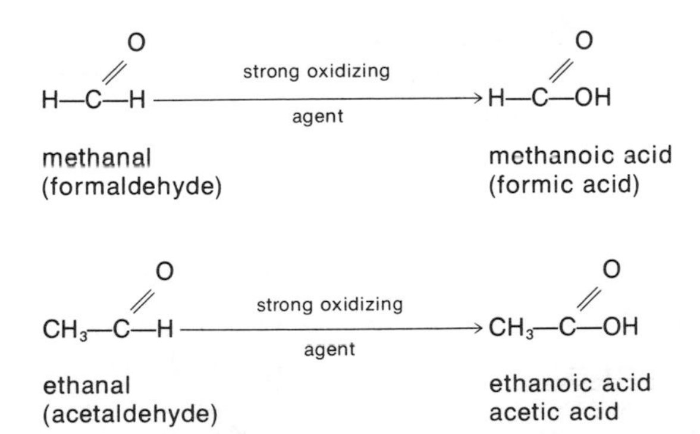

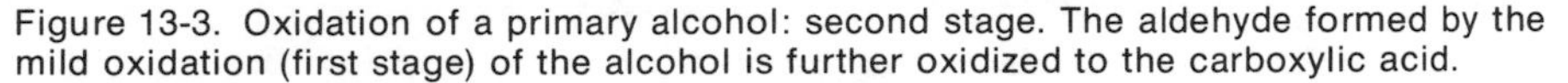
Figure 13-3. Oxidation of a primary alcohol: second stage. The aldehyde formed by the mild oxidation (first stage) of the alcohol is further oxidized to the carboxylic acid.

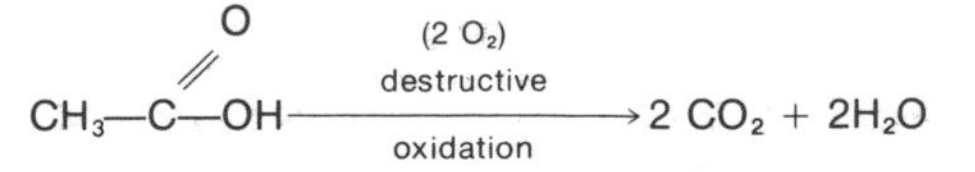

Figure 13-4. Oxidation of a primary alcohol: final stage. Complete oxidation of an alcohol yields CO_2 and H_2O; the carboxylic acid formed in the second stage is oxidized.

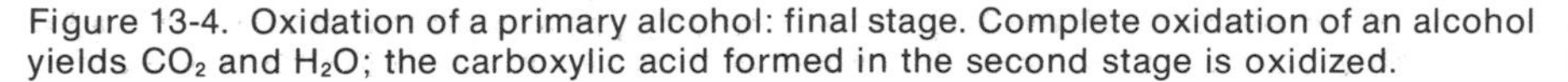

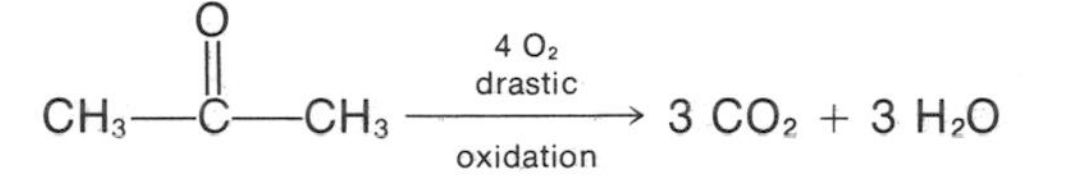

Tertiary alcohols resist oxidation under normal conditions. However, under strong oxidizing conditions, they are destroyed and yield CO_2 and H_2O.

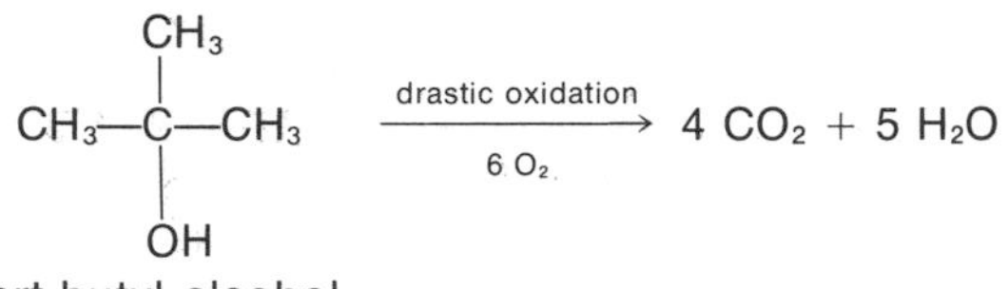

DEHYDRATION OF ALCOHOL

When primary or secondary alcohols are heated with dehydrating agents, such as sulfuric acid, they form alkenes, eliminating a molecule of water. Tertiary alcohols do not dehydrate to form the corresponding alkene (Fig. 13-6).

REACTION OF ALCOHOLS WITH ACTIVE METALS

Alcohols react with active metals such as metallic sodium and potassium to form salts, the metal replacing the H atom of the —OH group (R is an alkyl radical):

$$2\ R\text{—}OH + 2\ Na \longrightarrow 2\ RONa + H_2$$

REACTION OF ALCOHOLS WITH ORGANIC ACIDS

Alcohols react with organic acids (which have the functional group —COOH) to form a series of compounds called esters.

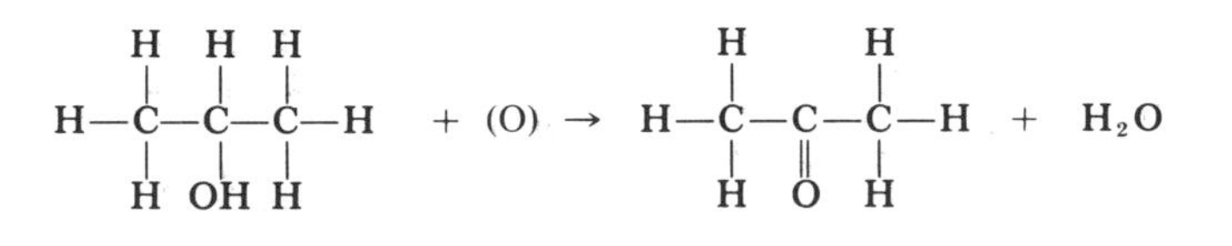

Figure 13-5. Oxidation of a secondary alcohol to yield ketones.

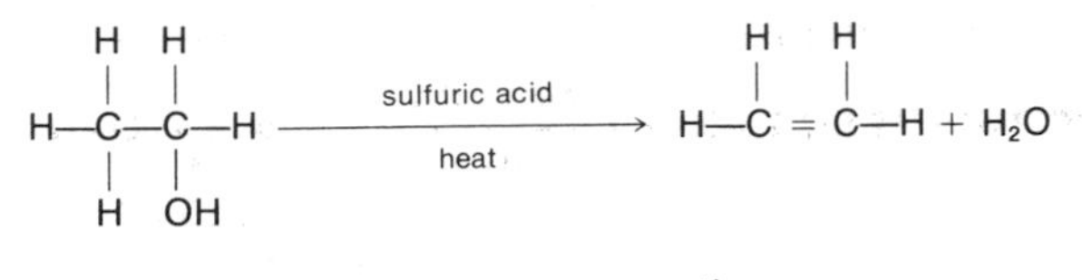

ethanol ethene

Dehydration of a primary alcohol to form the corresponding alkene.

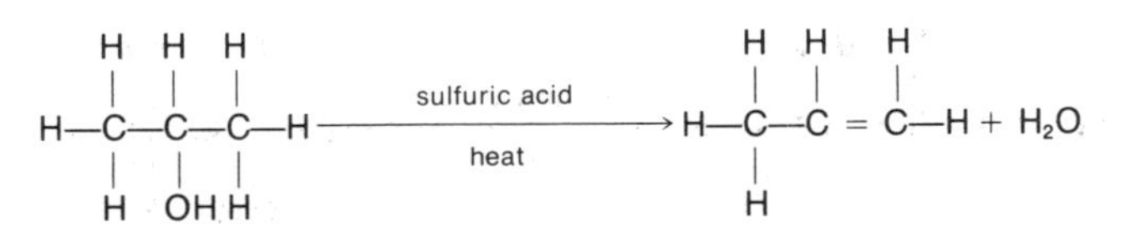

2-propanol
(isopropyl alcohol

propene

Dehydration of a secondary alcohol to form the corresponding alkene.

Figure 13-6. Dehydration of alcohols to form alkenes.

REACTION OF ALCOHOLS WITH SULFURIC ACID

Alcohols react with inorganic acids to form compounds called esters. With sulfuric acid, sulfate esters are formed. The alcohol functional group reacts with sulfuric acid to form the alkyl sulfate:

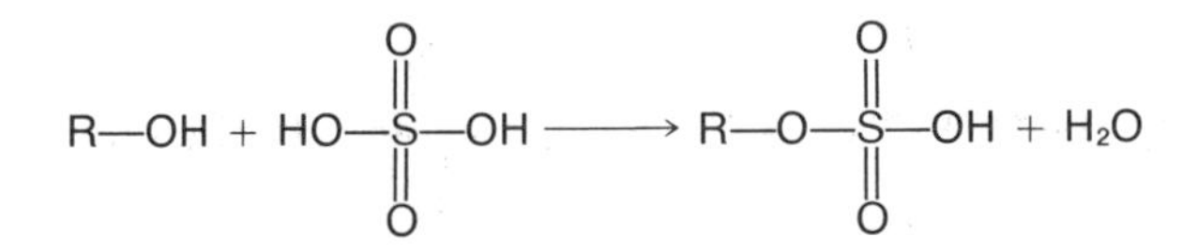

When the alcohol has 12 or more carbon atoms in the alkyl radical, compounds called sulfuric acid esters are formed, which form the soapless detergents found in the supermarkets, when neutralized by NaOH. They are superior to the ordinary soaps, because they lather even in hard water and do not leave the scum and rings that the ordinary soaps leave in hard water.

$$CH_3—CH_2—CH_2—CH_2—CH_2—CH_2—CH_2—CH_2—CH_2—CH_2—CH_2—CH_2—O—\overset{O}{\underset{O}{\overset{\|}{\underset{\|}{S}}}}—OH$$

The sulfate esters can react further and undergo a variety of reactions depending on the substances present and the conditions of the reaction. Normally these reactions are competitive and a mixture of products result.

REACTIONS OF ALKYL SULFATES

The alkyl sulfates regenerate the alcohol by reaction with water:

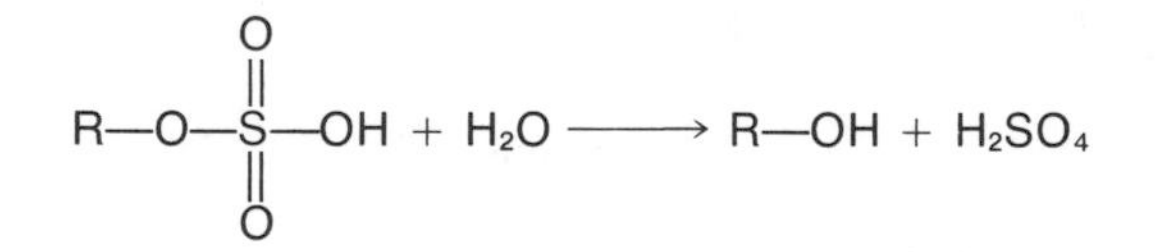

They also form compounds called ethers by reacting with another alcohol:

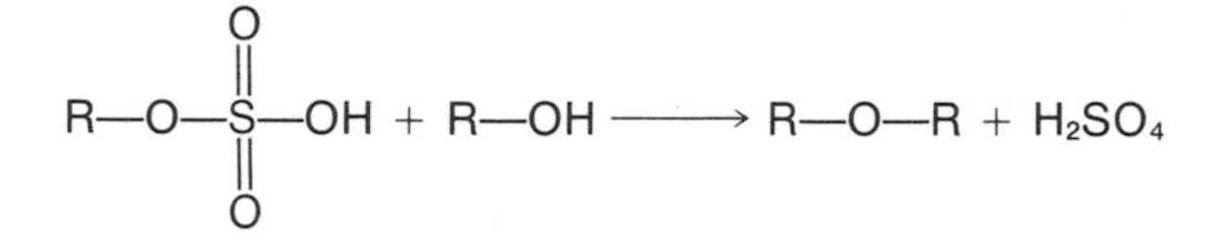

Alkylsulfates form an alkene when heated:

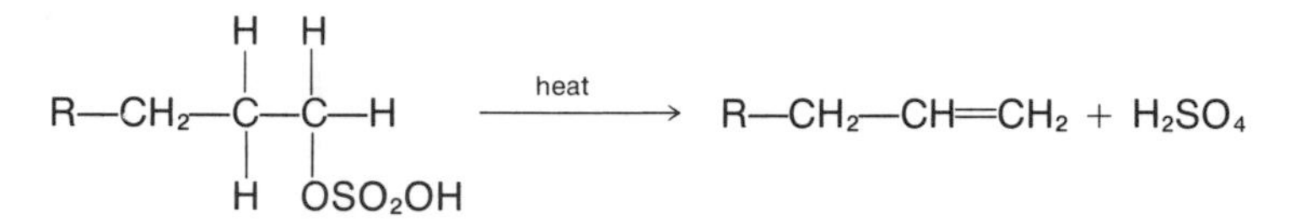

Only the sulfate esters of primary alcohols and secondary alcohols can be dehydrated to form alkenes, because of the availability of a hydrogen atom on the neighboring carbon atom. Alkyl sulfate esters of tertiary alcohols do not form alkenes when heated, but can form ethers.

Phosphoric acid reacts with alcohols to form phosphate esters that are extremely important in biological systems. The phosphate esters for both provide the storage and transfer of chemical energy in the metabolic processes of living systems. The reaction of an alcohol and phosphoric acid can be represented as follows:

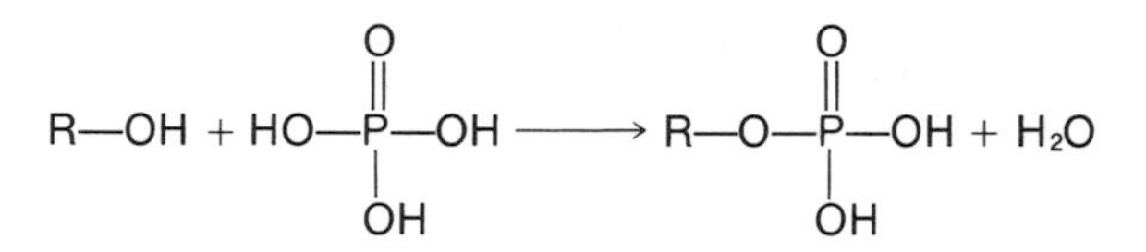

This is a monophosphate ester, and you can see that the ester has 2 additional —OH groups on it which can further react to form the diphosphate and triphosphate ester.

Summary of Alcohols

The characteristic functional group of alcohols is —C—OH. Alcohols can be primary, secondary or tertiary, depending on the carbon atom to which the group is attached. They are organic hydroxides but they are not basic and do not ionize in water to yield —OH ions. They react with oxidizing agents; the products formed depend on the carbon atom to which the —OH group is attached and the strength of the oxidizing agent. They also react with organic acids to form esters, with sulfuric acid to form alkenes, ethers, or soapless detergents depending upon the reaction conditions, and with active metals to form salts.

MOLECULAR ISOMERS

Isomers are not limited to the hydrocarbons, but are proliferated by the introduction of functional groups into the molecules. As you know, isomers are compounds that have the same number and kind of atoms, but those atoms are arranged differently within the molecule.

In the case of the dihydric alcohols, the glycols, the positions of the hydroxyl groups lead to isomers. For example:

1,4-butanediol has the formula: $CH_2—CH_2—CH_2—CH_2$ (OH on the first and fourth carbons)

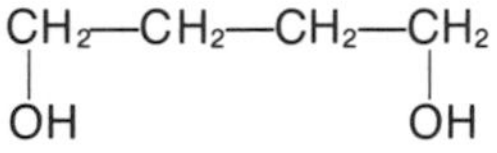

1,3-butanediol has the formula: $CH_2—CH_2—CH—CH_3$ (OH on the first and third carbons)

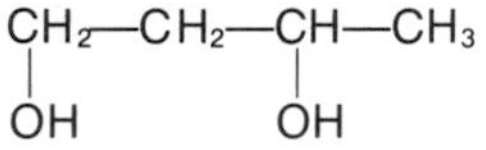

2,3-butanediol has the formula: $CH_3—CH—CH—CH_3$ (OH on the second and third carbons)

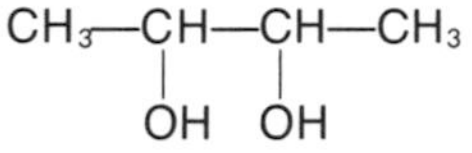

However, the existence of isomers is not limited to compounds that have the same functional groups placed in different locations. The isomers can be entirely different compounds because they have different functional groups, yet having the same *molecular formula* (Fig. 13-7).

The classic example of isomers is easily seen when you examine the formula, C_2H_6O, which can be written as ethyl alcohol ($CH_3—CH_2—OH$) or as dimethyl ether ($CH_3—O—CH_3$). These compounds are entirely different, but have the same molecular formula. It is for this reason the specific structural formula must be considered to identify the particular compound.

Methyl ether (dimethyl ether) is an anesthetic. Ethyl alcohol is the alcohol contained in wines, beers, liquors, and champagnes. As you can see in Table 13-5 they have the same molecular formula.

$CH_3—CH(OH)—CH_2—C(=O)—H$

hydroxy aldehyde

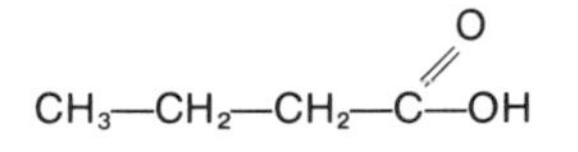

acid

$CH_3—C(=O)—CH_2—CH_2—OH$

keto alcohol

Figure 13-7. Isomers that have different functional groups. The molecular formula $C_4H_8O_2$ can represent an acid, an aldehyde with a hydroxyl group and a ketone with a hydroxyl group.

Table 13-5. Comparison of ethyl alcohol and dimethyl ether.

Name	*Molecular formula*	*Structural formula*	*Boiling point*	*Freezing point*	*State*
Ethyl Alcohol	C_2H_6O	CH_3CH_2OH	78°C	−117°C	Liquid
Dimethyl Ether	C_2H_6O	$CH_3—O—CH_3$	−24.9°C	−138°C	Gas

ETHERS

Ethers are characterized by the functional group —O—, which is an oxygen atom acting as a bridge between 2 carbon atoms of alkyl radicals, as in the general formula, R—O—R′ (where R and R′ can be the same or different alkyl radicals).

Physical Characteristics

Ethers are highly flammable, volatile liquids which can form explosive mixtures with air or oxygen. They are used as solvents for fats, greases, oils, resins, and organic compounds, and certain ones are used as anesthetics.

Chemical Characteristics

They are relatively stable to acids, bases, and reactive metals, but they do have a tendency to form peroxides, which are unstable and explosive. When the low molecular weight ethers are evaporated to dryness, they have a tendency to form the explosive peroxides, and therefore should be handled with care.

Naming of Ethers

The ethers are very simple to name. The names of the two alkyl groups are used followed by the term ether.

$$R—O—R'$$

where R and R′ are alkyl radicals, such as

$CH_3—$ methyl
$CH_3—CH_2—$ ethyl
$CH_3—CH_2—CH_2—$ propyl

Step 1. Identify the alkyl radicals.
Step 2. Combine the names of the alkyl radicals and add the term ether to them.

EXAMPLES:

$CH_3—O—CH_3$	dimethyl ether
$CH_3—O—CH_2—CH_3$	methyl ethyl ether or ethyl methyl ether
$CH_3—CH_2—O—CH_2—CH_3$	ethyl ethyl ether or, preferably, diethyl ether

Diethyl ether is the most commonly encountered ether. The vapors of diethyl ether have a higher density than air and fall to the floor where they can accumulate and form an explosive mixture with air that can be ignited by a chance spark. As an anesthetic, it is easy to administer, giving an excellent relaxation of the muscles, only slightly affecting blood pressure, pulse rate, and rate of respiration. Its disadvantage is that ether vapors are irritating to the respiratory passages and give aftereffects of nausea. In some cases ether may cause pneumonia after surgery. Due to the combustibility of ether, nylon clothing is prohibited in the operating room, because of the tendency for nylon to accumulate an electrostatic charge and give off a spark.

Synthesis of Ethers

Ethers are made from alcohols by the elimination of 1 molecule of water between 2 molecules of the alcohol. H_2SO_4 can be used as the dehydrating agent.

$$R{-}OH + HO{-}R \xrightarrow{H_2SO_4} R{-}O{-}R + H_2O$$

When one alcohol is used as the reactant, only one ether is formed, because both of the molecules of alcohol involved are identical.

When a mixture of two alcohols is used, three different ethers can be formed, two symmetrical ethers and one mixed ether:

EXAMPLE: If methyl alcohol (methanol) $CH_3{-}OH$ is mixed with ethyl alcohol (ethanol) $CH_3{-}CH_2{-}OH$ and heated in the presence of sulfuric acid, three ethers can be formed:

$$CH_3{-}OH + CH_3{-}OH \longrightarrow CH_3{-}O{-}CH_3$$
dimethyl ether

$$CH_3{-}OH + CH_3{-}CH_2{-}OH \longrightarrow CH_3{-}O{-}CH_2{-}CH_3$$
methyl ethyl ether

$$CH_3{-}CH_2{-}OH + CH_3{-}CH_2{-}OH \longrightarrow CH_3{-}CH_2{-}O{-}CH_2{-}CH_3$$
diethyl ether

ALDEHYDES

The functional group of the aldehyde is the $-\overset{\overset{O}{\|}}{C}-H$, which can be written more conveniently as —CHO. When that group is attached to an alkyl radical the corresponding aldehyde is formed.

The aldehydes are important to health care personnel because their presence in urine indicates the presence of blood sugar (glucose). Glucose is a sugar that contains an aldehyde functional group, and its presence in urine may indicate diabetes.

Nomenclature of Aldehydes

The common usage names of the simple aldehydes are gradually being replaced by the IUPAC system; however, it would be good practice to be familiar with both systems (Table 13-6).

Table 13-6. Nomenclature of aldehydes.

Compound	*Common name*	*IUPAC name*
H—CHO	Formaldehyde	Methan*al*
$CH_3{-}CHO$	Acetaldehyde	Ethan*al*
$CH_3{-}CH_2{-}CHO$	Propionaldehyde	Propion*al*
$CH_3{-}CH_2{-}CH_2{-}CHO$	Butyraldehyde	Butan*al*

Note: The IUPAC ending for an aldehyde is al and the ending for an alcohol is ol. You should be extremely careful of your pronunciation of these names and in writing them to avoid any possibility of confusion.

Chemical Reactions

The aldehyde functional group is a good reducing agent, and the principle of Benedicts' solution and Fehling's solution for identifying the presence of blood sugar in the urine is based upon this fact. Both Benedicts' and Fehling's solutions contain the copper II ion (the cupric ion) in solution, which has a blue color. When a solution containing an aldehyde functional group (as blood sugar in diabetic urine) is added to the blue test solution, the copper II (Cu^{++}) is reduced to copper I (cuprous ion, Cu^{+}) which causes the formation of an orange-red to red cuprous oxide, Cu_2O, precipitate. This precipitate formation indicates the presence of the aldehyde group which in turn signifies the presence of the blood sugar, according to the reaction:

$$\text{Radical—CHO} + 2\ Cu^{++}(OH)_2 \longrightarrow \text{Radical—COOH} + Cu_2O + H_2O$$

Tollens' test is another procedure for detecting aldehydes that uses a solution of silver nitrate as the test reagent. When the silver nitrate in solution is reduced by the aldehyde functional group, a silver mirror is deposited on the walls of a test tube. The silver ion (Ag^{+}) of the silver nitrate is reduced to metallic silver by the reducing action.

$$R\text{—CHO} + Ag_2O \longrightarrow R\text{—COOH} + 2\ Ag\downarrow$$

Oxidation of Aldehydes

Aldehydes are easily oxidized to acids, which have the functional group, $\text{—C}(=O)\text{—OH}$. This was the reaction that the aldehyde group underwent in the test for blood sugar, and it is evident that the aldehyde group has an affinity for oxygen. When aldehyde is oxidized, the corresponding acid is formed.

Synthesis of Aldehydes

When primary alcohols are oxidized in the presence of a suitable catalyst, 2 hydrogen atoms are removed in the process and an aldehyde is formed (Fig. 13-8).

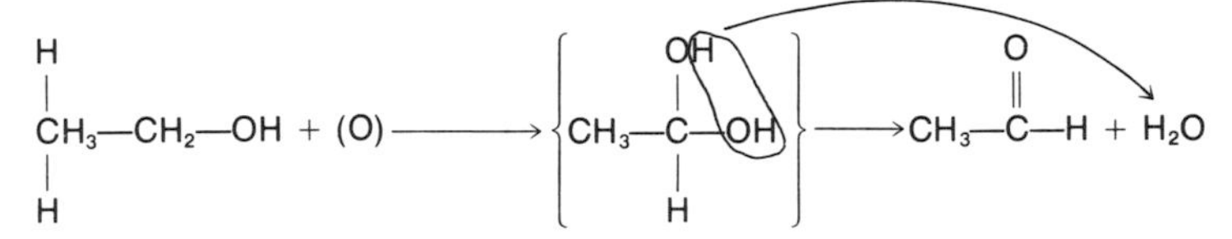

Ethyl alcohol + oxygen → unstable compound → acetaldehyde + water

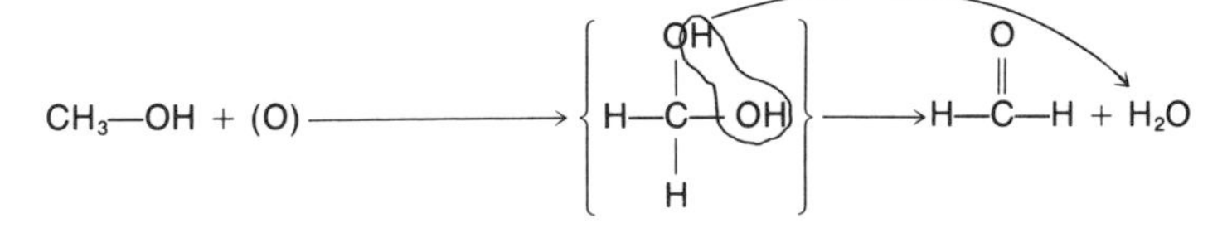

methyl alcohol + oxygen → unstable compound → formaldehyde + water

Figure 13-8. Synthesis of aldehydes.

FORMALDEHYDE

Formaldehyde is the simplest aldehyde and is a colorless gas with a sharp and penetrating odor. It is usually found as a 40 percent solution in water called formalin, which is used to preserve specimens in the laboratory, as a germicide, and as a disinfectant. It hardens tissues and coagulates proteins. Its presence is readily detectable because the fumes of formaldehyde are irritating to the mucous membranes. When formaldehyde is polymerized, several molecules joining together, paraformaldehyde, an antiseptic, is formed.

ACETALDEHYDE

Acetaldehyde is a colorless liquid having a sharp, penetrating, characteristic odor. It is an important chemical raw material, and when it is polymerized, the polymer paraldehyde is formed, which is much more stable than the acetaldehyde. Paraldehyde is a relatively nontoxic hypnotic (a sleep inducer) although it does have an unpleasant taste and an irritating odor.

UROTROPINE

When formaldehyde reacts with ammonia, urotropine, a urinary antiseptic is formed. Urotropine decomposes under the influence of acids in the body into formaldehyde and ammonia, and the formaldehyde destroys microorganisms in the urinary tract.

KETONES

Ketones are characterized by the carbonyl group, $-\overset{\overset{\displaystyle O}{\|}}{C}-$, which is the ketone functional group. Each of the 2 other valence bonds of the carbon are satisfied by organic radicals. Ketones are significant because dimethyl ketone, commonly called acetone, is normally found in the blood in small quantities. When blood analyses show larger than normal amounts of acetone in the blood, this is indicative of faulty fat metabolism, which accompanies diabetes mellitus. The presence of acetone in diabetic patients can also be detected in their breath, because excessive amounts are respired and acetone has a characteristic sweetish odor.

Nomenclature of Ketones

The ketones are named by using the names of the 2 organic radicals attached to the carbonyl functional group followed by the term ketone. However, the newer IUPAC system of nomenclature names the ketone by changing the name of the corresponding organic compound to include the ending one. The position of the ketone group on the carbon chain is indicated by a number preceding the name (Table 13-7).

Table 13-7. Naming of ketones.

Formula	*Common Usage*	*IUPAC name*
$CH_3-CO-CH_3$	Acetone	Propanone
$CH_3-CO-CH_2-CH_3$	Methyl ethyl ketone	Butanone
$CH_3-CH_2-CO-CH_2-CH_3$	Diethyl ketone	3-pentanone
$CH_3-CO-CH_2-CH_2-CH_2$	Methyl propyl ketone	2-pentanone

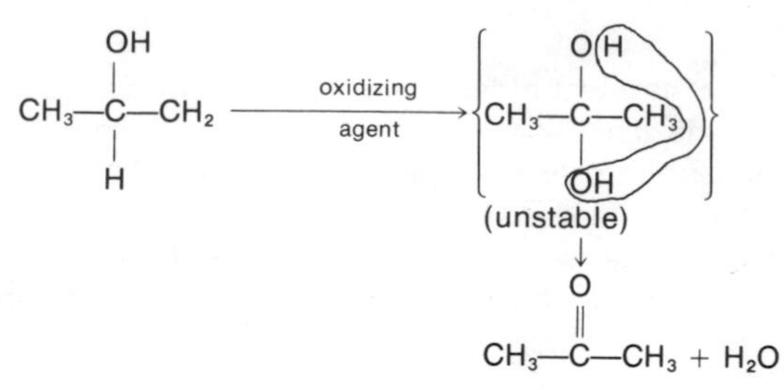

Figure 13-9. Synthesis of ketones.

Synthesis of Ketones

Ketones are formed by the controlled oxidation of secondary alcohols by oxidizing agents. Strong oxidizing agents destroy the molecule to form carbon dioxide and water. They differ from the aldehydes in that each of the valence bonds of the carbonyl carbon atom is attached to organic radicals in the ketone, whereas in the aldehydes, one of the valence bonds is attached to a hydrogen atom (Fig. 13-9).

ACETONE

Acetone, dimethyl ketone or propanone, is the most important simple ketone to health care personnel because of its excessive presence in diabetic patients. It is very volatile, extremely flammable, and an excellent solvent. It is found in nail polish removers because it readily dissolves nail polish lacquer.

ACIDS

Organic carboxylic acids have the characteristic functional group, —C(=O)—OH, a carboxyl group, written in condensed form as —COOH. They are related to the aldehydes and the ketones and can be considered as being a derivative of aldehydes, with an —OH group attached instead of merely a —H atom. This functional group is responsible for the acid qualities of acid compounds, because the H atom of the —OH group ionizes in water to yield H^+ ions, which are characteristic of inorganic acids. They differ from the inorganic acids in that they ionize to a much lesser extent than the strong inorganic acids, such as HCl and H_2SO_4. However, the fact that they are acids and do ionize gives the solution the characteristic properties of an acid.

NOMENCLATURE OF ACIDS

The organic acids have common names, which in most cases are used preferentially over the newer IUPAC names. They include such acids as acetic, propionic, butyric, stearic and oleic. Many times the name reflects the source of the acid, such as palmitic from palm oil (Table 13-8). When the acid has a chain of 12 or more carbon atoms, it is called a fatty acid, and these will be covered in detail in the chapter on lipids.

Table 13-8. Naming acids.

Formula		
Saturated acids	*Common usage*	*IUPAC name*
H—COOH	Formic acid	Methanoic acid
CH_3—COOH	Acetic acid	Ethanoic acid
CH_3—CH_2—COOH	Propionic acid	Propanoic acid
CH_3—CH_3—CH_2—COOH	Butyric acid	Butanoic acid
CH_3—$(CH_2)_{14}$—COOH	Palmitic acid	Hexadecanoic acid
CH_3—$(CH_2)_{16}$—COOH	Stearic acid	Octadecanoic acid
Unsaturated fatty acids		
CH_3—$(CH_2)_7CH{=}CH(CH_2)_7COOH$	Oleic acid	9-octadecanoic acid
CH_3—$(CH_2)_4CH{=}CH$—$CH_2CH{=}CH$ —$(CH_2)_7COOH$	Linoleic acid	
$CH_3CH_2CH{=}CH$—$CH_2CH{=}CH$ —$CH_2CH{=}CH(CH_2)_7COOH$	Linolenic acid	
$CH_3(CH_2)_3$—$(CH_2CH{=}CH)_4$—$(CH_2)_7COOH$	Arachidonic acid	

Physical Characteristics of Acids

Acids with up to 5 carbon atoms are clear liquids, between 5 and 9 carbon atoms they are waxy, and above 9 carbon atoms they are solids. The low molecular weight acids are soluble in water, but the solubility decreases as the molecular weight increases. Organic acids have a characteristic smell. Formic, acetic and propionic acids have sharp odors. Butyric acid has the smell of rancid butter. Higher molecular weight acids have disagreeable odors when heated to high temperatures.

Chemical Characteristics of Acids

They are acids because the hydrogen atom that is part of the —OH group of the carboxyl functional group ionizes. The other hydrogen atoms that are attached to the organic radical do not ionize. The ionization of the acid can be represented by the equation:

$$CH_3\text{—}COOH \longrightarrow CH_3COO^- + H^+$$

Neutralization of Acids

Acids can be neutralized by either organic or inorganic bases to form the corresponding salts.

$$CH_3\text{—}COOH + NaOH \longrightarrow CH_3\text{—}COONa + H_2O$$

Since the acids ionize to yield the H^+ ion and the corresponding anion, they react with bases to form salts and water, just as the inorganic acids react to form the corresponding salt and water:

$$R\text{—}COOH + NaOH \longrightarrow R\text{—}COONa + H_2O$$

The salt is named by using the name of the metal followed by the stem of the organic acid plus the ending ate.

EXAMPLES:

Acid	*Base*	*Salt*
Formic	NaOH	Sodium formate
Acetic	KOH	Potassium acetate
Propionic	NH_4OH	Ammonium propionate

Acetic acid will form soluble salts with almost all bases. The sodium salt of propionic acid, sodium propionate (CH_3—CH_2—COONa) is added as a preservative to pastry products to inhibit spoilage.

The Common Acids

FORMIC ACID

Formic acid is the simplest acid, causing the skin irritation from insect bites. If the bite area is washed with dilute base such as ammonia water or sodium bicarbonate, the acid is neutralized. Formic acid is a stronger acid than other alkyl carboxylic acids that have higher molecular weights.

ACETIC ACID

Acetic acid, ethanoic acid, is the acid of vinegar at a 5 percent concentration. When it is 99.5 percent pure, it is called glacial acetic acid, freezing at around 17°C and resembling ice in its solid state. Glacial acetic acid can be liquified by gradually immersing the bottle of solidified acid in a container of lukewarm water. The sodium salt, sodium acetate, is important as part of the acetic acid-acetate buffer system.

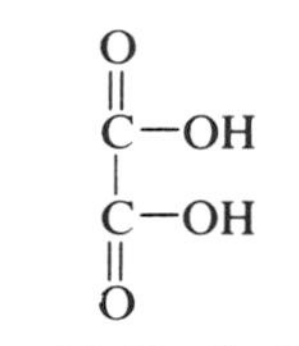

Figure 13-10. Oxalic acid.

Polycarboxylic Acids

There are a number of polycarboxylic acids (acids that have more than one carboxyl group attached to the carbon chain) that health care personnel will commonly encounter either in the laboratory or in the biochemical processes. These poly acids behave as carboxylic acids in most reactions, either one or more of the carboxyl groups being involved.

Oxalic acid is the simplest dicarboxylic acid (Fig. 13-10). It is highly poisonous and cannot be taken internally. It is used to remove iron rust stains and its salts are added to blood samples to be analyzed in the laboratory because they prevent the clotting of blood. It should be noted that these salts cannot be administered intravenously because they are poisonous.

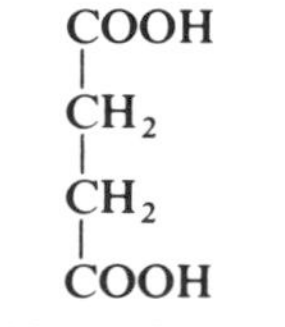

Figure 13-11. Succinic acid.

Succinic acid is a dicarboxylic acid containing 4 carbon atoms, with carboxyl groups at each end of the carbon chain (Fig. 13-11). It is an intermediate compound found in the series of biochemical reactions involved in the final phase of the oxidation of blood sugar in the Krebs cycle.

Fumaric acid is a dicarboxylic acid containing 4 carbon atoms. It differs from succinic acid in that it contains a double bond (Fig. 13-12). In the Krebs cycle, succinic acid is converted to fumaric acid by the removal of 2 hydrogen atoms and the formation of the double bond.

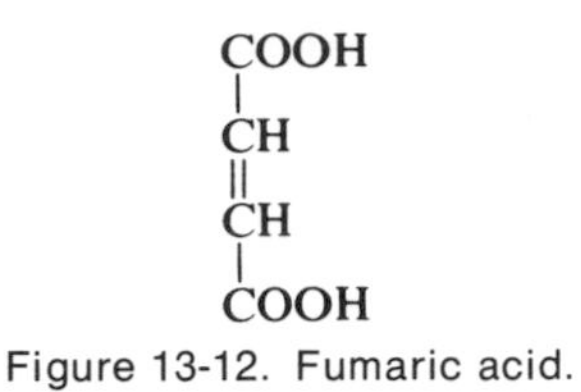

Figure 13-12. Fumaric acid.

HYDROXY AND KETO ACIDS

The introduction of an —OH group (alcohol) or a $\overset{O}{\overset{\|}{—C—}}$ group (ketone or carbonyl) on the carbon chain of the simple alkyl carboxylic acids leads to a series of derivatives that play a major role in the metabolic processes of the body. Acids such as lactic, citric, tartaric, pyruvic, ketoglutaric, oxaloacetic, and aspartic are intermediates in the various biochemical reactions of the catabolism (breakdown of the molecules) and anabolism (the building of molecules) of the body. It is essential that health care personnel understand the chemistry of the reactions of the functional groups and be able to relate why the body is able to interconvert these substances through the chemical reactions of oxidation, reduction, addition and dehydrogenation.

Nomenclature of Hydroxy and Keto Acids

The common practice is to identify the location of the hydroxyl group or the ketone group on the carbon chain by labelling the carbon atoms alpha (α), beta (β) and gamma (γ) starting with the *first* carbon on the chain *after* the carboxyl functional group. This system is also used to designate the locations of other functional groups.

The following compounds, therefore, would be named:

$CH_3—CH_2—\underset{\alpha}{\overset{O}{\overset{\|}{C}}}—COOH$	alpha keto butyric acid
$CH_3—\underset{\beta}{\overset{O}{\overset{\|}{C}}}—CH_2—COOH$	beta keto butyric acid
$CH_3—CH_2—\underset{\alpha}{CHOH}—COOH$	alpha hydroxy butyric acid
$CH_3—\underset{\beta}{CHOH}—CH_2—COOH$	beta hydroxy butyric acid

Important Hydroxy and Keto Acids

Lactic acid is 2-hydroxy propionic acid (2-hydroxy propanoic acid) and it is found in the muscle tissue as a result of the anaerobic oxidation of glucose (Fig. 13-13). It is found in sour milk resulting from the fermentation of lactose (milk sugar).

```
   H  H
   |  |
H—C—C—COOH
   |  |
   H  OH
```

Figure 13-13. Lactic acid.

Pyruvic acid is 2-keto propionic acid (2-keto propanoic acid) and it is formed on the muscle tissue as a result of the aerobic oxidation of blood sugar (Fig. 13-14). A comparison of pyruvic acid with lactic acid shows that the difference between them is that one is a ketone acid while the other is a secondary alcohol

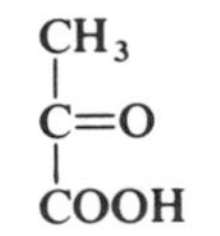

Figure 13-14. Pyruvic acid.

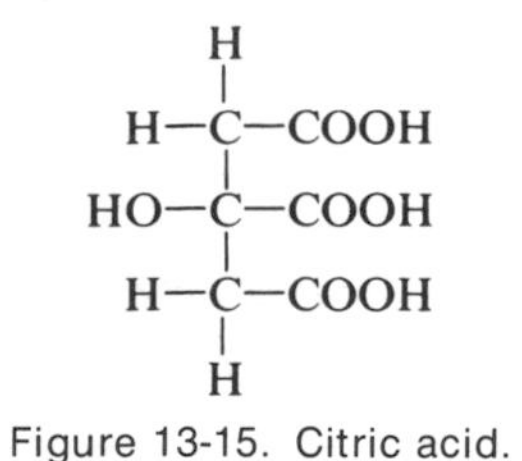
Figure 13-15. Citric acid.

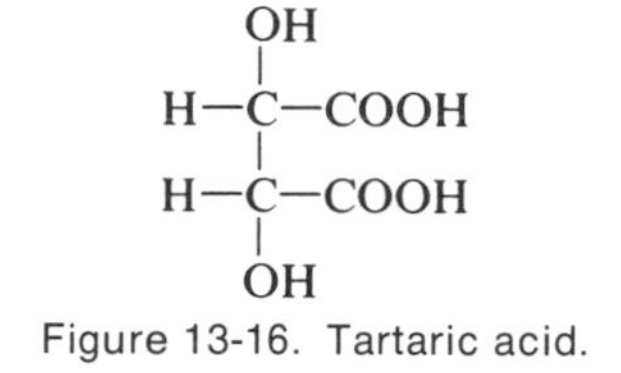
Figure 13-16. Tartaric acid.

acid. Since secondary alcohols can be oxidized to ketones, and ketones can be reduced to secondary alcohols, their relationship to each other can be understood if the conditions of oxidation (aerobic or anaerobic) are known.

Citric acid is a tricarboxylic acid (containing 3 carboxyl groups) with 1 hydroxyl group (Fig. 13-15). It is found in the citrus fruits such as lemons, limes and grapefruits, but more importantly it is found in the body in the final phase of the oxidation of glucose in a series of reactions liberating energy known as the Krebs cycle, citric acid cycle or tricarboxylic acid cycle.

Tartaric acid is a dihydroxy dicarboxylic acid found in grapes (Fig. 13-16). Its potassium salt is known as cream of tartar and it is used to make baking powders. Its mixed salts, where the 2 carboxyl groups are neutralized by different bases, are sodium potassium tartrate (Rochelle salts), a cathartic, and potassium ammonium tartrate (tartar emetic), an expectorant.

Hydroxybutyric acid is normally found in the blood in small quantities. As a secondary alcohol, it can easily be oxidized to a ketone, forming keto butyric acid, which is called a ketone body. The ketone bodies are normal intermediate products of the fatty acid metabolism, but when the number of ketone bodies rises above normal, a condition called ketosis exists. Since these ketone bodies are acid, the blood pH falls. This condition can result from starvation, low carbohydrate diets, metabolic abnormalities, and most commonly, diabetes mellitus.

Replacement of the —OH Group with a Carboxyl Functional Group

Many substances will react with carboxylic acid and will replace the —OH group with the other groups or atoms, forming different compounds having another functional group. One of the most important acid derivatives is called an ester, and it results from the reaction of the carboxylic acid with an alcohol. The products of this reaction are the ester and water, and it is an elimination reaction because part of the molecule is removed to form water. A catalyst, such as a trace of sulfuric acid, is used, and the overall reaction is represented in Figure 13-17.

ESTERS

Esters are colorless liquids with pleasant fruit-like odors. Amyl acetate has the odor of banana oil, ethyl butyrate smells like pineapples, amyl butyrate like

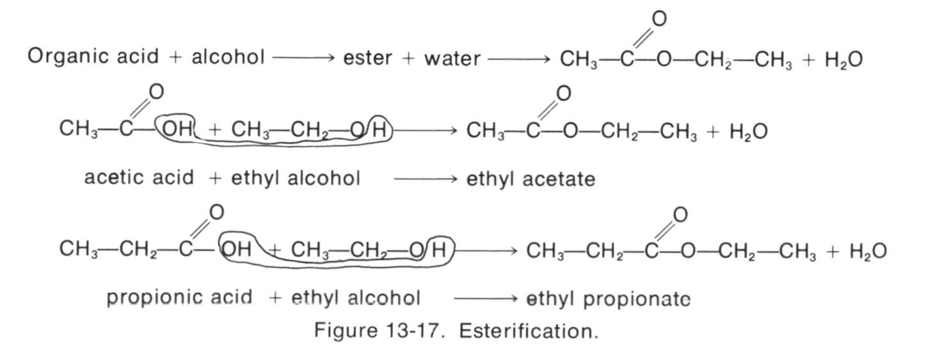

Figure 13-17. Esterification.

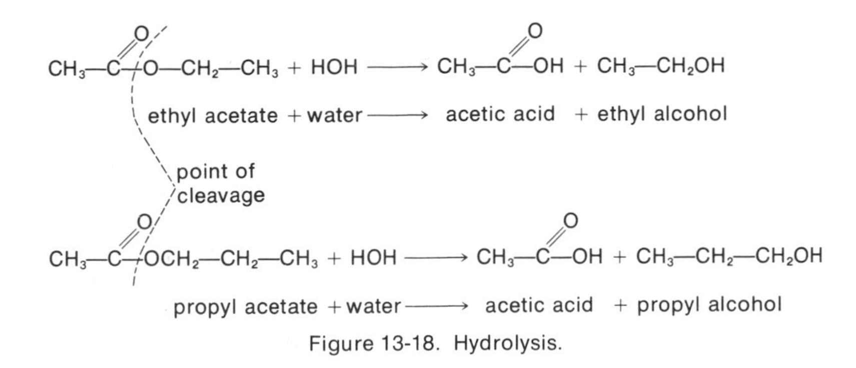

Figure 13-18. Hydrolysis.

apricots, octyl acetate like oranges, and isoamyl valerate like apples. The esters are used as perfumes and flavoring agents. In addition, the lower molecular weight esters are excellent solvents for cellulose nitrate lacquer (used in fingernail polish). Ethyl acetate is often used as a component of fingernail polish thinner and remover. Perfumes can easily go bad when exposed to moisture and heat over a period of time because of their hydrolysis back to the acid and alcohol from which they are formed (Fig. 13-18).

Alcohols react with organic acids to form organic esters, and the process of esterification results from the reaction of the hydroxyl group of the acid with the hydrogen atom of the alcohol eliminating water to form the ester. This reaction is a reversible one, the ester reacting with water to form the corresponding acid and alcohol. Catalysts, such as sulfuric acid, enable the reaction to reach the equilibrium point more rapidly.

Nomenclature of Esters

The name of the ester combines both the name of the alcohol and the acid that formed it. The name of the alcohol is used first, omitting the term alcohol, followed by the name of the acid, changing the ending of the acid from ic to ate (see Table 13-9).

Esters differ from organic salts of alcohols because they do not ionize, and they react with water very slowly (hydrolyze) unless a catalyst is present. Their

Table 13-9. Naming of esters.

Organic acid	*Organic alcohol*	*Ester*
Acetic acid CH_3COOH	Methyl alcohol CH_3—OH	Methyl acetate CH_3COOCH_3
Propionic acid CH_3CH_2COOH	Methyl alcohol CH_3OH	Methyl propionate $CH_3CH_2COOCH_3$
Butyric acid $CH_3CH_2CH_2COOH$	Methyl alcohol CH_3OH	Methyl butyrate $CH_3CH_2CH_2COOH$
Butyric acid $CH_3CH_2CH_2COOH$	Ethyl alcohol CH_3CH_2OH	Ethyl butyrate $CH_3CH_2CH_2COOCH_2CH_3$
Butyric acid $CH_3CH_2CH_2COOH$	Propyl alcohol $CH_3CH_2CH_2OH$	Propyl butyrate $CH_3CH_2CH_2COOCH_2CH_2CH_3$

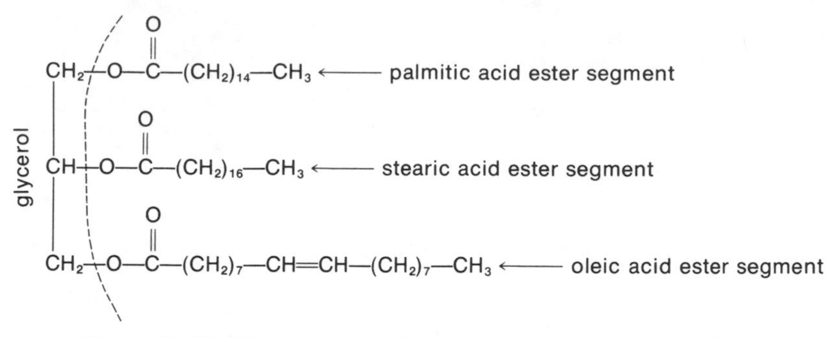

Figure 13-19. The triester of glycerol and long chain fatty acids.

most important reaction in the living system is hydrolysis, reacting with water to form the corresponding alcohol and organic acid. Most of the fats are triesters of glycerol and long chain fatty acids (Fig. 13-19). Drying oils and linseed oils contain esters of linoleic and linolenic acids (unsaturated fatty acids) and those unsaturated acids are responsible for the drying qualities of those oils.

Saponification of Esters

When esters are heated with water in the presence of a base, such as NaOH or KOH, the acid liberated appears as a sodium or potassium salt. The saponification of fats (containing high molecular weight fatty acids) yields the sodium or potassium salt, which are soaps (Fig. 13-20). The sodium salt is a solid soap, while the liquid soap found in washrooms is a potassium salt.

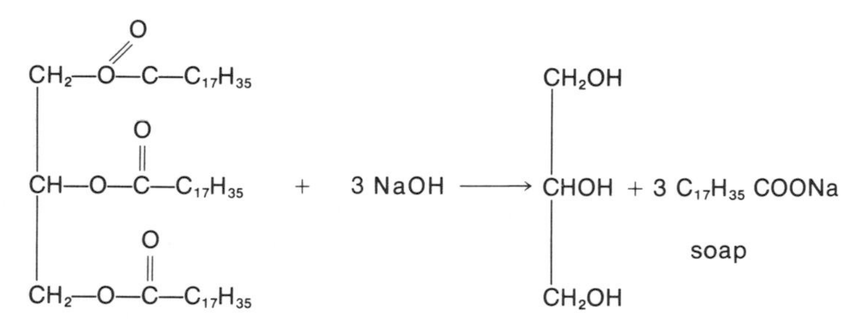

Figure 13-20. Saponification.

EPOXIDES

Epoxides are characterized by a three membered ring structure that is composed of 2 carbon atoms and 1 oxygen atom. This is a heterocyclic ring ether (Fig. 13-21).

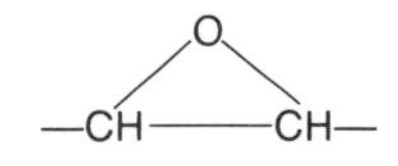
Figure 13-21. Epoxide structure.

Naming of Epoxides

The length of the carbon chain determines the root of the name, and the alkene term is used with the suffix oxide (Fig. 13-22).

Reactivity of Epoxides

Although they are internal ethers, epoxides are extremely reactive as compared to normal alkyl ethers, which are relatively inert. They react with water to form glycols (Fig. 13-23).

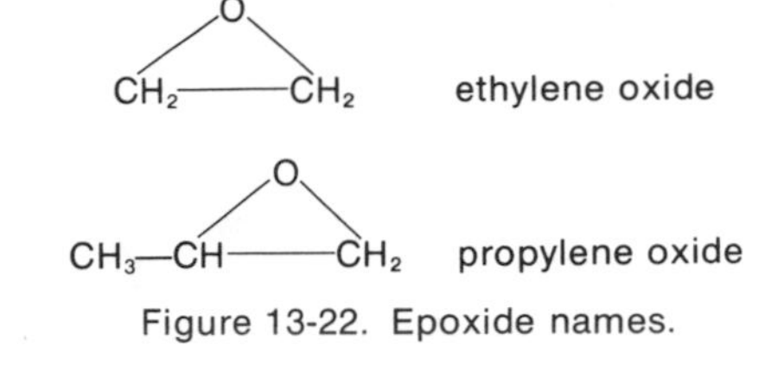

Figure 13-22. Epoxide names.

Figure 13-23. Hydrolysis of epoxides.

Important Epoxides

Ethylene oxide is used in the sterilization of both large and small equipment and instruments. The gas has the ability to penetrate extremely small orifices, cloth wrappings, and certain plastics. Ethylene oxide at concentrations of about 1 g/L is very effective against all types of microorganisms.

SUMMARY

The introduction of oxygen atoms in molecules forms a functional group. Different groupings form different functional groups, each having its own characteristic properties. In the molecule, an —OH yields an alcohol, 2 —OH groups on different C atoms a glycol, an oxygen atom bridging 2 C atoms an ether, a —CHO group an aldehyde, a —CO— group a ketone, a —COOH group an acid, a —COO— group an ester, and a three membered ring with 2 C atoms and 1 O atom an epoxide. Methyl, ethyl, and propyl alcohols are the important alcohols, but only ethyl alcohol can be taken internally. Glycerol is a skin lubricant and is used to make nitroglycerin. Diethyl ether is used as an anesthetic, but is very flammable and can form peroxides that explode. Formaldehyde is used as a disinfectant, and acetaldehyde as a raw material for paraldehyde, a hypnotic. The oxidation of the aldehyde group is the basis for the detection of blood sugar. Acetone is found in abnormal amounts in diabetic patients. Acetic acid, 5 percent, is the acid of vinegar, and is important in buffer systems. Polycarboxylic acids are part of the biochemical metabolic processes. Hydroxy and keto acids are products of cell metabolism. Organic acids react with alcohols to form esters. The esters can be split apart with water (hydrolysis) or with alkaline solutions (saponification) to yield the alcohols and the acids and their salts. Natural fats and oils are triesters of glycerol and high molecular weight fatty acids. Epoxides are used to sterilize supplies and equipment.

EXERCISE

1. What is the functional group of the alcohols, and what are the formulas, common names and IUPAC names for the first three alkyl alcohols?
2. Which alcohol can be taken internally?
3. What is denatured alcohol and why is methanol added as a component of the mixture?
4. Name and draw the formulas for a common diol and a triol.
5. List at least two uses of ethylene glycol and glycerin.
6. What is the primary oxidation product of methanol, ethanol and propanol?
7. Write the equation for the oxidation of primary alcohols to the stable acid.
8. What is the stable oxidation product of secondary alcohols and what is the importance of that reaction in the body?
9. Draw the structural formula for diethyl ether and list its chemical, physical and physiologic properties.
10. What is the oxidation product of an aldehyde? State how that reaction is used to identify the presence of blood sugar in the urine.
11. Describe the Tollens', Benedicts' and Fehling's tests and what they indicate.
12. Give the common usage names and the IUPAC names for the first three carboxylic acids and state the importance of those acids in the health care field.
13. What is an ester composed of, and how are esters named?
14. What is the difference between saponification and hydrolysis?
15. What are the products of saponification of an ester?
16. What are the products of the hydrolysis of esters? Relate the hydrolysis of esters to the catabolism of fats in the body.
17. Draw an epoxide and give the name and uses of one epoxide that is used to sterilize instruments and equipment.

OBJECTIVES

When you have completed this chapter you will be able to:

1. Write the names and formulas of the alkyl halides that are important in the health care field and state the uses of each.
2. Differentiate between the three types of amines, draw their formulas, and give the rule for naming them.
3. Show by chemical reactions why amines yield basic solutions in water.
4. Relate the formation of amine salts to ammonia salts and illustrate by example an important application.
5. Define quaternary ammonium salts, draw their formulas, and give two important uses of them.
6. Name acid amides and draw the functional group, showing its relationship to acids.
7. Discuss in detail the structure of amino acids and how the functional groups affect the chemical behavior.
8. Name and draw the formulas for common amino acids.
9. Explain by means of a chemical reaction the body synthesis of proteins from amino acids.
10. Define peptide, dipeptide and polypeptide, drawing the peptide linkage.
11. Show how proteins act as buffering agents.
12. Name an important nitroso and nitro derivative and state where they are used in medicine.
13. Name four heterocyclic nitrogen ring structures and list at least six types of compounds in which they are found.
14. Differentiate chlorophyll and hemoglobin.
15. Explain why sulfur derivatives generally behave like their oxygen analogues.

HALOGENS, NITROGEN, SULFUR DERIVATIVES AND POLYMERS

16. Give the general terms that identify sulfur compounds.
17. Distinguish between condensation and polymerization polymers, writing a general reaction to show how the polymers are formed.
18. Construct a table summarizing the simple functional groups found in organic compounds, giving the common and/or IUPAC name for the functional group.

The introduction of other elements into the organic molecule as atoms or as parts of functional groups create many new organic compounds. The chemical nature and behavior of the molecule depends on the presence of the various functional groups, each contributing their characteristic properties. Some groups have chemical properties resembling their oxygen counterparts, such as groups where sulfur atoms have been substituted for oxygen atoms.

THE HALOGENS

The halogen elements are fluorine, chlorine, bromine and iodine, each having an outer shell of 7 electrons and each needing only 1 electron to achieve a stable state. The halogen atoms can satisfy their outer shell requirements by sharing 1 of their electrons with 1 electron of a carbon atom, forming halogenated derivatives.

The carbon atom has 4 electrons in its outer shell, and therefore needs 4 more electrons to achieve stability (Fig. 14-1). It can do this, as you have seen previously, by sharing its electrons with hydrogen atoms to form methane. It also can share 1 or more of its electrons with halogen atoms (Fig. 14-2).

```
   H
   ..
H : C : H
   ..
   H
```

Figure 14-1. Alkane

NAMING HALOPARAFFINS

As you have seen with other organic compounds, many have common usage names and the newer IUPAC names. The common usage names are generally used for the simpler compounds, whereas the IUPAC name is used for the more

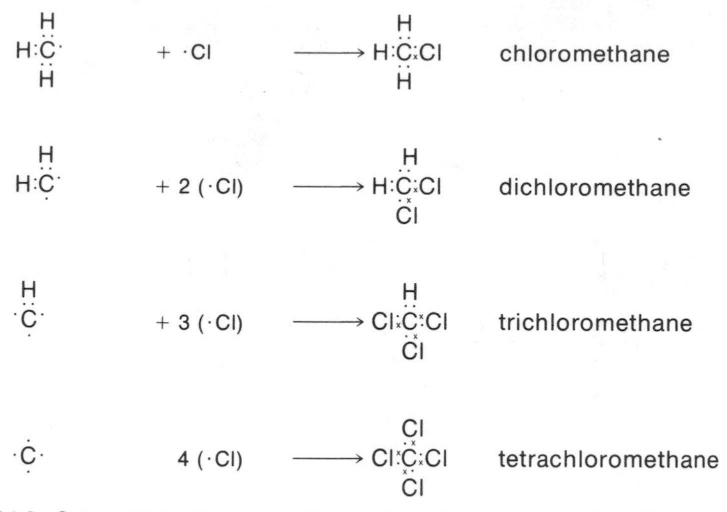

Figure 14-2. Chloro derivatives of methanes. The electrons shared by the chlorine atom is shown as x.

complicated long chain molecule. The IUPAC method uses the longest straight chain of carbon atoms as the base name (root) and identifies the element and its carbon atom location by prefix numbers in front of the base name (Table 14-1).

Chloroform is used as a general anesthetic, is not flammable and easily administered. (CAUTION: Exercise care in using chloroform. Too much can be lethal and the difference between the proper amount of chloroform and a lethal dose is not great.) Chloroform should not be used near an open flame because it oxidizes to form phosgene, a toxic gas.

$$CHCl_3 \xrightarrow[\text{conditions}]{\text{oxidizing}} COCl_2 + HCl$$

Chloroform that has been stored for a long time undergoes the same reaction. However, since HCl is liberated on decomposition, and HCl ionizes in water to

Table 14-1. Names of common halogenated alkanes.

Compound	*Common usage*	*IUPAC*
CH_3Cl	Methyl chloride	Chloromethane
CH_2Cl_2	Methylene chloride	Dichloromethane
$CHCl_3$	Chloroform	Trichloromethane
CCl_4	Carbon tetrachloride "carbon tet"	Tetrachloromethane
$CH_2Cl—CH_2Cl$	Ethylene dichloride	1,2 dichloroethane
CH_3CH_2Cl	Ethyl chloride	Chloroethane
$CH_2Cl—CHCl—CH_2Cl$	—	1,2,3 trichloropropane

yield chloride ions, the addition of silver nitrate can be used to test for the decomposition. A turbid AgCl precipitate shows the presence of chloride ions and, of course, decomposition. Discard any containers that give a positive test result.

Carbon tetrachloride is a heavy liquid, is not flammable and is an excellent dry cleaning fluid. It should not be used in confined spaces because its vapors will dissolve the protective oils in the respiratory tract and lead to respiratory infections. Furthermore, its vapors are lethal in high concentrations.

Trichloroethylene has the formula $CHCl{=}CCl_2$. Its base is ethylene, and the compound is used as an industrial solvent and as a substitute for carbon tetrachloride.

Iodoform, CHI_3 (tri-iodomethane) is the iodine equivalent of chloroform. It is a pale yellow solid which has been used as an antiseptic in ointments or as a powder dust for open wounds. In the laboratory the iodoform test is used to indicate the presence of ethyl alcohol. Its antiseptic action may be due to the gradual liberation of iodine as the molecule decomposes.

Dichloro-difluoromethane is commonly called freon, a gas that is used in home and industrial air conditioning and refrigeration units.

Vinyl chloride (chloroethene) $CH_2{=}CHCl$ is the raw material used in the manufacture of vinyl plastics, called PVC plastics. The Food and Drug Administration has prohibited its use as a propellant in aerosol cans because they have found that it is a very active carcinogenic (cancer producing) agent and that it was responsible for an outbreak of cancer among workers in vinyl chloride manufacturing plants. Other safe propellants have been substituted for vinyl chloride.

Tetrafluoro-ethene, $CF_2{=}CF_2$, is the monomer (single molecule) that is used to produce Teflon plastic products. Teflon plastics are composed of tens of thousands of these molecules linked together to give high molecular weight plastics that have the unique property of not sticking.

AMINES

Amines are characterized by the presence of the functional group $—NH_2$, and can be considered as derivatives of ammonia (NH_3), where one or more of the hydrogen atoms have been replaced by an alkyl group. As you know, ammonia dissolves in water to form a basic solution yielding hydroxyl ions. The amines react in the same way yielding basic solutions that can react with either inorganic or organic acids.

Types of Amines

When 1 hydrogen atom of ammonia is replaced by an alkyl group, a primary amine is formed. If 2 hydrogen atoms of ammonia are replaced by 2 alkyl groups, a secondary (diamine) amine is formed, and when all 3 are replaced, a tertiary amine results. Amines are characterized by an ammonia-like obnoxious odor, and they are formed by the decomposition and putrefaction of proteins.

Naming Amines

Amines are identified by naming the alkyl groups that are attached to the nitrogen atom and following those names with the term amine (see Table 14-2).

Table 14-2. Naming amines.

Name	*Formula*
Methyl amine	$CH_3—NH_2$
Dimethyl amine	$CH_3—N(H)—CH_3$ or $(CH_3)_2NH$
Trimethyl amine	$CH_3—N(CH_3)—CH_3$ or $(CH_3)_3N$
Methyl ethyl amine	$CH_3—N(H)—C_2H_5$ or $(CH_3)\ (C_2H_5)NH$
Ethylene diamine	$NH_2—CH_2—CH_2—NH_2$

Chemical Reactions of Amines

REACTION WITH WATER

Amines react with water to form ionizable bases just as ammonia does:

$$NH_3 + H_2O \longrightarrow NH_4OH \xrightarrow[\text{in water}]{\text{ionizes}} NH_4^+ + OH^-$$

$$CH_3—NH_2 + H_2O \longrightarrow CH_3NH_3OH \xrightarrow[\text{in water}]{\text{ionizes}} CH_3NH_3^+ + OH^-$$

REACTION WITH ACIDS

Ammonia reacts with acids to form the corresponding ammonium salts:

$$NH_3 + HCl \longrightarrow NH_4Cl \xrightarrow[\text{in water}]{\text{ionizes}} NH_4^+ + Cl^-$$

Amines react with acids to form the corresponding alkyl ammonium salts:

$$CH_3—CH_2—NH_2 + HCl \longrightarrow CH_3—CH_2—NH_3Cl \xrightarrow[\text{in water}]{\text{ionizes}} CH_3—CH_2—NH_3^+ + Cl^-$$

Ammonium salts are far more stable than the corresponding amines and much more water soluble. This factor is of great use in the preparation of many of our potent drugs. By treating the amino group present in the molecule with the equivalent amount of inorganic acid, a salt is formed which has greater stability and is easier to handle than the parent amine.

Morphine is insoluble in water but morphine sulfate (made by adding sulfuric acid to morphine) is very soluble. By converting the morphine to the salt it can be easily formulated for parenteral injection, and the compound is in a more stable state.

REACTION WITH ALKYL HALIDES TO FORM QUATERNARY AMMONIUM SALTS

Amines react with alkyl halides to form quaternary ammonium salts just as the amines react with acids to form salts. These quaternary ammonium salts are so named because there are 4 alkyl radicals attached to the nitrogen atom. The salts are excellent bacteriostatic agents (inhibiting the growth of bacteria) and germicides (killing bacteria). They are commonly called "quats" and are used in dilute concentrations as cold sterilization solutions. Zephiran at a 1:1,000 concentration sterilizes instruments and at a 1:20,000 concentration is used in urinary irrigations.

Naming Quaternary Ammonium Salts

The name of the amine and the alkyl halide are combined to give the name of the quaternary ammonium halide. Examples are:

trimethyl amine + methyl chloride gives tetramethyl ammonium chloride
diethyl-butyl amine + butyl bromide gives diethyl-dibutyl ammonium bromide

AMIDES

Amides are characterized by the functional group $-\overset{O}{\overset{\|}{C}}-NH_2$, or more conveniently written $-CONH_2$. They are related to the carboxylic acid $-\overset{O}{\overset{\|}{C}}-OH$, and are also called acid amides, the $-NH_2$ group replacing the $-OH$ group of the carboxylic acid.

Naming Amides

Amides have both common usage names and IUPAC names. The IUPAC name uses the name of the alkane root with the suffix amide (see Table 14-3).

Important Amides

The most important single molecule compound to health care personnel is carbamide (urea), which is the principle end product of the body's protein metabolism. As the carbon dioxide waste given off by the cells reacts with the NH_3 available from the amino groups, urea is formed and excreted.

Table 14-3. Names of amides and their relationship to carboxylic acids.

Acid	*Acid formula*	*Acid amide*	*Common name*	*IUPAC name*
Formic	$HCOOH$	$HCONH_2$	Formamide	Methanamide
Acetic	CH_3COOH	CH_3CONH_2	Acetamide	Ethanamide
Propionic	CH_3CH_2COOH	$CH_3CH_2CONH_2$	Propionamide	Propanamide
Carbonic	$\{H_2CO_3$, $CO\ (OH)_2$	$CO(NH_2)$	Urea	Carbamide

THE AMINO ACIDS

Amino acids are characterized by having a molecule containing both an amino group (—NH_2) and an acid group (—COOH). They are very important because they are the building blocks of protein molecules. Amino acids are encountered in diet therapy, nutrition, and in the study of enzymes and the physiology of the body. Since proteins are long chain molecules of amino acids they can be degraded to the amino acids by hydrolysis (the addition of water molecules to the protein molecule under special conditions). Of the approximately 20 amino acids known, 10 of them are considered to be absolutely essential to life.

Amino acids are not especially complex molecules, but they do differ from each other by the location of the methyl groups (or other radicals), by the presence of sulfur atoms in the molecule and by the presence of certain ring structures of carbon and other atoms.

NAMING AMINO ACIDS

The location of the amino group and other groups on the carboxylic chain is designated by the Greek letters: alpha α, beta β, gamma γ, and delta δ. The carbon atom that is attached to the carboxyl group is the alpha carbon atom, the next carbon atom is beta, the third is gamma, and the fourth is delta (see Fig. 14-3).

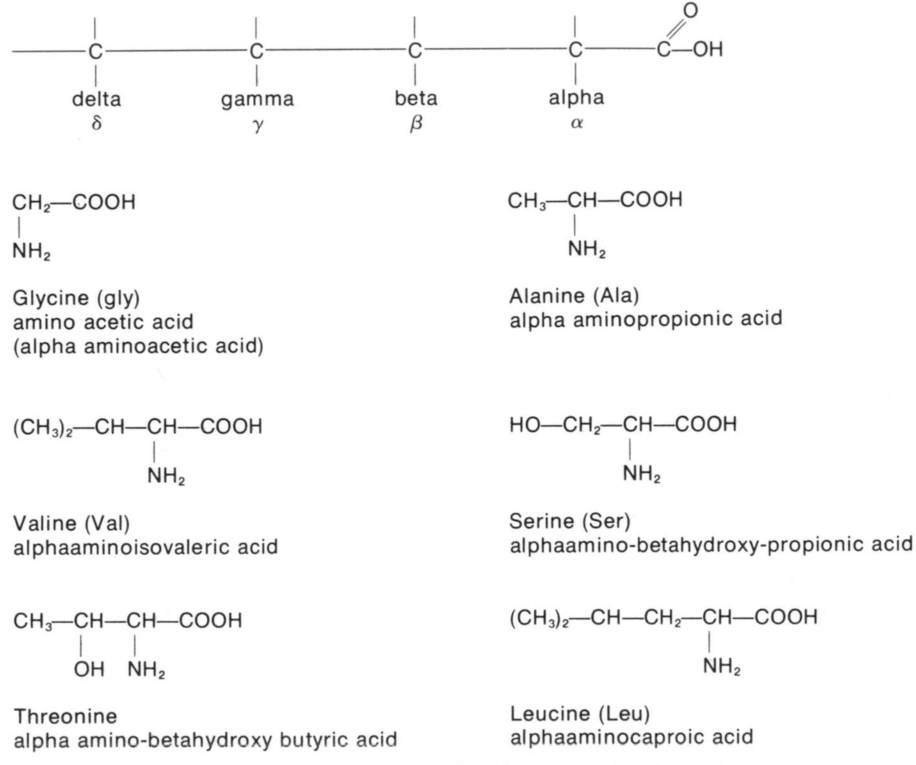

Figure 14-3. Common and IUPAC names of amino acids.

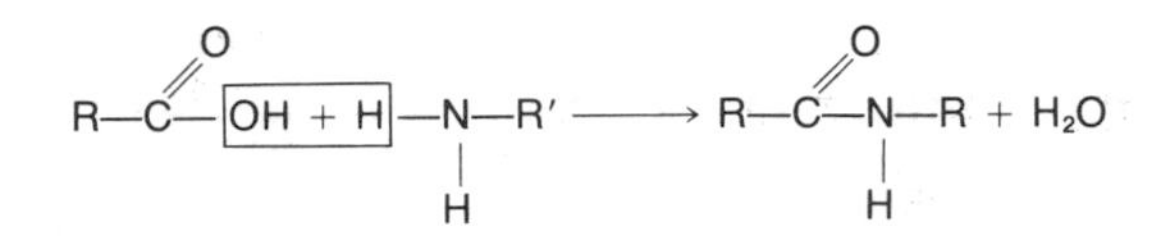

Figure 14-4. Reaction between a carboxyl group and an amino group.

Carboxyl groups can react with amino groups to form an acid amide, eliminating H_2O in the reaction and joining the 2 molecules together (Fig. 14-4).

Therefore amino acids react with one another in the body as they are supplied to the cells requesting them to form long chain molecules. The functional groups that are formed as a result of the reaction are called peptides and when many molecules react to form long chains, they are called polypeptides, the functional grouping $-\overset{O}{\overset{\|}{C}}-NH-$, being called the peptide linkage.

When 2 amino acids react, a dipeptide is formed (Fig. 14-5). As more and more amino acids react to lengthen the chain the result is a polypeptide. As the length of the molecule (and its complexity) increase, peptones, then proteoses, and finally proteins result.

This dipeptide is the beginning of a protein molecule. You can imagine the infinite and limitless number of possible sequence combinations by using all 20 amino acids and then building molecules that contain thousand upon thousand of amino acids per molecule. When the position of one amino acid in the sequence is altered a new molecule, an isomer, is formed and its chemical activity is different.

Chemical Properties of Amino Acids

AMPHOTERIC NATURE

Amino acids are amphoteric in nature, which means that they can act either as acids or bases, because the molecule contains both an acid group (—COOH) and a basic group ($-NH_2$). Because of these 2 functional groups that can react with each other amino acids can join together in extremely large and complex molecules; acid groups reacting with basic groups of another molecule and basic groups reacting with acid groups of another molecule. You might compare the molecule to an arrow that has a head and a tail. The head can be related to the —COOH group and the tail to the $-NH_2$ group (Fig. 14-6). Therefore the long molecule formed by the reaction of amino acids with each other can be represented as the formation of a series of head to tail couplings.

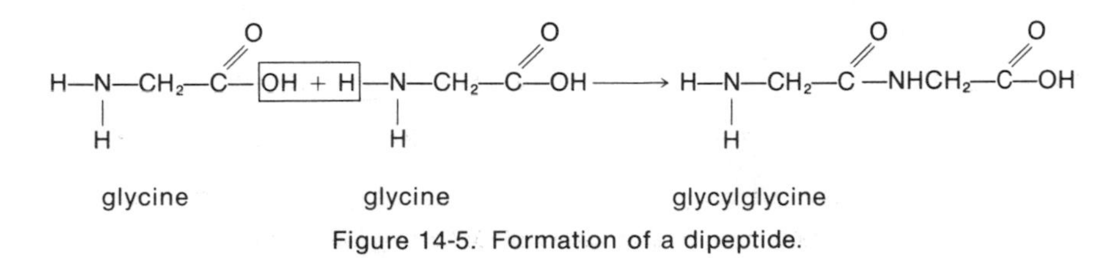

Figure 14-5. Formation of a dipeptide.

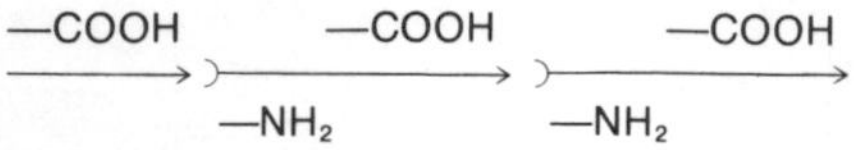

—COOH —COOH —COOH

—NH$_2$ —NH$_2$

Figure 14-6. A protein molecule.

It is the variation in the number, type and sequence of the amino acids in the protein molecule that gives rise to the vast number of possible combinations that are beyond the comprehension of the human mind.

For convenience and clarity protein molecules can be represented using the symbols R, R′, R#, R*, and R″ to designate the various radicals involved in the amino acids (see Fig. 14-7). Even with only 5 different amino acids, you can readily see how many variations are possible merely by resequencing the R's.

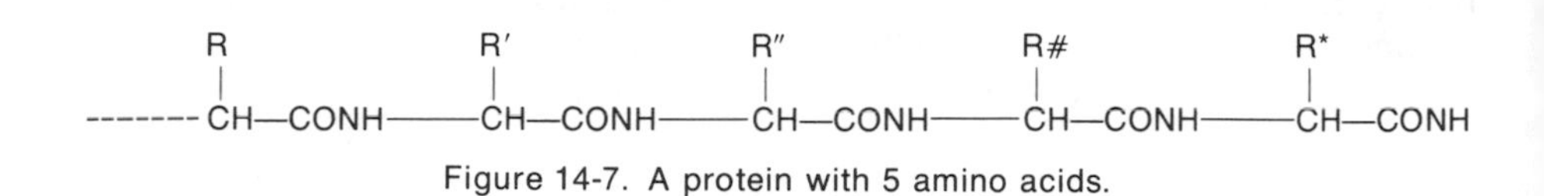

Figure 14-7. A protein with 5 amino acids.

Internal Neutralization of Amino Acids: Zwitterions

Since the molecule contains both an acid functional group and a basic functional group, the amino acid molecule can internally neutralize itself to form a dipolar ion that is called a Zwitterion (see Fig. 14-8).

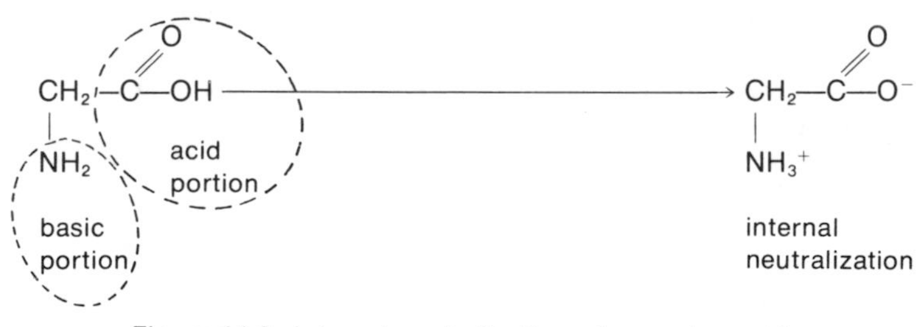

Figure 14-8. Internal neutralization of an amino acid.

The amino acid molecule is electrically neutral because it contains an equal number of positive and negative charges. And because it is both an acid and a base, it can react with both acids and bases (Fig. 14-9). It is this amphoteric nature of amino acids that is responsible for the buffering action by proteins in the blood to counter acidity and alkalinity and maintain the desired pH.

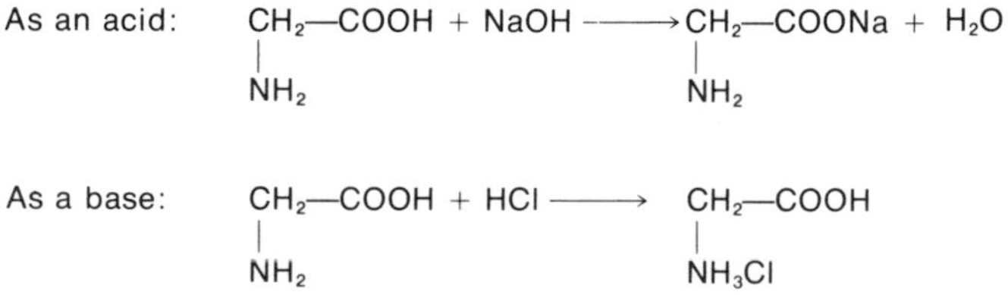

As an acid: $CH_2(NH_2)—COOH + NaOH \longrightarrow CH_2(NH_2)—COONa + H_2O$

As a base: $CH_2(NH_2)—COOH + HCl \longrightarrow CH_2(NH_3Cl)—COOH$

Figure 14-9. Action of amino acids as acids and bases.

$$\text{alcohol} + HNO_2 \longrightarrow \text{nitroso compound}$$

$$\text{Amyl alcohol} + HNO_2 \longrightarrow \text{amyl nitrite}$$

$$CH_3—CH_2—CH_2—CH_2—CH_2—OH + HNO_2 \longrightarrow CH_3—CH_2—CH_2—CH_2—CH_2—NO_2 + H_2O$$

Figure 14-10. Synthesis of a nitroso derivative.

NITROSO COMPOUNDS

The functional nitroso group is $—NO_2$ and is formed by the reaction of an alcohol with nitrous acid.

Amyl nitrite is a valuable medication used in the treatment of angina pectoris, an abnormal heart condition. It acts as a vasodilator to relax the blood vessels, increase blood flow, and thus reduce the pain associated with this condition (Fig. 14-10).

NITRO DERIVATIVES

The characteristic functional group of nitro compounds is the $—NO_3$ group. Esters of nitric acid with polyhydric alcohols, such as glycerin, are used as heart stimulants. The alcohol functional group reacts with the nitric acid to form the nitro ester, a nitrate. Glycerol trinitrate, commonly called nitroglycerin, is one of the most commonly used (Fig. 14-11).

$CH_2—OH$		$HONO_2$	$CH_2—ONO_2$	H_2O
$CH—OH$	+	$HONO_2$ ⟶	$CH—ONO_2$	H_2O
$CH_2—OH$		$HONO_2$	$CH_2—ONO_2$	H_2O
glycerine		nitric acid	glycery trinitrate	water

Figure 14-11. Nitro esters of alcohols.

NITROGEN HETEROCYCLICS

Heterocyclic ring structures contain atoms of other elements in addition to the carbon atoms. Those atoms may be nitrogen or sulfur, and ring structures that contain nitrogen atoms form the basis of many compounds that are essential to normal body processes. These compounds are the alkaloids, hormones and enzymes. The most important ring structures are given below, each ring structure having a specific name assigned to it.

Pyrrole

Pyrrole is a five membered ring that contains 4 carbon atoms, 1 nitrogen atom, and 2 double bonds (Fig. 14-12). This ring is found in the hemoglobin of the blood and in the chlorophyll of plants. The similarity of these two essential compounds becomes apparent when you realize that in hemoglobin there are

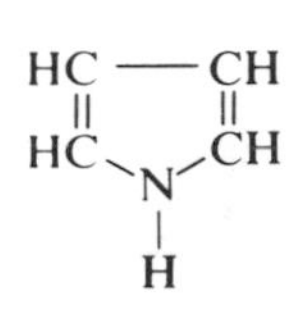

Figure 14-12. The pyrrole ring.

4 pyrrole rings surrounding 1 atom of iron, and in chlorophyll there are 4 pyrrole rings surrounding 1 atom of magnesium.

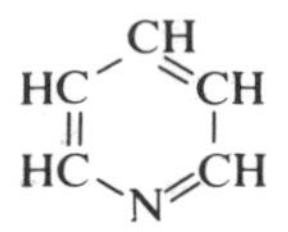

Figure 14-13. The pyridine ring.

Pyridine

Pyridine is a six membered ring (Fig. 14-13) that contains 5 carbon atoms, 1 nitrogen atom, and 3 double bonds. This ring structure is found in niacin (nicotinic acid) and in nicotinamide, members of the B vitamin family (Fig. 14-14).

Purine

Purine is composed of 2 rings fused together (Fig. 14-15), a pyrimidine ring (Fig. 14-16) and an imidazole ring. The purines are extremely important because purine bases are found in RNA (ribonucleic acid) and in DNA (deoxyribonucleic acid). In the DNA molecule, the 2 purines (adenine and guanine) link up with the pyrimidines (thymine and cytosine), respectively (Figs. 14-17 and 14-18).

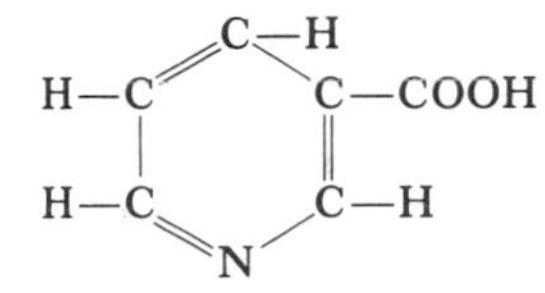

Figure 14-14. Niacin (nicotinic acid).

Alkaloids

Alkaloids are very complex nitrogeneous compounds which are found in plants. Some are liquids while others are crystalline solids, the liquid ones having a disagreeable odor and the solids having a bitter strong taste. Most of them are insoluble in water, but are made water soluble by the formation of a salt with the addition of HCl or H_2SO_4 to form the hydrochloride or sulfate salt. The alkaloids are composed of heterocyclic rings containing nitrogen atoms, and are of extreme interest because they can act as tranquilizers, stimulants, muscle relaxers, mild pain killers, powerfull sedatives, anesthetics, vasodilators, antimalarials, and hallucinogenics. Table 14-4 lists sources and uses for various alkaloids.

DRUG OVERDOSE

In the past decade the use of drugs by people of all ages throughout the world has led to a marked increase in the number of drug fatalities, drug addictions and emergency room treatment of drug overdose. Health care personnel should

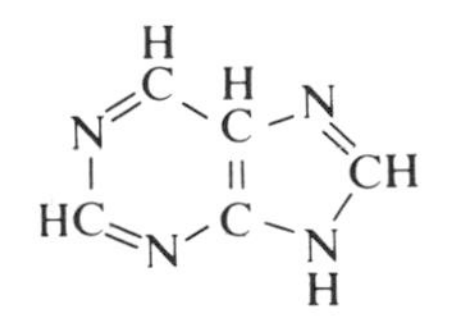

Figure 14-15. The purine ring.

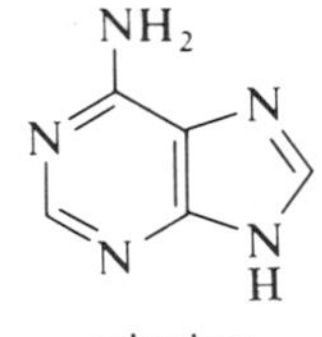

adenine

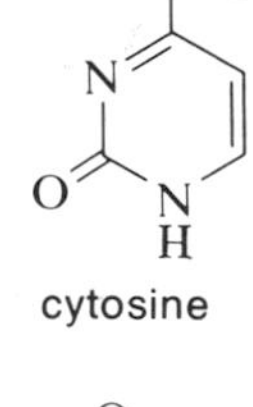

cytosine

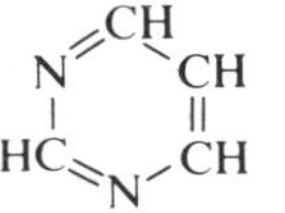

Figure 14-16. The pyrimidine ring.

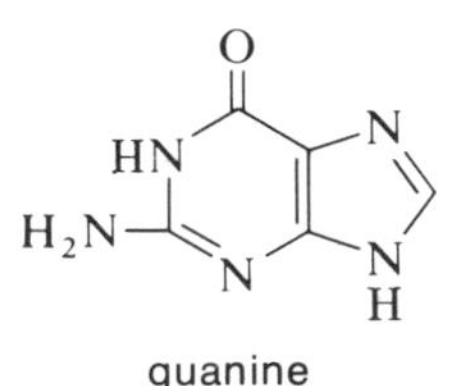

guanine

Figure 14-17. The purines.

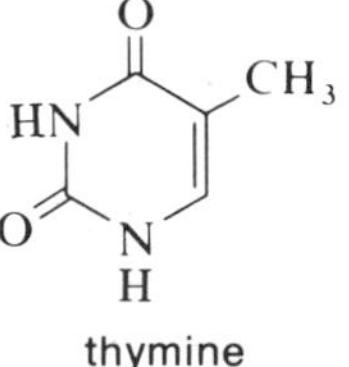

thymine

Figure 14-18. The pyrimidines.

Table 14-4. Sources and uses of alkaloids.

Alkaloid	*Source*	*Uses*
Reserpine	Rauwolfia roots	Tranquilizer, induces sleep.
Atrychnine	Nux vomica tree	Tonic and stimulant in small doses. Overdose can cause convulsions and death.
Morphine	Opium poppy	Powerful sedative, pain reliever.
Papaverine	Opium poppy	Relaxes involuntary muscles, anti-spasmodic, dilates blood vessels.
Codeine	Opium poppy	Similar to morphine but not as strong, habit forming, used in cough medi-cines.
Paregoric	Opium extract	Alcoholic, camphorated solution, eases pain, slows persistaltic action.
Caffeine	Tea, coffee	Mild stimulant, vasodilator.
Nicotine	Tobacco plant	A spastic poison, causes spasm of pe-ripheral arteries.
Cocaine	Coca leaves	Local anesthetic, toxic, has now been replaced by novocain.
Quinine	Cinchona bark	Specific for malaria, reduces fever, used as tonic to stimulate appetite.
Atropine	Deadly nightshade	Dilates pupils, secretory depressant, used by ophthalmologists
Emetine	Ipecac	Used in treatment pyorrhea and ame-bic dysentery.
LSD	Lysergic acid	Hallucinogenic.

learn to recognize the symptoms of drug overdose, regardless of the drug, and the emergency procedures that can be rendered to save and help the patient.

SULFUR COMPOUNDS

The sulfur atom can usually be found in organic molecules wherever oxygen atoms are found because sulfur is in the same periodic table group as oxygen. Its chemical reactions resemble those of oxygen, and the nomenclature of the sulfur analogs to oxygen is modified by the use of the terms thio, sulfhydryl, sulfide or mercapto.

Thioalcohols

The sulfur atom replaces the oxygen atom in the functional group —OH to become —SH (Table 14-5). The thioalcohols are highly odoriferous and have disagreeable odors, resembling that of hydrogen sulfide. When thioalcohols are

Table 14-5. Thioalcohols.

Oxygen alcohol	*Oxygen alcohol formula*	*Sulfur alcohol formula*	*Names of sulfur alcohols*
Methanol	CH_3—OH	CH_3—SH	Methyl mercaptan, methanethiol
Ethanol	CH_3—CH_2—OH	CH_3—CH_2—SH	Ethyl mercaptan, ethanethiol

$$CH_2—CH_2—SH + CH_3—CH_2—SH \xrightarrow[\text{agent}]{\text{oxidizing}} CH_3—CH_2—S—S—CH_2—CH_3 + H_2O$$

Figure 14-19. Oxidation of thioalcohols.

oxidized disulfides are formed; a linkage that is most important in maintaining the structure of proteins (Fig. 14-19).

Thioethers

The sulfur atom replaces the oxygen atom and acts as the bridge between the alkyl groups instead of the oxygen atom. The thioethers are also called sulfides (Table 14-6).

Table 14-6. Thioethers.

Oxygen ether	*Sulfur ether*	*Name of sulfur ether*
$CH_3—O—CH_3$	$CH_3—S—CH_3$	Dimethyl sulfide, methyl thiomethane
$CH_3—CH_2—O—CH_2—CH_3$	$CH_3—CH_2—S—CH_2—CH_3$	Diethyl sulfide, ethyl thio-ethane

Thioacids

The thioacids have the functional group $—\overset{O}{\overset{\|}{C}}—SH$, in which the sulfur atom replaces the oxygen atom of the —OH part of the group. Acetic acid is $CH_3—\overset{O}{\overset{\|}{C}}—OH$; thioacetic acetic acid is $CH_3—\overset{O}{\overset{\|}{C}}—SH$. The thioacids serve as raw materials for the synthesis of the thioamides.

Thioamides and Sulfonamides

The thioamides have the functional group $—\overset{S}{\overset{\|}{C}}—NH_2$, the oxygen amides being $—\overset{O}{\overset{\|}{C}}—NH_2$. The sulfonamides are derived from sulfonic acid $—SO_2OH$ which is analogous to the carboxylic acid group $—\overset{O}{\overset{\|}{C}}—OH$. The drug sulfanilamide is not a true amide but a sulfonamide, $—SO_2NH_2$ (the acid amide of a sulfonic acid). This compound was accepted in the medical field in the middle 1930s to treat many infectious diseases. Since that time, many modifications of the structure of sulfanilamide have been tested, and some have fewer of the undesirable side effects of the original compound and greater bactericidal properties. Sulfanilamide was the first of the so-called "wonder" drugs.

Sulfones

Sulfones have the functional group $—\underset{\underset{O}{\|}}{\overset{\overset{O}{\|}}{S}}—$ where the $—SO_2—$ group bridges 2 radicals. A commonly used pH colorimetric indicator is phenolsulfonphthalein (PSP), commonly called phenol red. The kidneys have the capability to excrete this compound, and when it is intravenously injected, samples of urine are analyzed for the amount of dye present, which is a measure of the excretory function of the kidneys. Another important sulfone is red dye #2, a dye that has recently been banned for use in foods because it is a suspected carcinogen.

PLASTICS AND HIGH MOLECULAR WEIGHT POLYMERS

In today's health care, the use of disposable plastic items is increasing at a very high rate. Hypodermic syringes, irrigation components, tubing, food and dispensing items, and many more plastic items are commonplace. After being manufactured, they are sealed and sterilized, offering health care personnel added assurance of sterility, the chance of nonsterility and contamination being extremely minute if the container has not been opened or broken. To use these plastic items to their best advantage you should know something about them and how they behave under the influence of heat and pressure. They are lumped under the term plastics but there are many different plastic products, each having its own characteristic advantages and disadvantages.

In general, a plastic is a substance which under the proper conditions of heat and pressure can be molded into a desired form and shape. Some plastic products will deform or change their shape if strain, stress or heat is applied whereas others will not. *Thermoplastic* substances will flow under the effect of heat. Combs, novelties, toys, and ornamental items usually are manufactured from these plastic substances. *Thermosetting* plastics, once formed under the influence of heat and pressure, resist any further change by heat or pressure.

Most synthetic plastic products, polymers, are very large molecules that have been made from single small molecules called monomers. The large molecules may be formed by two methods: *Polymerization* takes place when double bonds of small molecules open up and the molecules link together to form long chains. *Condensation* takes place when functional groups between molecules react, usually eliminating water. The formation of esters (polyesters) and amides (polyamides) are examples of such reactions. See Table 14-7 for a summary of common plastic materials.

The Polyolefins

The carbon atom has the ability to covalently bond with other carbon atoms to form chain-like structures containing tens of thousands of atoms. This is the reason why we have polyethylene, polypropylene, and polybutylene plastic products. They are manufactured from the corresponding olefin, which as you know has a double bond, ethylene ($CH_2{=}CH_2$), or propylene ($CH_3{-}CH{=}CH_2$). That double bond is unstable, and under the proper conditions, the double bond of the olefin can open up, causing the molecule to have 2 free valence electrons. Molecules can then join together to form extremely long chains, now containing stable single bonds between the carbon atoms (Fig. 14-20). They are called polyethylene and polypropylene because they were made from the corresponding olefin.

Polyethylene is a soft and inert plastic substance used to make plastic squeeze bottles. It resists deformation by springing back into shape, and is used to make plastic tubing, plastic bottles, waterproof paper, and food wrap.

Other polyolefines have been prepared, and one of the most widely known

Table 14-7. Summary of the various common plastic materials.

Type	Trade Name	Method of formula	Uses
Vinyls	Vinylite Koroseal Orlon	P	Floor tiles, rain coats, shower curtains, upholstery.
Phenolics	Bakelite	C	Electrical fixtures, electronic components, insulators, equipment knobs.
Acrylics	Lucite Plexiglas	P	Bomber noses, nonshattering glass acrylic lacquers, auto paints, optical lenses, safety glass laminate.
Cellulose acetate	Cellophane Tenite	N	Fabrics, radio cabinets, clear protective sheetings, signs, auto plastic parts.
Urea	Beetle Plaskon	C	Buttons, bottle closures.
Celluose Nitrate	Celluloid Pyroxylin		Decorative plastic, eyeglass frames, cutlery handles, buttons, toilet seats, novelties, screw driver handles.
Isoprene	Buna Butyl rubber	P	Synthetic rubber for tires, hoses, fan belts, power transmission belting, foul weather gear.
Polyamides	Nylon	C	Hosiery, parachutes, brush bristles.

P = Polymerization.
C = Condensation.
N = Cellulose is a natural polymer, and it is modified by acetic acid reacting with the hydroxyl groups to give an acetate (an ester).

is polytetrafluoroethylene (Teflon). Instead of polymerizing ethylene ($CH_2{=}CH_2$), tetrafluoroethylene ($CF_2{=}CF_2$) is used. The Teflon polymer is unique. Nothing will stick to it, and it is used for coating cookware and to make bearings for equipment.

Lucite and Plexiglas are trade names for polymers of acrylic acid esters. The double bonds open up to form long chain polymers that are crystal clear, thermoplastic and very strong. Infant incubators, dentures, windshields for air-

$CH_2{=}CH_2$ $\quad$ $CH_2{=}CH_2$ $\quad$ $CH_2{=}CH_2$ $\quad$ Molecules of ethylene

$(—CH_2—CH_2—)$ $(—CH_2—CH_2—)$ $(—CH_2—CH_2—)$ reactive form of molecule

$(—CH_2—CH_2—CH_2—CH_2—CH_2—CH_2—)$ $\quad$ part of the chain of the high molecular weight polymer

Figure 14-20. Simplified reactions involved to form polyethylene plastic.

Polymethylmethycrylate.

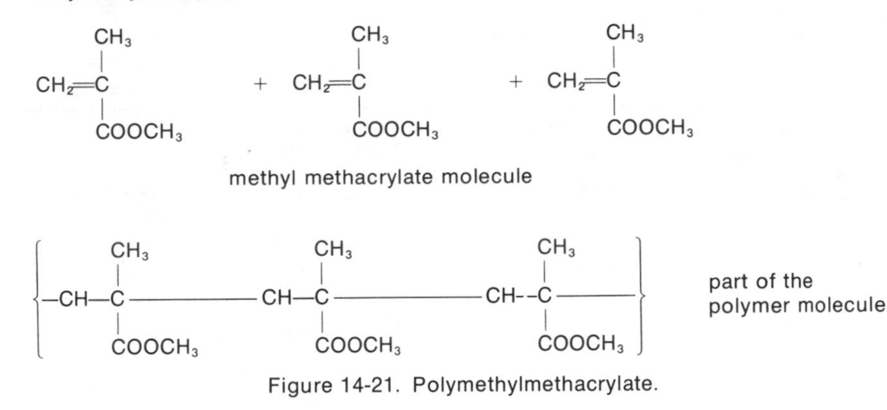

Figure 14-21. Polymethylmethacrylate.

planes and optical instruments are made from polymethylmethacrylate, the methyl ester of acrylic acid (Fig. 14-21). Polymethylmethacrylate has the unique property of conducting light just as a water pipe conducts water, and therefore it can be used to "see around corners" and is used in endoscopes for gastrointestinal diagnosis.

Polyesters and Polyamides

Organic acids react with alcohols to form esters, and polyacids react with polyalcohols to form chain-like molecules called polyesters. In the reaction, water is split out between the —OH group of the acid and the —OH group of the alcohol (Fig. 14-22).

Polyamides result from the condensation of a dicarboxylic acid (a polyacid) with a diamine (a polyamine). Nylon 66 is a polyamide made from a 6 carbon dicarboxylic acid and a 6 carbon diamine. Water is eliminated between the acid and the amine to form the amide linkage.

Sources of Raw Materials for Plastics

Most of the raw materials for making various plastic products come from the petroleum industry, and are formed when large oil molecules are broken apart under the influence of heat and pressure.

Care and Use of Plastic Products

WASHING: You can wash all plastic products just as you would wash glassware, except that you should not use harsh abrasive cleansers.

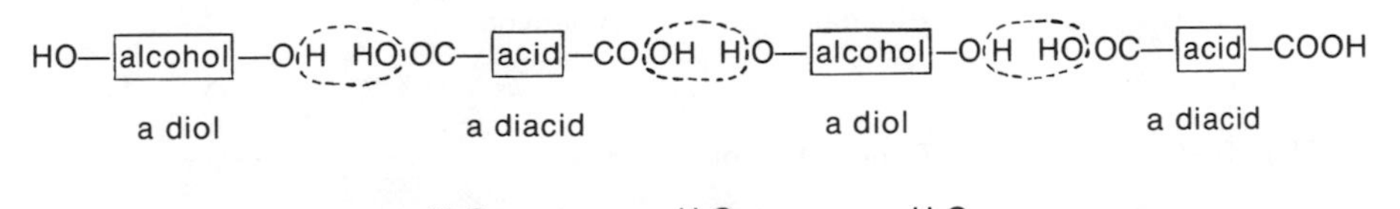

—alcohol—acid—alcohol—acid—alcohol—acid—

Figure 14-22. Reaction to form polyester.

Table 14-8. A summary of the simple functional groups of organic compounds

Formula	*Name*	*Representative compound*	*Functional group*
$CH_3—CH_3$	Eth*a*ne	Alk*ane* (saturated hydrocarbon)	None
$CH_2=CH_2$	Eth*e*ne	Alk*ene* (unsaturated hydrocarbon)	Double bond $C=C$
$CH≡CH$	Eth*y*ne	Alk*y*ne (unsaturated hydrocarbon)	Triple bond $C≡C$
$CH_2—CH_2$ / $CH_2—CH_2$ (ring, vertical bonds)	*Cyclo*butane	Cyclic hydrocarbon	None
$CH_3—CH_2—OH$	Ethano*l*	Primary alcohol	—OH
$CH_3—CH(OH)—CH_3$	Isopropyl alcohol	Secondary alcohol	—OH
$Ch_3—C(CH_3)(CH_3)—OH$	Tertiary butyl alcohol	Tertiary alcohol	—OH
$CH_2(OH)—CH_2(OH)$	Ethylene Glycol	Glycol	—OH
$CH_3—C(=O)—OH$	Ethan*oic* Acid	Carboxylic acid	$—C(=O)—OH$
$CH_3C(=O)—O—CH_2—CH_3$	Ethyl ethan*ate*	Ester	$—C(=O)—O—$
$H_3—CH_2—O—CH_2—CH_3$	Diethyl *Ether*	Ether	—O—
$CH_2—CH_2$ with bridging O	Ethylene oxide	Epoxide	$—C—C—C$ with O bridging the first two C
$CH_3—C(=O)—H$	Ethanal	Aldehyde	$—C(=O)—H$
$CH_3—C(=O)—CH_3$	Dimethyl ketone	Ketone	$—C(=O)—$
$CH_3—CH_2—Cl$	Ethyl chloride	Alkyl halide	—Cl

Table 14-8. Continued

Formula	*Name*	*Representative Compound*	*Functional group*
$CH_3—NH_2$	Methyl amine	Primary amine	$—NH_2$
CH_3 | $CH_3—N—H$	Dimethyl amine	Secondary amine	$—NH$ |
CH_3 | $CH_3—N$ | CH_3	Trimethyl amine	Tertiary amine	| $—N$ |
O // $CH_3C—NH_2$	Ethamide	Amide	O // $—C—NH_2$
O // $CH_2—C—OH$ | NH_2	Aminoethanoic acid	Amino acid	| O | // $—C—C—OH$ | NH_2
$CH_3—CH_2—NO_2$	Ethyl nitrite	Nitroso compound	$—NO_2$
$CH_2—ONO_2$ | $CH—ONO_2'$ | $CH_2—ONO_2$	Trinitroglycerol	Nitro compound	$—ONO_2$
HC—CH ‖ ‖ HC CH \\ / N | H	Pyrrole	Nitrogen heterocyclic	—C—C— | | —C C— \\ / N |
$CH_3—SH$	Methanethiol	Thioalcohol	$—SH$
$CH_3—S—S—CH_3$	*D*imethyl	Disulfide	$—S—S—$
O // $CH_3—C—SH$	Thioethanoic acid	Thioacid	O // $—C—SH$
S // $CH_3—C—NH_2$	Thioacetamide	Thioamide	S // $—C—NH_2$
$R—SO_2NH_2$	Radical + sulfonamide	A sulfonamide	$—SO_2NH_2$

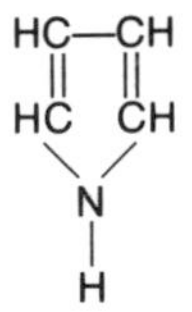

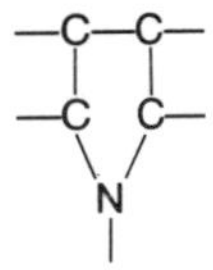

STERILIZING: The polyolefins, (linear polyethylene, polypropylene polycarbonate, and Teflon can be repeatedly autoclaved at 121°C for 20 minutes (see product instructions). You normally clean items before sterilizing. The following plastics *cannot be sterilized with heat,* because they will deform: PVC (polyvinyl chloride), polystyrene, styrene-acrylonitrile, and conventional polyethylene. You can sterilize them with the quatenary ammonium halides. Always check the instruction sheets.

AUTOCLAVING: The closures on all containers must be loose during and after the autoclave procedure. If they are closed tightly they may rupture on expansion or collapse because of a partial vacuum.

MARKING: Use conventional grease or wax marking pencil.

Table 14-8 summarizes the organic substances discussed in the previous chapters.

SUMMARY

When halogen, nitrogen and sulfur atoms are introduced into the molecule, they may be present as single atoms or in functional groups. Alkyl halides are used as anesthetics and solvents for grease and oil. Amines can be considered as alkyl derivatives of ammonia. Their solutions are basic, and they react with inorganic acids to form salts. Teritiary amines react with alkyl halides to form quaternary ammonium halides, called "quats," which are cold disinfecting solutions. When the —OH group of acids is replaced with an $—NH_2$ group, acid amides result. The amino acids are the building blocks of proteins containing both an amino and an acid group; therefore they are amphoteric. They can act as either acids or bases and are components of proteins. Nitrogen atoms as nitroso or nitro functional groups are used to treat abnormal heart conditions, and nitrogen atoms form important heterocyclic ring structures in the body. Sulfur, being in the same periodic group as oxygen, can yield sulfur analogs of the oxygen compounds. They are usually identified by the terms thio, sulfhydryl, sulfide, and mercapto. The disulfide linkage is a very important protein bond linkage. Table 14-8 summarizes the common functional groups found in organic compounds along with their formulas and common names.

EXERCISE

1. Name three important alkyl halides used in medicine and write their formulas.
2. Write the reaction between water and a primary, secondary, and tertiary amine.
3. Why are inorganic acid salts of amines important? Illustrate with the compound morphine.
4. What are quaternary ammonium halides and what are they used for? Draw the formula for tetraethyl ammonium chloride.
5. Show the relationship of the acid amide functional group to the carboxylic acid functional group and name a very important amide.
6. Give the general structure for amino acids and state the rules for naming them.
7. How do amino acids react to form dipeptides and polypeptides?
8. What is a zwitterion and why can amino acids act as buffers?
9. What is the importance of the nitroso and nitro derivatives of organic compounds to health care personnel?
10. What is the difference between chlorophyll and hemoglobin?
11. Give the general rules for naming the sulfur analogs of oxygen alcohols, ethers, acids, and amides.
12. What is the importance of the disulfide linkage?
13. What is meant by the term sulfa drugs and what is their functional group?
14. Define monomers and polymers and distinguish between polymerization and condensation reactions.
15. What is the source of most of the monomers used to manufacture plastics, and name one that tends to cause cancer.
16. Complete the following chart:

Name	Formula	Functional group
Propane		
Ethene		
Ethyne		
Cyclopropane		
1-propanol		
2-propanol		
Tert-butanol		
Ethylene glycol		
Glycerol		

Propanoic acid ____________________

Methyl acetate ____________________

Dimethyl ether ____________________

Propylene oxide ____________________

Propanal ____________________

Methyl ethyl ketone ____________________

Ethyl chloride ____________________

Ethyl amine ____________________

Dimethyl amine ____________________

Trimethyl amine ____________________

Acetamide ____________________

Amino-acetic acid ____________________

Ethyl nitrite ____________________

Glycerol trinitrate ____________________

Ethanethiol ____________________

Diethyl disulfide ____________________

Thioacetamide ____________________

OBJECTIVES

When you have completed this chapter you will be able to:

1. Describe the basic structure of the aromatic compounds and draw formulas to explain the delocalization of the electrons in the molecule.
2. Give both the common usage and IUPAC names of the simple aromatic hydrocarbons and their derivatives.
3. Explain by drawing formulas why there is only one monosubstituted derivative of benzene.
4. Identify compounds that are ortho, meta and para derivatives.
5. List the properties of the aromatic hydrocarbons and their uses.
6. Draw and name the polynuclear aromatic hydrocarbons and state their uses and importance.
7. Describe why halogen derivatives of the aromatic compounds are important to health care personnel.
8. Distinguish the difference between an alcohol and a phenol.
9. Discuss phenol coefficients.
10. Draw the formulas for the three isomeric cresols, for resorcinol and hexylresorcinol, giving the uses of these compounds.
11. Give the formulas for the two most commonly encountered aromatic carboxylic acids and their uses.
12. Summarize the uses of methylsalicylate, phenylsalicylate, and acetylsalicylic acid and discuss how they act in the body through hydrolysis.
13. Name and draw the formulas of trinitroglycerine, trinitrotoluene, and trinitrophenol and state where they are used.
14. Illustrate the formulas of the simplest aromatic aldehyde and ketone and list uses of these compounds.

AROMATIC HYDROCARBONS AND DERIVATIVES

15. Write the chemical reaction showing the formation of an amine salt and state the importance of this reaction.
16. Illustrate the formation of an amide and draw the formula for phenacetin.
17. Show the basic structure of the sulfa drugs and state how they act to stop the growth of pathogenic microorganisms.

A completely different series of hydrocarbons are formed when a six membered carbon ring contains 3 double bonds. These bonds are not the same type of double bonds that you found in the alkenes, and their presence in the ring structure markedly affects the chemical reactivity of the molecules. These compounds are called the aromatic series and the compounds must have a ring structure containing 6 carbon atoms with 3 double bonds attached to *alternate* carbon atoms.

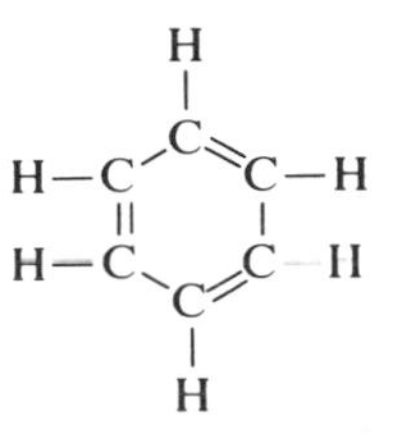

Figure 15-1. Benzene.

AROMATIC COMPOUNDS

The simplest aromatic compound is benzene (Fig. 15-1). It can be represented as shown in Figure 15-2. At each corner of the hexagon there is a —CH= grouping. Regardless of which figure is shown they both mean the same thing, that an aromatic nucleus (a benzene ring) is present.

The one important fact to remember is that these 3 double bonds are not the ordinary double bonds that you studied in the alkenes, and they do not react as the alkene double bonds. The electrons are delocalized (spread out) over the whole ring, and the double bond oscillates and changes positions in the molecule. This oscillation is indicated by double headed arrow (see Fig. 15-3).

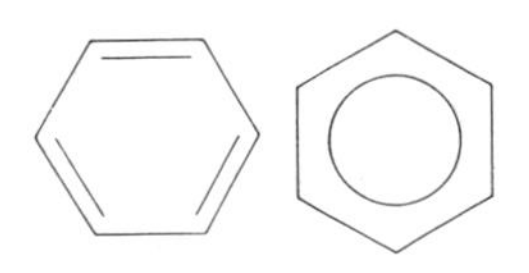

Figure 15-2. Drawings representing benzene.

Figure 15-3. Representation of oscillating double bonds.

Naming Derivatives of Aromatic Compounds

Carbon atoms in the straight and branched chain hydrocarbons and their derivatives were numbered to identify the location of radicals and functional groups. In the aromatic compounds the carbon atoms are also numbered to

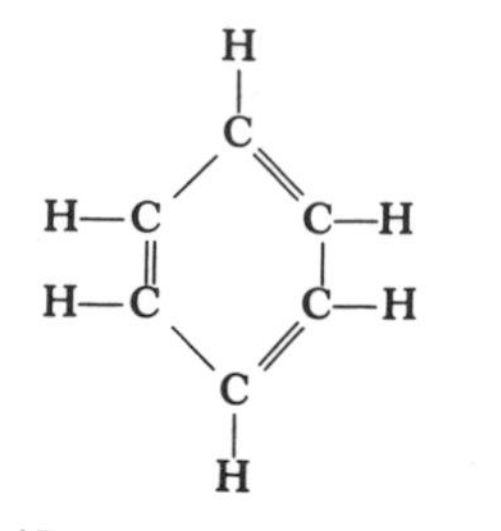

Figure 15-4. Hydrogen and carbon bonds of benzene.

locate the positions of any substituents for the single hydrogen atom attached to each carbon atom in the six membered ring. You should take special notice that there is only 1 hydrogen atom attached to each carbon atom in benzene, and that hydrogen atom can be replaced with any radical or functional group (Fig. 15-4).

IMPORTANT: it does not matter which carbon atom is designated as number one as long as all other substituting groups and radicals are located with respect to that particular carbon atom. The usual practice is to start with a radical, such as methyl, or an ethyl.

These are substitution reactions. They *are not* addition reactions, and since the 6 carbon atoms are in a ring, each carbon atom is identical to every other carbon atom. When only 1 substituting radical or functional group is inserted in the ring, the location of that radical does not have to be specified by a number because the molecule is always randomly rotating (see Fig. 15-5).

When 2 radicals are substituted for 2 hydrogen atoms in the ring, a series of isomers are obtained, depending upon the relative position of the substituting groups in the ring with respect to each other. The naming of these isomers is governed by the following rules:

Rule 1: When the 2 groups are adjacent to each other (on neighboring carbon atoms) the compound is termed an *ortho* isomer. If 1 group is labeled as being on carbon atom #1, the second group would be on carbon atom #2 (Fig. 15-6).

Rule 2: When the substituting radicals or groups are separated by *1* carbon atom, the compound is termed a *meta* isomer, and if 1 group were on carbon atom #1, the second group would be on carbon atom #3 (Fig. 15-7).

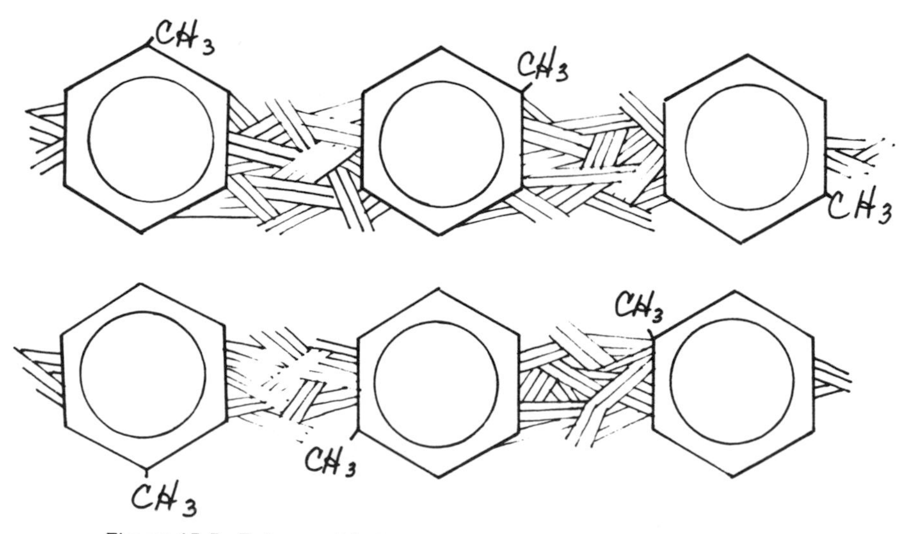

Figure 15-5. Toluene. All structures represent the same compound.

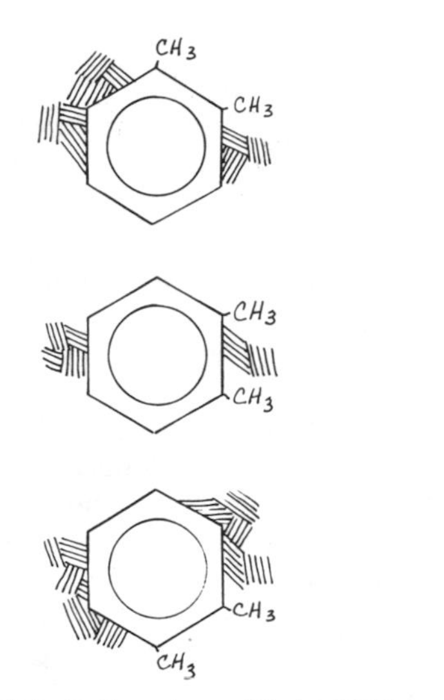

Figure 15-6. Orthoxylene. All structures represent same compound.

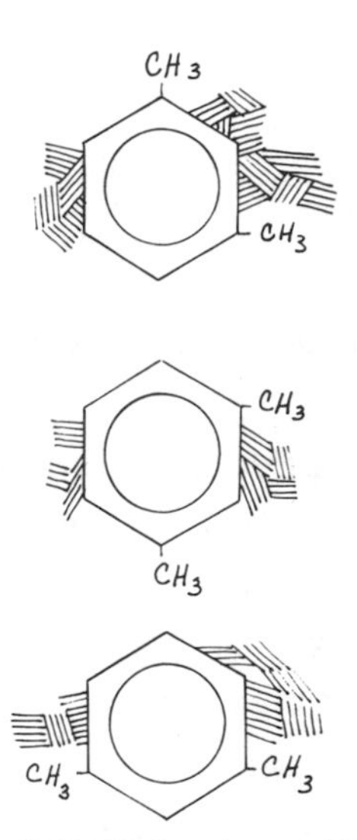

Figure 15-7. Metaxylene. All structures represent same compound.

Rule 3: When the substituting radicals or groups are separated by *2* carbon atoms, the compound is termed a *para* isomer, and if 1 group were on carbon atom #1, the other would be on carbon atom #4 (Fig. 15-8).

Properties of Aromatic Hydrocarbons

The aromatic hydrocarbons are relatively nonreactive hydrocarbons. Their main use is as solvents and as raw materials for the manufacture of drugs and chemicals that have the aromatic ring structure. They are insoluble in water and are extremely flammable, burning with a smoky flame. This smoky flame combustion is a characteristic property of the aromatic hydrocarbons. Extreme care should be exercised when using the volatile compounds because their vapors are toxic, and some of the higher molecular weight aromatic derivatives, especially the polynuclear (polybenzene ring) ones, are carcinogenic.

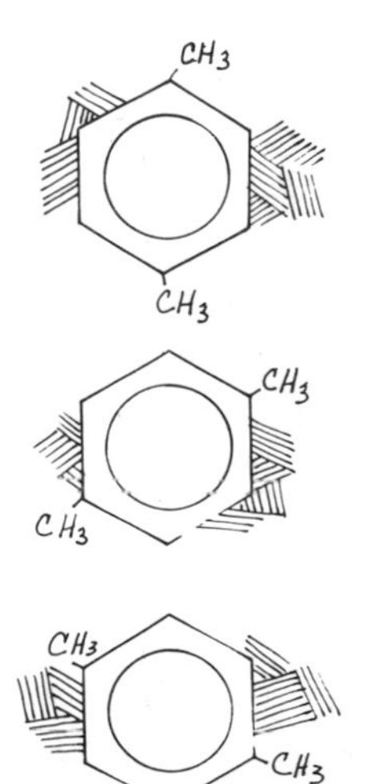

Figure 15-8. Paraxylene. All structures represent same compound.

BENZENE

Benzene, C_6H_6, should not be confused with the substance called *benzine,* which is the common name for a mixture of straight chain hydrocarbons, pentane and hexane, and their isomers. Benzene is the basic structure for many single ring compounds in which the hydrogen atoms have been replaced by alkyl groups and functional groups previously studied (the —OH, —COOH, —CHO, —NO_3, and halogens). Continued inhalation of benzene (and other volatile aromatic compounds) leads to a reduction of the red and white corpuscles of the blood, which could be fatal.

TOLUENE

Toluene, C_7H_8, is a solvent for grease and oils and is the basis for manufacturing trinitrotoluene. Structurally it is methylbenzene, and since it is a mono substituted benzene, there is only one toluene.

XYLENE

Xylene is a solvent for greases and oils. Because it is structurally dimethylbenzene there are 3 possible isomers of xylene.

Polynuclear Aromatic Compounds

Two or more benzene rings can be fused together, forming polynuclear benzenoid ring structures. As in benzene, the electrons are delocalized and the compounds resemble benzene in their chemical reactivity regarding the double bonds. These polynuclear compounds are important as compounds and as raw materials for the synthesis of drugs and chemicals.

NAPHTHALENE

Naphthalene, $C_{10}H_8$, is the simplest polynuclear aromatic compound. Its structure is represented in Figure 15-9, and the location of any substituting radical or functional group is designated by the number of the carbon atom to which it is attached. Naphthalene is the substance commercially used in moth balls which prevent damage to clothing. It is a white solid that slowly sublimes (changing directly from a solid to a vapor).

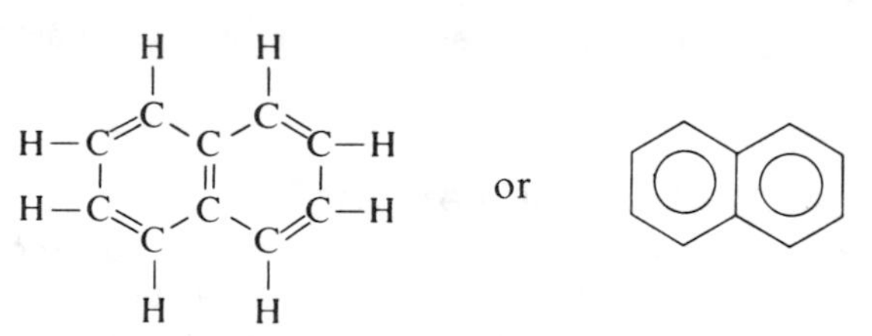

Figure 15-9. Structures representing naphthalene.

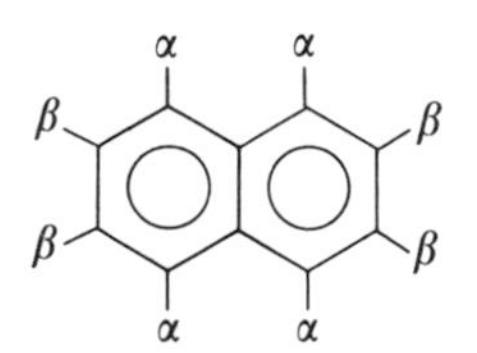

Figure 15-10. Alpha and beta carbon atoms of naphthalene.

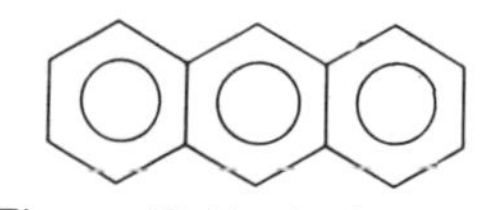

Figure 15-11. Anthracene.

Naming of Naphthalene Derivatives

Substituted naphthalene can be named according to common usage or by the IUPAC system. Carbon atoms 1,4,5, and 8 are the ALPHA carbon atoms, and 2,3,6, and 7 are the BETA carbon atoms (Fig. 15-10).

Anthracene, $C_{14}H_{10}$, is a tribenzene fused ring with the rings in a straight line. It is used for the manufacture of clothing dyes and indicators for chemical procedures (Fig. 15-11).

Phenanthrene, a tribenzene fused structure, has 1 ring located at an angle (Fig. 15-12). This structure is very important as it is the basic structure of all male and female sex hormones, bile acids, vitamin D, cholesterol, and some alkaloids in modified degrees of saturation.

Benzopyrene is a polynuclear benzeneoid structure known to cause lung cancer and genetic mutations (Fig. 15-13). It is found in the incomplete combustion of coals and oils and in the atmosphere of urban centers (caused by factory furnaces and automobile pollution). It has also been found in charcoal broiled steaks and chops, due to the decomposition of fats and greases falling on the hot charcoal. Lung cancer can result from smoking and inhalation of cigarette smoke.

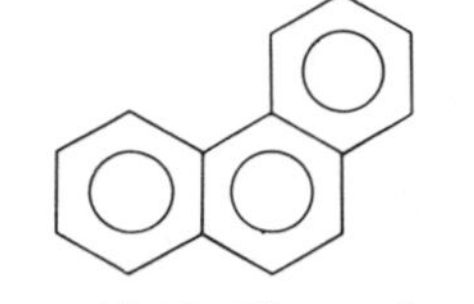

Figure 15-12. Phenanthrene.

The Aromatic Radicals

Just as the alkyl radicals were formed by the removal of a hydrogen atom from the corresponding alkane, the aromatic radicals are formed by the removal of 1 or more hydrogen atoms from the corresponding aromatic compounds. The radicals cannot exist by themselves, needing another radical or functional group attached to the free valence bond (Table 15-1).

Table 15-1. Names of the aromatic hydrocarbon radicals.

Hydrocarbon	*Radical*
Benzene	Phenyl
Naphthalene	Naphthyl

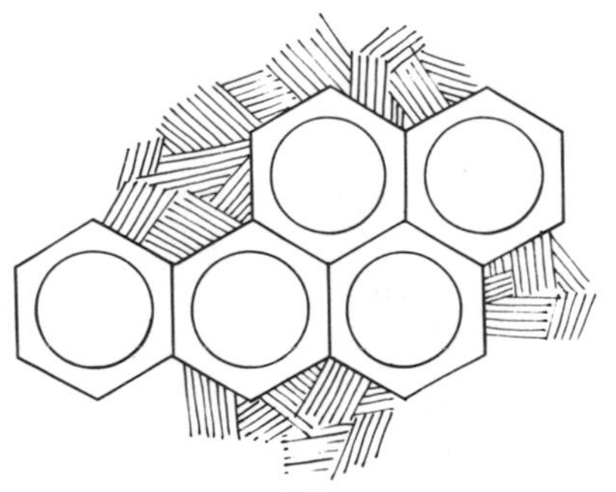

Figure 15-13. Benzopyrene.

FUNCTIONAL GROUP DERIVATIVES OF AROMATIC COMPOUNDS

All of the functional groups that you studied in the aliphatic compounds, the alkanes and alkenes, can be substituted for hydrogen atoms attached to the carbon atoms of the benzene and polynuclear benzene ring structures. They are named according to the IUPAC system or by their common usage names.

Halogen Derivatives

Chlorobenzene (phenylchloride) is the simplest halogen derivative of benzene (Fig. 15-14). When 2 chlorine atoms are substituted, 3 isomeric dichlorobenzenes can be formed. More complicated halogenated aromatics such as hexachlorophene (Fig. 15-15) are used as germicides; hexachlorophene is marketed under the trade name of pHisoHex. It consists of 2 benzene rings coupled by a methylene group. It has remarkable bactericidal properties, and until recently, was incorporated into soaps, shampoos, surgical scrubs, deodorants, cleansing creams, and preparations used by physicians and dentists. However, the Food and Drug Administration has now prohibited its use except under physician's orders, because tests reported that it was absorbed through the skin, especially in infants, and had an adverse effect on brain tissue. Another very powerful halogenated derivative that has been banned for general public use because of its poisonous effects is dichlorodiphenyl trichloroethane, DDT.

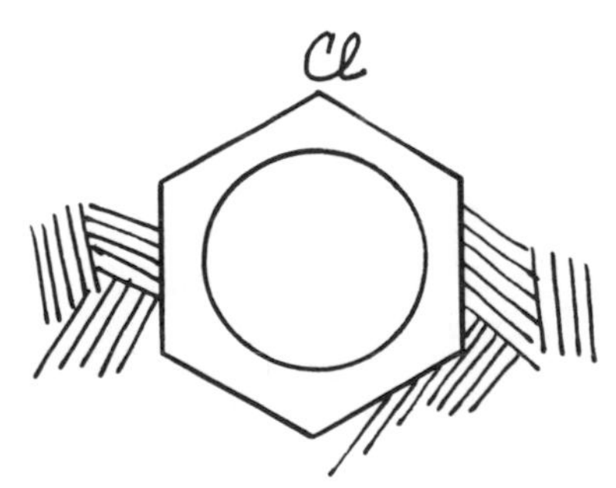

Figure 15-14. Phenylchloride.

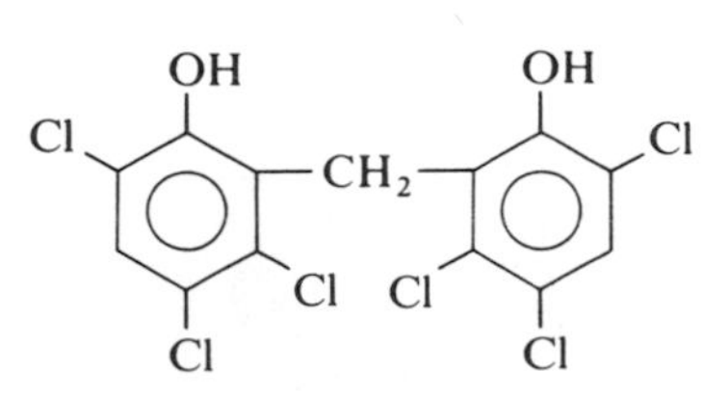

Figure 15-15. Hexachlorophene.

CHEMICAL REACTIONS OF THE HALOGEN COMPOUNDS

The halogen aromatics react similarly to the alkyl halides. They do not ionize in water to yield the halide ion, being covalently bound to the carbon atom.

The Hydroxy Aromatics

The substitution of an —OH group onto the benzene ring yields a hydroxybenzene, which behaves completely differently than the —OH group bound to an alkyl radical (as in methanol or ethanol). These hydroxy aromatics are called phenols, ionize weakly to yield hydrogen ions and therefore are acidic in nature (Fig. 15-16). They are extremely corrosive, causing severe burns, are poisonous when taken internally, and are bactericidal. There may be more than a single —OH group attached to the benzene, naphthalene or polynuclear ring structure.

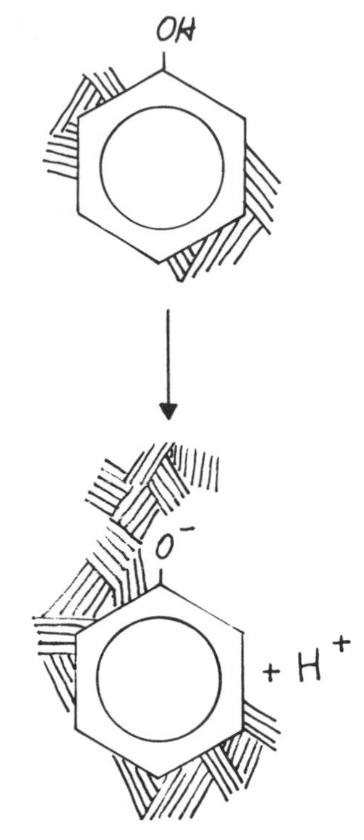

Figure 15-16. Ionization of phenol to yield acidic solutions.

PHENOL

Phenol (hydroxybenzene) is the standard to which all other germicides are compared (Fig. 15-17). In fact, Lister began antiseptic surgery in 1867 with the use of phenol. Phenol is sometimes called carbolic acid, and you should never confuse it with carbonic acid, H_2CO_3, which is the component of carbonated water and is harmless.

Other germicides are rated and measured in arbitrary units called *phenol coefficients.* For example, a 1 percent solution of a germicide being tested that has the identical germicidal effect as a 5 percent solution of phenol in water is designated as having a phenol coefficient of 5. Phenol itself is too corrosive to be applied to the skin as it will cause blisters. However, it is marketed in dilute solutions. Pure phenol *should never be handled* with your hands. It is a white crystalline solid having a low melting point, 41°C. It darkens on exposure to light. Should any pure or highly concentrated phenol or phenol solution come in contact with your skin, you should wash the area thoroughly with water and soap.

Alcoholic solutions of phenol are even more dangerous to handle than water solutions because the alcohol and the phenol are absorbed even through unbroken skin. Do not use an alcohol wash to remove phenol from your skin.

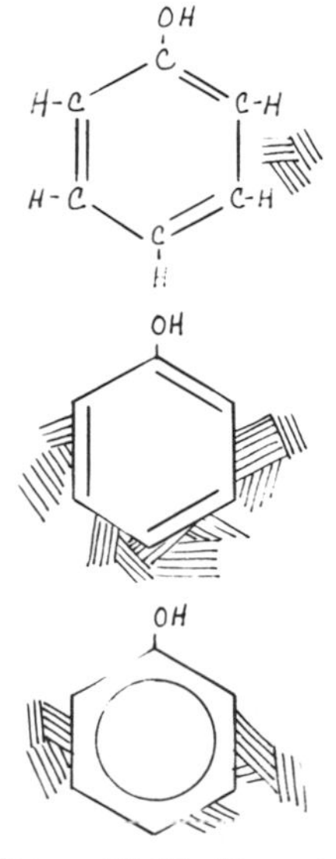

Figure 15-17. Phenol.

CRESOLS

Cresols are hydroxytoluenes (hydroxymethyl benzenes). Since there are 2 groups attached to the benzene ring, there are 3 isomers (Fig. 15-18). Cresols are less corrosive than phenol, and they are better antiseptics. Lysol, a commercial preparation, is a mixture of the 3 isomeric cresols in a dilute soap and water solution.

RESORCINOL

Resorcinol is dihydroxybenzene, and it is used as an antiseptic although it is not as effective a germicide as phenol (Fig. 15-19).

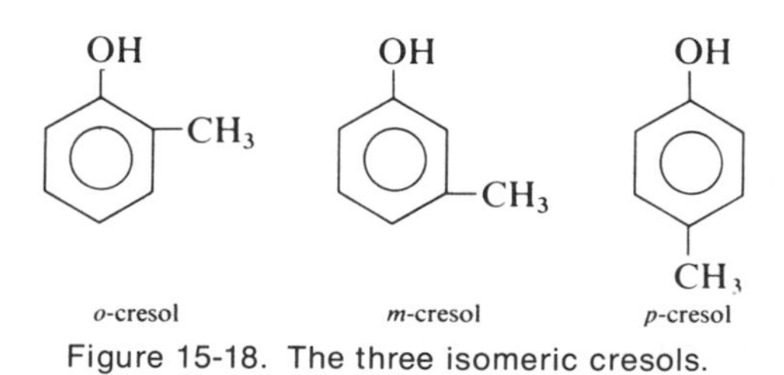

Figure 15-18. The three isomeric cresols.

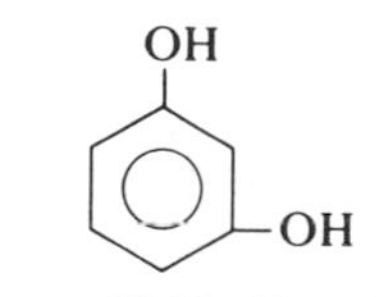

Figure 15-19. Resorcinol.

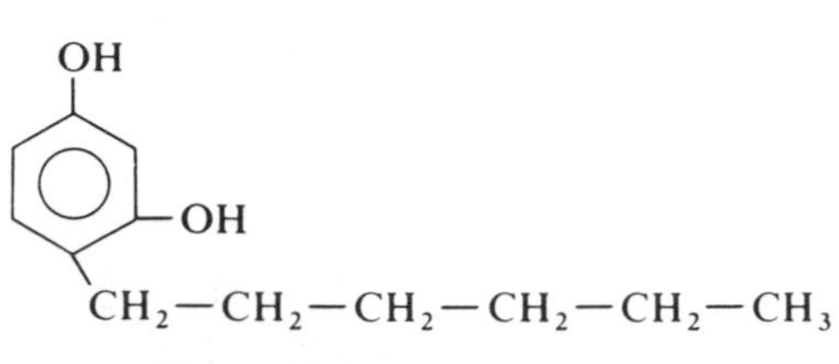

Figure 15-20. Hexylresorcinol.

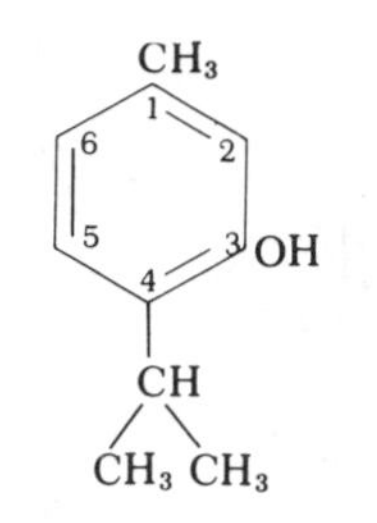

Figure 15-21. Thymol.

HEXYLRESORCINOL

Hexylresorcinol is resorcinol with a hexyl radical attached to the 4-position carbon atom. It is an excellent bactericidal compound (Fig. 15-20).

THYMOL

Thymol, which is 3-hydroxy, 4-isopropyltoluene, is an excellent antiseptic that finds applications in dental proprietary formulations. It is used to treat infections caused by hookworm in the intestinal tract (Fig. 15-21).

Figure 15-22. Benzoic acid.

Carboxylic Acid Derivatives

The substitution of the —COOH group for a hydrogen atom attached to ring carbon atoms yields the corresponding carboxylic acid derivatives.

BENZOIC ACID

Benzoic acid is found in gum benzoin and in cranberries (Fig. 15-22). It is used as a stimulant, a diuretic, and an antiseptic in the intestinal and urinary tracts. It is a colorless solid, soluble in hot water, and its sodium derivative, sodium benzoate, is used as a preservative, as you will notice from the labels of foods and soft drinks.

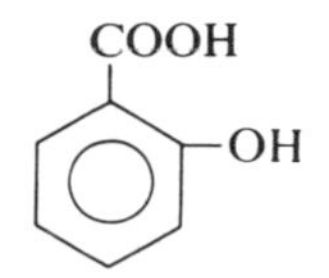

Figure 15-23. Salicylic acid.

SALICYLIC ACID

Salicylic acid is 2-hydroxybenzoic acid, being both a phenol and a carboxylic acid (Fig. 15-23). The sodium salt, sodium salicylate, has been used universally as a treatment for rheumatism, but because it is absorbed mainly in the stomach it sometimes causes gastric irritation and nausea. Sodium salicylate reduces fever and relieves pain caused by bursitis, arthritis, and headaches. It depresses the pain centers in the thalamus regions of the brain, and therefore raises the threshold of pain. The acid itself is too irritating to be used, so the sodium salt or other derivatives are used medicinally (Fig. 15-24).

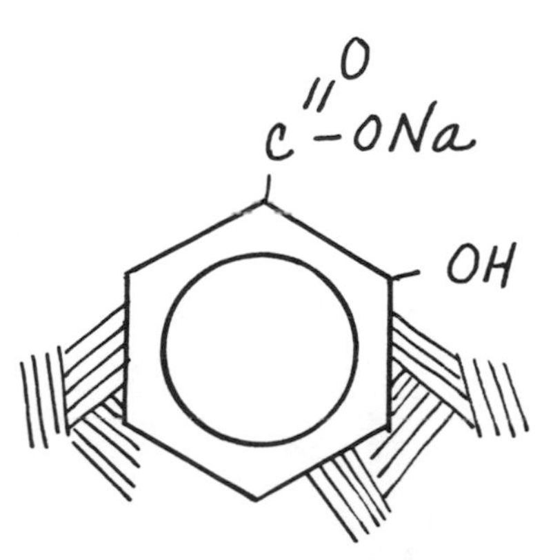

Figure 15-24. Sodium salicylate.

Esters of Carboxylic Acids and Their Derivatives

Esters of aromatic acids are made by reacting the acid with an alcohol. The methyl ester of benzoic acid, methylbenzoate, is used in the manufacture of perfumes and flavoring agents and has the odor of new mown hay.

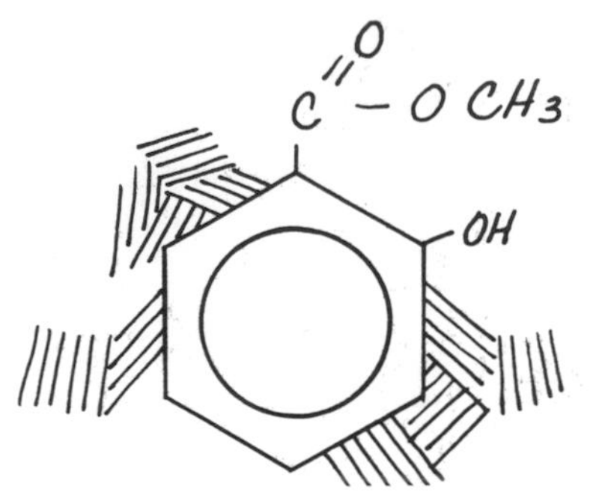

Figure 15-25. Methylsalicylate.

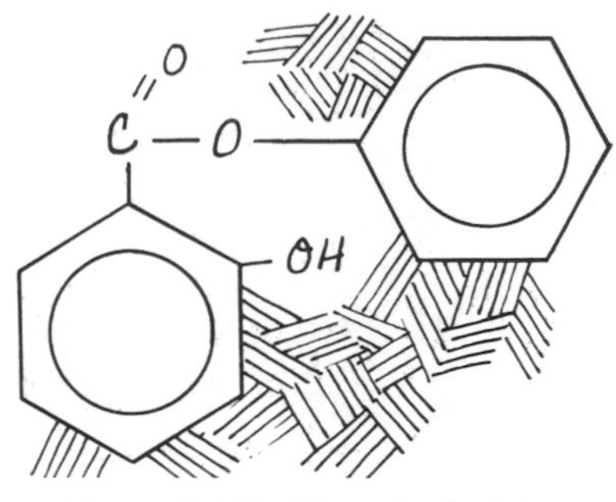

Figure 15-26. Phenylsalicylate.

METHYLSALICYLATE

Methylsalicylate, the methyl ester of salicylic acid, is commonly called oil of wintergreen (Fig. 15-25). The natural oil of wintergreen is gaultheria oil, and liniments commonly use methylsalicylate in their formulations. It is readily absorbed through the skin, and when applied on painful joints and muscles it hydrolyzes to yield salicylic acid.

PHENYLSALICYLATE

Phenylsalicylate, the ester of phenol and salicylic acid, is commonly called salol (Fig. 15-26). It finds applications as an intestinal antiseptic because it is not hydrolyzed by the gastric juices. It passes through the stomach unchanged, and in the intestines under basic conditions is hydrolyzed to yield both phenol and salicylic acid. Because phenylsalicylate does not hydrolyze in the stomach it makes an ideal coating for medications that need to act in the intestines. It protects drugs that might be destroyed by the gastric juices and is therefore used as an enteric coating.

ACETYLSALICYLIC ACID

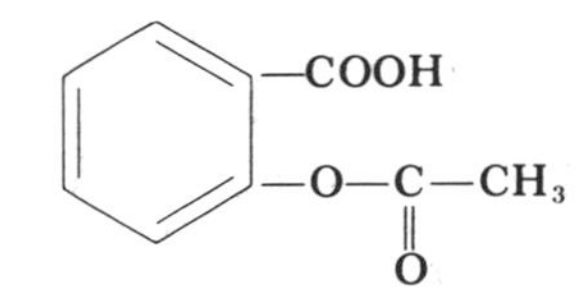

Figure 15-27. Acetylsalicylic acid (aspirin).

Acetylsalicylic acid is the acetic acid ester of the —OH group of salicylic acid, made by the reaction of acetic acid and salicylic acid, and is commonly called aspirin (Fig. 15-27). It is a fever reducer (antipyretic) and an analgesic (pain reducing agent). Since it is a carboxylic acid it may cause stomach irritation, and many commercial preparations combine it with antiacids or buffering agents to form the sodium salt. The sodium salt is supposed to be less irritating and is said to be absorbed more rapidly from the intestinal tract to give more rapid relief from pain. Aspirin is used to treat the symptoms of the common cold, headaches, minor muscle aches and pains, and for the treatment of rheumatic fever. It retards the clotting of blood, and its use after surgery is contraindicated, as it can cause hemorrhaging. Since the salicylates hydrolyze to form salicylic acid, they are midly antiseptic.

Aromatic Nitro Compounds

The substitution of the nitro functional group (—NO_2) for hydrogen atoms in the carbon ring yield the corresponding nitro derivatives.

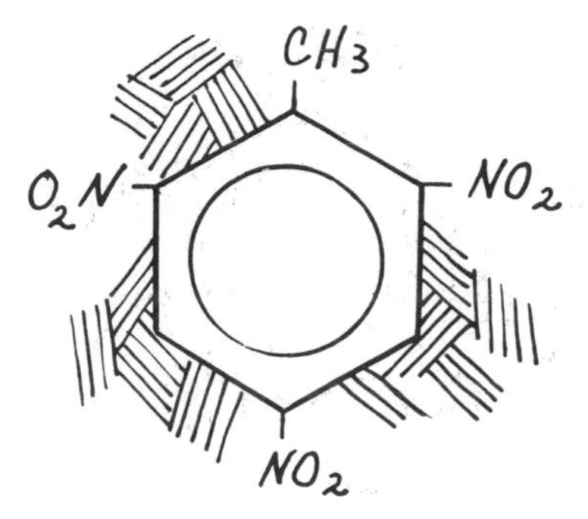

Figure 15-28. 2,4,6 trinitrotoluene (TNT).

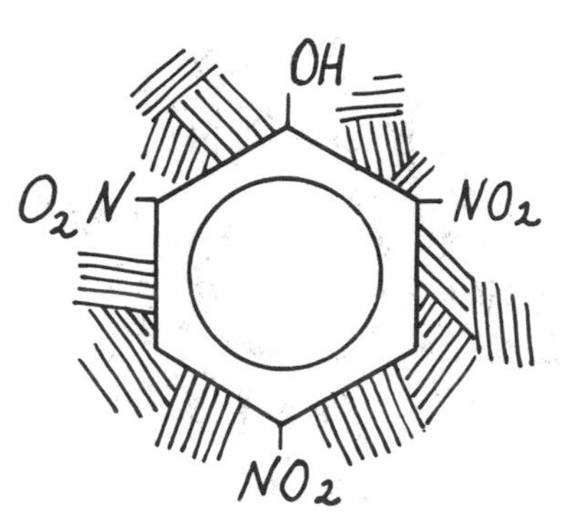

Figure 15-29. Trinitrophenol.

TRINITROTOLUENE

Trinitrotoluene is the trinitro derivative of toluene (methylbenzene) with the nitro groups in the 2, 4 and 6 positions (Fig. 15-28). Commercially it is called TNT and is an extremely powerful explosive.

TRINITROPHENOL

Trinitrophenol is the 2, 4, 6 trinitro derivative of phenol (hydroxybenzene), and its common name is picric acid (Fig. 15-29). Picric acid is used in the laboratory to precipitate proteins from solution. Albumin in the urine is coagulated (the Esbach's test) by picric acid and is therefore detected. Picric acid is also used in the treatment of burns. It coagulates the proteins on the burn surface, thus preventing the loss of blood serum. It is marketed as Butesin picrate salve, containing the local anesthetic Butesin to relieve the pain of burns. Structurally, it is similar to TNT, and because it contains such a high percentage of nitro groups, it is also used as an explosive.

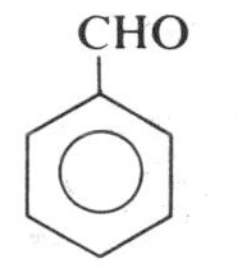

Figure 15-30. Benzaldehyde.

Aromatic Aldehydes

The substitution of the aldehyde functional group (—CHO) for a hydrogen atom yields aldehyde derivatives, and the simplest aromatic aldehyde is benzaldehyde (Fig. 15-30). Benzaldehyde is used as a perfume and flavoring agent, having the odor of bitter almonds. As an aldehyde, it serves as a raw material for the manufacture of other compounds such as drugs, dyes, and chemicals.

Vanillin is probably the most widely encountered aldehyde, imparting the flavor of vanilla to foods, ice cream, and candies. It is found in the vanilla bean, sugar beets, balsams and resins. Structurally it is an aldehyde with an ether and phenolbenzene ring structure, 3-methoxy-4-hydroxybenzaldehyde (Fig. 15-31).

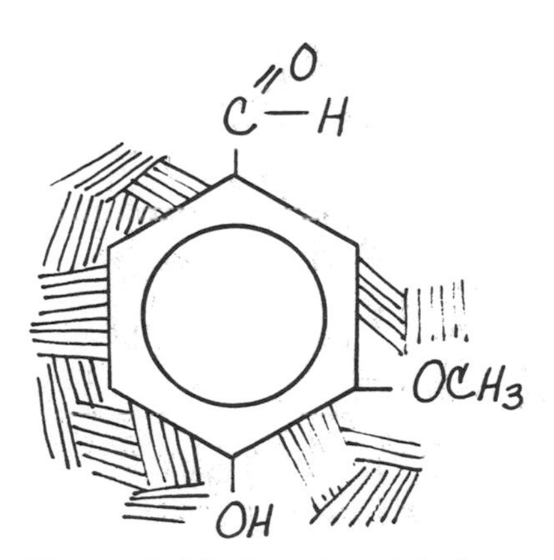

Figure 15-31. 3-methoxy-4-hydroxy-benzaldehyde (vanillin).

Aromatic Ketones

Acetophenone is the simplest aromatic ketone, the ketone group ($—\overset{O}{\overset{\|}{C}}—$) linking a benzene ring and a methyl group (Fig. 15-32). It is used as a hypnotic (a drug that tends to produce sleep) and as a raw material for the manufacture of tear gas, chloroacetophenone, where a chlorine atom substitutes for a hydrogen atom of the methyl group (Fig. 15-33).

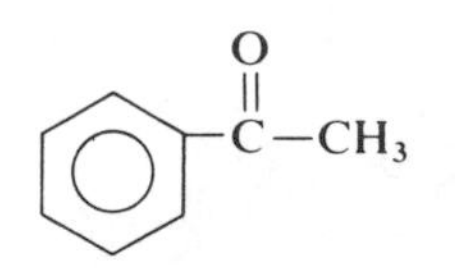

Figure 15-32. Acetophenone.

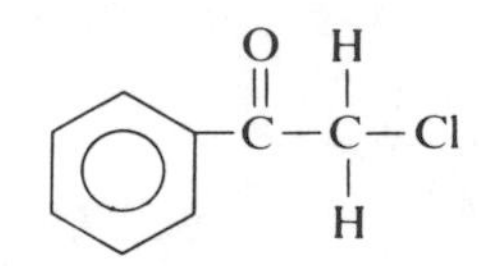

Figure 15-33. Chloroacetophenone (tear gas).

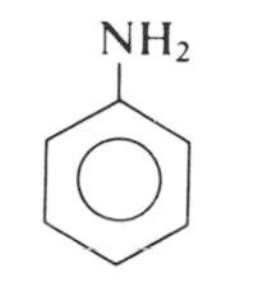

Figure 15-34. Aniline.

Aromatic Amines

The substitution of the amino functional group (—NH_2), either in the form of a primary, secondary or tertiary amine, yields the extremely important amino derivatives.

Aniline is the simplest aromatic amine, being aminobenzene (Fig. 15-34). It is much weaker as a base than ammonia, and the diphenyl amine is even weaker. Amines are made by reducing the corresponding nitro compounds, which can be made by nitrating the benzene ring with nitric acid.

Aromatic amines form salts with inorganic acids, just as ammonia, and the solubility of the salt formed in water is much greater than that of the amine (Fig. 15-35).

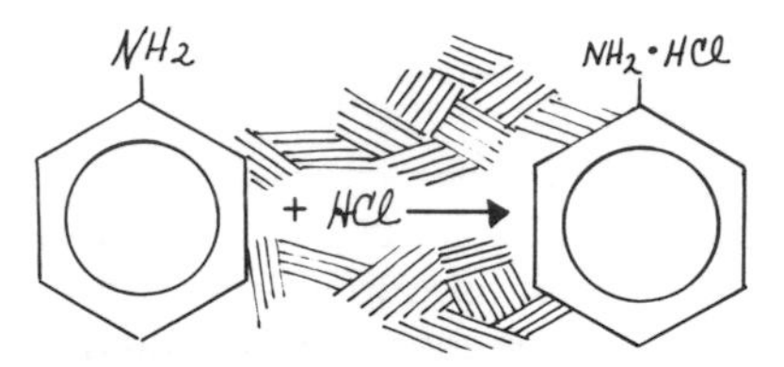

Figure 15-35. Aromatic amines form salts with inorganic molecules.

Aromatic Amides

Amines react with carboxylic acids to form the corresponding amides. When aniline is reacted with acetic acid, the corresponding acetanilide is formed (Fig. 15-36). Acetanilide is used as an antipyretic and an analgesic, but it is toxic. Because of this, other derivatives of acetanilide have been prepared that are less toxic. In the laboratory acetanilide prevents the rapid decomposition of hydrogen peroxide into oxygen and water, acting as a catalytic agent to stabilize the peroxide solution.

Phenacetin is a derivative of acetanilide, being the paramethyl ether (Fig. 15-37). It is a very effective antipyretic and analgesic and is sold commercially in headache preparations that advertise no aspirin content and as relief against neuralgia.

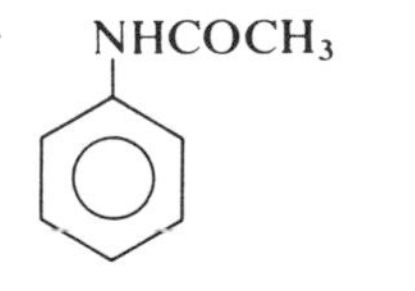

Figure 15-36. Acetanilide.

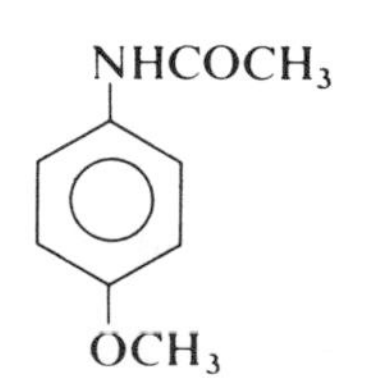

Figure 15-37. Phenacetin.

THE SULFA DRUGS—ANTIMETABOLITES

When benzene is treated with sulfuric acid, benzene sulfonic acid results, which can then react with ammonia (or amines) to form the corresponding sulfonamides (Fig. 15-38). These sulfonamides are not true amides because they contain the —SO_2NH_2 group instead of the —$CONH_2$ group. These are the sulfa drug compounds used to treat a wide variety of infections.

As antimetabolites, the sulfa drugs act in this manner. Pathogenic organisms require paraaminobenzoic acid (PABA), one of the Vitamin B's, to synthesize folic acid, a coenzyme. Coenzymes are small organic molecules, nonproteins, which can be separated from enzymes, and which are needed as part of the whole molecule to enable the molecule to act as an enzyme. The sulfa drugs contain a molecular structure closely resembling PABA which is the paraaminobenzene sulfonic acid (see Fig. 15-39). Apparently pathogenic organisms use the sulfa compound instead of the PABA in their coenzyme formation. The altered coenzyme fails and the normal metabolism of the pathogen is inhibited. Thus the drug serves its purpose by inhibiting the growth of the pathogen due to vitamin deficiency. The various sulfa drugs are made by using different amines and amino derivatives to react with the amino sulfonic acid. The structure of some sulfa drugs is shown in Figure 15-40.

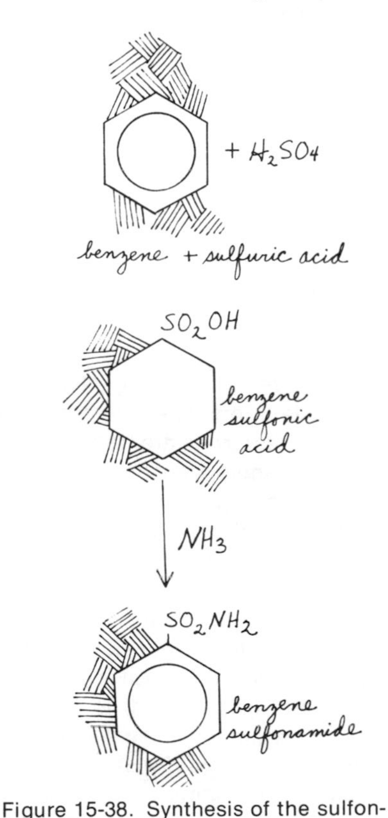

Figure 15-38. Synthesis of the sulfonamides.

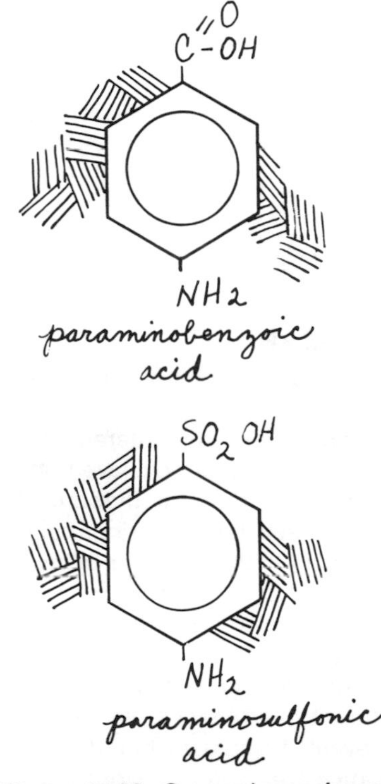

Figure 15-39. Comparison of paraaminobenzoic acid and paraaminosulfonic acid.

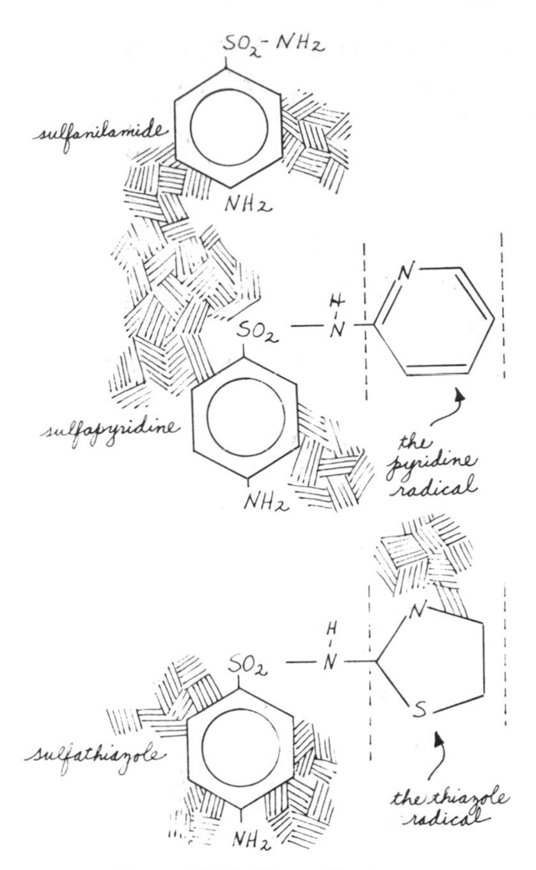

Figure 15-40. The sulfa drugs.

SUMMARY

Aromatic compounds are characterized by the benzene type ring structure which contains 3 pairs of double bonds. These double bonds are different from the previously encountered alkyl double bonds as the electrons are delocalized. Aromatic compounds have both common and IUPAC names, benzene being the simplest aromatic compound. Polynuclear aromatic compounds can contain many fused benzene rings: some are carcinogenic and the skeleton structure of others is the basis for many hormones, bile acids and alkaloids. Phenols are compounds that have 1 hydroxyl group on the benzene ring, while cresols have 2 hydroxyl groups. When the carboxyl group is attached to the aromatic ring, carboxylic acids result, and their derivatives are used to reduce fever, treat headaches and minor aches. Aspirin and phenacetin are common examples. The nitro aromatics are explosives and trinitrophenol (picric acid) is used to prevent the loss of blood serum in burn victims. The aldehyde and keto aromatics are used as flavoring agents and perfumes. the nitrogen derivatives (the amides) are antipyretics and analgesics. The sulfa drugs, amides of benzene sulfonic acid and its derivatives act as antimetabolites to alter the coenzyme formation of pathogenic microorganisms and inhibit their growth.

EXERCISE

1. Draw three representations of benzene that designate the benzene ring.
2. Draw and name (common and IUPAC names) the three dimethyl benzenes.
3. What are the aromatic hydrocarbons used for?
4. Draw the structures for naphthalene, anthracene and phenanthrene.
5. What is the importance of the anthracene structure to health care personnel?
6. Where are benzopyrenes found and why are they important?
7. What is the formula for hexachlorophene and what is it used for?
8. What is the difference between the —OH group of an alcohol and that of a phenol?
9. What is meant by phenol coefficient?
10. What is another name for orthohydroxybenzoic acid?
11. What is the common name for acetylsalicylic acid and why is the sodium salt sometimes used?
12. Draw the simplest aromatic aldehyde and aromatic ketone and state where they are used.
13. What is TNT? Draw its formula and state where it is used.
14. What is picric acid and what is it used for?
15. What is the simplest aromatic amine, and how can it be stabilized?
16. What results from the reaction of an aromatic amine and a carboxylic acid? Name a common compound that possesses this functional group.
17. Compare a carboxylic acid amide with a sulfonic acid amide structure.
18. How do the sulfonamides act as antimetabolites in the body?

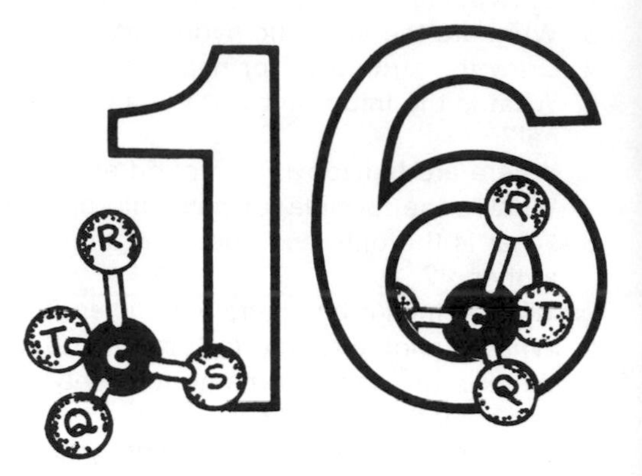

OBJECTIVES

When you have completed this chapter you will be able to:

1. Define the term carbohydrate.
2. Define the term hydrolysis.
3. Define the terms monosaccharide, disaccharide and polysaccharide.
4. Trace the path of carbon in nature.
5. Describe an asymmetric carbon.
6. Differentiate between dextrorotatory and levorotatory isomers.
7. Explain the purpose of the polarimeter.
8. State the meaning of the term superimposability.
9. Name the carbohydrates according to functional group present and number of carbons.
10. Classify carbohydrates.
11. Write the straight chain formula and ring structure of ribose and dexyribone.
12. Write the straight chain formula and ring structure for glucose.
13. Differentiate between α and β glucose.
14. Write the straight chain formula and ring structure of galactose.
15. Write the straight chain formula and ring structure of fructose.
16. Write the structure of sucrose.
17. Identify the $1 \rightarrow 2$ glycoside linkage.
18. Define invert sugar.
19. Write the structure of lactose.
20. Write the structure of maltose.

THE CARBOHYDRATES

21. Identify the β $(1 \rightarrow 4)$ glycoside linkage in cellulose.
22. Identify the α $(1 \rightarrow 4)$ glycoside linkage in starch.
23. Distinguish amylose from amylpectin.
24. Describe the hydrolysis steps of starch.
25. Define the term reducing sugar.
26. Explain the energy pathway of glucose.

The sugars or saccharides are commonly called carbohydrates, which may suggest to you that the compound is made up of carbon and water (hydrate). Basically this is the case. The ratio of the hydrogen atoms to the oxygen atoms in the carbohydrate molecule is the same as that for water. In fact, if you were to heat ordinary table sugar (also called sucrose) in a container over a flame, you would observe (as the sugar turns brown, melts and decomposes) that water condenses on the cool sides of the container. Continued heating leaves only pure carbon as the residue (see Fig. 16-1).

However, it is impossible for you to take 12 parts of carbon and 11 parts of water and recombine them to form the sucrose molecule. The water in the molecule was not water of hydration, because the hydrogen and hydroxl groups of water were bound by their valence bonds to the carbon atoms. The carbohydrates are molecules in which many hydrogens and hydroxyl groups are attached to the carbon atoms. In addition, there is also either a ketone or aldehyde functional group present (see Fig. 16-2).

ORIGIN OF THE CARBOHYDRATES: THE CARBON CYCLE

Carbohydrates are synthesized by plants and constitute an important part of their nutrition. This process is called *photosynthesis.* The plant absorbs carbon dioxide from the air and absorbs water from the soil. These materials are syn-

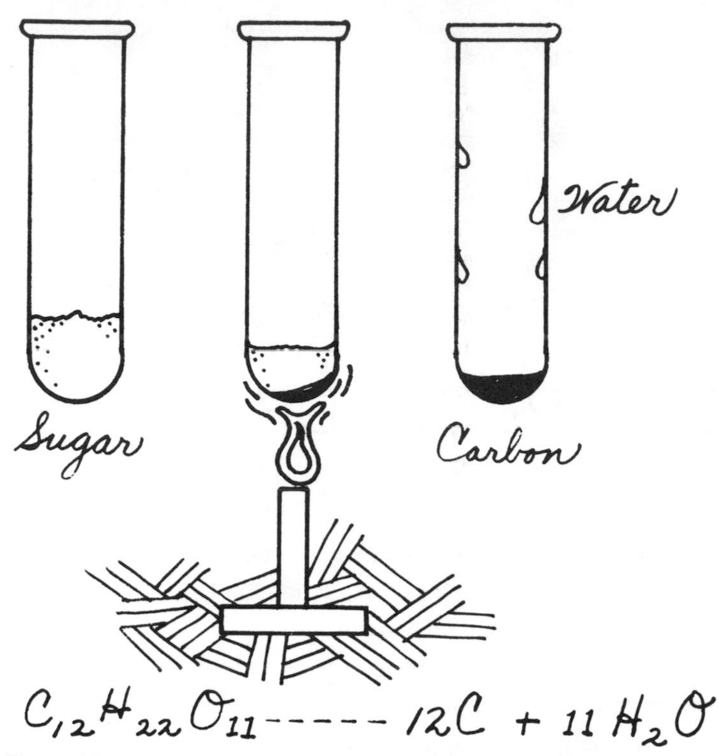

Figure 16-1. As sugar is heated it decomposes into carbon and water.

thesized into carbohydrates by means of energy in the form of sunlight with the aid of chlorophyll as a catalyst. The overall reaction can be represented as:

$$6\ CO_2 + 6\ H_2O \xrightarrow{\text{chlorophyll \& light}} C_6H_{12}O_6 + 6\ O_2$$

The above reaction is a simple summary of a very complex series of biochemical reactions in the plant cells.

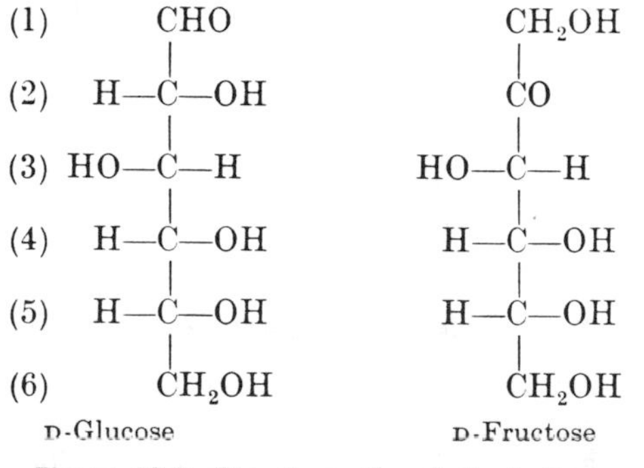

Figure 16-2. Structure of carbohydrates.

Carbohydrates can be classified as monosaccharides (which means 1 sugar molecule), as disaccharides (which means 2 sugar molecules) and as polysaccharides (which means many sugar molecules). The monosaccharides are the simplest sugars which cannot be broken down or *hydrolyzed* into more simple sugars. The disaccharides can be broken down or *hydrolyzed* to yield 2 simple sugars. The polysaccharides can be hydrolyzed to yield many simple sugars. Plants have the ability to convert monosaccharides into disaccharides and polysaccharides by the removal of 1 or more molecules of water.

Cellulose, which gives rigidity to the leaves, stalks and stems of plants, is a polysaccharide. Plants store *starch,* another polysaccharide, as a reserve source of food in their roots (e.g. potatoes) and in their seeds (e.g. beans). Fruit bearing plants store monosaccharides and disaccharides in their fruits (e.g. glucose in grapes).

Plants have the ability to convert carbohydrates to fats and proteins. Corn can synthesize the fat corn oil. Flax can synthesize the fat linseed oil. Certain plants, such as the bean, can synthesize protein from carbohydrates by making use of atmospheric nitrogen and nitrogen-containing substances in the soil.

Animals cannot synthesize carbohydrates as plants do from CO_2 and H_2O and consequently, animals depend upon plants for their supply of carbohydrates. In the process called respiration, these carbohydrates are oxidized in the animal cells to yield carbon dioxide, water and energy. The net process is essentially the reverse of photosynthesis:

$$C_6H_{12}O_6 + O_2 \longrightarrow 6\ CO_2 + 6\ H_2O + \text{energy}$$

Again, this is a simple summary of a complex series of biochemical reactions. The high energy carbohydrate molecule is converted into low energy carbon dioxide and water. The energy thus released is used by animals for normal physiological processes. Some of this energy is used to convert carbohydrates to fats and proteins. The carbon dioxide and water are returned to the environment and are reconverted by plants to carbohydrates and oxygen. This is called the carbon cycle (Fig. 16-3).

THE ASYMMETRIC CARBON ATOM AND OPTICAL ISOMERISM

Isomerism is a common occurrence in organic chemistry. Two types have already been discussed: structural isomerism (as in butane and isobutane) and geometric isomerism (as in *cis* and *trans* isomers).

Optical isomerism is frequently encountered in organic compounds of biological interest and is now included in our study of the carbohydrates. This type of isomerism results from the presence of 1 or more asymmetric carbon atoms as part of the molecular structure. An *asymmetric carbon atom* is defined as one that is bonded to 4 different kinds of atoms or groups:

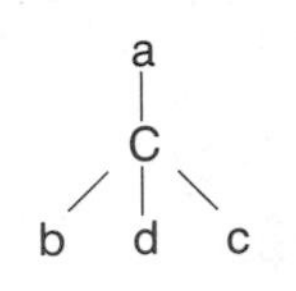

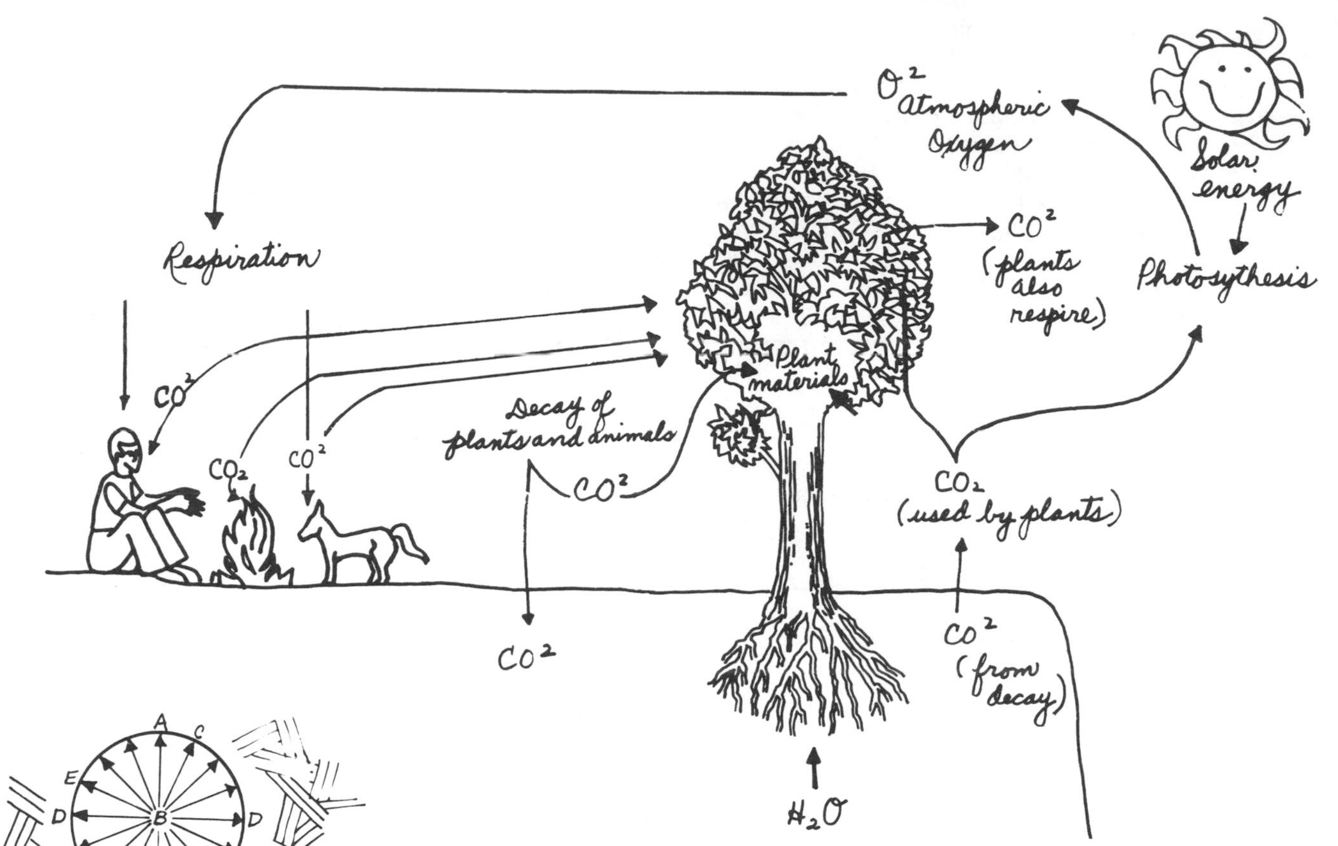

Figure 16-3. The carbon cycle.

A C E D B D E C A

Unpolarized light vibrates in all directions (planes)

A B A

Polarized light vibrates only in one direction (plane)

Figure 16-4. Vibration of normal and polarized light.

Carbohydrates in solution show the property of optical rotation, which means that a beam of polarized light is rotated as it passes through the solution. Normal light vibrates in all directions; polarized light vibrates only in one direction (see Fig. 16-4). The extent to which the beam of polarized light is rotated (defined as the angle of rotation) is measured by an instrument called a polarimeter.

If the solution of a substance rotates the beam of polarized light to the right (clockwise) it is said to be *dextrorotatory.* If the solution of a substance rotates the light to the left (counterclockwise) it is said to be *levorotatory.* Glucose rotates light to the right and is sometimes called *dextrose.* Fructose rotates light to the left and is sometimes called *levulose.* Dextrorotatory is symbolized by (D-) and levorotatory by (L-).

Let us consider D-glyceraldehyde and its mirror image, L-glyceraldehyde (see Fig. 16-5). These mirror images are called optical isomers or *enantiomers.* Note that these two isomers have the same number and kinds of atoms and the same grouping of atoms but they have *different spatial arrangements.* The D-glyceraldehyde isomer rotates the light from —H to —CHO to —OH, in a clockwise direction to the right. The L-glyceraldehyde isomer also rotates the light from —H to —CHO to —OH, but in a counterclockwise direction (to the left) around the asymmetric carbon atom. These mirror image isomers have the same relationship to each other as a person's right and left hands, or a pair of shoes. The isomers are not *superimposable* on each other (Fig. 16-6). You cannot rotate the D-glyceraldehyde so that its atoms will have a point to point correspondence with the atoms of L-glyceraldehyde. As an example, put your right hand face down on the table. Now try to put your left hand *face down* over your right hand so that your left thumb is over your right thumb, your left index finger is over your right index finger, your left ring finger is over your right ring finger, and your left pinky is over your right pinky. You will find it impossible to *superimpose* your left hand over your right hand. Again, you will find it difficult to put your right shoe on your left foot and your left shoe on your right foot, at least not if you plan to do much walking. The shoes are not *superimposable.*

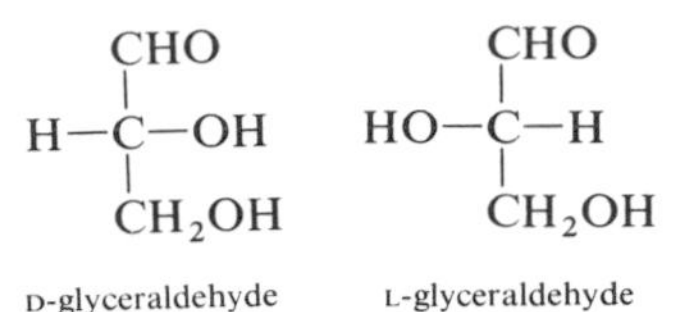

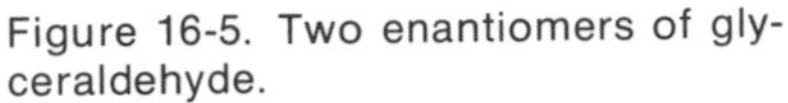

Figure 16-5. Two enantiomers of glyceraldehyde.

Enantiomers have the same physical and chemical properties. They differ only in the direction, not the magnitude, of the rotation of polarized light. Because of superimposability, they also differ in their biochemical or physiological activity. It should be noted in passing that even though a molecule contains asymmetric carbon atoms, it will be optically inactive if it contains a plane of symmetry, as in tartaric acid (Fig. 16-7).

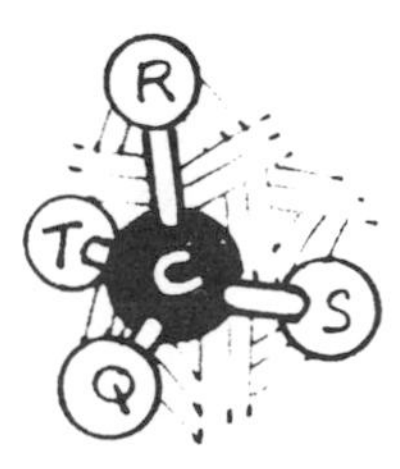

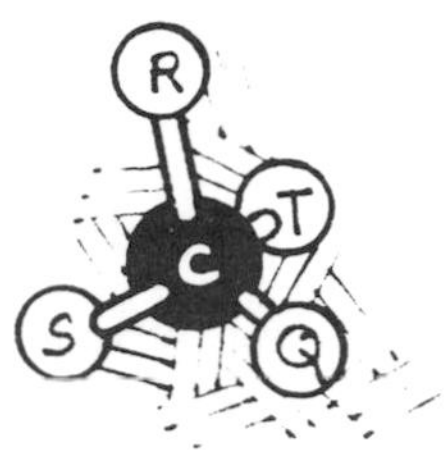

R : -CHO

T = -H

Q = $-CH_2OH$

S = -OH

Figure 16-6. Nonsuperimposibility of glyceraldehyde enantiomers.

Biochemical Significance of Optical Isomerism

Molecular orientations are of very great importance in biochemistry. For example, when lactic acid is obtained from different natural resources it exhibits different optical activities. The lactic acid obtained from contraction of muscle tissue is in the dextro form, whereas lactic acid obtained from the fermentation of cane sugar is in the levo form. But the lactic acid obtained from sour milk consists of an equal mixture of the D- and L- forms. Therefore, they cancel out each other's rotation and the mixture does not rotate polarized light.

In general, when a compound that exhibits optical activity is synthesized in the laboratory, a mixture of equal parts of the dextro and levo forms result. This is called a *racemic mixture.*

Reactions carried out in the body or in the presence of microorganisms often produce optically active isomers since the reactions are catalyzed by enzymes which are themselves optically active. The enzyme reactions are often specific for the D- or L- component of a racemic mixture.

Almost all of the organic compounds that occur in living organisms are only one enantiomer of a pair. Foods and medicines must have the proper molecular orientation or *configuration* if they are to be useful to the organism. A widely used meat flavoring agent (which has been ruled as being possibly dangerous when used in baby foods) is L-monosodium glutamate. The D- form does not change the flavor of the meat but the L- form does. Epinephrine (Adrenalin) is levorotatory and has about 15 to 20 times the physiological effect of its D-enantiomer.

As you study the naturally occurring optically active compounds such as the carbohydrates and the amino acids, you will see that the structure is closely related and extremely important to the physiological activity of these substances.

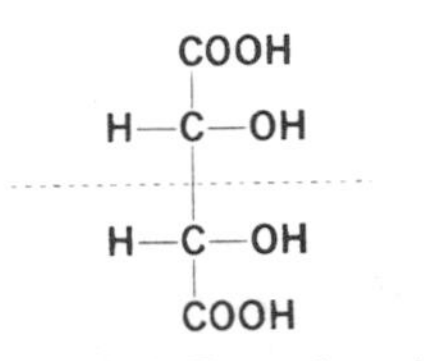

Figure 16-7. Tartaric acid.

Table 16-1. Summary of carbohydrate nomenclature

Classification	*Compounds*
Monosaccharides	
Trioses $C_3H_6O_3$	Glyceraldehyde
Tetrose $C_4H_8O_4$	
Pentoses $C_5H_{10}O_5$	Ribose
	Deoxyribose
Hexoses $C_6H_{12}O_6$	Glucose (dextrose)
	Galactose
	Fructose
Monosaccharides cannot by hydrolyzed to simpler units.	
Disaccharides $C_{12}H_{22}O_{11}$	Sucrose (glucose + fructose)
	Lactose (glucose + galactose)
	Maltose (glucose + glucose)
Disaccharides can be hydrolized to give monosaccharides.	
Polysaccharides $(C_6H_{12}O_6)_n$	Cellulose (β-glucose)$_n$
	Starch (α-glucose)$_n$
	Dextrin (α-glucose)$_n$
	Glycogen (α-glucose)$_n$
	Heparin (glucuronic acid-glucosamine)$_n$
Polysaccharides can be hydrolyzed to disaccharides and finally to monosaccharides.	

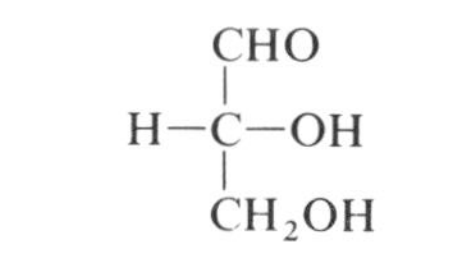

Figure 16-8. Glyceraldehyde.

NOMENCLATURE OF CARBOHYDRATES

The structure of the carbohydrate determines its name (see Table 16-1). The structure depends upon:

1. The length of the carbon chain.
2. The functional group present, a ketone or an aldehyde group.
3. The optical isomerism of the molecule.

The smallest sugar molecule has 3 carbon atoms and is called a *triose.* A sugar molecule with 4 carbon atoms is called a *tetrose.* One with 5 carbon atoms is called a *pentose,* and one with 6 carbon atoms is called a *hexose.*

If a sugar molecule contains a ketone group, it is called a *ketose.* If the sugar molecule contains an aldehyde group, it is called an *aldose.* For example, glucose has 6 carbon atoms and an aldehyde group. It is an *aldohexose.* Fructose has 6 carbon atoms and a ketone group. It is a *ketohexose.* Glyceraldehyde has 3 carbon atoms and an aldehyde group; it is an *aldotriose* (Fig. 16-8).

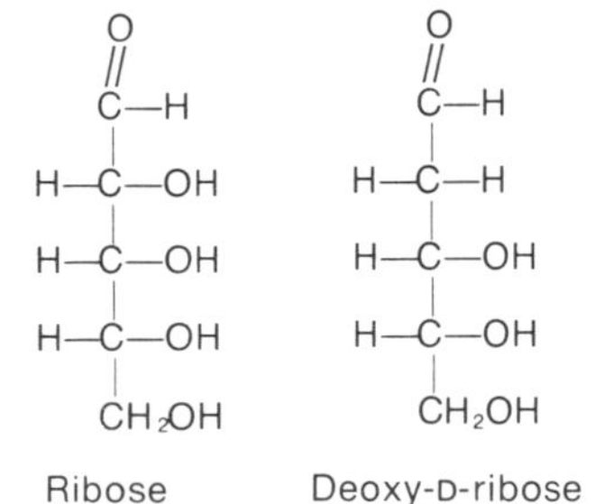

Figure 16-9. Consecutive numbering of carbon atoms.

CLASSIFICATION OF CARBOHYDRATES

The carbohydrates are grouped into three principal classes: the monosaccharides, the disaccharides, and the polysaccharides.

The Monosaccharides

THE TRIOSES: $C_3H_6O_3$

Of all the trioses, the most important for our purposes is D-glyceraldehyde. It has an aldehyde group and 3 carbons. Therefore, it is an aldotriose. The oxidation of glucose in animal cells yields D-glyceraldehyde.

THE TETROSES: $C_4H_8O_4$

There are no tetroses of any special interest to health care personnel.

THE PENTOSES: $C_5H_{10}O_5$

The most important pentoses we will study are ribose and deoxyribose. Ribose is a constituent of ribonucleic acid (RNA), and deoxyribose is a constituent of deoxyribonucleic acid (DNA). Since deoxyribose has one less oxygen atom than ribose, it has the prefix deoxy. Both pentoses are necessary components of the cytoplasm and nucleus of every cell. They are also found in the structure of viruses.

The pentoses (and the hexoses) can be represented as either a straight chain or a ring structure. The carbon atoms of ribose and of deoxyribose are shown in Figure 16-9. The carbon atom with the aldehyde group is numbered 1 and the other carbon atoms are numbered consecutively. In the ring structure (which is the way the sugars exist in nature) the aldehyde bonds to the OH group on carbon atom number 4. This gives a five membered ring consisting of 1 oxygen atom and 4 carbon atoms (see Fig. 16-10). The ring is called a *furanose.*

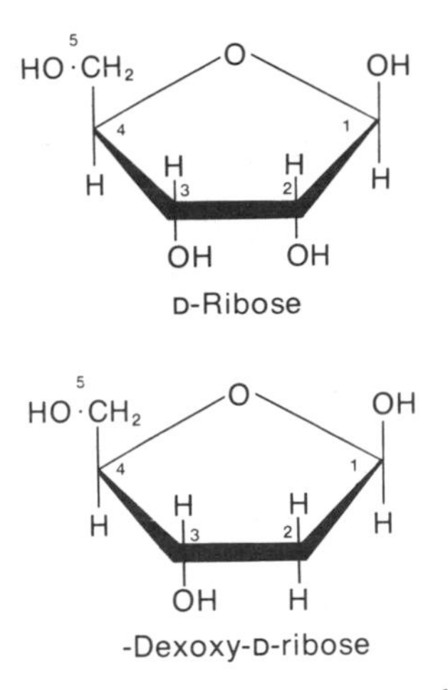

Figure 16-10. Ring structures of ribose and deoxyribose.

THE HEXOSES: $C_6H_{12}O_6$

The three most important hexoses to health care personnel are glucose, galactose and fructose. All are isomers of each other.

Glucose

Glucose is a white powder, which is not as sweet or as soluble in water as table sugar. In nature fruits are a major source of glucose. In the human body blood has a maximum glucose concentration of about 0.1 percent by volume. This concentration tends to rise after heavy meals. Glucose is the principle carbohydrate which can pass through the cell membrane. In the process of respiration glucose is oxidized to yield energy for cellular activities. This is why glucose is used in intravenous therapy. After strenuous exercise, honey, which contains 40 percent glucose, is given to replenish depleted blood sugar.

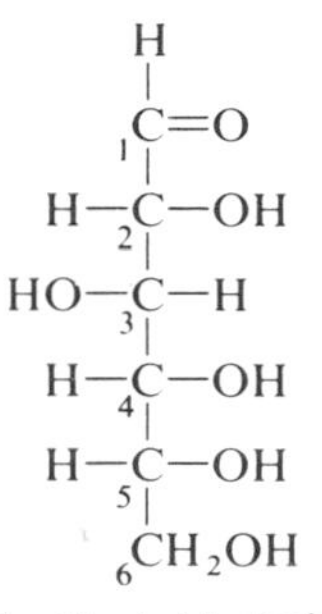

Figure 16-11. Straight chain structure of glucose.

Like the pentoses, glucose can be represented as a straight chain or as a ring structure (Fig. 16-11). The carbon atoms in glucose are numbered in the same way as ribose. In the ring structure for glucose the aldehyde group bonds to the OH group on carbon atom number five. This gives us a six membered ring containing 1 oxygen and 5 carbon atoms. This is called a *pyranose ring* (see Fig. 16-12).

When glucose forms the ring structure, the OH group on carbon atom number one can be either above the plane of the ring or below the plane of the ring. If the OH group is above the plane of the ring we have a β-glucose. If the OH group is below the plane of the ring we have an α-glucose. See Figure 16-13.

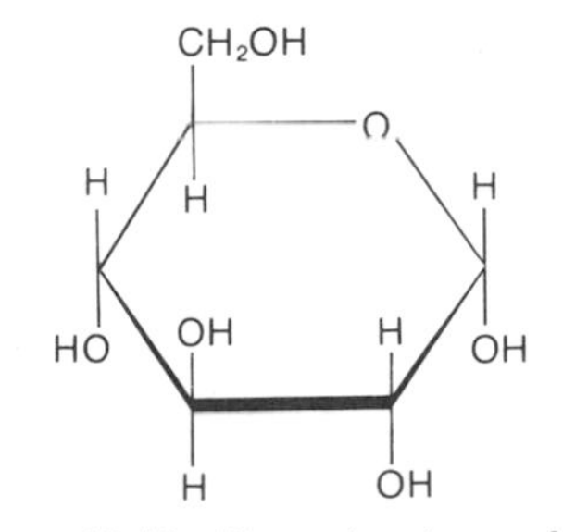

Figure 16-12. Ring structure of glucose.

Galactose

Galactose is an isomer of glucose. Th only difference between the two is the position of the OH group on carbon atom number four. See Figure 16-14. The principle source of galactose is from the hydrolysis of milk sugar or lactose. In the body the mammary glands convert glucose into galactose.

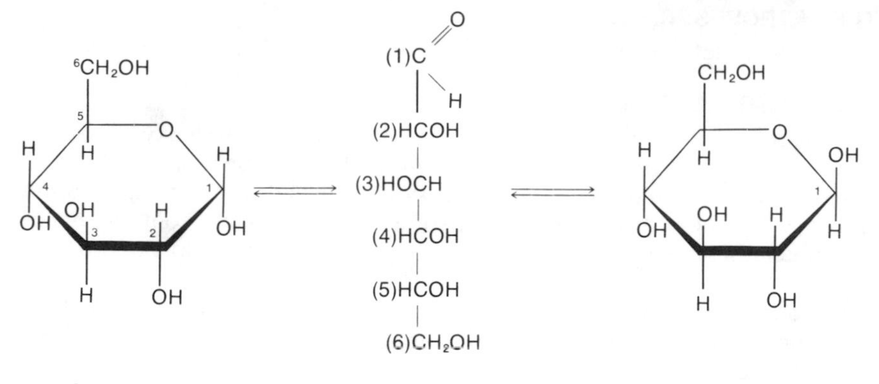

Figure 16-13. The difference between alpha and beta glucose.

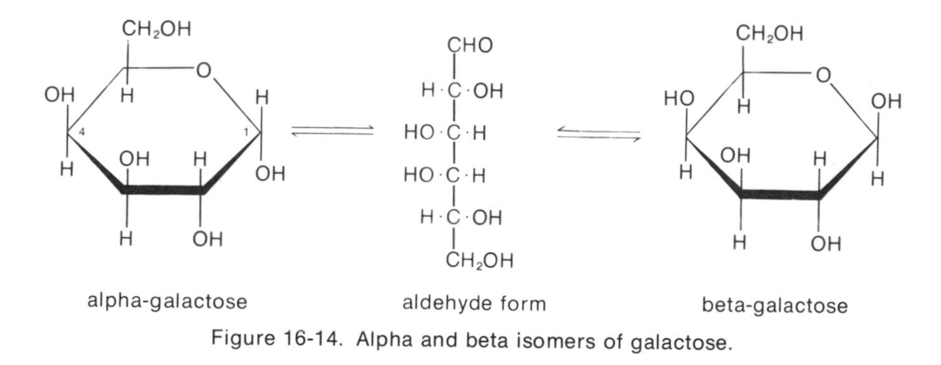

Figure 16-14. Alpha and beta isomers of galactose.

Fructose

Fructose is a ketohexose (see Fig. 16-15). In nature, fructose is found, along with glucose, principally in fruits and honey. It is also formed by the hydrolysis of table sugar. Fructose forms a furanose ring and is the sweetest of all the sugars.

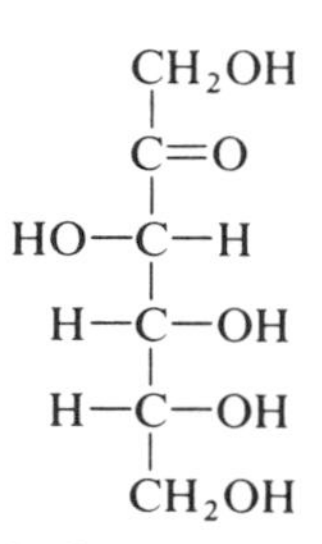

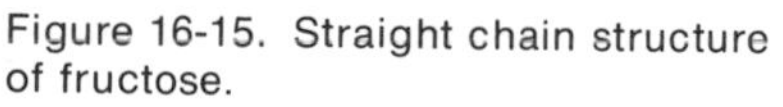

Figure 16-15. Straight chain structure of fructose.

The Disaccharides

We shall confine our studies to the three disaccharides which are important in the human body: sucrose, lactose and maltose. All are isomers of each other and have the empirical formula $C_{12}H_{22}O_{11}$. Upon hydrolysis they form 2 molecules of the simple sugars.

SUCROSE

Sucrose is the technical name for table sugar (also called cane or beet sugar). Upon heating, sucrose will partially decompose to yield a caramel brown amor-

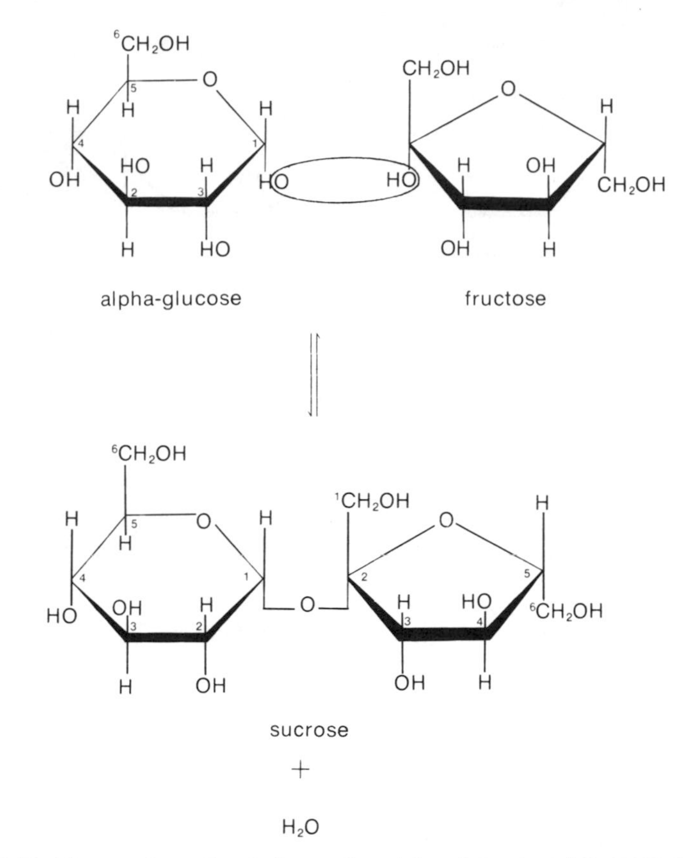

Figure 16-16. The condensation between the alpha glucose and fructose results in the formation of sucrose and elimination of 1 molecule of water.

phous substance. Upon hydrolysis, sucrose yields the monosaccharides glucose and fructose.

$$\text{sucrose} + H_2O \longrightarrow \text{glucose} + \text{fructose}$$

The structure of sucrose is shown in Figure 16-16. Sucrose is composed of an α-glucose plus a fructose. When these 2 molecules are joined together a water molecule is released. Essentially this is the reverse of the above reaction. The OH group on carbon atom number one of α-glucose is bonded to the OH group of carbon atom number two of fructose. This is called a $1 \rightarrow 2$ linkage. The oxygen atom bridging the 2 monosaccharides is known as *glycoside bond.*

The mixture of glucose and fructose, formed from the hydrolysis of sucrose, is called *invert sugar.* In the body, the enzyme *sucrase* hydrolyzes sucrose. Until sucrose has been hydrolyzed, it cannot be used by the body. This is another reason why glucose and sometimes fructose are used in I.V. therapy.

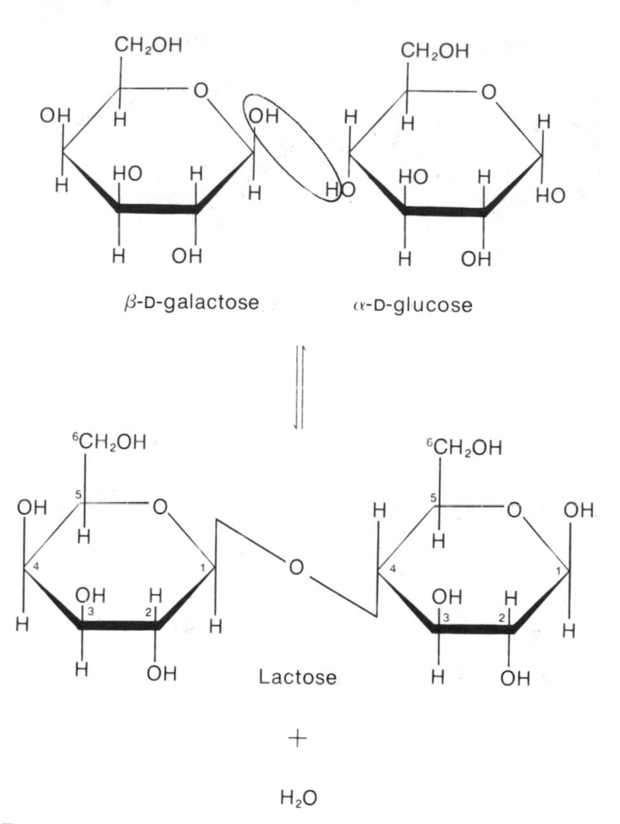

Figure 16-17. The condensation of beta galactose and alpha glucose results in the formation of lactose and 1 molecule of water.

LACTOSE

Lactose, also called milk sugar, is found in the milk of all mammals. Upon hydrolysis, lactose yields glucose plus galactose:

$$\text{Lactose} + H_2O \longrightarrow \text{glucose} + \text{galactose}$$

The structure of lactose is shown in Figure 16-17. The OH group on carbon atom number one of β-galactose is bonded to the OH group on carbon atom number four of α-glucose. This is a $1 \rightarrow 4$ linkage.

Cow's milk usually contains about 5 percent lactose by volume whereas human milk contains about 7 percent. Cow's milk is fortified with carbohydrates to make it more like human mother's milk. Certain bacteria can ferment lactose to produce lactic acid, causing sour milk. Unless the milk is refrigerated, the air temperature will hasten the fermentation reaction. Lactose is more difficult to ferment than the other sugars. This is important in infant feeding, because there is less of a tendency for the infant to have intestinal disturbances.

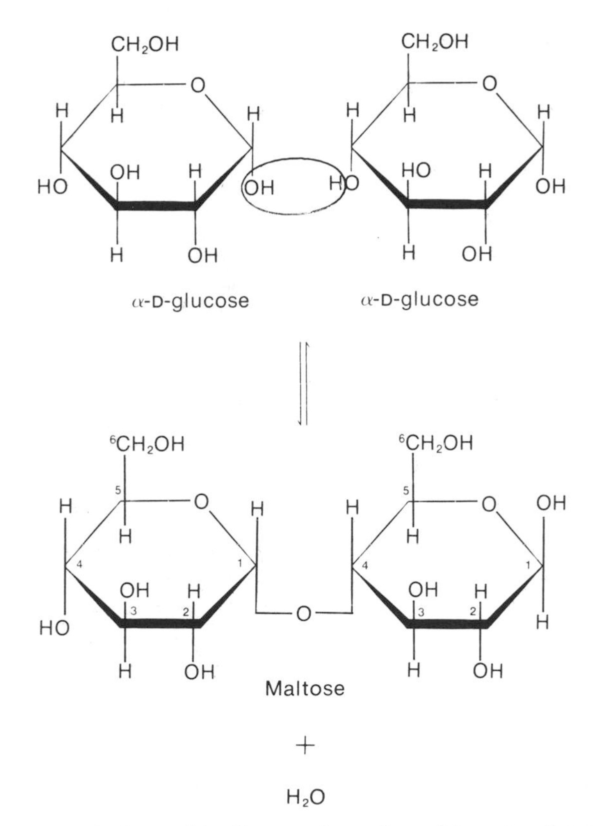

Figure 16-18. Maltose is formed by the condensation of 2 molecules of alpha glucose with the elimination of 1 molecule of water.

Lactose is not as sweet as sucrose. Therefore, the high caloric content of lactose in milk is not excessively sweet. This is an important consideration in diet therapy. In the body the enzyme *lactase* hydrolyzes lactose. Only the monosaccharides can be utilized by the body.

MALTOSE

Maltose, called malt sugar, is found in the germinating seeds of barley or malt. Maltose is also produced in the hydrolysis of starch. Upon hydrolysis, maltose yields 2 molecules of glucose:

$$\text{Maltose} + H_2O \longrightarrow \text{glucose} + \text{glucose}$$

The structure of maltose is shown in Figure 16-18. Maltose has a $1 \rightarrow 4$ linkage between two α-glucose molecules. In the body the enzyme *maltase* hydrolyzes maltose to glucose. Like sucrose, maltose is easily fermented into ethyl alcohol and CO_2, as evidenced by the sale of malt liquors and beer.

The Polysaccharides

The polysaccharides are complex polymers of the monosaccharides. The exact number of monosaccharides which are joined together will vary with the origin of the polysaccharide. Therefore, the letter subscript n is used to designate that unknown value.

CELLULOSE: $(C_6H_{12}O_6)_n$

Cellulose is a polymer of β-glucose. The molecules are joined by $1 \rightarrow 4$ linkages. This combination is called a β $(1 \rightarrow 4)$ linkage. See Figure 16-19. Since man does not have an enzyme capable of breaking the β $(1 \rightarrow 4)$ linkage, he cannot digest cellulose. Ruminants, such as cows, have the ability to hydrolyze cellulose. So they can transform cellulose (grass) to protein (steaks), which man can eat. In the laboratory, cellulose can be hydrolized by acid to give the glucose monomer. A cellulose molecule can yield from 2,000 to 9,000 glucose molecules.

Cotton, linen and hemp (used in ropes) are primarily composed of cellulose, since they are all plant products. Cellulose, which is also found in fruits and vegetables, gives bulk to foods. This bulk causes peristaltic action in the intestines, preventing constipation. Commercially, cellulose is the raw material used in the manufacture of nitrocellulose (an explosive) and in the manufacture of cellulose acetate and cellulose propionate (which are used to make plastics and fibers). When cellulose is treated with NaOH, the solution formed is regenerated back into a fiber in dilute acid solution to form rayon and cellophane.

STARCH: $(C_6H_{12}O_6)_n$

Starch is found in plants. There are various numbers which may be assigned to n, depending upon the source of the starch. Since starch is insoluble in water, it forms a colloidal dispersion.

Starch hydrolyzes to form a number of α-glucose molecules. Starch is not homogeneous. It consists of 95 percent *amylose*, which is a straight chain, and *amylopectin*, which has a great deal of branching. The amylose polymer consists of a number of α-glucose molecules joined by α $(1 \rightarrow 4)$ linkages (see Fig. 16-20). Amylopectin consists of a number of amylose chains joined by α $(1 \rightarrow 6)$ linkages (see Fig. 16-21). Starch hydrolyzes through the following stages:

$$\text{starch} \longrightarrow \text{dextrin} \longrightarrow \text{maltose} \longrightarrow \text{glucose}$$

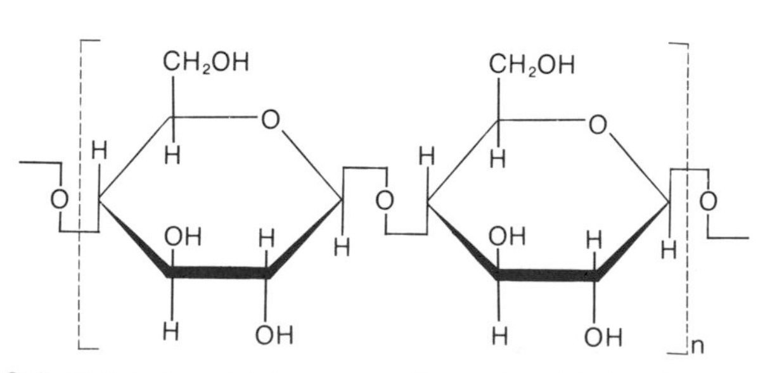

Figure 16-19. Cellulose is formed from repeating units of beta glucose monomers that are joined by beta $(1 \rightarrow 4)$ bonds.

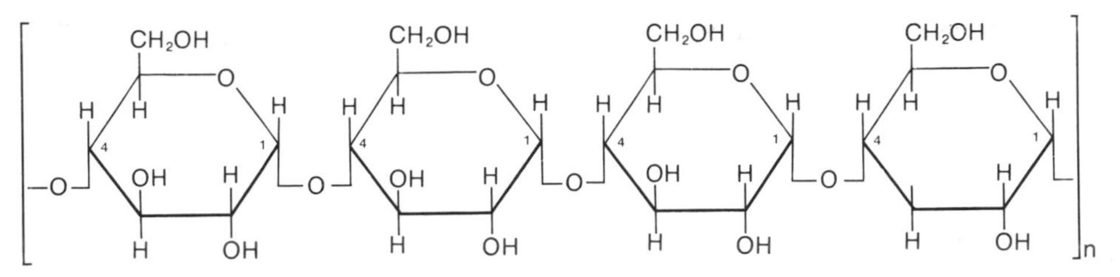

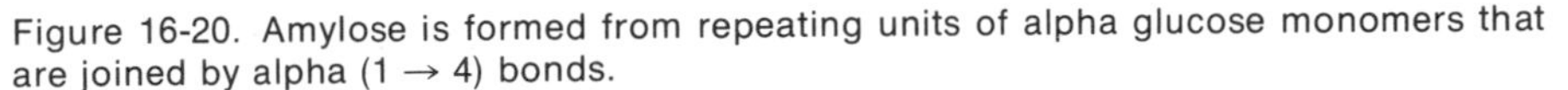

Figure 16-20. Amylose is formed from repeating units of alpha glucose monomers that are joined by alpha (1 → 4) bonds.

In the laboratory, one can observe the process and the degree of hydrolysis by testing the solution with *iodine reagent.* Starch reacts with iodine reagent to produce a *deep blue-black color.* Dextrin produces a *red color* with iodine reagent, while maltose and glucose produce *no color* change.

The starch content of potatoes is 20 percent, wheat 65 percent, corn 65 percent, and rice 80 percent. As fruits ripen the starch is hydrolyzed to glucose and the fruit becomes sweet. Unlike cellulose, starch can be hydrolyzed by the body enzymes. You can detect the sweetness of the sugar formed from the hydrolysis of starch in bread if you chew it thoroughly without swallowing.

DEXTRIN: $(C_6H_{12}O_6)_n$

Dextrin is formed from the partial hydrolysis of starch. It is also formed when bread is toasted. Since dextrin forms sticky and colloidal suspensions, most starch based adhesives are dextrins, for example, postage stamp glue. Dextrin is sweet in taste, is used to impart smoothness to candy, and causes corn syrup to be sticky.

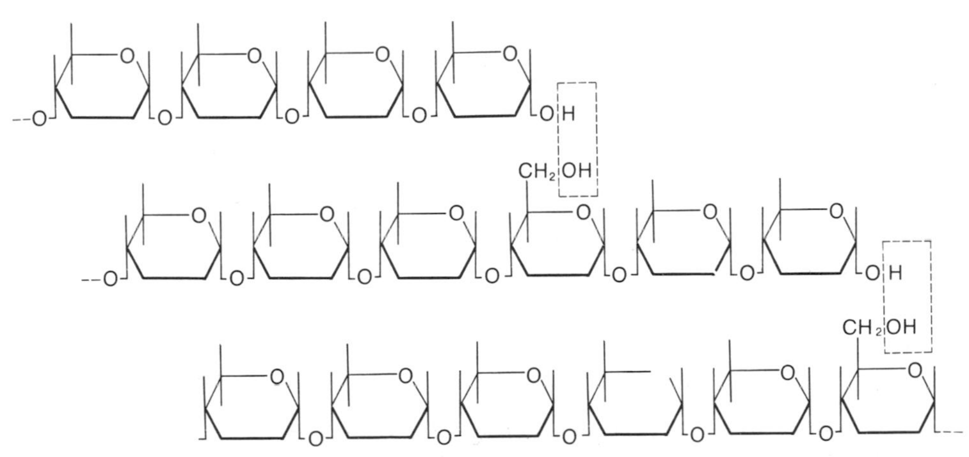

Figure 16-21. Amylopectin is formed from chains of amylose that have joined together by the elimination of a water molecule between chains by alpha (1 → 6) bonds.

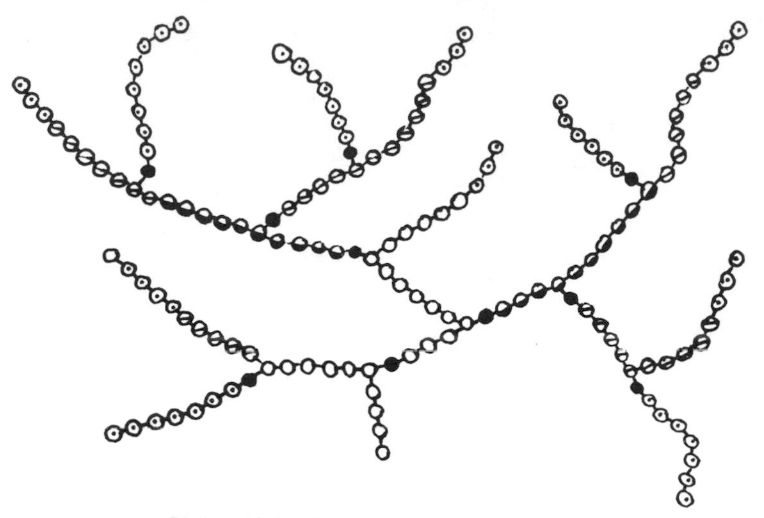

Figure 16-22. The structure of glycogen.

GLYCOGEN: $(C_6H_{12}O_6)_n$

Glycogen is a polymer of α-glucose molecules. It has the same structure as amylopectin except that it is more branched. See Figure 16-22.

Glycogen is produced in the liver. It is stored in both the liver and muscle cells. An adult liver can contain as much as one pound of glycogen. Glycogen serves the same role in the animal world as starch does in the plant world: it serves as a reserve food supply. When exercise depletes blood sugar, the hydrolysis of liver glycogen maintains the normal glucose content of the blood.

DEXTRAN: $(C_6H_{12}O_6)_n$

Dextran is a polysaccharide of glucose obtained from bacterial sources. The molecule is large, having a molecular weight of 70,000. It is very highly branched. Hence, dextran cannot be hydrolyzed by the body enzymes that can hydrolyze starch and glycogen. Therefore, dextran is used as a plasma substitute to alleviate the effects of shock.

When a dextran solution is intravenously injected, the dextran remains in the blood stream for several days and is slowly excreted in the urine. While in the vascular system, dextran increases the osmotic pressure of the blood. As a result water passes into the blood vessels from the extracellular spaces. The increase in blood volume raises the blood pressure and in this way provides relief to a patient suffering from shock.

HEPARIN

Heparin is a mixed polysaccharide. It consists of a repeating of glucuronic acid plus glucosamine. Both are derivatives of glucose. Its structure is shown in Figure 16-23.

Heparin is a blood anticoagulant. It inhibits the conversion of prothrombin into thrombin. Since thrombin catalyzes the formation of blood clots, heparin prevents the clotting of blood.

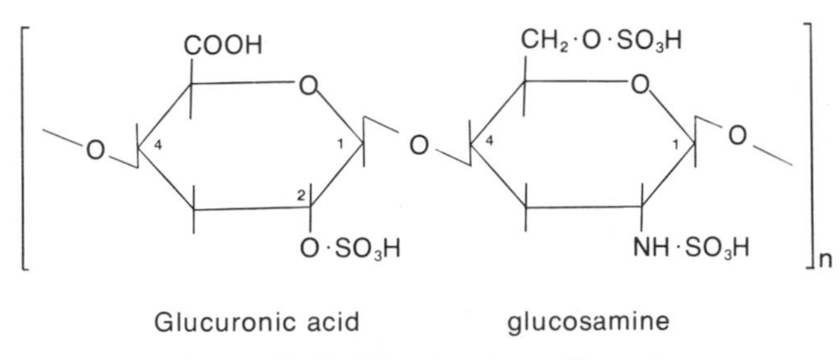

Glucuronic acid glucosamine

Figure 16-23. The structure of heparin.

CHEMICAL PROPERTIES OF CARBOHYDRATES

From a chemical standpoint, the reactions of monosaccharides and disaccharides may be divided into those involving: 1) the carbonyl group (aldehyde or ketone); 2) the alcoholic hydroxyl group; and 3) the carbon skeleton of the entire molecule.

Carbonyl Group Reaction

When a sugar is classified as a *reducing sugar,* it refers to the fact that the particular sugar can be oxidized by such oxidizing agents as Benedict's solution, Fehling's solution or Tollen's reagent. Glucose gives positive tests with any of these reagents because of its easily oxidizable aldehyde group. Both Benedict's and Fehling's solutions contain cupric ion. In a positive test, the blue solution first turns to green and then to an orange-red precipitate. The more sugar present, the deeper the color becomes. The reaction is:

$$\text{carbonyl group} + \underset{\text{blue}}{2\,Cu(OH)_2} \longrightarrow \underset{\text{red}}{Cu_2O} + \text{acid} + H_2O$$

All of the monosaccharides and disaccharides we have studied, *except for sucrose,* give a positive test.

Tollen's reagent contains a silver compound. With a reducing sugar, Tollen's reagent gives a precipitate of metallic silver, the silver mirror test. The reaction is:

$$\text{carbonyl group} + 2AgOH \longrightarrow \underset{\substack{\text{silver}\\\text{mirror}}}{2\,Ag} + \text{acid} + H_2O$$

Again, *except for sucrose*, all of the monosaccharides and disaccharides we have studied give a positive test.

Sucrose does not give a positive test because the aldehyde group on its component glucose is tied up to the fructose. When sucrose is hydrolyzed into its two components, the glucose is free to give the positive test.

Lactose is sometimes found in the urine of pregnant women. It is a reducing sugar. The linkage in the lactose does not tie up the aldehyde as the linkage in sucrose. Therefore, one should use caution in interpreting the urine analyses of pregnant women.

Hydroxyl Group Reactions

By careful oxidation the —CH_2OH group can be oxidized to the carboxyl group without affecting the aldehyde group. With nitric acid both the aldehyde and —CH_2OH group are oxidized to the carboxyl group. Also, the hydroxyl group of sugars reacts with inorganic or organic acids to produce esters. The most important is phosphoric acid, which is involved in the formation of *adenosine triphosphate* (ATP), a high energy compound that releases and stores energy for biochemical reactions.

Reactions of the Carbon Skeleton

The sugars may be oxidized in the body to form water and carbon dioxide, and to release energy. Glucose and fructose are fermented by the catalytic action of yeast enzymes. Other monosaccharides are fermented by bacterial enzymes. Some fermentations produce acids; acetic acid is formed in the preparation of vinegar from apple juice and commercial lactic acid is formed from glucose or molasses. Sucrose and maltose readily undergo fermentation to form a variety of compounds. Lactose is not fermented by yeast, while sucrose and maltose are. Acids are the most common products of these fermentations. Hence, fermentation of sucrose by a fungus is a commercial method of making citric acid. Butter, cheese, and sauerkraut result from lactic acid fermentation. See Figure 16-24.

CARBOHYDRATES AND BIOENERGETICS

Bioenergetics is defined as the study of energy transformations in living organisms. Energy is transformed from one type to another at three major stages in the biological world.

The first stage is photosynthesis. Chlorophyll uses the energy of the sun to convert CO_2 and H_2O to high energy carbohydrates.

The second stage is respiration. Animals oxidize carbohydrates and convert the chemical energy of carbohydrates to other forms of energy. This is the reverse of photosynthesis. The stored energy is released along with the waste products CO_2 and H_2O.

In the third stage, the released chemical energy is used by animal cells to do work either within the cell or upon the environment. This may comprise the mechanical work of muscular contraction, osmotic work, electrical work in the

Alcoholic Fermentation:

$$C_6H_{12}O_6 \longrightarrow 2\ C_2H_5OH + 2\ CO_2$$

Acetic acid Fermentation:

$$C_6H_{12}O_6 + 2O_2 \longrightarrow 2CH_3COOH + 2H_2O + 2CO_2$$

Lactic Acid Fermentation:

$$C_6H_{12}O_6 \longrightarrow 2\ CH_3CHOHCOOH$$

Figure 16-24. Fermentation reactions.

nervous system, or the chemical work involved in growth. As these functions are performed waste in the form of heat energy flows into the environment. The reactions involved are:

$$6\ CO_2 + 6\ H_2O + 686{,}000\ \text{cal} \underset{\text{respiration}}{\overset{\text{photosynthesis}}{\rightleftharpoons}} C_6H_{12}O_6 + 6\ O_2$$

The term *metabolism* describes the various chemical processes used by a living organism to transform food to provide energy, growth materials, and cell repair. Any chemical compound that is involved in a metabolic reaction is called a *metabolite.* Carbohydrates are the primary metabolites of the animal world, since over one half of the food we consume consists of carbohydrates.

Returning to the photosynthesis reaction, we notice that 686,000 cal are consumed. Every mole of glucose formed contains 686,000 more calories of energy than the molecules from which it was formed. In the reverse reaction, respiration, this same energy is released as free energy. Since the molecular weight of glucose is 180 g/mole, one gram of glucose releases approximately 4,000 cal. This energy is available for physiological work. However, because of various inefficiencies in the body, only about 50 percent of the maximum possible energy is actually available as free energy.

These caloric values are obtained by burning the substance in a bomb calorimeter. Investigators have found that carbohydrates and proteins but not fats have the same caloric value when burned in a bomb calorimeter or the human body.

SUMMARY

On decomposition carbohydrates yield carbon and water. They are important as energy sources. They are named by common names, according to length of the chain, according to the functional group present, and according to how they rotate polarized light. Carbohydrates can be represented by straight chain formulas or ring structures. Glucose is the most important aldohexose in the body. Glucose is found in the blood, is converted to glycogen and is stored in the liver. Fructose is a ketohexose found in fruits and honey. Galactose is an aldohexose that is important in infant feeding. The important disaccharides are sucrose, lactose and maltose. Sucrose, table sugar, hydrolyzes to glucose and fructose. Sucrose *does not* give a positive aldehyde test. Lactose, milk sugar, hydrolyzes to glucose and galactose. Maltose, malt sugar, hydrolyzes to give 2 molecules of glucose. Cellulose is a polysaccharide, made of many β-glucose units. Human beings cannot hydrolyze cellulose for food as do cows and horses. Starch hydrolyzes to α-glucose and forms a store of energy for plants. Upon partial hydrolysis, starch forms dextrin, a lower molecular weight polysaccharide. Glycogen, also formed from α-glucose, is produced and stored in the body. Glycogen serves as a store of energy for animals. Heparin is a mixed polysaccharide that prevents the clotting of blood.

EXERCISE

1. What is the origin of the term carbohydrates?
2. What is the hydrogen to oxygen ratio in carbohydrates?
3. Draw the general formula for a carbohydrate.
4. What three facts are considered in animal carbohydrates?
5. Draw the straight chain and ring structures for glucose.
6. What is the most important pentose and where is it found?
7. What is another name for glucose?
8. What does the prefix D- in front of a carbohydrate name mean?
9. What is the formula for fructose, and where is it usually found?
10. What is invert sugar?
11. How do glucose and fructose combine to form sucrose?
12. What is the source of lactose and why is it important?
13. Give the hydrolysis products of sucrose.
14. Why will sucrose not give a positive aldehyde test?
15. What is the source of lactose, and what does it hydrolyze to give?
16. Will lactose give a positive aldehyde test? Why or why not?
17. What are the hydrolysis products of maltose? Where is it found?
18. Write the general formula of cellulose. What is the ultimate hydrolysis product? For what purpose is cellulose used?
19. What is heparin and why is it important?
20. What is the general formula of starch and what are the stages of its hydrolysis?
21. How is dextrin used in foods?
22. How is glucose stored in the body and where?
23. How would you distinguish sucrose from glucose?
24. How would you distinguish maltose from starch?

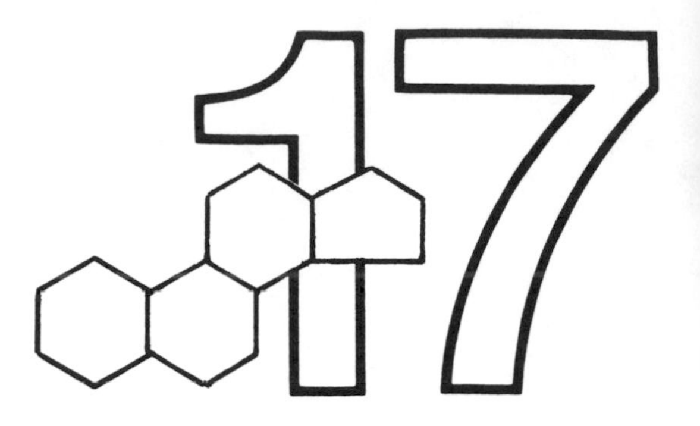

OBJECTIVES

When you have completed this section, you will be able to:

1. Summarize the physical and chemical characteristics of lipids.
2. State where lipids are found in nutrients and where they are found in the body.
3. List the various functions of lipids in the body.
4. Classify lipids according to their hydrolysis products.
5. State the physical properties of lipids.
6. Enumerate the names and structures of the saturated and unsaturated fatty acids.
7. Distinguish between the fatty acid composition of fats and oils and the degree of unsaturation of the unsaturated fatty acids.
8. Write general and specific formulas of simple fats, knowing the fatty acid components.
9. Write the chemical reactions that the fats undergo in the body and in the laboratory.
10. Understand what the term polyunsaturated means.
11. Explain how liquid fats are changed into solid fats.
12. Detect fats by a simple chemical test.
13. Distinguish between the various classifications of lipids, where the various lipids are found in the body, and their physiological activities in the body.
14. Discuss the steroids and name some of the compounds that make up this classification.
15. Relate the presumed relationship between cholesterol and certain types of heart disease.

LIPIDS

16. Show the structural relationship of the bile acids to the steroids.
17. Discuss the prostaglandins and how they are biosynthesized from unsaturated fatty acids.

Lipids are oily, greasy fatty substances which have a glistening appearance. They are also called fats, oils or waxes. All have a specific gravity less than 1, so they will float on water. The *fats* (butter, lard or tallow) are solid at room temperature, while the *oils* (olive oil, peanut oil, or corn oil) are liquid at room temperature. Lipids are soluble in nonpolar solvents (dry cleaning fluids or chloroform) but are insoluble in polar solvents (water). Lipids are composed of the elements carbon, hydrogen and oxygen, but they can also contain phosphorus and nitrogen as a part of a complex molecule.

Even though lipids do not dissolve in water, they can be broken down into extremely small droplets that will remain suspended in water. This is called an emulsion. Milk, salad dressings, mayonnaise, cosmetic creams, and body lotions are examples of such emulsions.

Mineral oil, paraffin waxes, and essential oils are not classified as lipids. Mineral oils and paraffin waxes are high molecular weight hydrocarbons. Essential oils are liquids that have characteristic odors and are used as flavoring agents and perfumes. All of these substances have different chemical properties than the lipids.

Fats and oils are the most abundant lipids found in nature. Generally, solid fats are obtained from animal sources, and liquid oils from vegetable sources. Thus, lard is obtained from hogs, tallow from beef and sheep, butter from milk, and lanolin from sheep. Olive oil, corn oil, cotton seed oil, and safflower oil are obtained from the fruit or seed of the respective vegetable sources.

In animal organisms fats are found in all tissues, between the muscles, around all organs, in bone marrow, and in very large amounts of adipose tissue directly beneath the skin. In the human body all cells contain small amounts of fat and there are specialized adipose tissues which are known as depot fat. Here fats are especially abundant and they are stored as sources of energy. Lipids also exist in cell membranes, in brain and nerve tissue, and in blood and tissues as lipoproteins (lipids combined with proteins).

FUNCTIONS OF FATS IN THE BODY

Fats are used as:

1. Sources of energy: More energy is liberated from fat (9 Kcal per gram) than from either proteins or carbohydrates (4 Kcal per gram). Because of this high caloric value, fats should be eaten sparingly in warm weather, but in cold weather the diet can contain a higher fat content in order to supply extra heat energy.
2. Reserve supply of energy: The body accumulates fats when the food intake exceeds energy requirements. Excess fat, in the form of adipose tissue, is distributed under the skin and around the abdominal organs as a reserve store of energy, available when needed during illness or a period of starvation. Some adipose tissue is synthesized from glucose and consequently, a diet rich in carbohydrates results in an increase of stored fat in the body. Generally speaking, individuals with too much adipose tissue are less healthy and often die at a younger age than persons of normal weight. This is partially due to the added weight and an additional network of blood vessels that may cause a strain on the heart, leading to heart attacks.
3. Insulation: Layers of fat under the skin serve to protect against cold because fat is a poor conductor of heat, and prevents the loss of heat from the body through the skin.
4. Protection of nerve endings and delicate organs: Fat deposits at nerve endings and at the ganglia of nerve cells serve to protect them against irritation by stimuli. They also support abdominal organs and protect them from shock and mechanical injuries.
5. Constituents of protoplasm: Fat-like substances in cells can absorb large quantities of water without themselves being dissolved and passing through cell membranes. The lipids are found in biological membranes and at interfaces where they can permit the entry of nutrients and at the same time allow the exit of waste products from the cell.

CLASSIFICATION OF LIPIDS

Lipids are usually classified by making use of their solubilities and their hydrolysis products. The classifications are:

1. Simple lipids which are fats, oils, and waxes. They are esters of fatty acids with various alcohols.
   a. Neutral fats are solids (lard), esters of fatty acids and glycerol.
   b. Neutral oils are liquids (olive oil), esters of fatty acids and glycerol.
   c. Neutral waxes (beeswax) are esters of fatty acids and monohydroxy alcohols having a higher molecular weight than glycerol.
2. Compound lipids are esters of fatty acids, with an alcohol and other additional groups or compounds. These include:

a. Phospholipids (lecithin, cephalins and sphingomyelins) that yield glycerol, fatty acids, phosphoric acid and a nitrogen compound upon hydrolysis.
b. Glycolipids that yield glycerol, fatty acids, phosphoric acid, a carbohydrate, and a nitrogen compound upon hydrolysis.
c. Lipoproteins are higher molecular weight compounds in which lipids exist in conjunction with proteins.

3. Derived lipids are compounds obtained by hydrolysis of simple and compound lipids. They are the:
   a. Fatty acids, monoglycerides, and diglycerides.
   b. Organic nitrogen compounds (choline, sphingosine, ethanolamine).
   c. The steroids (sterols, corticoids, and sex hormones).

PHYSICAL PROPERTIES

Lipids have a slippery and greasy feeling, forming a transparent grease spot when placed on a sheet of paper. Pure lipids are colorless, odorless, and tasteless, but in the natural form their color is yellow, because of the presence of a yellow fat soluble pigment, and these do have a characteristic odor and taste.

Melting Point

Pure compounds have a sharp melting point, but because fats and oils are mixtures of glycerides of different fatty acids their melting points are not very sharp. Relatively low melting points of fats and oils and fatty acids indicate the presence of glycerides of unsaturated fatty acids.

Specific Gravity

Solid fats have a specific gravity of about 0.87 while oils have a specific gravity ranging from 0.91 to 0.94. They all float on top of water, as their specific gravity is less than 1.0.

Solubility

The largest portion of the fat or oil molecule is mostly hydrocarbon, therefore the lipids resemble the higher hydrocarbons in their lack of solubility in water. But they are soluble in either petroleum solvents or halogenated hydrocarbons.

Emulsification

When lipids are vigorously shaken or mixed with water for a prolonged period of time they are broken up into minute droplets and therefore a temporary emulsion is formed. This emulsion will eventually break and the oil and water layer will separate. However, if an emulsifying agent such as a soap or detergent is added, the emulsion will persist for much longer periods of time. In an emulsion, there is a film formed around the surface of the tiny dispersed fat globule that prevents them from forming larger globules and separating out on top of the water. In the digestive process, the emulsion of fats in the intestines is accomplished with the help of bile, an emulsifying agent. This emulsification of the lipids is a prerequisite for digestion and absorption. The blood carries the fats in an emulsified form to the individual body cells.

FATTY ACIDS

The fatty acids are usually long, straight chain carboxylic saturated or unsaturated acids. Those found in nature invariably have an even number of carbon atoms in the molecule, including the carbon atom of the carboxylic acid. Fatty acids are classified as derived lipids, because they are constituents of lipids and are obtained by the hydrolysis of lipids. Most fatty acids have only one carboxyl group, and the hydrocarbon chain is usually straight and unbranched. Table 17-1 lists the more common saturated fatty acids. The most important saturated fatty acids are palmitic and stearic acids, which are found in animal fats and vegetable oils.

Table 17-1. Common fatty acids.

Saturated acid	*Formula*	*Melting point °C*	*Common source*
Acetic acid	CH_3COOH	16.7	Vinegar
Butyric acid	$CH_3(CH_2)_2COOH$	−8	Butter
Caproic acid	$CH_3(CH_2)_4COOH$	−1.5	Butter
Caprylic acid	$CH_3(CH_2)_6COOH$	16.5	Butter
Capric acid	$CH_3(CH_2)_8COOH$	31.3	Butter, coconut oil
Lauric acid	$CH_3(CH_2)_{10}COOH$	44	Coconut oil
Myristic acid	$CH_3(CH_2)_{12}COOH$	58	Coconut oil
Palmitic acid	$CH_3(CH_2)_{14}COOH$	63	Animal & vegetable fats
Stearic acid	$CH_3(CH_2)_{16}COOH$	70	Animal & vegetable fats
Arachidic acid	$CH_3(CH_2)_{18}COOH$	75	Peanut oil
Lignoceric acid	$CH_3(CH_2)_{22}COOH$	81	Brain & nerve tissue
Cerotic acid	$CH_3(CH_2)_{24}COOH$	82.5	Beeswax & wool fat

Unsaturated fatty acids may contain one or more carbon-to-carbon double bonds. The greater the number of double bonds the greater the degree of unsaturation and the lower the melting point of the acid or the fat. This is why animal fats (they have fewer double bonds) are solid, while vegetable oils (they have more double bonds) are liquid at room temperature. From Table 17-2 you will see the most common unsaturated fatty acids.

Table 17-2. Common unsaturated fatty acids.

Unsaturated acids	*Formula*	*Melting point °C*	*Common source*
Palmitoleic acid	$CH_3(CH_2)_5CH{=}CH(CH_2)_7COOH$	–	Animal & vegetable fats
Oleic acid	$CH_3(CH_2)_7CH{=}CH(CH_2)_7COOH$	16	Animal & vegetable fats
Linoleic acid	$CH_3(CH_2)_4CH{=}CHCH_2CH(CH_2)_7COOH$	−5	Linseed & vegetable oils
Linolenic acid	$CH_3CH_2CH{=}CHCH_2CH{=}CHCH_2CH{=}CH(CH_2)_7COOH$	−11	Linseed oil
Arichodonic acid	$CH_3(CH_2)_4(CH{=}CHCH_2)_3CH{=}CH(CH_2)_3COOH$	−49.5	Brain & nerve tissue

The most common unsaturated acid is oleic acid which contains 1 carbon-to-carbon double bond. Linoleic has 2 double bonds, linolenic has 3, and arichodonic acid has 4. These 4 unsaturated fatty acids are especially important in nutrition; they are known as essential fatty acids. They are indispensable because they cannot be synthesized in the body rapidly enough to satisfy the body's requirements and therefore they must be included in the diet.

FATTY ACID COMPOSITION OF FATS AND OILS

From a nutritional standpoint there has been a great deal of interest in polyunsaturated fats and oils. The term polyunsaturated refers to an oil that is highly unsaturated, containing a large number of carbon-to-carbon double bonds in the molecule. Recent research has shown that the body is less able to synthesize cholesterol from unsaturated fatty acids than from saturated fatty acids, and cholesterol is believed to be an important factor in heart attacks. There appears to be a relationship between excess deposits of cholesterol (a derived lipid) in the blood and subsequent stiffening of the blood vessels (hardening of the arteries) and severe heart disease. Therefore, if a person tends to synthesize too much cholesterol from a diet high in saturated fats this condition may possibly be helped by modifying the diet to contain more liquid fats (polyunsaturated) and less solid fats (saturated).

In underdeveloped countries where the diet contains insufficient polyunsaturates skin diseases occur frequently. This is caused by a dietary imbalance between saturated and unsaturated fats, leading to an accumulation of excess cholesterol deposits in the circulatory system.

An examination of Table 17-3 will reveal that the vegetable oils contain a higher percentage of unsaturated fatty acids while animal fats contain a lower percentage of unsaturated fatty acids. For example, animal fat (butter) contains 10 to 13 percent stearic acid (saturated) and 4 to 5 percent linoleic acid (unsaturated). Vegetable oil (corn oil) contains 39 to 42 percent linoleic acid (unsaturated) and 3 to 4 percent stearic acid (saturated).

Table 17-3. Fatty acid composition of fats and oils.

		Saturated		*Unsaturated*		
	% Myristic	% Palmitic	% Stearic	% Oleic	% Linoleic	Iodine number
Animal Fats						
Butter	7-10	23-26	10-13	30-40	4-5	26-40
Lard	1-2	28-30	12-18	41-48	6-7	46-66
Tallow	3-6	24-32	14-32	35-48	2-4	46-66
Vegetable Oils						
Olive	0-1	5-15	1-4	49-84	4-12	79-90
Peanut	-	6-9	2-6	50-70	13-26	84-102
Corn	0-2	7-11	3-4	43-49	34-42	116-130
Soybean	0-2	6-10	2-4	21-29	50-59	127-138

Chemical Composition

You learned earlier in the chapter on alcohols and acids that alcohols react with organic carboxylic acids to form compounds called esters. Natural fats and oils are triesters of the alcohol glycerol ($CH_2OH—CHOH—CH_2OH$) and high molecular weight fatty acids and are called triglycerides. The fats of animals and plants are mixed triglycerides because they contain two or three different fatty acids in each molecule.

Since the esterification of alcohols with organic carboxylic acids is reversible, the hydrolysis of esters (splitting them with water) or the saponification of esters (splitting them with alkali such as NaOH or KOH) will yield the original alcohol and the corresponding fatty acids (in the case of water) or the salt of the acid (in the case of the alkali). The sodium or potassium salt of the acid is a soap, and splitting fats with alkali is the method of making soaps. The sodium salt yields a solid soap while the potassium salt yields a liquid soap.

A simple triglyceride contains only one type of fatty acid component, and upon hydrolysis will yield 1 glycerol and 3 identical fatty acid molecules. A mixed triglyceride contains as many as three different fatty acids and will yield them upon hydrolysis.

Chemical Properties

HYDROLYSIS

Hydrolysis is the most characteristic and important reaction of fats and oils. In the presence of a catalyst (such as an acid, a base or an enzyme) a fat or oil will react with water to split the molecule and form glycerol and fatty acids (see Fig. 17-1).

$$\text{Fat} + \text{water} \xrightarrow{\text{catalyst}} \text{glycerol} + \text{fatty acids}$$

SAPONIFICATION

Saponification is the splitting of a glyceride with a base, such as NaOH or KOH, to form the glycerol and the sodium or potassium salt of the fatty acids (soaps). See Figure 17-2.

$$\text{Fat} + \text{base} \longrightarrow \text{glycerol} + \text{salt of fatty acid (soap)}$$

The saponification number of a fat is the number of milligrams of KOH required to saponify 1 gram of fat or oil. The higher the saponification number, the smaller the fatty acid chains found in the molecule.

RANCIDITY

Upon long exposure to air many unsaturated fats turn yellow and develop objectionable tastes and odors. This is known as *rancidity,* and is the result of 3 chemical reactions:

a. Hydrolysis of the ester, liberating foul smelling and volatile fatty acids.
b. Oxidation of the unsaturated fatty acids to yield aldehydes and ketones.

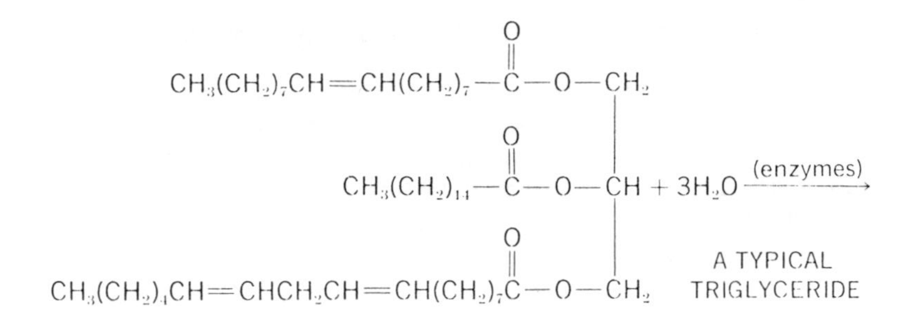

$$CH_3(CH_2)_7CH{=}CH(CH_2)_7{-}\overset{O}{\overset{\|}{C}}{-}OH \qquad HO{-}CH_2$$

OLEIC ACID

$$CH_3(CH_2)_{14}{-}\overset{O}{\overset{\|}{C}}{-}OH + HO{-}CH$$

PALMITIC ACID

$$CH_3(CH_2)_4CH{=}CHCH_2CH{=}CH(CH_2)_7\overset{O}{\overset{\|}{C}}{-}OH \qquad HO{-}CH_2$$

LINOLEIC ACID — GLYCEROL

Figure 17-1. Hydrolysis in the presence of enzymes.

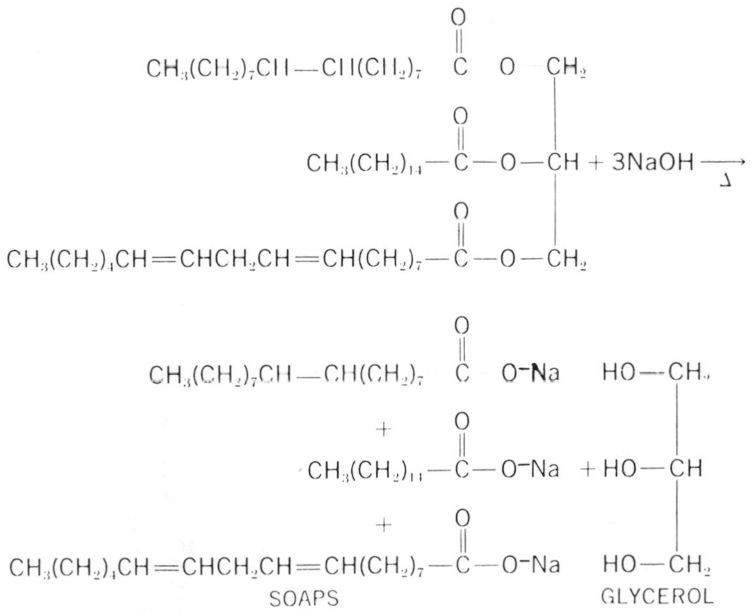

$$CH_3(CH_2)_7CH{=}CH(CH_2)_7{-}\overset{O}{\overset{\|}{C}}{-}O{-}CH_2$$

$$CH_3(CH_2)_{14}{-}\overset{O}{\overset{\|}{C}}{-}O{-}CH + 3NaOH \xrightarrow[\Delta]{}$$

$$CH_3(CH_2)_4CH{=}CHCH_2CH{=}CH(CH_2)_7{-}\overset{O}{\overset{\|}{C}}{-}O{-}CH_2$$

$$CH_3(CH_2)_7CH{=}CH(CH_2)_7{-}\overset{O}{\overset{\|}{C}}{-}O^-Na \qquad HO{-}CH_2$$

$$+ \; CH_3(CH_2)_{14}{-}\overset{O}{\overset{\|}{C}}{-}O^-Na + HO{-}CH$$

$$+ \; CH_3(CH_2)_4CH{=}CHCH_2CH{=}CH(CH_2)_7{-}\overset{O}{\overset{\|}{C}}{-}O^-Na \qquad HO{-}CH_2$$

SOAPS — GLYCEROL

Figure 17-2. Saponification.

c. Bacterial growth upon impurities present and the subsequent formation of butyric acid.

Two methods are used to prevent foods rich in fats and oils from becoming rancid: lowering the temperature and adding antioxidants. When you lower the temperature, you decrease the rate of reaction. Adding antioxidants inhibits the auto-oxidation of the unsaturated fatty acids. Most vegetable shortenings used in cooking and even certain brands of lard have antioxidants added to protect them from becoming rancid.

HYDROGENATION

Liquid vegetable oils have carbon-to-carbon double bonds, and hydrogen atoms can be added to those double bonds to form saturated compounds. This process is called *hydrogenation,* in which 1 molecule of hydrogen (H_2) adds to each double bond. For example, triolein can be hydrogenated to tristearin by adding 3 molecules of hydrogen to the 3 double bonds in the molecule.

Many of the solid vegetable shortenings that you purchase in the supermarket are *partially* hydrogenated vegetable oils. Completely hydrogenated oils result in brittle products similar to mutton tallow, which is unsuitable for cooking purposes. In the manufacture of margarine (which is a mixture of partially hydrogenated vegetable oils), skim milk, and salt, Vitamin A and D are added to make it taste like butter. Today more margarine is produced than butter.

IODINE NUMBER

Just as hydrogen will add to a carbon-to-carbon double bond in an unsaturated fat or oil, so will iodine. Therefore, the iodine number is a measure of the unsaturation of a fat or oil and it is defined as the number of grams of iodine that will react with 100 grams of the fat or oil. The higher the iodine number, the greater the degree of unsaturation, as you can see by an examination of Table 17-3 and Table 17-4.

Table 17-4. Iodine number.

Fat or Oil	*Iodine number*
Animal fats	
Beef tallow	35-45
Butter	25-40
Human fat	65-70
Lard	45-70
Vegetable oils	
Castor oil	85-90
Coconut oil	6-10
Corn oil	115-130
Cottonseed oil	100-120
Linseed oil	175-205
Olive oil	75-95
Soybean oil	125-140

ACROLEIN

When fats or oils are heated to high temperatures, *acrolein* is liberated from the decomposition of the fat or oil to yield glycerol. Glycerol then dehydrates to form acrolein and unsaturated aldehyde with a sharp, penetrating and unpleasant odor. This reaction is characteristic for fats and oils. You produce acrolein every time you burn a fat or oil in cooking.

SIMPLE LIPIDS

Fats and Oils

Simple lipids include animal fats (butter and lard) that predominately contain ester of saturated fatty acids and vegetable oils (corn oil) that predominately contain unsaturated fatty acids. Upon hydrolysis they yield glycerol and the fatty acids. The fatty acid composition varies widely. This variable composition of the lipids results from several factors: the plant or animal species, the diet of the plant or animal, and the environmental climate. Lard made from corn-fed hogs is more highly saturated than lard produced from peanut-fed hogs.

Waxes

Waxes are esters of fatty acids and high molecular weight alcohols other than glycerol. They contain only 1 molecule of fatty acid because the alcohol contains only 1 hydroxyl group. The fatty acid obtained on hydrolysis is generally one of the fatty acids usually involved with fats and oils, but the alcohol has a high molecular weight. Waxes are not as easily hydrolyzed as fats and oils, and usually occur in nature as protective coatings on plants (see Table 17-5).

Table 17-5. Naturally occurring waxes.

Name	*Alcohol*	*Acid*	*Source*	*Uses*
Beeswax	$CH_3(CH_2)_{29}OH$, Myricyl alcohol	Palmitic acid	Honeycomb	Cosmetics, ointment, candles
Lanolin	Mixture	Mixture acids	Wool	Ointments, salves, creams
Spermaceti	$CH_3(CH_2)_{15}OH$, Cetyl alcohol	Palmitic acid	Sperm whale	Ointments, creams, candles
Carnauba	$CH_3(CH_2)_{29}OH$, Myricyl alcohol	Palmitic acid	Palm leaves	Floor & auto wax

COMPOUND LIPIDS

Compound lipids (the phospholipids, glycolipids, and lipoproteins) yield glycerol, fatty acids and other substances that contain either nitrogen or phosphoric acid or both upon hydrolysis.

Phospholipids

Phospholipids are subdivided into the lecithins, the cephalins, and the sphingomyelins, which yield upon hydrolysis a nitrogen containing compound and phosphoric acid. They are found in all living organisms, and are particularly abundant in egg yolk, brain tissue, spinal tissue, liver, and in the outer membrane of most cells. Because phospholipids are large molecules and because they contain both polar and nonpolar components they act as emulsifying agents at the surface of cell membranes, where water insoluble lipids must come in contact with water soluble substances. They act to transport lipids in the blood stream, and play an important part in processes involving secretions, transport of ions across cell membranes, and in systems involving the transport of electrons.

LECITHINS

Lecithins are probably the most common of the phospholipids, hydrolyzing to yield glycerol, 2 molecules of a fatty acid, choline (a nitrogen compound) and phosphoric acid. Fats are partially converted to lecithin in the body and therefore are transported as lecithins from one part of the body to another. The lecithins are excellent emulsifying agents and are used extensively in commercial food preparations. They vary in composition depending upon the assortment and identity of the fatty acid component.

The venom of some snakes contain certain enzymes that hydrolyze phospholipids forming compounds known as lysolecithins (lysocephalins) and the unsaturated fatty acid. The lysolecithins have the ability to destroy red blood cells by their hydrolytic action and if this hemolysis is sufficiently extensive in the victim of a snake bite death may result.

CEPHALINS

Cephalins yield glycerol, 2 fatty acids, phosphoric acid and either ethanolamine ($HO—CH_2—CH_2—NH_2$) or serine ($HO—CH(NH_2)—COOH$) instead of choline. They differ from each other by the identity of the fatty acids contained in the molecule. Cephalins play an important part in the clotting of blood. Because they are released upon the disintegration of the blood platelets, they have been used as hemostats to arrest the flow of blood during surgery. Cephalins are important in the metabolism of fats in the liver and as a source of phosphoric acid in the building of new tissue.

SPHINGOMYELINS

Sphingomyelins are derivatives of the amine sphingosine ($CH_3\ (CH_2)_{12}—CH{=}\underset{\underset{OH}{|}}{C}H—\underset{\underset{NH_2}{|}}{C}H—CH_2OH$). The sphingomyelins consist of a fatty acid (saturated or unsaturated) joined by an amide linkage to the primary $—NH_2$ group of the amine. They are essential constituents of protoplasm of cells and seem to be concentrated in the brain, liver, and spleen, but they are found in kidney, liver, egg yolk, blood, and muscle. The pathologic accumulation of sphingomyelins in the brain, liver, and spleen during infancy or early childhood causes mental retardation and possible early death.

GLYCOLIPIDS

Glycolipids upon hydrolysis yield a fatty acid, an alcohol, a carbohydrate and a nitrogen containing compound, but no glycerol or phosphoric acid. Glycolipids are also called cerebrosides because they are found in the cerebrum. They resemble the sphingomyelins but instead of having a choline and a phosphoric acid, they contain a carbohydrate (galactose or occasionally glucose) in addition to the sphingosine and a fatty acid. Young children need milk in their diet because it contains lactose, which is needed for the formation of the cerebrosides.

LIPOPROTEINS

Lipoproteins are lipids combined with proteins. The fibrin in the blood is an example of the type, and the various types of lipids found in the blood are believed to be carried as lipoprotein complexes.

DERIVED LIPIDS

Compounds that are obtained from the hydrolysis of simple and compound lipids are called derived lipids and these include the fatty acids, glycerol, long chain alcohols, amine compounds and a special class of compounds called the steroids. Steroids are compounds that have a skeleton structure with a cyclopentyl-phenanthrene polycyclic nucleus (Fig. 17-3). The skeleton of the phenanthrene molecule is completely saturated with hydrogen atoms, and a five membered carbon ring is attached.

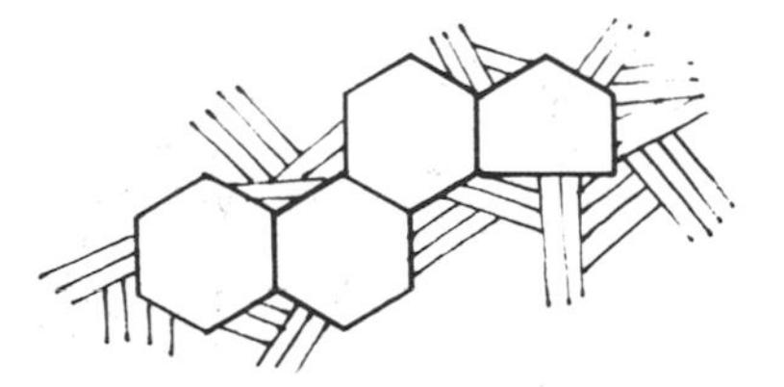

Figure 17-3. Perhydrocyclopentanophenanthrene.

The steroids include sterols, corticoids, bile acids, certain drugs, vitamins and hormones. Over 30 different steroidal compounds have been found in nature, occurring in plant and animal tissue, yeasts and molds, but not in bacteria. They may exist in the free state or combined with acids or carbohydrates. They resemble high molecular weight hydrocarbons because of the preponderantly large hydrocarbon portion of the molecule, and therefore are soluble in fat solvents. However, unlike other lipids, they cannot be hydrolyzed or saponified.

When the steroid structure contains one or more hydroxyl groups it is called a *sterol*. Cholesterol is the best known and most abundant steroid in the body and it is present in all living animal cells but not vegetable cells. It can exist as the free form or as the ester.

Cholesterol

Cholesterol is the chief constituent of gallstones, and is found in the brain, nerve tissue, and blood plasma. In the blood plasma, along with the lecithins and other lipids, it is thought to take part in the transportation of fats to the tissues and in the metabolism of fats. It is found in the kidneys of nephritis patients. It is used by certain endocrine glands for the synthesis of the steroid hormones and it is chemically related in structure to the bile acids, the sex hormones, and some carcinogenic compounds.

High cholesterol levels in the blood are believed to be related to certain types of heart disease because of the suspected correlation between high cholesterol levels in the blood and arteriosclerosis (hardening of the arteries). Cholesterol is believed to be responsible for destroying the elasticity of the walls of arteries, and this causes high blood pressure. The administration of anticholesterol drugs may help to prevent heart disease by reducing cholesterol level in the blood. Another method is to substitute foods that are high in polyunsaturated fatty acids for those foods that are highly saturated. A typical advertisement may suggest that you substitute highly unsaturated margarines or cooking oils for the more saturated ones to aid in reducing high cholesterol levels in the blood.

Cholesterol is detected by color tests, one of which is the Lieberman-Burchard test. In this test, cholesterol reacts with acetic anhydride in the presence of sulfuric acid to give a color change from pink to blue to green.

BILE ACIDS

Bile acids are among the most important metabolic excretion products of cholesterol. The salts of the bile acids (sodium salts of peptide-like combinations of bile acids and amino acids) act as emulsifying agents in the digestive process and in the absorption of lipids and their hydrolysis products. Bile acids are produced in the liver, stored in the gall bladder, and secreted in the intestine. The most abundant of the bile acids is cholic acid, which closely resembles the cholesterol steroid structure.

Hormones and Prostaglandins

Sex hormones and adrenocortical hormones also have a steroid skeleton. Recently a new class of high molecular weight fatty acids called prostaglandins have been isolated and studied. They have been termed lipids because of their solubility characteristics. These prostaglandins are simple nonsaponifiable lipids and hormone-like compounds that have profound biological effects even in trace amounts. They are formed by the cyclization of a 20-carbon unsaturated fatty acid such as arichodonic acid or from some closely related essential polyunsaturated fatty acid. The prostaglandins (about 14 of them are known) act to induce labor, bring about therapeutic abortion, lower blood pressure and induce smooth muscle contractions.

SUMMARY

Lipids are oily, greasy substances that are water insoluble but soluble in nonpolar solvents. They are found in all body tissues, and serve as sources of energy and insulation. They may be classified as simple lipids, compound lipids and derived lipids. Lipids have fatty acids (saturated or unsaturated) as part of their composition. The greater the number of double bonds, the greater the degree of unsaturation. The simple lipids are triglycerides of varying fatty acid composition. Upon hydrolysis they yield fatty acids and glycerol. The compound lipids are the phospholipids, the glycolipids and the lipoproteins. The derived lipids cannot be hydrolyzed. The most important derived lipids are the steroids such as cholesterol and cholic acid.

EXERCISE

1. How can you tell by physical examination if a substance is a lipid?
2. Where are lipids found in nature?
3. Where are lipids found in the body?
4. Which has more energy per gram, a fat or a protein?
5. What are the functions of fat in the body?
6. State the classifications of lipids and give an example of each.
7. What is the basis for the classification of lipids?
8. What do you mean by emulsification, and why is it important in fat metabolism in the body?
9. What are the names of the fatty acids that contain 12, 14, 16, and 18 carbon atoms (saturated acids)? Write their formulas.
10. Write the names and formulas for the four most important unsaturated fatty acids.
11. Which contain more unsaturated fatty acids, oils or solid fats? Explain.
12. What is the general formula for the simple neutral fats and oils?
13. What is the hydrolysis product of a simple fat?
14. What is the difference between hydrolysis and saponification of a fat?
15. What happens to fats when they become rancid?
16. How can you change an unsaturated fat to a saturated fat?
17. What is the iodine number? What does it signify?
18. Where does acrolein come from when fats are heated?
19. What is a simple fat? Name two.
20. What is a wax and what would be its general formula?
21. What substances do phospholipids yield upon hydrolysis?
22. What are some of the activities of the phospholipids?
23. What does the body use lecithins for?
24. What are derived lipids?
25. What is the presumed effect of high cholesterol levels in the blood?
26. What is the most abundant bile acid?
27. What are prostaglandins and what do they do in the body?

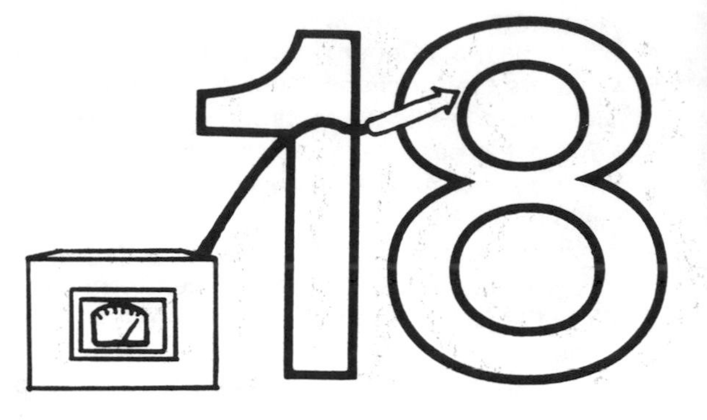

OBJECTIVES

When you have completed this chapter, you will be able to:

1. State the function of proteins in the body.
2. Recognize what complete protein foods and incomplete protein foods are, and to name the deficiencies of incomplete protein foods.
3. Explain why proteins must be a part of your daily diet.
4. Illustrate the hydrolysis of proteins through a reaction.
5. Define amino acids and list ten essential amino acids and other representative amino acids.
6. Discuss the optical activity of amino acids.
7. Draw the four bondings of a protein molecule, and explain the: peptide bond, disulfide bond, salt bridge and the hydrogen bond.
8. Discuss the four structures of a protein molecule and their effect upon the protein molecule.
9. Discuss the classification of the proteins and what those classifications are based upon.
10. Discuss the components of nucleic acids and distinguish between DNA and RNA.

PROTEINS AND NUCLEIC ACIDS

FUNCTION OF PROTEINS

Every living cell contains proteins and requires a continuous supply of protein for its existence. Proteins are involved in all life processes and are present in all body fluids except bile and the urine. The important functions of proteins are as follows:

1. Proteins are needed to build new tissue, muscles, organs, and to form protoplasm and nucleoproteins.
2. Proteins are needed to repair damaged and worn out cells.
3. Proteins act as buffers to control and maintain the normal pH in the cells, body fluids, blood, and lymph.
4. Proteins aid in maintaining the water balance between blood, lymph, and body tissues.
5. Proteins supply the building block components for the (a) formation of hemoglobin, essential to the respiratory process, (b) for enzymes needed in metabolism, (c) for hormones, required to regulate the secretion of specific fluids that maintain normal and healthy body growth, (d) for antibodies, which are the body's protection against invading pathogenic microorganisms.
6. Proteins are needed to supply amino acids for the synthesis of other proteins and protein materials that contain nitrogen (such as creatine, bile salts, and coloration pigments).
7. Proteins can furnish energy when the body stores of carbohydrates and fats are too low or cannot be used.

THE AMINO ACIDS

Proteins are composed of structural units known as amino acids. There are 21 different amino acids known to exist in the various proteins. The simplest amino acid is glycine (Fig. 18-1). Proteins are condensation polymers of the amino acid monomers. Amino acids which occur in nature have an amino group (NH_2—) attached to the alpha carbon atom (the first —CH_2— group attached to

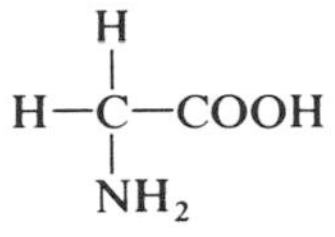

$$\begin{array}{c} \mathrm{H} \\ | \\ \mathrm{H{-}C{-}COOH} \\ | \\ \mathrm{NH_2} \end{array}$$

Figure 18-1. Glycine.

Table 18-1. Nonessential amino acids.

Name	*Abbreviation*	*Structure*
Alanine	Ala	$CH_3—CH—COOH$ $\quad\quad|$ $\quad\;NH_2$
Aspargine	Asp-NH_2	$H_2N—C—CH_2—CH—COOH$ $\quad\quad\;\|\quad\quad\quad\;|$ $\quad\quad\;O\quad\quad\;NH_2$
Aspartic Acid	Asp	$HOOC—CH_2—CH—COOH$ $\quad\quad\quad\quad\quad|$ $\quad\quad\quad\quad NH_2$
Cysteine	Cys	$HS—CH_2—CH—COOH$ $\quad\quad\quad\quad|$ $\quad\quad\quad NH_2$
Cystine	Cys-Cys	$HOOC—CH—CH_2—S—S—CH_2—CH—COOH$ $\quad\quad\quad|\quad\quad\quad\quad\quad\quad\quad\quad|$ $\quad\quad NH_2\quad\quad\quad\quad\quad\quad NH_2$
Glutamic Acid	Glu	$HOOC—CH_2—CH_2—CH—COOH$ $\quad\quad\quad\quad\quad\quad\quad|$ $\quad\quad\quad\quad\quad\quad NH_2$
Glutamine	Glu-NH_2	$H_2N—C—CH_2—CH_2—CH—COOH$ $\quad\quad\;\|\quad\quad\quad\quad\quad\;|$ $\quad\quad\;O\quad\quad\quad\quad\;NH_2$
Glycine	Gly	$CH_2—COOH$ $|$ NH_2
Proline	Pro	$CH_2—CH_2$ $|\quad\quad\quad|$ $CH_2\quad CH—COOH$ $\quad\diagdown\;\;\diagup$ $\quad\;NH$
Serine	Ser	$HO—CH_2—CH—COOH$ $\quad\quad\quad\quad|$ $\quad\quad\quad NH_2$
Tyrosine	Tyr	HO—⟨benzene ring⟩—$CH_2—CH—COOH$ $\quad\quad\quad\quad\quad\quad\quad\quad|$ $\quad\quad\quad\quad\quad\quad\quad NH_2$

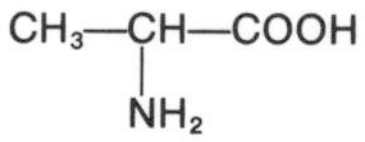

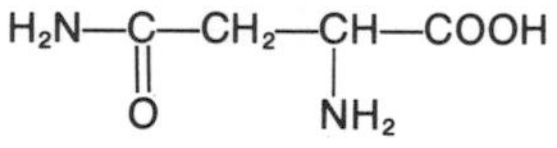

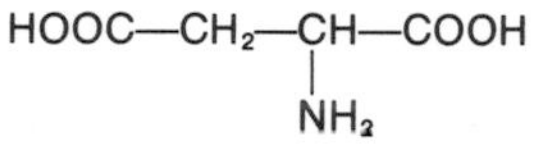

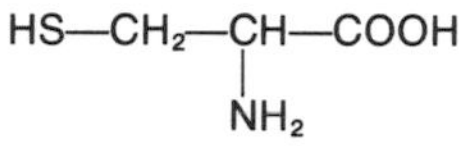

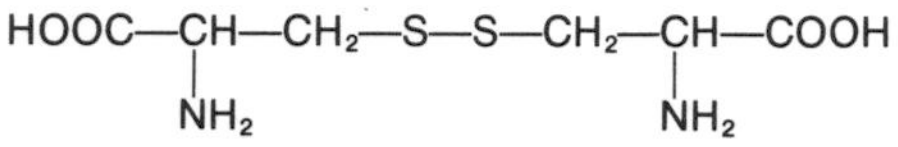

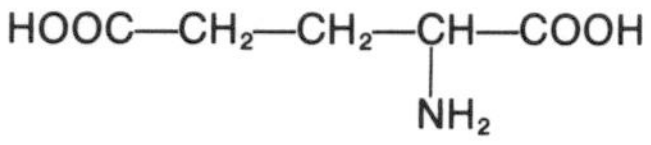

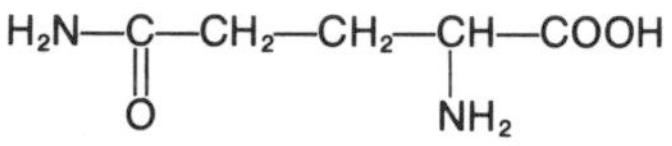

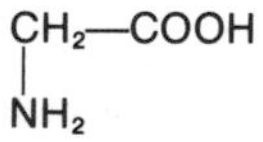

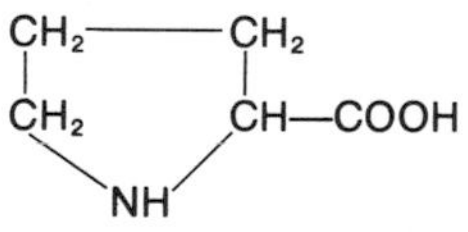

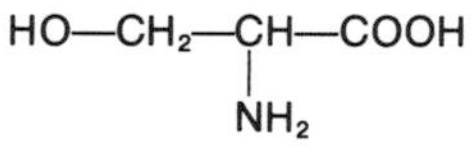

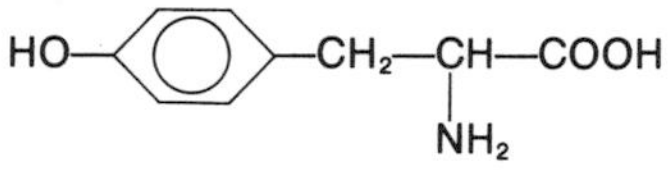

the —COOH). These amino acids differ from each other by distinctive side chains that are attached to the alpha carbon atom (replacing H atom).

Nutritionally we can divide the amino acids into two groups: the nonessential amino acids (Table 18-1) and the essential amino acids (Table 18-2). The body can synthesize the nonessential amino acids. But the body cannot synthesize the essential amino acids or can only synthesize them in small amounts. Since the body requires the essential amino acids for growth and bodily function it is absolutely imperative that they be included in the diet. A food source that contains all of the essential amino acids is called a complete protein but if one or more amino acids are missing it is an incomplete protein.

Table 18-2. Essential amino acids.

Name	*Abbreviation*	*Structure*
Arginine*	Arg	$H_2N—C(=NH)—NH—CH_2—CH_2—CH_2—CH(NH_2)—COOH$
Histidine*	His	$H—C{=}C(—N{=}C(H)—N—H)—CH_2—CH(NH_2)—COOH$ (imidazole ring: C—H, N, N—H, H—C=C)
Isoleucine	Iso	$CH_3—CH_2—CH(CH_3)—CH(NH_2)—COOH$
Leucine	Leu	$(CH_3)_2—CH—CH_2—CH(NH_2)—COOH$
Lysine	Lys	$CH_2(NH_2)—CH_2—CH_2—CH_2—CH(NH_2)—COOH$
Methionine	Met	$CH_3—S—CH_2—CH_2—CH(NH_2)—COOH$
Phenylalanine	Phe	(benzene ring)$—CH_2—CH(NH_2)—COOH$
Threonine	Thr	$CH_3—CHOH—CH(NH_2)—COOH$
Tryptophan	Try	(indole ring, N)$—CH_2—CH(NH_2)—COOH$
Valine	Val	$(CH_3)_2—CH—CH(NH_2)—COOH$

*Needed during childhood.

Optical Activity of the Amino Acids

Like the monosaccharides, all the amino acids, except for glycine, contain an asymmetric carbon atom. Amino acids will also rotate polarized light when viewed through a polarimeter. As with the monosaccharides (which in nature produces a primarily D-configuration) the amino acid can have a D- or L-configuration, depending on whether they resemble D-glyceraldehyde or L-glyceraldehyde (Fig. 18-2). Naturally occurring amino acids from plant and animal sources have the L configuration and are designated as L(+) or L(−) depending upon their plane of polarized light. Most are weakly dextrorotary in neutral solution, but the optical rotation of any amino acid is controlled by the pH of

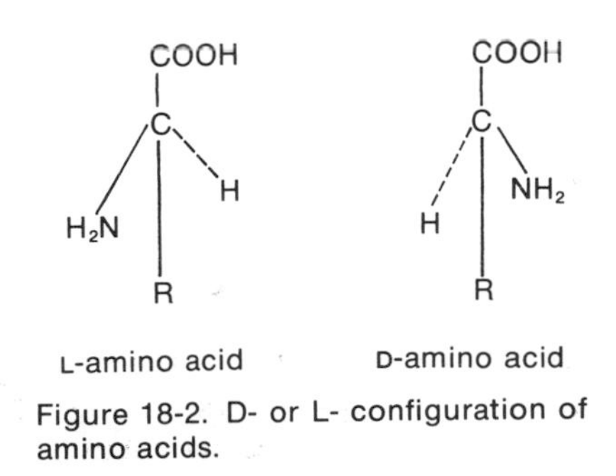

Figure 18-2. D- or L- configuration of amino acids.

the solution. Some D amino acids (D-alanine and D-glutamic acid obtained from the cell walls of microorganisms) are used as antibiotics.

Amphoteric Nature of Amino Acids

The various amino acids contain both a basic amino group and an acid carboxyl group. Therefore the amino acids are dipolar ions and can neutralize both acids and bases even though the molecule is neutral. Because of this, the amino acids can act as a buffer system to neutralize excess acids or bases, and provide an important buffering mechanism for the biological control of pH. The amino acids are called zwitterions.

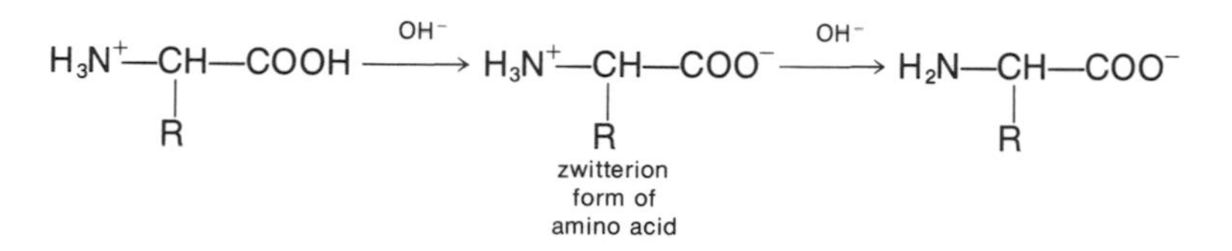

At very low pH both the amino and acid groups are protonated. As base is added, first the acid and then the amino group is deprotonated. If the R group contains an —NH_2 (as in lysine or arginine) or a —COOH (as in aspartic acid or glutamic acid) then as the pH changes, these groups also will be protonated and deprotonated. When the number of negative charges on the amino acid equals the number of positive charges we have reached the isoelectric point of the amino acid.

Chemistry of the Amino Acids

THE PEPTIDE BOND

Proteins are formed by the union of amino acids, in which the amino group of one molecule reacts with the carboxyl group of another molecule with the elimination of a molecule of water. When two amino acids combine, they form a dipeptide:

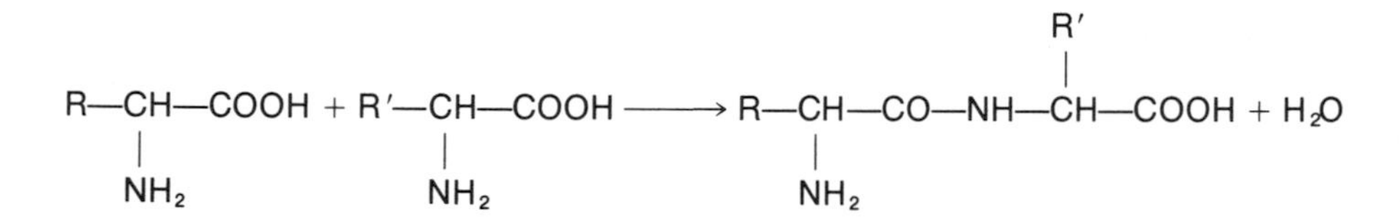

If three amino acids combine, they form a tripeptide, and when many combine, you have a polypeptide. It is that —CO—NH— grouping, the amide grouping, called the peptide linkage, which joins amino acids together to form proteins. A polypeptide molecule might have the following structure:

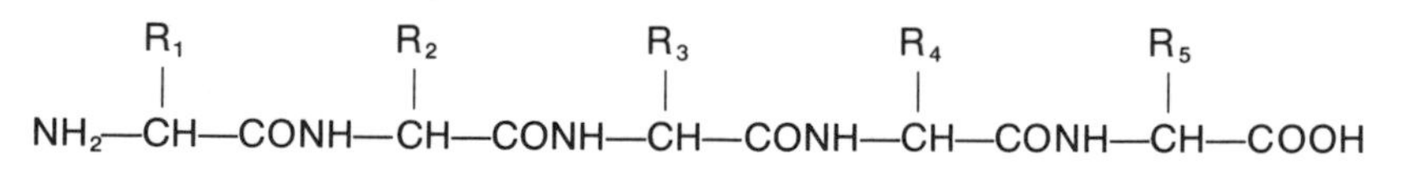

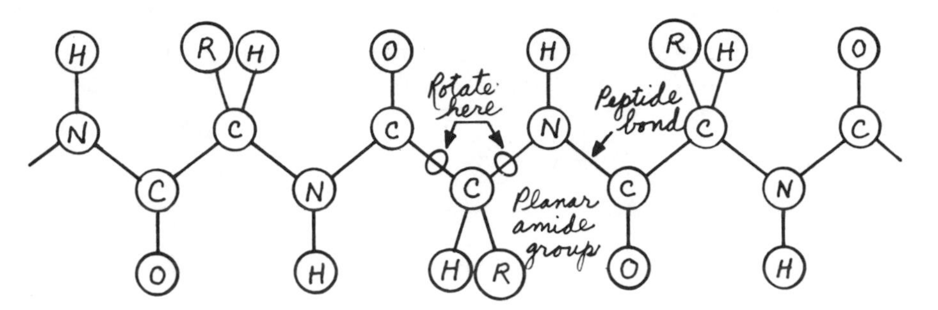

Figure 18-3. The polypeptide chain with the R groups extending from the chain.

where the R's are any amino acid. A polypeptide or protein will have 1 free —NH_2 at one end of the chain and 1 free —COOH group at the other end. By convention we start numbering the amino acid residues (the R groups) at the —NH_2 end of the polypeptide chain (Fig. 18-3). In a polypeptide or protein, the peptide bond is the strongest chemical bond.

THE DISULFIDE BOND

This is an oxidation reaction between 2 cysteine residues to produce a cystine residue plus water. Reduction will break this bond (see Fig. 18-4). The 2 cysteine residues can be on different polypeptides or on the same polypeptide chain. This is not as strong a bond as the peptide bond.

THE SALT BRIDGE

This is an ionic bond between an acid group and a base or amino group to form a salt. They may be from different chains or the same chain. This bond is weaker than the peptide bond (Fig. 18-5).

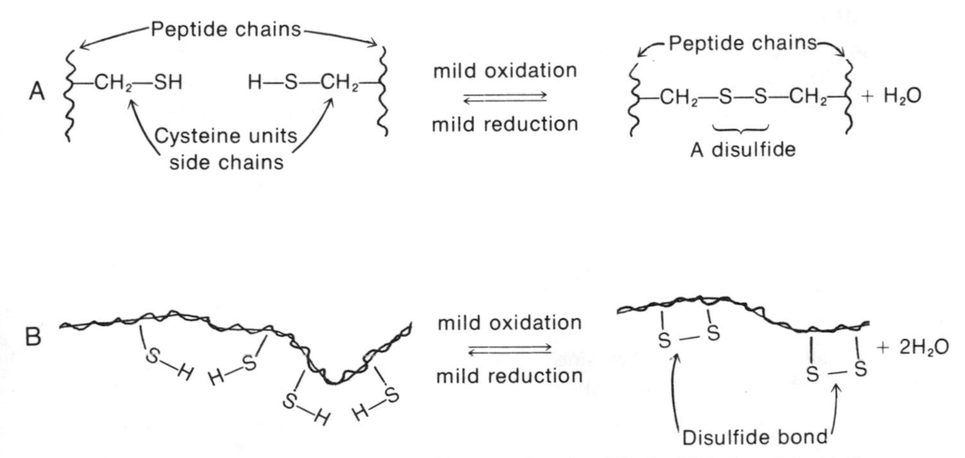

Figure 18-4. Intermolecular (A) and intramolecular (B) disulfide bond formation.

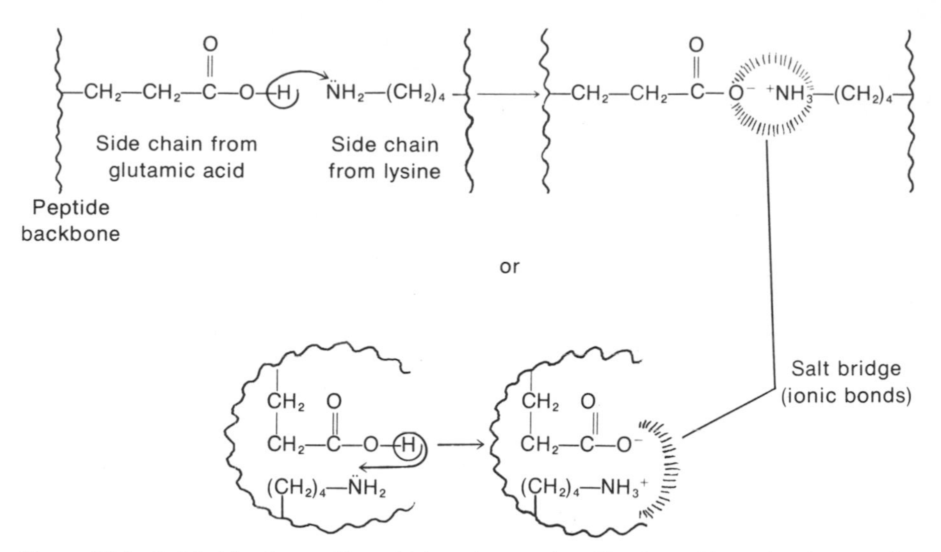

Figure 18-5. Salt bridge formed by acid-base interaction. The 2 peptide molecules acquire opposite charges that act to bond them together.

THE HYDROGEN BOND

This is a weak bond between hydrogen and oxygen or hydrogen and nitrogen atoms. The bond may occur between side chains of the polypeptide chains or between the amide nitrogen, oxygen and hydrogen of the chain (Fig. 18-6).

STRUCTURE OF PROTEINS

There are four structures in proteins that are recognized by biochemists: the primary structure, the secondary structure, the tertiary structure and the quaternary structure.

Primary Structure

The sequence of amino acids in a protein is called the primary structure of the protein. The peptide bonds of the amino acids form the backbone of the chain while the radical groups extend from the chain like charms on a charm bracelet.

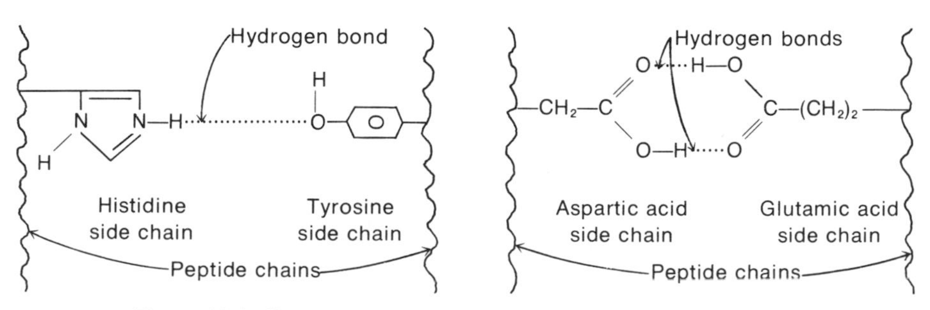

Figure 18-6. The hydrogen bond linkage between peptide chains.

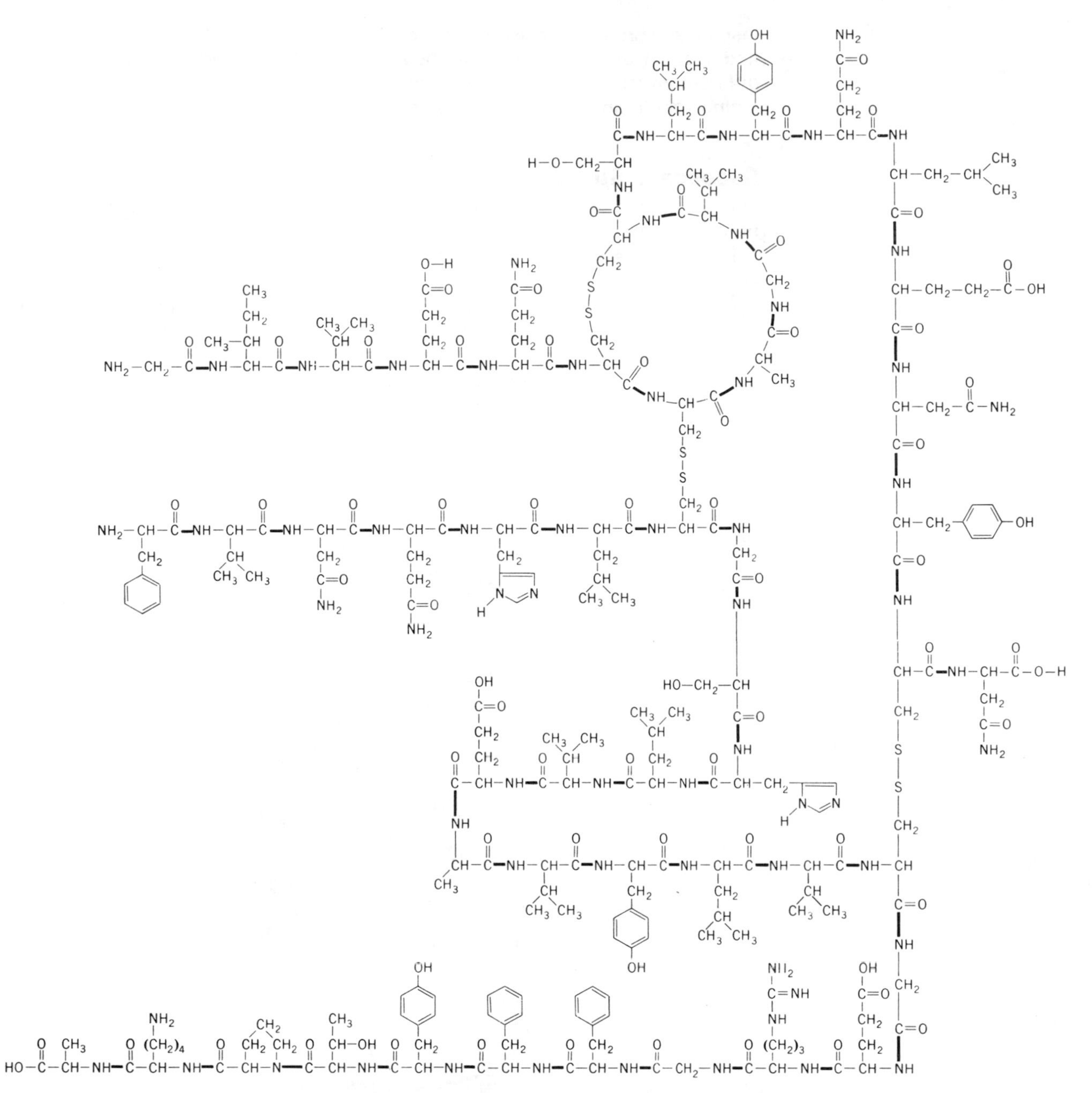

Figure 18-7. Primary structure of a sheep insulin molecule.

Insulin is a protein consisting of 2 chains held together by 2 disulfide bonds and 1 chain also has a loop formed by a disulfide bond. The insulins from different animals differ slightly in the amino acid sequence. The sequence of amino acid in sheep insulin can be seen in Figure 18-7.

Secondary Structure

The secondary structure depends upon the type of hydrogen bonding in the polypeptide chain. We have two types.

The first type is called the alpha helix structure. It occurs when hydrogen bonding within the chain twists it into a coil (see Fig. 18-8). Hydrogen bonding occurs between the =C=O group of one turn to the —N—H group of the turn below. Examples of alpha helixes are wool and hair keratins or proteins.

The second type is called the pleated sheet. In this case, we have hydrogen bonding between different chains. The chains are then arrayed in the form of sheets of proteins (see Fig. 18-9). This type of structure is found in silk.

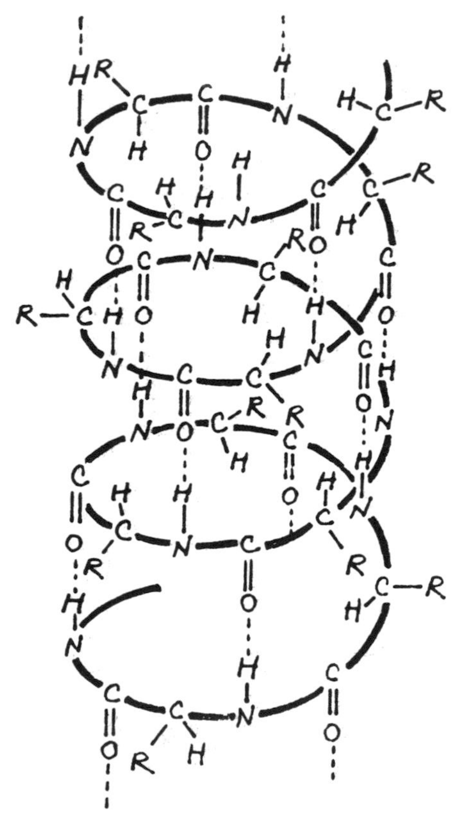

Figure 18-8. Secondary structure of a protein.

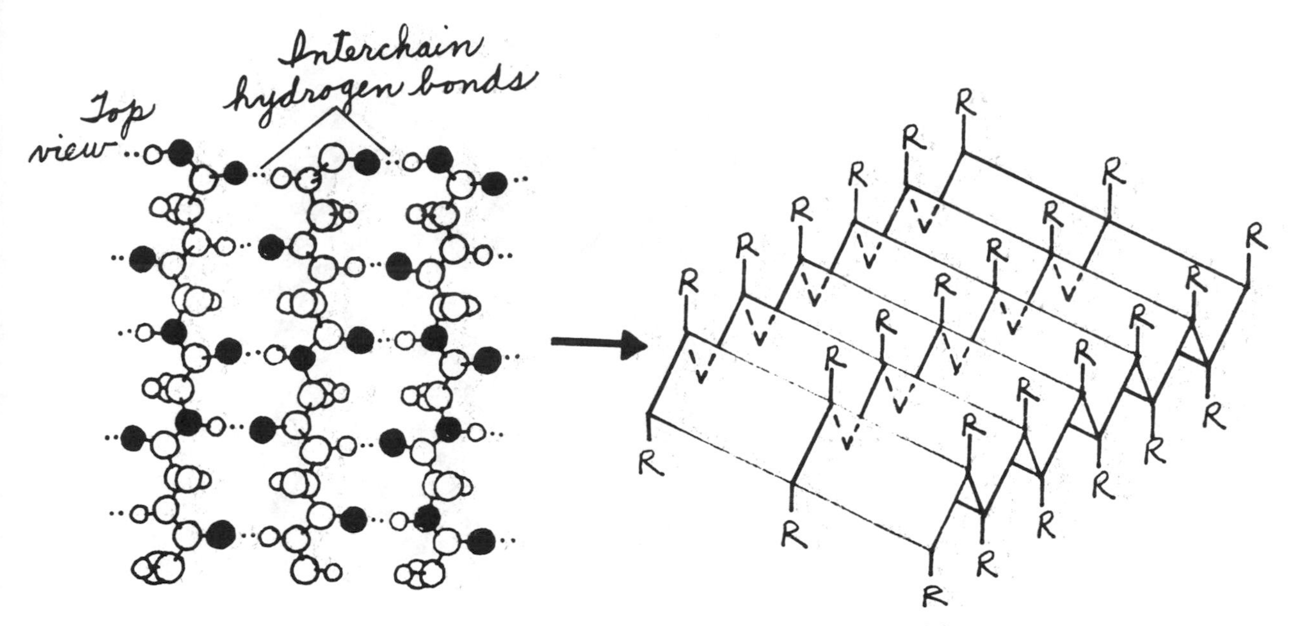

Figure 18-9. Secondary structure of a protein (beta-keratin).

Tertiary Structure

The tertiary structure depends on how the protein molecule is folded upon itself. Generally, tertiary structure applies to globular proteins while secondary structure applies to fibrous proteins. The folded molecule is held together by hydrogen bonding between side chains, salt bridges, disulfide bonds and other weak bonds. Myoglobin (Fig. 18-10), which is related to hemoglobin, has an alpha helical coil that is folded in upon itself.

Quaternary

The quaternary structure depends on how many chains must be brought together to form the protein. Hemoglobin, for example, is composed of 4 protein chains that are held together by hydrogen bonding, salt bridges and other weak bonds (Fig. 18-11).

A STUDY OF TWO PROTEINS

Hemoglobin

Hemoglobin is a tetramer of 2 chains called alpha (α) chains and 2 chains called beta (β) chains. Each of the chains has a heme group attached to it and a structure similar to myoglobin (Fig. 18-12). The α-chain contains 141 amino acid residues while the β-chain contains 146 residues.

Normal hemoglobin, designated HbA, has the structure $\alpha_2\beta_2$. Fetal hemoglobin, HbF, has the structure $\alpha_2\beta_2$ where the gamma (γ) chain has a structure similar to β but has different amino acids in certain positions.

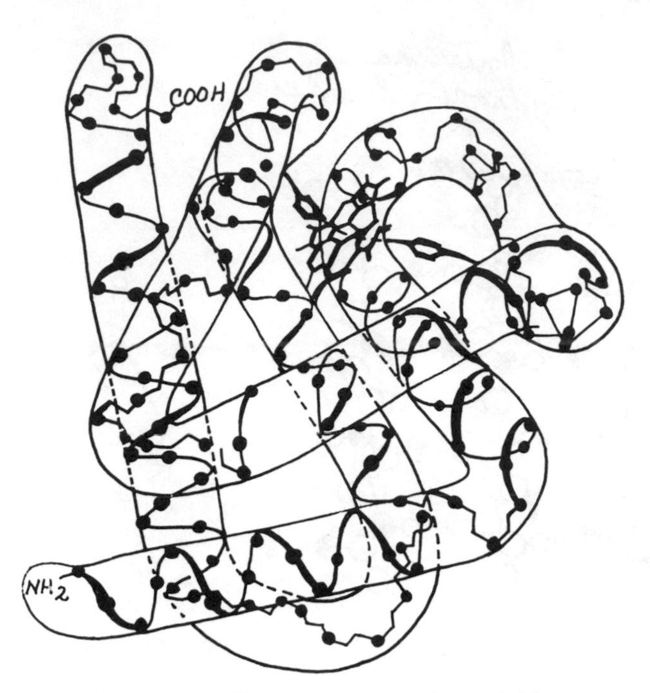

Figure 18-10. Tertiary structure of myoglobin.

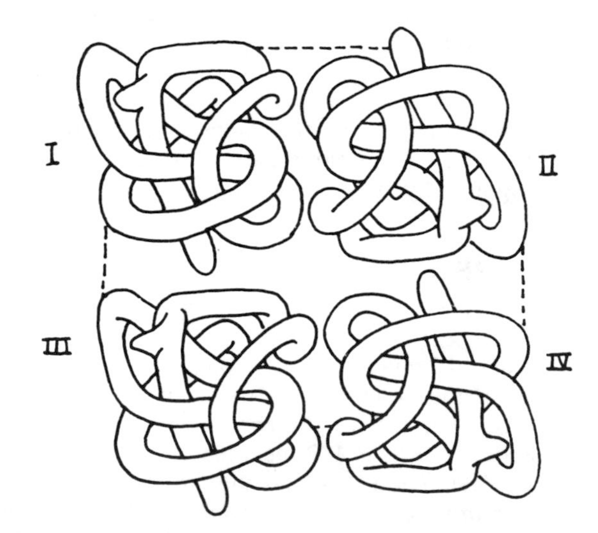

Figure 18-11. Quaternary structure of a protein tetramer.

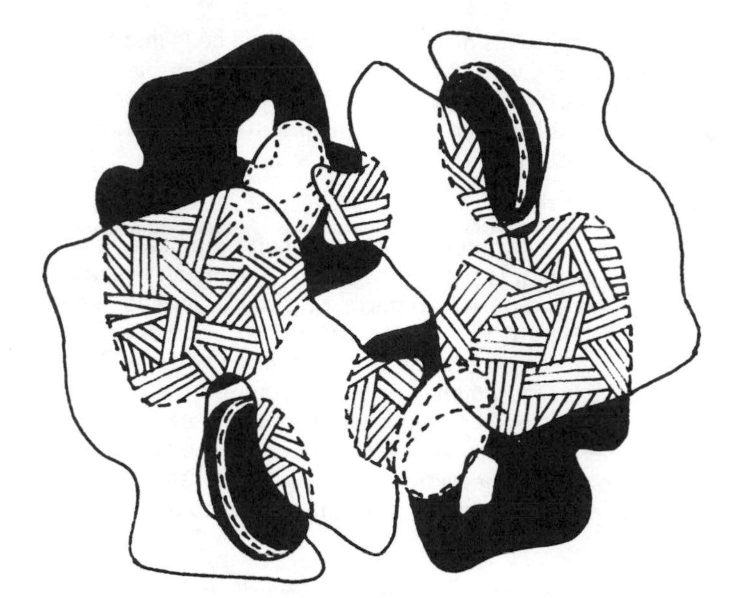

Figure 18-12. Normal hemoglobin is a tetramer of 2 alpha chains (light color) and 2 beta chains (dark color).

It is vital to life that a cell can produce the same protein as its parent cell since it is the sequence of amino acids within the protein molecule that enables the molecule to function as it should. The substitution of even *one* amino acid in a large protein molecule can completely alter its biological function (see Table 18-3). A startling example of this is the hereditary disease sickle cell anemia, designated HbS where the hemoglobin (a protein) molecule in the blood of an afflicted individual contains a valine unit in place of a glutamic acid unit in the β chain. This substitution involves only *one* of approximately 300 amino acid units of the hemoglobin yet it causes a drastic change in the properties of the molecule. The abnormal hemoglobin tends to precipitate in the red

Table 18-3. First nine amino acids in hemoglobin.

	Position number								
Chain	1	2	3	4	5	6	7	8	9
HbA									
α chain	Leu	Ser	Ala	Leu	Ser	Asp	Leu	His	Ala
β chain	Phe	Ala	Thr	Leu	Ser	Glu	Leu	His	Cys
γ chain	Phe	Ala	Glu-Nh_2	Leu	Ser	Glu	Leu	His	Cys
HbS									
β chain	Phe	Ala	Thr	Leu	Ser	Val	Leu	His	Cys

NOTE: 1. HbA = normal hemoglobin, HbS = sickle cell hemoglobin.
2. The heme group is attached to His at position eight.
3. The difference between HbA-chain and HbS chain is at position six.

blood cells containing it. This distorts the red blood cells into the characteristic shape of a sickle, thus giving the disease its name. The distorted cells clump together, sometimes blocking the veins and arteries, causing (in some cases) the cell to break open. If children inherit this trait from both parents they seldom live beyond the age of two years.

Collagen

Collagen is a fibrous protein composing the major element of skin, bone, tendon, cartilage and teeth. It is present in nearly all cells and serves to hold cells together. Collagen is composed primarily of glycine, proline, 4-hydroxyproline and hydroxylysine (Fig. 18-13). The latter two are derivatives of proline and lysine respectfully and are not found in many other proteins. The basic unit of collagen is tropocollagen, which is a triple stranded helical coil (Fig. 18-14). Tropocollagen has a molecular weight of 285,000 and each of the 3 chains has 1,000 residues. The tropocollagen molecules are cross linked with each other by means of hydroxylysine or hydroxyproline. The greater the degree of cross linking, the greater the mechanical strength of the collagen fiber (Fig. 18-15).

Hydroxylysine Lys-OH

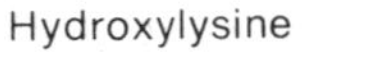

4-Hydroxyproline Pro-OH

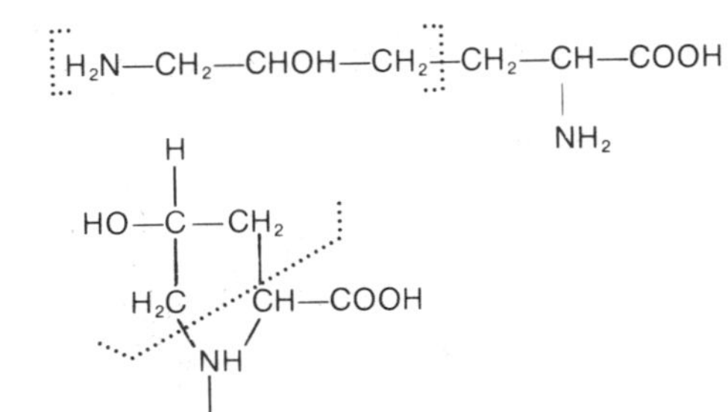

Figure 18-13. Components of collagen.

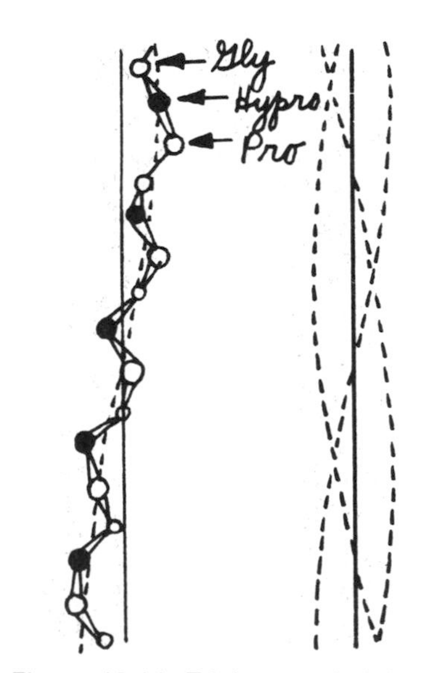

Figure 18-14. Triple stranded tropocollagen molecule.

SIZE OF PROTEIN MOLECULES

Proteins are fantastically large molecules having molecular weights from 30,000 to as high as 50,000,000. There are tens of thousands of different protein molecules which vary in their size, shape, and configuration as a result of the interactions of the amino acids, their sequence and occurrence in the molecule, the relative number of each amino acid in the molecule, and the spatial arrangement of the amino acids. Sucrose has a molecular weight of 342, glucose, a molecular weight of 180, while acetic acid has a molecular weight of 60. Obviously, it is impossible to conceive of molecular structures having molecular weights in the millions, but we can schematically relate their size in Figure 18-16.

CLASSIFICATION OF PROTEINS

Proteins are extremely complex and it is very difficult to classify them, but the most common method is based upon what groups are formed upon the hydrolysis of the protein material. Essentially there are three types of proteins: simple, conjugated and derived (see Table 18-4).

Table 18-4. Classification of proteins.

Simple proteins	*Chief characteristics*	*Examples*
Albumins	Soluble in water	Egg albumin, blood serum albumin
Globulins	Insoluble in water, soluble in dilute salt solutions	Blood serum globulin
Glutelins	Insoluble in water or salt solutions, soluble in dilute acids and alkalies	Glutenin of wheat
Prolamins	Insoluble in water, soluble in 70% alcohol	Zein in grain
Albuminoids	Insoluble in water, salt solutions, dilute acids, or alkalies	Keratin of nails and hair
Histones	Soluble in water and dilute acids; strongly basic and insoluble in ammonia	Globin in hemoglobin
Protamines	Strongly basic, soluble in water; small molecule	Clupein in fish sperm
Conjugated proteins		
Nucleoproteins	Protein combined with nucleic acid	Nuclein in cell nuclei; glandular tissue
Glycoproteins	Combined with carbohydrate group	Mucin of saliva
Phosphoproteins	Combined with a phosphate group	Casein of milk
Chromoproteins	Combined with a color producing nucleus	Hemoglobin of blood
Lecithoproteins	Combined with lecithin	Fibrinogen of blood
Derived proteins		
Proteans	First hydrolysis products of water, acids, or enzymes	Edestan
Metaproteins	Further hydrolytic products of acids or alkalies	Acid or alkali metaprotein
Coagulated protein	Insoluble products of action of heat or alcohol	
Proteoses	Hydrolysis products; soluble in water; precipitated by saturated ammonium sulfate solution; very slightly diffusible	
Peptones	Products of further hydrolysis; soluble in water; not precipitated by saturated ammonium sulfate; readily diffusible.	
Peptides	Products of further hydrolysis; soluble in water; smaller molecule than peptones; more diffusible than peptones.	

Simple proteins, upon complete hydrolysis yield only alpha amino acids. Examples of simple proteins are albumin, globulin, glutelins, prolamins, albuminoids, histones, and protamins.

Conjugated proteins are combinations of a simple protein with an additional molecule. Examples of conjugated proteins are nucleoproteins, glycoproteins, phosphoproteins, lipoproteins, chromoproteins and lecithoproteins.

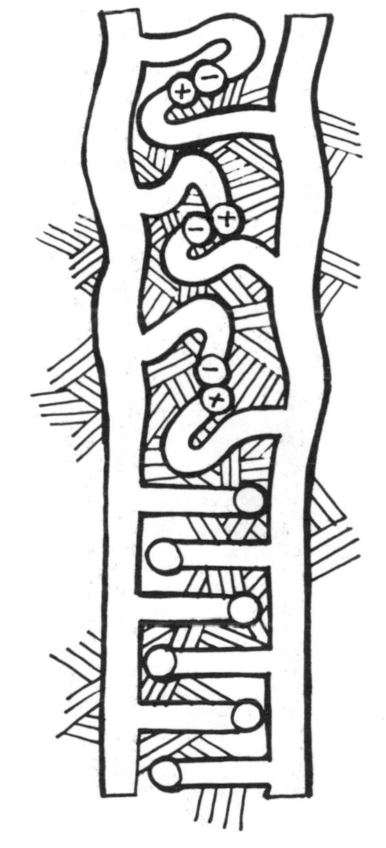

Figure 18-15. Two tropocollagen molecules cross-linked to form collagen.

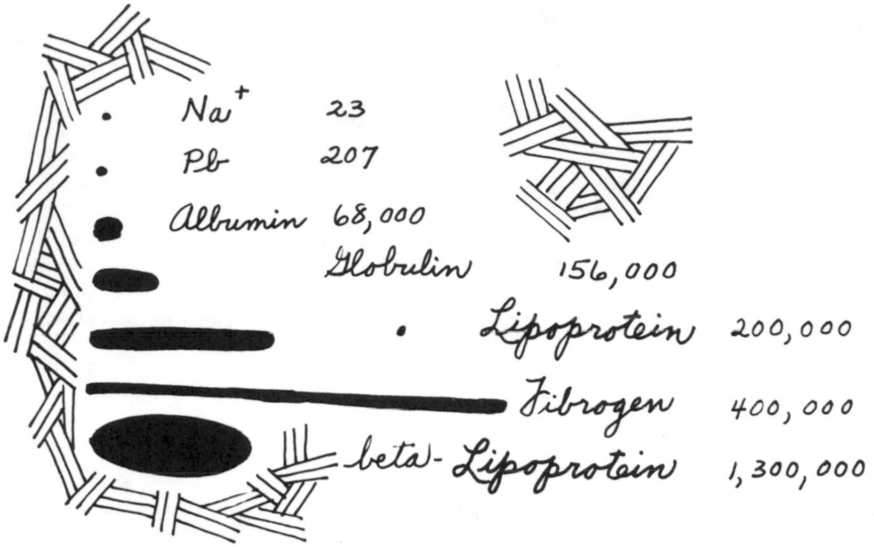

Figure 18-16. A comparison of the sizes and shapes of protein molecules. Numbers represent the molecular weight.

Derived proteins are so named because they are the hydrolysis products of proteins or are obtained in some way from other proteins. This group includes proteins, metaproteins, coagulated proteins, proteoses, peptones and peptides.

Functional Classification of Proteins

Another classification of proteins can be made according to the function of the protein. Those functions are:

1. *Structural proteins:* More than one half of the total protein of the mammalian body is collagen, which is found in skin, cartilage, and bone, performing the same function as cellulose in plants.
2. *Contractile protein:* Actin and myosin isolated from muscular tissue.
3. *Enzymes:* Biochemical catalysts that promote specific physiologic reactions.
4. *Hormones:* Organic compounds produced by the endocrine glands and carried by the blood to other glands, organs, and tissues where they perform their biological function.
5. *Antibodies:* Proteins that act to destroy foreign materials (antigens) released by infectious agents into the body. Gamma globulins in the blood are antibodies.
6. *Blood proteins:* Albumins, hemoglobin, and fibrogen are essential protein components of the blood.

PHYSICAL PROPERTIES OF PROTEINS

Amino acids are colorless, crystalline compounds that melt above 200°C with decomposition, and are very soluble in water. They are, however, insoluble in organic solvents such as ether. Some have a slightly sweetish taste, while others are tasteless or bitter. The characteristic size of the protein molecule, because of its high molecular weight, leads to a number of fundamental properties such as:

1. Proteins form colloidal dispersions rather than true solutions.
2. Many osmotic and amphoteric properties depend on the particle size of the proteins.
3. The large size of the molecule decreases the proteins' permeability through the pores of natural and artificial membranes.
4. Proteins can carry electrical charges and can attract smaller molecules.

Normally the proteins of the blood, lymph, and tissue cells will not diffuse through the membranes. For example, the blood plasma proteins do not readily diffuse through the membranes in the glomerulus, the filtration apparatus of the kidney. Therefore, there is normally no protein in the urine. However, should the glomerulus membrane become damaged by disease, as in nephritis, protein will be found in the urine. Artificial membranes of collodion can be prepared through which the smaller crystalloid materials will pass, but which will retain the large protein molecules. This process, called dialysis, is used in the artificial kidney machine.

Solubility in Water and Basic Solutions

Generally speaking, the simple proteins are more soluble than the conjugated or derived proteins. For example: albumins are water soluble, while globulins are insoluble. Histones are water soluble but insoluble in ammonium hydroxide, while protamins are soluble in both water and ammonium hydroxide.

ISOELECTRIC POINT OF PROTEINS

Even though proteins contain polypeptide linkages that tie up the majority of their amino and carboxyl groups, some amino acids (aspartic, lysine, arginine, histidine) have additional side chain amino and carboxyl groups. These groups in solution will ionize and therefore the proteins exist in solution as ions, carrying a net positive or negative charge depending upon the pH of the solution (Fig. 18-17), except at one definite pH. This pH is called the isoelectric point (Table 18-5). At the isoelectric point, the protein will not migrate in an electric field because its net electrical charge is zero. At pH values on the acid side, protein ions have a positive charge and will migrate to the negative pole. At pH values on the basic side of the isoelectric pH, protein ions have a negative charge and will therefore migrate to the positive pole (see Fig. 18-18).

Table 18-5. Isoelectric points of various proteins.

Protein	*Isoelectric* pH
Pepsin	1.0
Casein (in milk)	4.6
Insulin (a hormone)	5.3
Hemoglobin (in blood)	6.8
Ribonuclease	9.5

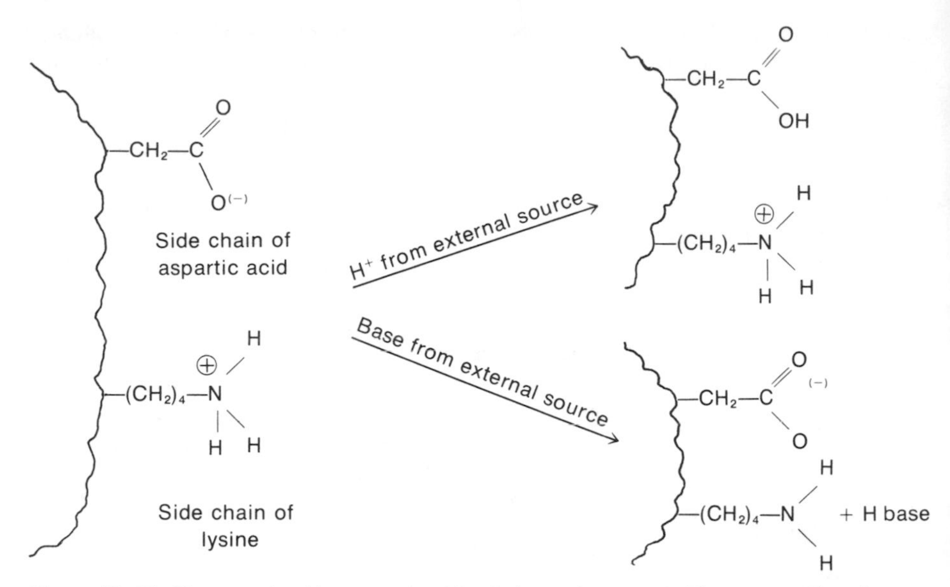

Figure 18-17. Changes in pH can cause side chains to have a positive or negative charge.

This principle is utilized in the method known as electrophoresis. Different isomeric forms of proteins can be separated from each other by using the difference in the number of charges each isomeric protein has. The more charges a protein has, the faster it will migrate to one pole or another.

This method is used to screen people with the sickle cell trait. If you remember, the difference between HbA and HbS is in the sixth position of the β chain where HbS has an uncharged valine substituted for the negatively charged glutamic acid. This means that HbS molecule has 2 less negative charges than the HbA molecule. In an electric field, the HbA will migrate more than the HbS. When proteins have an overall positive or overall negative charge, these like-charged particles repel each other. This prevents coagulation, or precipitation. However, once the isoelectric point is reached, then the protein molecules do not repel each other. In fact, because they now have an equal number of acid groups and basic groups, the molecules tend to cluster together and coagulate

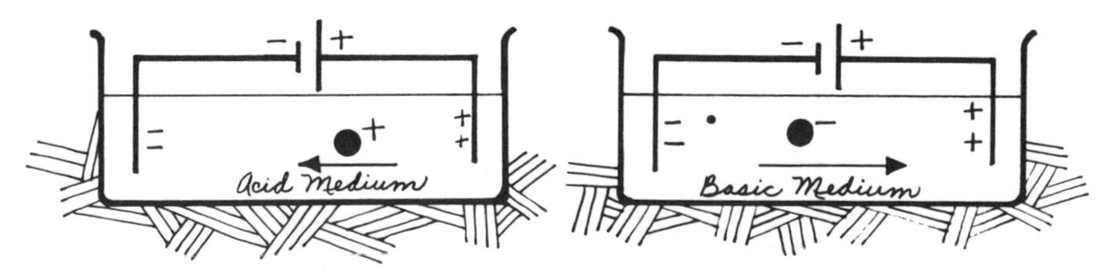

Figure 18-18. How pH affects the electrical charge of protein ions.

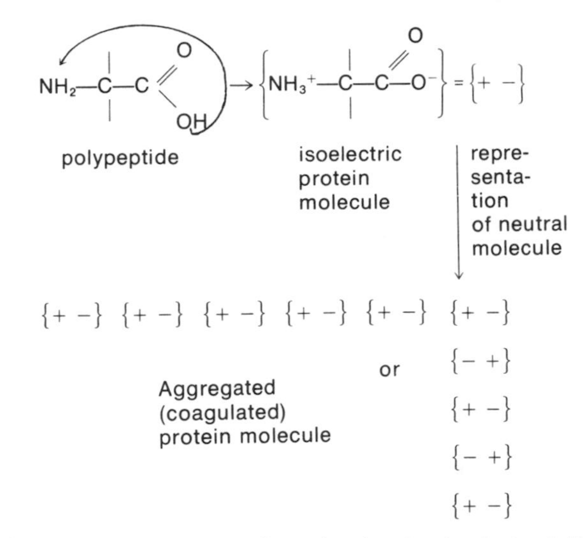

Figure 18-19. Coagulation of protein molecules due to electrostatic attraction.

because of the electrostatic attraction of the oppositely charged groups on the molecules for each other (Fig. 18-19).

This knowledge of the isoelectric point of proteins is extremely important to health care personnel because of the reactions of proteins in the body. For example, the curdling of milk illustrates the effect of pH on the solubility of protein. Normally, cow's milk is at a pH of 6.3, but the isoelectric point of casein is 4.6. When bacterial action produces lactic acid, the pH drops. The molecules become more and more isoelectric, and eventually they coagulate and precipitate: the milk curdles. The addition of acetic acid (vinegar) to milk will curdle it. Another example of protein coagulation is found in blood analysis for sugar content. Proteins are precipitated by tungstic acid and removed prior to the procedure analysis. One practical application is the use of milk or raw eggs as an antidote for mercury poisoning. The positively charged mercury ion combines with the negatively charged proteins to form insoluble compounds. An emetic forces the ejection of the precipitated compound. The use of an emetic after formation of the heavy metal precipitate is important, because if left in the digestive system the precipitate would eventually be digested, releasing the mercury or other heavy metal in the system.

CHEMICAL REACTIONS OF PROTEIN

Hydrolysis of Proteins

When boiled with dilute acids or alkalies or subjected to protein splitting enzymes (proteases) as are found in digestive processes, proteins are hydrolyzed (broken down) to amino acids. The digestion of proteins in the body occurs in a series of hydrolyses that are catalyzed by certain enzymes to form proteoses, peptones, polypeptides, dipeptides, and finally, amino acids.

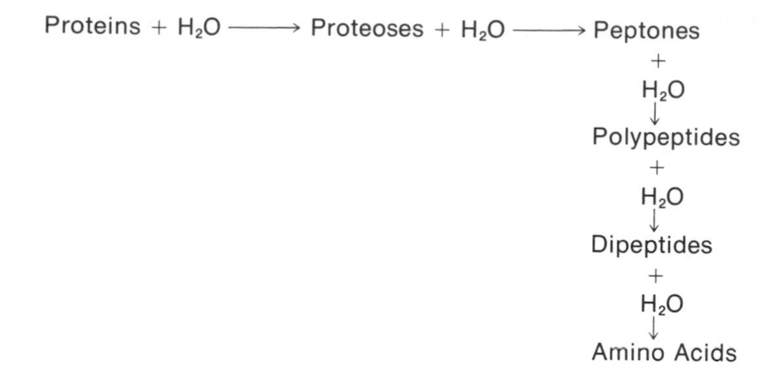

Denaturation

Denaturation is essentially a disorganization of the helical structure of the protein molecule caused by the breaking up of the cross-linked chains in the protein structure. Not only the hydrogen bonds, but also the disulfide (—S—S—) bond salt bridges and other weak bonds can be broken. And, as you know, this bond breaking results in the loss of biological activity because the unique three-dimensional structure involving secondary and tertiary structures is destroyed. This is often accompanied by precipitation and coagulation. You should especially note that only the weak bonds of the protein molecules are broken. None of the peptide linkages are affected (Fig. 18-20).

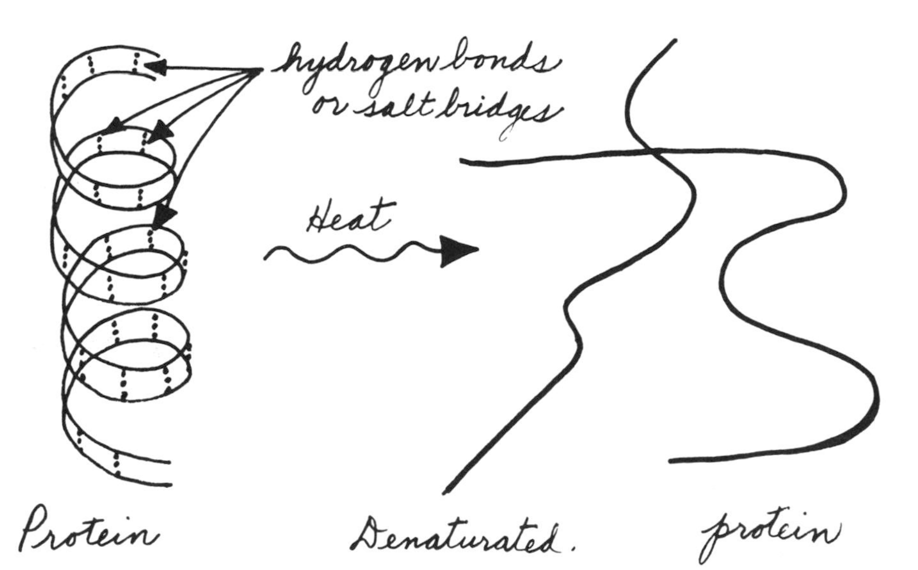

Figure 18-20. Denaturation.

PHYSICAL METHODS OF DENATURATION

1. Heat
   a. As in the coagulation of egg white when heated.
   b. As in the coagulation of protein of bacteria during sterilization.
   c. As in testing urine for the presence of proteins by heating to coagulate and precipitate the protein.
2. Ultraviolet radiation
   Ultraviolet radiations, which are electromagnetic waves similar to heat waves (infrared rays) but have different frequencies, work like heat does to disrupt the hydrogen bond salt bridges and weak bonds.
3. Violent Agitation
   The beating of an egg or the whipping of cream gives a surface film of denatured protein.

CHEMICAL METHODS OF DENATURATION

1. Alkaloidal Reagents
   Alkaloidal reagents precipitate proteins. Some are: trichloroacetic acid, phosphotungstic acid, tannic acid, picric acid, and phosphomolybdic acid. Tannic and picric acids have been used for burns because they precipitate the proteins on the surface to form a protective coating to exclude air from the burn and prevent the loss of water. In an emergency, strong tea may be used as a source of tannic acid for the treatment of burns.
2. Ethyl Alcohol
   Alcohol coagulates nearly all proteins including bacteria. If 95 percent alcohol is used, only the outer surface of the bacteria is coagulated which serves as a protective coating for the interior part, and consequently fails to kill them. However, a 70 percent alcohol acts more slowly and *penetrates* more effectively to completely coagulate the protein and thereby kill the bacteria.
3. Heavy Metals
   a. Heavy metals are Hg, Cu, Pb, Ag, and As. They are poisonous when taken internally but they will form insoluble precipitates with proteins. In the event of heavy metal poisoning, proteins in the form of milk or eggs are given so that the precipitate formed then can be ejected.
   b. Silver nitrate is toxic when taken internally, yet a silver nitrate protein preparation (Argyrol) is used in nose and throat infections, and silver nitrate is used to cauterize wounds and prevent gonorrheal infection in the eyes of newborn babies. Silver nitrate and mercuric chloride are good disinfectants because they combine with protein of bacteria, forming insoluble precipitates and in this way kill them.
   c. When heavy metal ions are swallowed the ions combine with the proteins of the mouth, esophogus, and stomach, precipitating them and causing local destruction. If ions are absorbed, they have a destructive action on all proteins, but their most serious effect is on the kidneys. The resulting inflammation interferes with urine excretion.
4. Inorganic Acids and Bases (change in pH)
   The casin protein of milk is coagulated by the lactic acid formed when it

sours. It is also coagulated by the HCl of the gastric juices. Proteins can be precipitated by compounds such as trichloroacetic acid, phosphomolybdic acid, and phosphotungstic acid. The Heller's ring test for albumin is based on the fact that concentrated HNO_3 precipitates proteins in the form of a white ring.

5. Reducing Agents
   The home permanent wave is an example of reversible denaturing. The disulfide linkage is responsible for the shape of the hair. Waving lotions contain reducing agents (ammonium thioglycolate) that ruptures the disulfide bond. The hair is put onto rollers, and then an oxidizing agent (potassium bromate, $KBrO_3$) is put on the hair. New disulfide bonds then hold the hair in the desired shape.
6. Salting Out
   Ammonium sulfate, $(NH_4)_2SO_4$, will salt out proteins that are soluble in water (egg and blood albumins), because these proteins become insoluble in this salt solution. Therefore, when you wish to isolate a protein from a solution without appreciably altering its chemical nature and properties, the protein can be precipitated by saturating the solution with ammonium sulfate. After filtration of the precipitate, the salt can be removed by dialysis.

SOURCES OF PROTEINS

As we stated before, some foods supply sufficient quantities of all essential amino acids, and they are called complete protein foods. These include milk, eggs, fish, poultry, and cheese. Other food sources are called incomplete protein foods because they have a deficiency of one or more essential amino acids. For example, gelatin lacks the amino acid tryptophan, and cereal grains lack lysine. However, you normally ingest varied protein-containing foods, providing you with a balanced diet; supplementing any incomplete protein food with a complete protein food that contains additional quantities of the deficient amino acid.

This does not mean that you need equal amounts of each amino acid, because the body does not require equal amounts of each in order to function normally. Therefore, it is important that you not just eat any protein, but a blend that contains the proper balance of amino acids. This is measured by the *biological value* of the protein, which is based both on the amino acid content and the digestibility of the protein. In general, animal proteins have greater biological value than plant proteins, which are lower in their lysine, methionine and tryptophan content, and contain protein in less digestible form. A nutritionist, dietician, or physician can analyze a diet and can determine whether it is a complete or incomplete protein diet.

Protein deficiency can have serious consequences. Kwashiorkor, a disease found to be prevalent in the economically deprived areas of Africa, Asia, and South America is the result of a deficiency of essential amino acids. The symptoms of this in children are dermatosis, decoloration, enlarged fatty liver, thinning hair, edema, weak and poorly developed musculature, distended abdomens due to abnormal fluid balance of the blood and interstitial fluids (caused by a lack of serum albumin to maintain osmotic pressure of the blood plasma), and apathy. The mortality is high but treatment is based on a milk diet. Therefore you can readily see that a protein deficient diet definitely has serious consequences because of the amino acid deficiency, which leads to growth failure, edema, anemia, peptic ulcers, low basal metabolism, and diminished cortical activity of the brain.

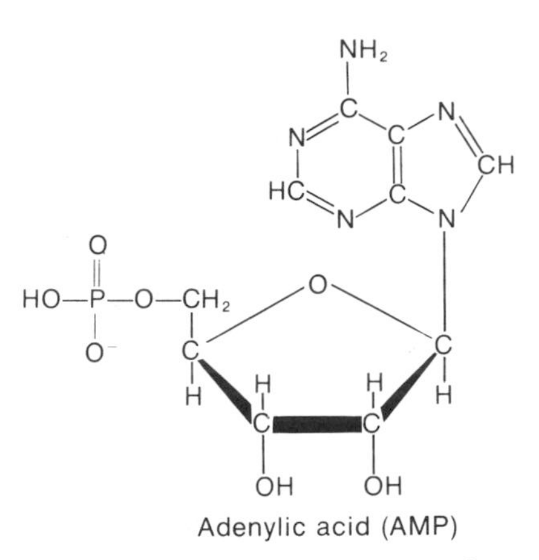

Figure 18-21. Adenylic acid; a nucleotide.

PROTEIN STORAGE IN THE BODY

Man cannot store either proteins or their component amino acids in his body as a reserve. Therefore, proteins must be continually obtained as a part of your diet, and to be effective, the proteins should be eaten daily because the amino acids cannot be stored. Of course, the body has the capability to synthesize some of the amino acids that it needs from the intermediates obtained in the metabolism of the carbohydrates, but the body *cannot synthesize* certain amino acids. They must be supplied by the intake of protein foods that contain them. Milk is a complete protein food, but a vegetable mixture can be formulated that has the same biological value as that of cow's milk. One such mixture contains corn meal, soy flour, and nonfat dried milk. In countries where malnutrition is a problem such mixtures are used to supplement the incomplete diet.

NUCLEIC ACIDS

The conjugated nucleoprotein consists of a protein part and a nonprotein part called a nucleic acid (Table 18-6). The nucleic acid is a very large polymer of nucleotides. Each nucleotide (Fig. 18-21) consists of a nucleoside and a phosphate. A nucleoside consists of a pentose sugar (either ribose or deoxyribose) attached to a purine ring or a pyrimidine ring (Fig. 18-22). The principal purines in nucleic acids are adenine and guanine. The principal pyrimidines ae cytosine, uracil and thymine (Fig. 18-23). Deoxyribonucleic acid (DNA) and ribonucleic acid (RNA) are the two fundamental nucleic acids found in all life on earth. DNA is usually found only in the nucleus of cells while RNA can be found in both the nucleus and cytoplasm. DNA is composed of the sugar deoxyribose and

Table 18-6. Nucleic acids.

nucleo protein ⟶ nucleic acid + protein
nucleic acid ↓ nucleotides; protein ↓ amino acids
nucleotides ↓ nucleosides + H_3PO_4
nucleosides ↓ pentose sugar + purine or pyrimidine

	Nucleosides	
	Ribose	*Deoxyribose*
Purines		
Adenine	Adenosine	Deoxyadenosine
Guanine	Guanosine	Deoxyguanosine
Pyrimidines		
Cytosine	Cytidine	Deoxycytidine
Uracil	Uridine	
Thymine		Thymidine

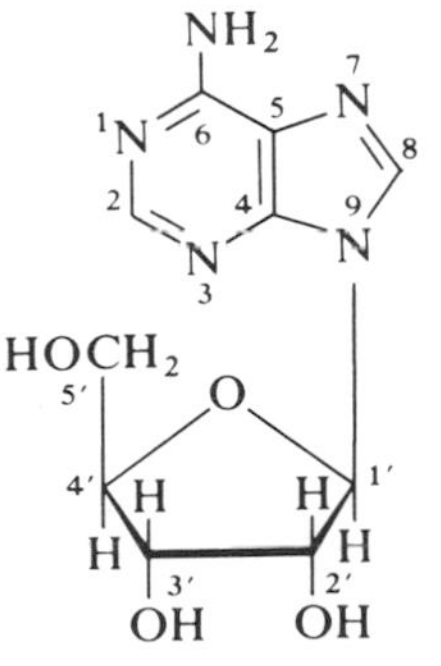

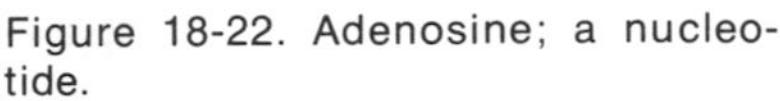
Figure 18-22. Adenosine; a nucleotide.

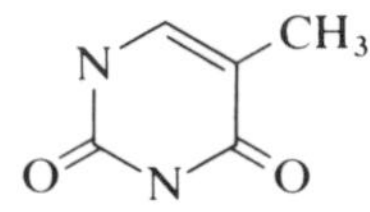

thymine

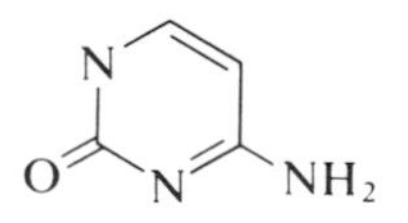

cytosine

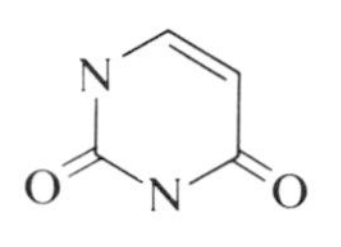

uracil

Figure 18-23. Pyrimidines.

Table 18-7. Nucleic acids.

	DNA	*RNA*
Pentose sugar	Ribose	Deoxyribose
Purines	Adenine	Adenine
	Guanine	Guanine
Pyrimidines	Cytosine	Cytosine
	Thymine	Uracil

adenine, guanine, cytosine and thymine whereas RNA is composed of ribose, adenine, guanine, cytosine and uracil (Table 18-7). Some nucleotides are important in the body. One of the most important is adenosine triphosphate (ATP) which is a source of high energy bonds in the body (Fig. 18-24). In another form as cyclic-AMP it is an important chemical messenger in the body.

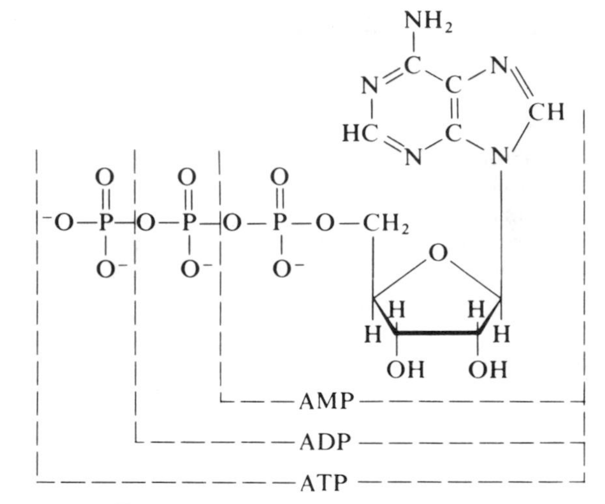

Figure 18-24. Adenosine phosphates.

SUMMARY

Every living cell contains proteins and requires a continuous supply of proteins for its existence. Proteins are used to build new tissues, act as buffers and are used to synthesize hormones, enzymes and antibodies. The basic building blocks of proteins are the amino acids. Some 21 amino acids are important to the body. Ten of these amino acids the body cannot synthesize and are called essential amino acids. The others the body can synthesize and are called non-essential amino acids. Amino acids are joined to form polypeptide chains by means of the peptide bond. Other weaker bonds (the disulfide bond, the salt bridge and the hydrogen bond) give each protein its own unique three-dimensional structure. The sequence of amino acids in a protein is called the primary structure. The configuration of the protein composes the secondary, tertiary and quatenary structures. Proteins can be classified according to their structure or their function. Because proteins have both acid and base groups, they are amphoteric. As pH rises, the protein can change from a primarily positive charge to a primarily negative charge. At the isoelectric point the positive charges equal the negative charges and the protein coagulates. The difference in charges of proteins is used in electrophoresis. When the peptide bonds of proteins are broken the protein is hydrolysed. If the weaker bonds are broken the three-dimensional configuration of the protein is destroyed and we have denaturation. Nucleic acids consist of polymers of phosphate, a pentose sugar and purines of pyrimidines. DNA consists of phosphate, ribose, adenine, guanine, cytosine and thymine. RNA consists of phosphate, deoxyribose, adenine, guanine, cytosine and thymine.

EXERCISE

1. Name seven functions of proteins.
2. Give the sources of proteins and distinguish between complete and incomplete proteins.
3. What is the difference between an essential and nonessential amino acid?
4. Are proteins stored in the body? What is the significance of your answer?
5. What are the intermediate and final products of hydrolysis of proteins?
6. What is the basic structure of an amino acid; what functional groups does it contain, and what is the spatial relationship of those groups?
7. Name ten essential amino acids and some other representative amino acids.
8. What do you mean by the optical activity of amino acids and why do they have this characteristic?
9. What is the basic primary structure of proteins?
10. Discuss the disulfide bond, the salt bridge, and the hydrogen bond with respect to the structure of the protein molecule.
11. Discuss the secondary structure of proteins.
12. What three factors stabilize the tertiary structure of proteins?
13. Give an example of a protein molecule that exhibits the quaternary structure.
14. What are the three classifications of proteins? Give three examples of each and their source.
15. What do you mean by the functional classification of proteins? Name five classes.
16. List the physical properties of proteins, and describe what part they play in the metabolic processes.
17. What is an amphoteric compound?
18. Discuss how the amino acids can act amphoterically.
19. What is meant by the isoelectric point, and what happens to proteins when they are brought to that point?
20. What is denaturation of proteins?
21. Name five methods of denaturing proteins and give examples of each.
22. When and how are proteins salted out?
23. What is an alkaloidal reagent and where does it find application?
24. Discuss the difference between nucleoproteins, nucleotides, and nucleosides.
25. What are the differences between DNA and RNA?

OBJECTIVES

When you have completed this chapter, you will be able to:

1. Define the term enzyme and state its function in the body.
2. List six major classes of enzymes.
3. Discuss enzyme structure.
4. Draw a diagram of how an enzyme works.
5. Explain factors that influence enzyme activity.
6. Define and discuss coenzymes.
7. Define inhibitors and explain how they work.
8. State the medical applications of enzymes.
9. Discuss some of the diseases caused by enzyme defects.
10. Discuss the role of vitamins in the metabolic processes of the body.
11. State the difference between vitamins and hormones.
12. Classify the vitamins and state their coenzyme function.
13. Define the term recommended daily allowance.
14. State what hormones are.
15. Describe how hormones work in the body.

ENZYMES, VITAMINS, AND HORMONES

BIOENERGETICS OF ENZYMES

Imagine that the bond holding 2 atoms together is like a tightly coiled spring. When the bond is broken the spring uncoils and energy is released. In some molecules the spring is tightly coiled: this is a high energy bond. In other molecules the spring is loosely coiled: this is a low energy bond. In chemical reactions chemical bonds break and reassemble.

In Figure 19-1 the bond energy of the reactants is more than the bond energy of the products. When the reaction proceeds from reactants to products the excess energy must be released. The reaction is *exergonic* (energy releasing) and proceeds spontaneously. In Figure 19-2 the bond energy of the reactants is less than the bond energy of the products. When the reaction proceeds from reactants to products energy must be put into the system. The reaction is *endergonic* (energy absorbing) and needs energy to proceed. We can say that an exergonic reaction proceeds downhill while an endergonic reaction proceeds uphill.

An example of an exergonic reaction is the burning of a safety match. The phosphorus in the match head reacts with oxygen in the air to form a phosphorus oxide. The excess energy is released as heat and ignites the match. You may argue that the reaction does not proceed spontaneously as you have to strike the match in order to light it. By striking the match we are putting energy into the system. After this initial input of energy, the reaction proceeds by itself: the match burns. This initial energy is called the energy of activation, abbreviated E_a (Fig. 19-3). All reactions need this input of energy to start the reaction. In exergonic reactions, after adding the E_a, the reaction proceeds spontaneously (Fig. 19-4). In endergonic reactions, after adding E_a, more energy has to be added to keep the reaction going (Fig. 19-5). An example of an endergonic reaction would be the roasting of a chicken. Roasting occurs only as long as the chicken is in the oven.

However, friction energy is not enough to ignite the safety match. If you rub

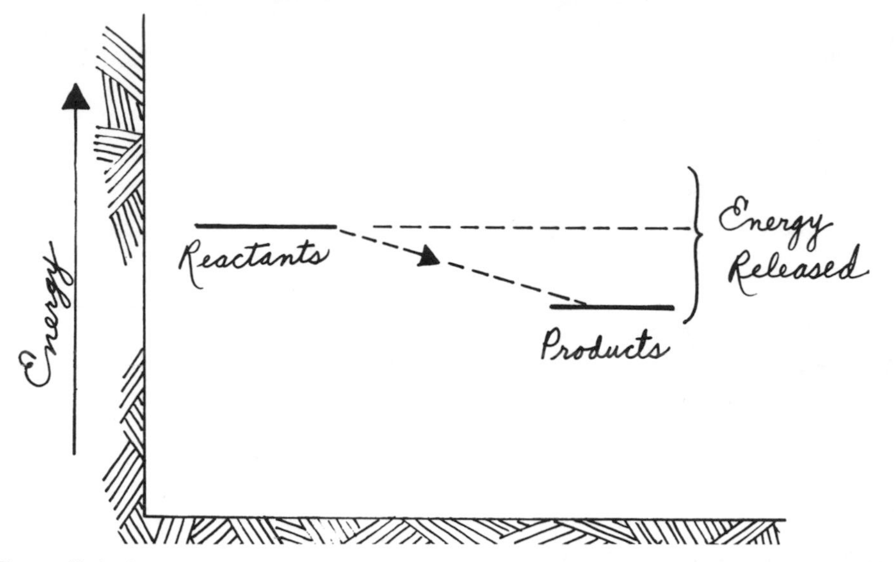

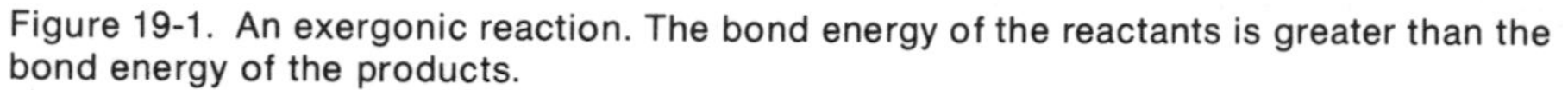
Figure 19-1. An exergonic reaction. The bond energy of the reactants is greater than the bond energy of the products.

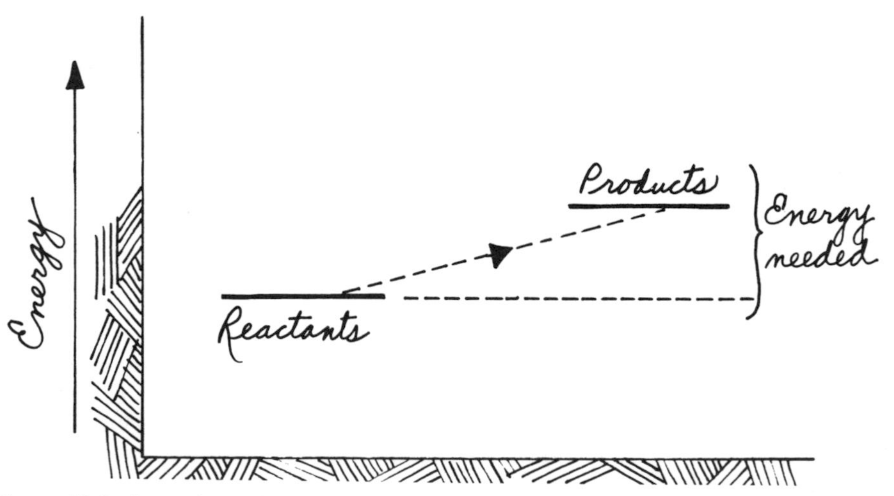

Figure 19-2. An endergonic reaction. The bond energy of the reactants is less than the bond energy of the products.

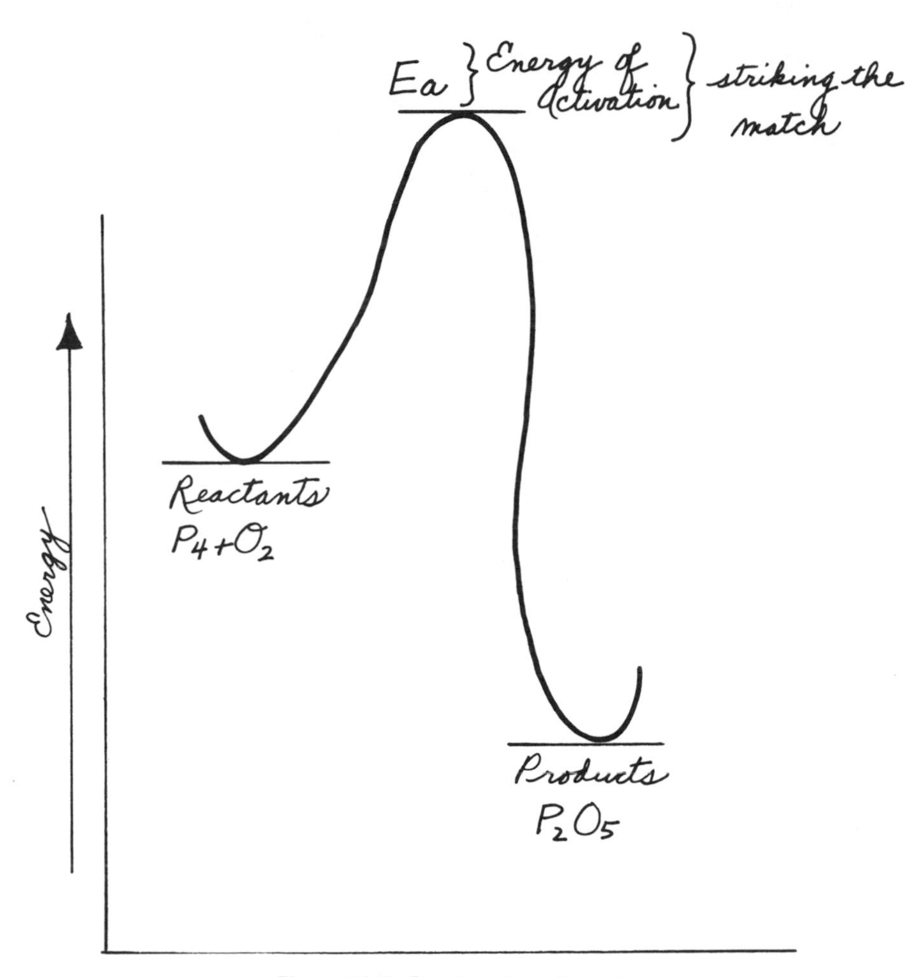

Figure 19-3. The burning of a match.

the match on the bottom of your shoe, on sandpaper, or on any rough surface, the friction heat will not be great enough to ignite the match. But if you strike the safety match on the special striker provided with the matchbook the friction heat is sufficient to ignite the match. Why is this so? It occurs because there is a substance in the striker, called a catalyst, that lowers the energy of activation of the reaction (Fig. 19-6). By lowering E_a, friction heat is sufficient to ignite the safety match. *A catalyst is a substance that lowers the energy of activation but does not take part in the reaction.* In the case of the safety match, red phosphorus in the striker acts as the catalyst.

Enzymes act as catalysts in reactions that occur in living organisms. Each enzyme is specific for a particular metabolic reaction. For example, the enzyme lipase will catalyze the hydrolysis of lipids but not carbohydrates or proteins;

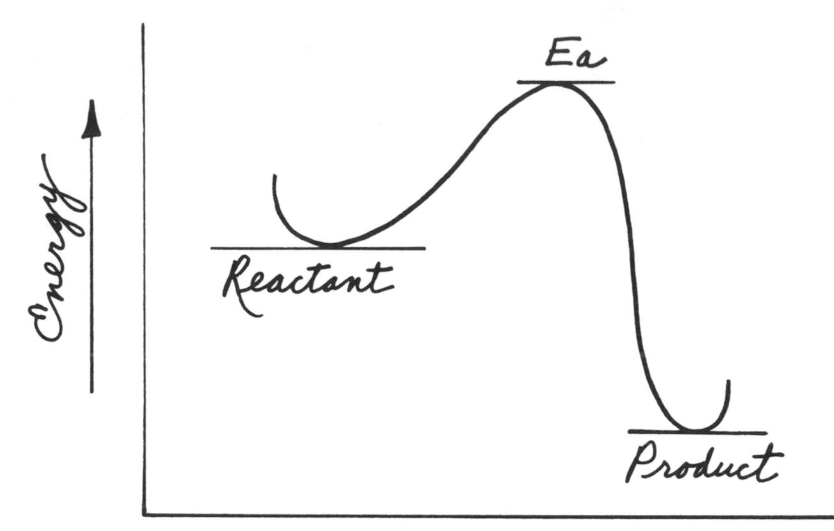

Figure 19-4. In an exergonic reaction, the reaction proceeds spontaneously after the energy of activation has been applied.

the enzyme sucrase will hydrolyze sucrose but not maltose or lactose. (Most enzymes have the ending ase.)

An enzyme is composed of two parts: a protein part called the apoenzyme and a nonprotein part called the coenzyme (see Fig. 19-7). If the nonprotein part is already part of the enzyme then it is called a prosthetic group. The apoenzyme and the coenzyme together form the enzyme (sometimes called the holoenzyme).

$$\text{Apoenzyme} + \text{Coenzyme} \longrightarrow \text{Enzyme}$$

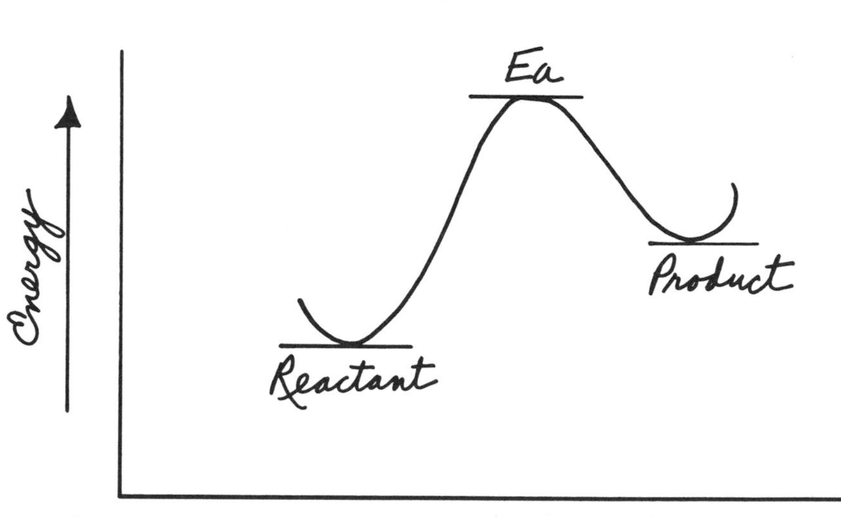

Figure 19-5. In an endergonic reaction, more energy has to be added continuously to keep the reaction going.

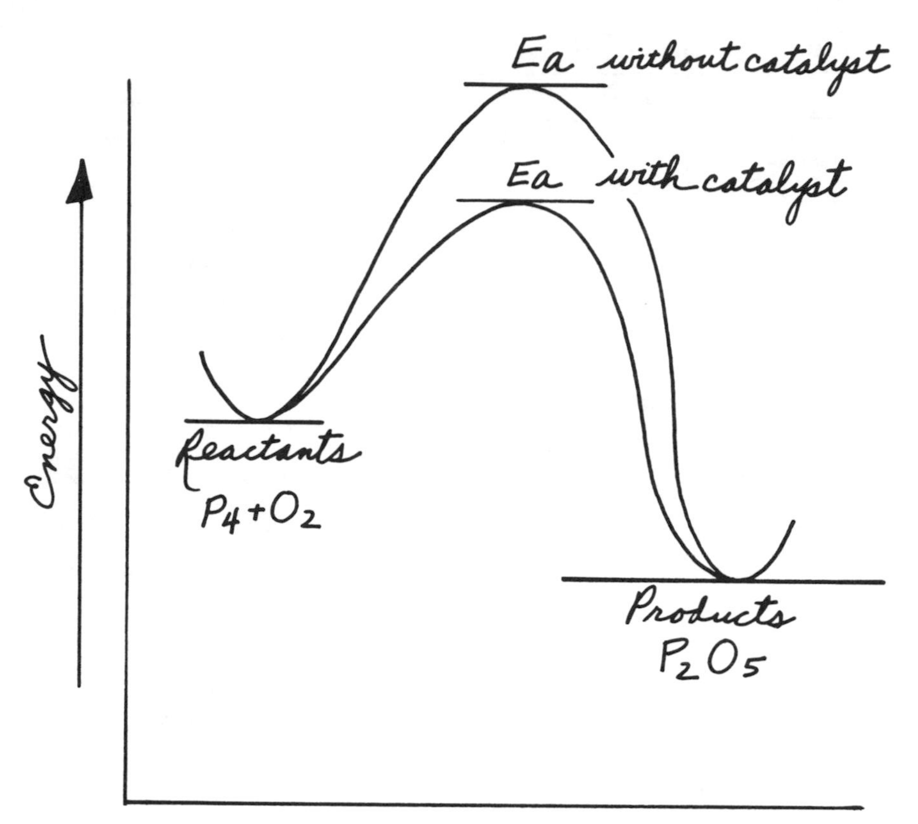

Figure 19-6. The action of a catalyst.

CLASSIFICATION OF ENZYMES

Enzymes are classified according to the type of reaction they catalyze (see Table 19-1). The six major classes of enzymes are as follows:

1. Oxidoreductases catalyze oxidation-reduction reactions.
2. Transferases catalyze the transfer of a characteristic chemical group from one molecule to another.
3. Hydrolases catalyze the reaction of the substrate with water.
4. Isomerases catalyze various types of isomerizations.
5. Lysases catalyze the formation of a double bond.
6. Ligases catalyze the coupling of two molecules.

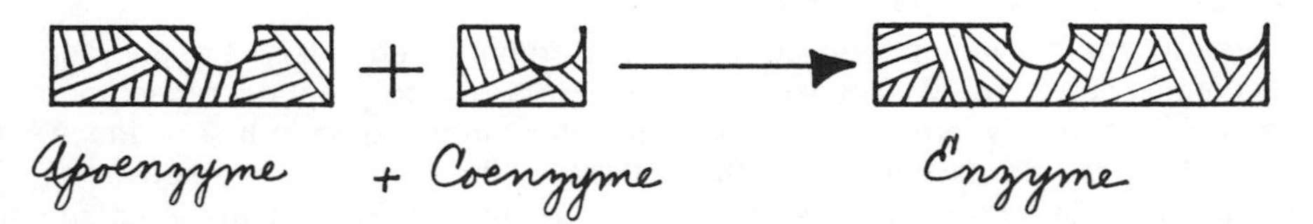

Figure 19-7. The composition of an enzyme. The nonprotein part is the coenzyme and the protein part is the apoenzyme.

Table 19-1. Classification of enzymes.

Group	*Enzyme*	*Source*	*Catalytic Action*
Oxido-reductase	Oxidases	Mitochondria	Adds oxygen to substrate
	Dehydrogenases	"	Removes hydrogen from substrate
	Peroxidases	"	Liberate O_2 and H_2O_2 from organic peroxides
	Catalases	"	Liberate O_2 from H_2O_2
Transferases	Transaminase	Heart muscle	Transfer of atoms or functional groups from one molecule to another
	Transacetylase		Transfer of acetyl group
Hydrolases	*Carbohydrases*		
	Ptyalin	Saliva	Starch → Dextrins → Maltose
	Sucrase	Intestinal juice	Sucrose → Glucose + Fructose
	Maltase	"	Maltose → Glucose + Glucose
	Lactase	"	Lactose → Glucose + Galactose
	Pancreatic amylase	Pancreatic juice	Starch → Dextrins → Maltose
	Lipases		
	Gastric	"	Fats → Fatty acids + Glycerol
	Steapsin	"	Fats → Fatty acids + Glycerol
	Esterases		Fats → Farry acids + Glycerol
	Phosphatases		Phosphate esters → Glycerol + H_3PO_4
	Proteases		
	Proteinases		
	Pepsin	Gastric juice	Protein → polypeptides
	Trypsin	Pancreatic juice	" "
	Chymotrypsin	"	" "
	Peptidases		
	Aminopeptidase	Intestinal	Polypeptides → aminoacids
	Carboxypeptidase	Pancreatic	" "
	Nucleases	Pancreatic and intestinal	Hydrolysis of nucleic acids
Lyases			Nonhydrolytic addition or removal of groups
Isomerases			Internal rearrangements
Ligases			Coupling of two molecules with breaking of pyrophosphate bond

MECHANISM OF ENZYME ACTIVITY

Since an enzyme is a protein it also has a primary, secondary, tertiary, and quaternary structure. We can describe the enzyme as a polypeptide folded upon itself to form a very large molecule with a cleft or indentation in it. This indentation is called the active site. At the active site the *substrate* (the substance upon which the enzyme acts) is changed to the product. The substrate fits into the active site like a key into a lock. Each enzyme has an active site that can fit only certain substrates. This is why enzymes are so specific. The enzyme com-

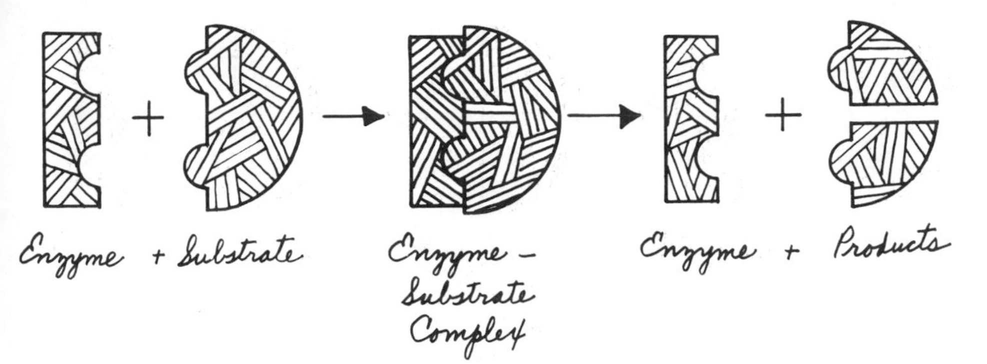

Figure 19-8. An enzyme-substrate complex.

bines with the substrate to form an enzyme-substrate complex. This complex intermediate then breaks apart into the products of the reaction, liberating the original enzyme and allowing it to combine with another molecule of substrate (Fig. 19-8). Because of this recycling only small amounts of enzyme are needed.

Coenzymes

A coenzyme is a nonprotein substance needed for enzyme activity that forms part of the enzyme. Vitamins frequently form part of the coenzyme molecule.

Factors That Influence Enzyme Activity

Since part of the enzyme is a protein, any of the factors that denature proteins will also destroy the activity of enzymes.

Enzymes in the body have their greatest activity at normal body temperature, 98.6°F or 37°C. Temperatures above normal reduce enzyme activity and extreme temperatures can destroy their activity. At below normal temperatures the activity is decreased but not destroyed. Each enzyme exerts its maximum activity at its own optimum pH. On either side of that particular pH (or range) its activity is markedly decreased. Amylase in the saliva is most active at a neutral pH and becomes inactive in the stomach where the pH is 1.6.

Some enzymes are initially in an inactive form. These are called proenzymes or zymogens. Before they can become active, another substance called an activator must be present. These activators may be organic substances secreted by the cell or inorganic ions (such as Ca^{+2}, Zn^{+2}, Mn^{+2}, or Mg^{+2}). Zinc activates the enzyme carbonic anhydrase which catalyzes the reaction: $CO_2 + H_2O \leftrightharpoons H_2CO_3$. Calcium activates prothrombin for normal blood clotting.

Inhibitors

Inhibitors decrease enzyme activity. They may be either organic or inorganic molecules. They react either directly with the enzyme or with the activator to prevent the activation of the enzyme (Fig. 19-9). Many poisons are fatal because they inhibit vital enzyme activity.

There are two types of inhibitors: competitive and noncompetitive. Competitive inhibitors compete directly with the substrate because they resemble the

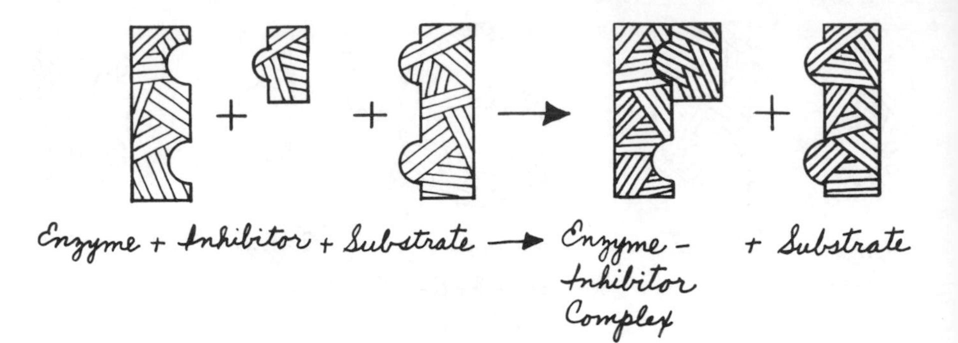

Figure 19-9. An enzyme-inhibitor complex.

substrate in their chemical nature. In this way the enzyme no longer performs its normal function. In other cases the competitive inhibitors are desirable. For example, sulfa drugs destroy bacteria through competitive inhibition.

Noncompetitive inhibitors react with certain nonspecific functional groups of the enzyme, altering the shape of the active site so that the substrate will not fit in. An example is the cyanide ion. It combines with iron to form a very stable complex, interfering with cytochrome oxidase, which is essential in respiration. Respiration ceases and the person dies.

Medical Applications of Enzymes

Protein splitting enzymes have been used to treat infected wounds, purulent (pus containing) bronchitis, and empyeme (pus in the body cavities). The enzymes liquify clotted blood, necrotic (dead) tissue, and pus, all of which aids the healing process.

Some diagnostic medical applications of enzymes are important in detecting various illnesses. Human infectious hepatitis causes a tenfold increase in the enzyme glutamic pyruvic transaminase (SGPT), which can be detected in the blood serum. Coronary occlusions with heart muscle damage are accompanied by an increase in the concentration of glutamic oxaloacetic acid transaminase (SGOT) in blood serum. Monocytic leukemia causes the presence of high concentrations of the enzyme lysozyme in body fluids. Arteriosclerosis is thought to be accompanied by high levels of the enzyme proline hydroxylase in the blood.

Diseases Caused by Enzyme Defects

There are a number of hereditary diseases that are caused by the lack of a certain enzyme. These diseases are referred to as inborn errors of metabolism.

Phenylketonuria (PKU) is characterized by excessive levels of phenylalanine in the blood and the presence of phenylketopyruvic acid in the urine. This disease affects the brain, causing mental retardation. Affected children are born lacking the enzyme phenylalanine hydroxylase, which converts phenylalanine to tyrosine. Early detection of this disease, followed by treatment with special diets low in phenylalanine, can prevent mental retardation.

Cystinuria results from the lack of certain enzymes in kidney tubules that aid the normal reabsorption of cystine from urine. Because cystine is relatively insoluble, cystine kidney stones result.

Galactosemia results from the lack of an enzyme that catalyzes the conversion of galactose to glucose. Damage to the central nervous system can result, causing mental retardation and death. The disease can be controlled by the elimination of galactose containing carbohydrates from the diet.

VITAMINS

Vitamins are organic compounds which are essential to normal growth in human beings but are not synthesized by the human body. Therefore, these substances must be part of the diet along with the carbohydrates, fats, proteins, and minerals. The vitamins (many of which have been synthesized in pharmaceutical laboratories) vary considerably in their chemical structure and in their physiological effects on the body.

Because vitamins have no similarity in chemical structures, biochemists have classified them according to their solubility. Vitamins are classed as being either water soluble or fat soluble. The B-complex vitamins and vitamin C are water soluble. Vitamins A, D, E, and K are fat soluble. Table 19-2 summarizes the role of vitamins in the human body.

Vitamins are important in many essential enzyme systems since they form part of many coenzymes. Vitamin C is a coenzyme in the conversion of lysine and proline to hydroxylysine and hydroxyproline. The latter two compounds are needed to cross-link collagen molecules, which form the connective tissue for muscles and gums. One of the symptoms of scurvy, caused by a lack of vitamin C, is bleeding gums followed by a loss of teeth. However, not all enzyme systems in the body need vitamins in order to function.

Many people are vitamin conscious, believing that they must take a pill containing all of the vitamins essential for growth. In general, a balanced diet of varied foods usually contains enough of each of the vitamins to maintain good health. The vitamins that are sold commercially are manufactured either in large scale plants by chemical synthesis or fermentation, or are extracted from their natural sources.

Vitamins act to regulate the myriad chemical reactions taking place in the cells. They promote growth and maintain health, although we still do not know exactly how they function. Extremely minute quantities of vitamins determine the difference between life and death, yet vitamins that are essential for one species are not essential to another species.

Recommended Daily Allowance (RDA)

Biochemists use a variety of methods to analyze the vitamin content of foods. In the early days there did exist a certain confusion regarding the units of potency of vitamins. As the chemical structures of the vitamins were determined, the biochemists were able to use the pure compounds in their procedures. Then the International Units (IU) were adopted. These units are identical to the United States Pharmacopoeia Unit (USP unit). Today, whenever possible, vitamin requirements are given in milligrams or in micrograms. The recom-

Table 19-12. The vitamins.

Vitamin	*Why it is needed*	*Recommended daily allowance*	*Foods it is found in*	*Deficiency symptoms*	*Overdose symptoms*
Vitamin A	Good vision, healthy skin and bones, pregnant women and fetus	Men: 5000 units women: 4000 units	Butter, milk, cream, eggs, liver, yellow fruit, vegetables	Poor night vision, rough-red skin, premature births, xerophthalmia, male sterility	Nausea, tension, weakness, constipation, retarded growth in children
Vitamin B_1	Keeps nervous system healthy, helps digestion, circulation, burns off excess carbohydrates	Men: 1.2mg women: 1.1mg	Pork, liver, brewer's yeast, whole grain cereals, milk, peanuts, soybean, wheatgerm	Beriberi, muscle weakness, gastroenteritis	None known
Vitamin B_2	Promotes normal cell growth (eyes & skin)	Men: 1.8mg women: 1.4mg	Liver, skim milk, green vegetables, brewer's yeast, organ meats, wheat germ	Impaired vision, dermatitus	None known
Vitamin B_3	Burns off certain carbohydrates and sugar	Men: 20mg women: 14mg	Lean meats, fish, poultry, peanuts, whole grain cereals, vegetables, brewer's yeast	Nervous disorders, intestinal trouble, pellagra, mental problems (headaches to hallucinations)	Liver damage, low blood pressure, swollen capillaries
Vitamin B_6	Enables body to use protein	adults: 2mg	Whole grain cereals, poultry, meat, shellfish, eggs, potatoes, nuts, prunes, raisins	Dry mucous membrane, dry, rashy skin	None known
Folic acid	Conditions the blood aiding in its clotting tendency	Adults: 400 micrograms	All foods, liver, kidneys, green leafy vegetables	Ruptured capillaries (blotchy red spots under skin), internal bleeding	None known
Vitamin B_{12}	Promotes healthy blood cells, maintains natural clotting of blood	Adults: 5 micrograms	All foods, cheese, eggs, milk	Anemia and internal bleeding (rare except for newborn babies & people taking antibiotics)	None known
Vitamin C	Combats stress, repairs cell damage, prevents scurvy, helps condition the blood	Adults: 45mg	Fruits, potatoes (notably the skin), most green vegetables (fresh)	Scurvy, extreme weakness, disintegrating gums, loose teeth, joint pains, longer time for wounds to heal	Possible kidney stones
Vitamin D	Healthy skin, essential for bone growth	Adults: 400 units	Sun's rays on the skin produce vitamin D; fortified milk	Bone disorders, rickets, cramps of muscles	Nausea, vomiting, appetite loss, high cholesterol, brittle bones, high blood pressure, Ca deposits in muscles & heart
Vitamin E	Maintains healthy cells, provides protection against chemical poisons in the system	Men: 15 units women: 12 units	Almost all foods	Almost impossible	None known
Vitamin K	Clotting of blood	adults: 1 microgram	Green leafy vegetables	Loss of blood	None known

mended daily allowances (RDA) are those values of nutrients judged by nutritionists to be adequate for the maintenance of good health. Vitamin deficiency is called hypovitaminosis; vitamin excess is called hypervitaminosis.

HORMONES

Most of the glands in the body have ducts, and their secretions pour out through these ducts, for example, the salivary glands. There are glands that have no ducts and these are called endocrine glands. These glands discharge their secretions directly into the lymph or blood stream. These secretions contain hormones that are synthesized by the endocrine glands. The hormones produce characteristic physiological effects on other cells.

Hormones resemble vitamins in their potent physiologic effect, and are essential in enzyme systems for normal and healthy growth. Hormones differ from vitamins in that they are synthesized in the body by the endocrine glands, whereas vitamins are synthesized by other organisms and are ingested in the diet.

The hormones are chemical messengers, initiating, regulating, and controlling a large number of metabolic and physiological activities. When the secreted substances stimulate activity they are called hormones; when they inhibit activity they are called chalones. It is possible that hormones have no effect upon the glands that secrete them, only affecting specific cells, which may be distant from the secreting glands.

Mechanism of Hormone Action

The specific action of hormones is not well understood. However, it is now supposed that cyclic AMP (cAMP) may act as a mediator between the hormone and the action it initiates inside a cell. Moreover, since hormones only act on specific tissues, the cells of these tissues must have a receptor site particular to that individual hormone (See Fig. 19-10).

The hormone attaches itself to a particular receptor site on the cell membrane. This releases or activates the enzyme adenyl cyclase. Adenyl cyclase changes ATP to cAMP, which initiates other reactions in the cell. After the reaction is completed, the hormone becomes inactivated and leaves the receptor site. There is some evidence that prostaglandins may also act as hormone mediators.

Summary of Hormone Actions

Hormones may:

1. affect the membrane transport of various substances into or out of the cell
2. affect the activity of genes
3. affect protein synthesis
4. change the amount or activity of various enzymes
5. change the amount or availability of a coenzyme
6. act as a coenzyme
7. change the structure of a large molecule, for example, an enzyme, protein or a nucleic acid
8. affect the structural elements of a cell to cause the above
9. stimulate or inhibit a hormonal mediator

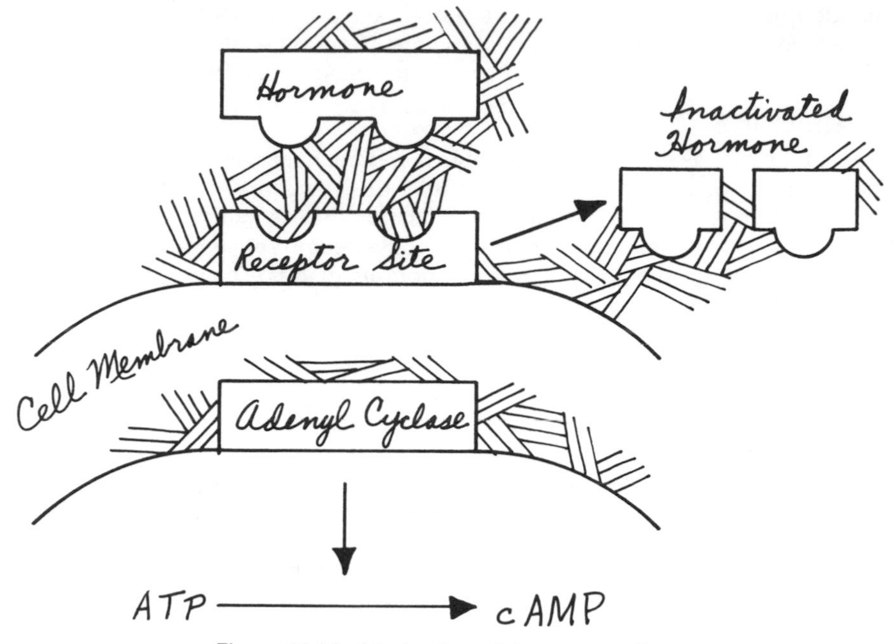

Figure 19-10. Mechanism of hormone action.

Dysfunction of the Endocrine Glands

When insufficient quantities of hormones are excreted the disorder is characterized by the prefix hypo, and when excessive quantities are excreted the disorder is characterized by the prefix hyper. The important hormone secreting glands are the thyroid, the parathyroid, the pancreas, the adrenal, the ovary, the testes, the pituitary glands and the intestinal mucosa. The malfunction of any of these glands that causes an abnormal excretion of hormones produces serious metabolic disturbances.

Classification of the Hormones

Hormones do not have a common structure or common functional groups. They may be either proteins, steroids or amino acid derivatives. The steroid groups are excreted by the testes, placenta, ovaries, corpus luteum, and the adrenal cortex. Protein hormones are produced by the pituitary, parathyroid, pancreas glands and by the cells of the gastrointestinal tract. The amino acid derivatives are produced by the thyroid and adrenal glands.

It is beyond the scope of this book to discuss the functions of all the hormones. They have been summarized in Table 19-3. However, four hormones are important in understanding the metabolism of the carbohydrates. These are epinephrine, ACTH, insulin and glucagon.

Epinephrine (previously called adrenalin) is secreted by the adrenal medulla. It is released into the blood stream only during emergencies (alarm, fright, emotional upset). Epinephrine produces an accelerated heart beat, raises the

Table 19-3. Hormones and their effects.

Gland	*Hormone*	*Site of action*	*Effect*
Pituitary	Growth hormone	General	Increases growth
	Thyroid stimulating hormone (TSH)	Thyroid	Increases thyroid hormone production
	Adrenocorticotropin (ACTH)	Adrenal cortex	Increases cortical hormone production
Thyroid	Thyroxine	General	Increases metabolic rate
	Calcitonin	Blood	Regulates calcium ion concentration
Parathyroid	Parathyroid hormone	Bone and kidney	Regulates calcium metabolism
Testis	Testosterone	General	Stimulates development of male characteristics
Ovary	Estrogen	Uterus	Prepares wall for pregnancy
		General	Stimulates development of female characteristics
	Progesterone	Uterus	Maintains wall
Adrenal medulla	Epinephrine	Heart	Increases pulse and blood pressure
		Liver and muscle	Controls breakdown of glycogen
Adrenal cortex	Corticosterone	General	Increases metabolic rate
Pancreas	Insulin	General	Reduces blood sugar
	Glucagon	General	Increases glycogenolysis
Intestinal cells	Secretin	Pancreas and liver	Releases pancreatic juice and bile

blood pressure, releases blood sugar (glucose) into the blood stream for energy, and dilates the air passages. All this enables the body to cope with stress situations (the flight or fight reaction).

Adrenocorticotropin (ACTH) normally is continuously secreted in small amounts. However, stress quickly increases its secretion. This increased secretion acts upon the adrenal cortex and causes an increased secretion of adrenal cortex hormones. Among other effects, ACTH increases blood sugar.

Throughout the pancreas there are scattered groups of specialized cells, called pancreatic islets or islands of Langerhans. These islets contain two types of cells called alpha cells and beta cells. The beta cells secrete insulin. Insulin is released when the blood sugar level is raised following carbohydrate ingestion. It interacts with the cell to permit the entrance of glucose. Insulin regulates the cellular intake of blood sugar.

Glucagon is secreted by the alpha cells of the pancreatic islets. Glucagon acts primarily on the liver to promote the breakdown of glycogen into glucose and stimulate its release into the bloodstream. Glucagon acts in opposition to insulin.

SUMMARY

Enzymes are specific catalysts for chemical reactions in the body. Enzymes are proteins and have the structural characteristics of proteins. The mechanism of enzyme reaction is known as the lock and key theory. Activators are needed to convert the inactive form of the enzyme to the active form. An important component of enzymes are coenzymes, the nonprotein part of the enzyme. Inhibitors are substances that decrease enzyme activity by competing with the substrate or altering the shape of the active site. Vitamins are organic compounds that are essential for normal growth and good health in the body. They act as part of the coenzyme systems and help regulate the many biochemical reactions taking place in the cell. Vitamins are classed as being either water soluble or fat soluble. The RDA requirements for vitamins are specified in micrograms, milligrams, or USP units. Hormones are secreted by the various endocrine glands. They are chemical messengers that initiate, regulate and control a large number of metabolic and physiological activities in the body. Hormones stimulate or reduce enzyme activity, are extremely potent in minute quantities, are excreted to meet the needs of the body, and have no common structure or functional group.

EXERCISE

1. What is an enzyme?
2. What are the six classes of enzymes?
3. What is the basic structure of enzymes and how do they act with substrates?
4. Describe the lock and key theory of enzymes.
5. What are coenzymes and why are they needed?
6. What are inhibitors and how do they act?
7. Give four medical applications of enzymes.
8. Name and discuss three diseases caused by enzyme defects.
9. Define vitamins as to their source and their function in the body.
10. What is meant by recommended daily allowance?
11. Define hypervitaminosis and hypovitaminosis.
12. How are vitamins classified?
13. What are hormones and where are they found?
14. Summarize the characteristics of hormones.

OBJECTIVES

When you have finished this chapter, you will be able to:

1. State how carbohydrates are digested in the body.
2. Discuss the digestion of lipids and their absorption in the body.
3. Discuss the digestion of proteins in the body.
4. List the enzymes and functions of the enzymes, gastric and intestinal juices.
5. Discuss the role of the liver in fat, carbohydrate, and amino acid metabolism.
6. Discuss the digestion of the nucleoproteins.
7. State the difference between liver and muscle glycogen.
8. Explain glycogenesis and glycogenolysis.
9. Write the sequence of reactions to show the formation of lactic acid from glycogen.
10. Discuss the role of the Emden-Meyerhof pathway and the hexose monophosphate shunt in the body.
11. Explain the cascade reaction.
12. Explain how the Kreb's cycle converts glucose to energy.
13. Differentiate between anaerobic and aerobic glycolysis.
14. Discuss the nitrogen balance in the body.

METABOLISM I

Foods are composed of carbohydrates, fats, proteins, and nucleoproteins, all of which are very large molecules. In order to be utilized by the body these molecules must be broken down into smaller molecules. This process is called digestion. This breakdown of large molecules is also called catabolism, while the reverse process, the joining of small molecules to form large molecules, is called anabolism. The combination of anabolism and catabolism is known as metabolism.

Enzymes hydrolize these large molecules into their component parts. Carbohydrates are hydrolized into monosaccharides, fats into glycerol and fatty acids, proteins into amino acids, and nucleoproteins into amino acids, phosphates, ribose, deoxyribose, purines and pyrimidines.

DIGESTION

Digestion is both a mechanical and chemical process. The process begins in the mouth where the food is masticated by the teeth into smaller particles. The food bole (masticated food) then goes to the stomach where further grinding takes place. The food bole, now called chyme, then goes into the small intestine where it is hydrolyzed and absorbed into the body (see Fig. 20-1). There are specific enzymes in the mouth, stomach and small intestine for the hydrolysis of carbohydrates, fats, proteins, and nucleoproteins.

Digestion of Carbohydrates

The digestion of carbohydrates begins in the mouth by the saliva, which contains 2 enzymes, salivary amylase and salivary maltase. This combination is known as ptylin. The amylase catalyzes the breakdown of starch to amylose and dextrin, while maltase hydrolyzes dextrin to maltose. Pytylin becomes inactive when the pH drops below 4. Hence its activity ceases when you swallow your food and it enters the stomach, where the pH is 1.6. The stomach does not contain any enzymes capable of hydrolyzing carbohydrates.

The major part of carbohydrate digestion occurs in the small intestine where

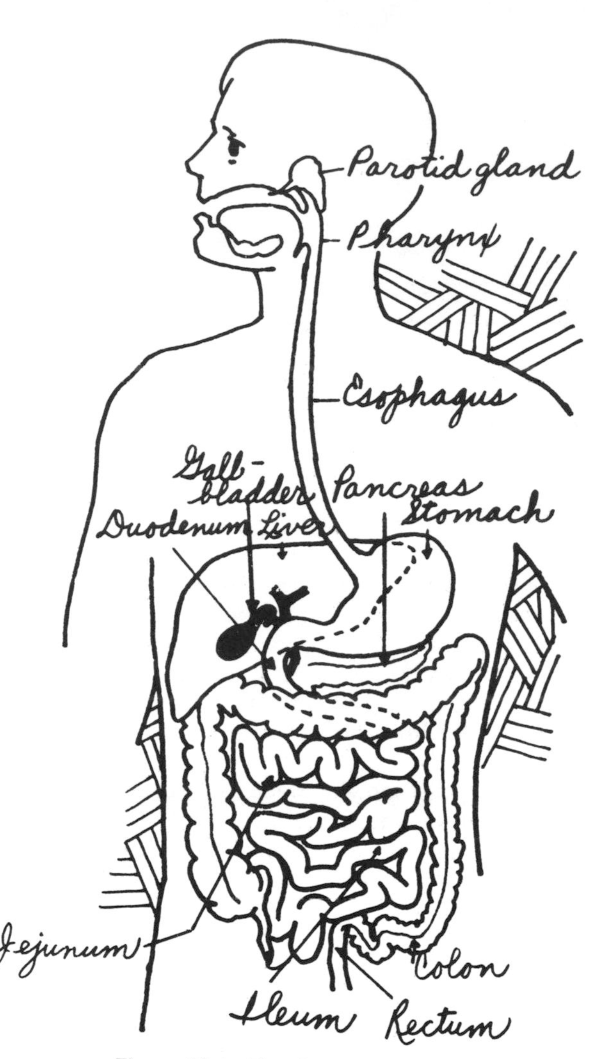

Figure 20-1. The digestive tract.

enzymes from the pancreas and the intestine hydrolyze polysaccharides to monosaccharides. The pancreas contains pancreatic amylase, while the intestinal juices contain maltase, sucrase, and lactase. Maltase hydrolyzes maltose into 2 glucose molecules; lactase hydrolyzes lactose into glucose and galactose; and sucrase hydrolyzes sucrose into glucose and fructose. Many adults lose their ability to form lactase and they cannot digest milk. They are plagued by gastrointestinal stress and diarrhea. All of the carbohydrates that we eat are hydrolyzed into the monosaccharides glucose, fructose, and galactose, which are absorbed from the small intestine (Fig. 20-2). The galactose is absorbed the most rapidly, followed by glucose, and finally fructose.

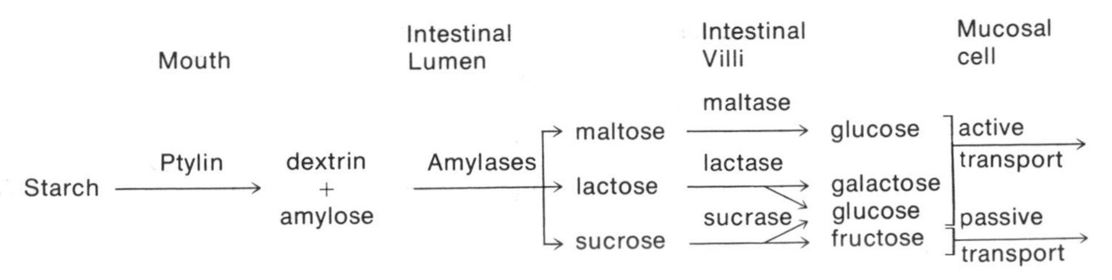

Figure 20-2. Digestion of carbohydrates.

Digestion of Fats

We know that fats are extremely insoluble in water and that body fluids are very watery, yet fats are transported throughout the body by the body fluids. This occurs because fats are emulsified when carried by the body fluids. The fats that are eaten can be either in a neutral fat condition, as pure fat-like salad oil, or in an emulsified fat condition, as in egg yolks.

When emulsified fats are eaten they are hydrolyzed in the stomach by gastric lipase to glycerol and fatty acids. Since glycerol is water soluble it can be readily absorbed through the stomach walls. The fatty acids are present in the emulsified form and are also absorbed.

The nonemulsified fats must first be emulsified by the bile acids of the intestinal juices. The bile acids emulsify fat by reducing the surface tension of the fat, in the same way that soaps and detergents work on grease. Once the neutral fats have been emulsified the pancreatic and intestinal lipases hydrolyze them. The resulting glycerol and fatty acids are absorbed through the intestinal wall.

The absorption of fats differs from that of the carbohydrates and proteins, both of which must be first hydrolyzed to their respective monomers before they are absorbed into the body. The fats and the fatty acids, on the other hand, after being emulsified pass right through the intestinal walls as milky microdroplets of oil. Once they have passed through the intestinal membrane, these microdroplets coalesce to form larger droplets of oil which are then carried by the lymphatic and blood systems to various parts of the body. Some of the fat is chemically bound to other substances such as proteins to form lipoproteins.

STORAGE OF FATS

The body stores fat in three places: the abdominal cavity, the subcutaneous connective tissue, and the intermuscular connective tissue. The body tends to store fat during surplus times, when the caloric intake of the body exceeds the energy requirements of the body. The composition and characteristics of stored body fats can be changed by changing the composition of the ingested fats, for example, changing a diet of saturated fats to unsaturated fats. The liver acts as a temporary depository for absorbed fats. The quantity stored in the liver depends upon the amounts received from ingestion and synthesis, the amount converted to other substances, and the rate of exchange between the liver and other depots of fats (Fig. 20-3).

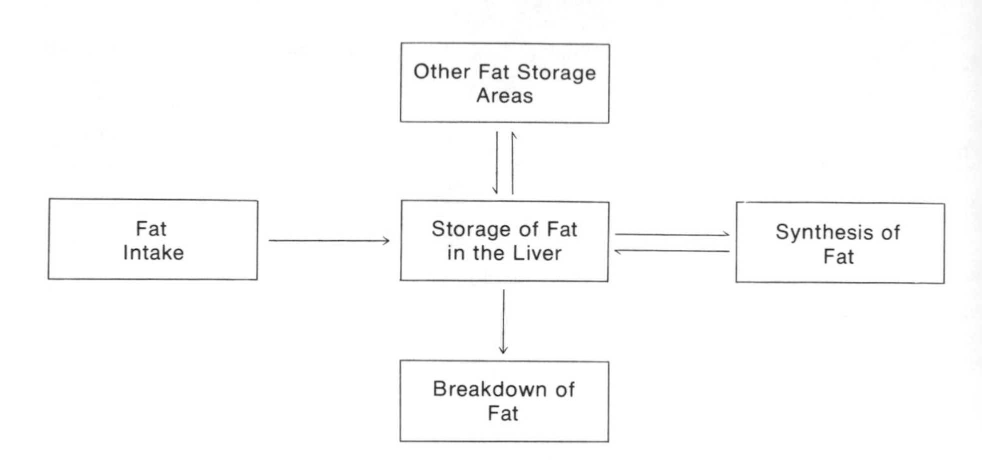

Figure 20-3. Fate of fat in the body.

Digestion of Proteins

After mastication by the teeth food passes into the stomach, where gastric juices begin the hydrolysis of proteins. The peristaltic action of the stomach (the wave-like motions of the stomach muscles) propels the food onward toward the small intestine. When in the stomach the proteins are hydrolyzed by the enzymes in the gastric juice and the hydrochloric acid secreted by the stomach.

ENZYMES IN THE GASTRIC JUICES

Pepsin is an active enzyme formed from the inactive proenzyme pepsinogen by the action of hydrochloric acid. Pepsin hydrolyzes proteins to proteoses, peptones, and polypeptides.

Renin is an enzyme that coagulates casein in milk to paracasein. Coagulated protein is easier to digest than noncoagulated protein. If the milk protein were not coagulated by renin, it would pass through the stomach too quickly to be acted on by the protein splitting enzymes. When renin coagulates casein, the paracasein then reacts with calcium to form precipitated curds.

Hydrochloric acid is present in the stomach at concentrations ranging from 0.2 percent to 0.5 percent. It acts as a germicide for any bacteria that have been swallowed and at the same time it provides the correct pH for the stomach enzymes.

INTESTINAL JUICE

The thick liquid mixture of partially digested food that passes into the small intestine is called chyme. There, the low pH of the mixture (from the acid of the stomach), is changed to a high pH by the alkaline juices of the small intestine, bile, and pancreas. These juices complete the digestive process (see Table 20-1).

Enzymes in the Intestinal Juices

Pancreatic juice is composed of 98.7 percent water and 1.3 percent solids and has a pH of 8. The solids consist of digestive enzymes, albumin, globulin, NaCl

Table 20-1. Important substances secreted into intestinal lumen.

Substance	*Origin*	*Function*
Pepsinogen	Stomach	To yield pepsin which preferentially splits peptide links adjacent to aromatic amino acids.
Amylase	Pancreas	To hydrolyze polysaccharides mainly to maltose.
Trypsinogen	Pancreas	To yield trypsin for the splitting of peptide links mainly between the carboxyl group of lysine or arginine and the amino group of another amino acid.
Chymotrypsinogen	Pancreas	To yield chymotrypsin for splitting peptide links mainly between carboxyl of aromatic acid and the amino group of an amino acid other than aspartate or glutamate.
Maltase	Pancreas	Splits maltose to glucose.
Carboxypeptidase	Pancreas	Removes amino acids from carboxyl end of a peptide.
Lipase	Pancreas	Splits triglycerides in stepwise fashion to fatty acid and glycerol at rates depending on the fatty acid composition
Maltase	Intestinal mucosa	Splits maltose to glucose.
Sucrase (invertase)	Intestinal mucosa	Splits sucrose to glucose and fructose.
Lactase	Intestinal mucosa	Splits lactose to galactose and glucose.
Amino peptidase	Intestinal mucosa	Removes amino acids from amino end of a peptide.
Dipeptidase	Intestinal mucosa	Splits dipeptides to amino acids.
Phosphatase	Intestinal mucosa	Removes phosphate from organic phosphates.
Lecithinase	Intestinal mucosa	Breaks down phospholipids to constituents.
HCl	Gastric mucosa	Stops enzymic actions in the material ingested; converts pepsinogen to pepsin; provides favorable pH for action of pepsin; hydrolyzes fructose-containing sugars.
$NaHCO_3$	Pancreatic juice	Brings material discharged by the stomach into the intestine to a roughly neutral pH suitable for the action of the digestive enzymes.
Bile salts	Liver	Emulsifies and solubilizes water-insoluble material.

and $NaHCO_3$. It contains not only protein splitting enzymes but also carbohydrate and fat hydrolyzing enzymes. Two protein hydrolyzing enzymes are trypsin and chymotrypsin. Both have an optimum pH between 8 and 9 and both hydrolyze proteins to proteoses, peptones, polypeptides, and amino acids.

Intestinal juice is secreted by the walls of the small intestine and contains peptidases (among other enzymes) which complete the digestion of the proteins. Dipeptidases hydrolyze dipeptides to amino acids. Aminopeptidases convert polypeptides to amino acids. Nucleases digest nucleoproteins and nucleic acids.

Digestion of the Nucleoproteins

The nucleoproteins are broken down into amino acids and nucleic acids. The latter is further split by nuclease into smaller units called nucleotides. The nucleotides are further split into nucleosides by phosphatase enzymes found in intestinal juice, bile and pancreatic juice. The nucleosides are further broken down into pentose sugars, purines and pyrimidines.

ABSORPTION OF PROTEINS

Amino acids are absorbed into the blood stream through the capillary blood vessels of the villi of the small intestine. The small intestine is very long (about 20 to 25 ft) and it presents a large surface area (about 100 sq ft) for absorption of nutrients. Further, about 5 million tiny, finger-like projections, called villi, increase the absorption area many times. Each villus has a network of blood capillaries which surround a central lymph capillary.

Occasionally, some traces of unchanged protein and partially hydrolyzed proteins may be absorbed in the intestine. This may cause allergic reactions in the individual. It may happen after the person has eaten certain food proteins such as eggs or fish. These people are said to be sensitive or allergic to the particular food. However, you should remember that allergies can be caused by any substance (clothing, animals, environment) or intangible agents (heat, cold, sunlight).

NITROGEN BALANCE IN THE BODY

Our bodies do not store any appreciable amount of protein, the major source of nitrogen for the human body. Hence, our bodies must use the nitrogen they need each day and excrete the excess. In normal individuals the nitrogen intake should equal the nitrogen output. Nutritionists do not know how much nitrogen an individual needs each day, but they concur that:

1. Growing children retain more nitrogen than they excrete and have a positive nitrogen balance.
2. Elderly people or people with a wasting disease excrete more nitrogen than they take in and have a negative nitrogen balance.

Amino acids are not only involved in the building of new tissue but also in the repair of damaged and worn out tissues. Actually most of the proteins in the body are in a state of constant turnover; they are constantly being broken down and reformed (Fig. 20-4). One cannot state that every protein is hydrolyzed and resynthesized each day as this action depends on the protein involved. Some are reformed daily, while others apparently remain unaffected by the rebuilding process.

Since the amount of nitrogen that we eat varies each day, our body must have some sort of control mechanism which will maintain the nitrogen balance at a normal level. That mechanism is the ability of the body to excrete urea, a nitrogen containing compound found in the urine. By varying the output of urea, the body can maintain the proper nitrogen balance.

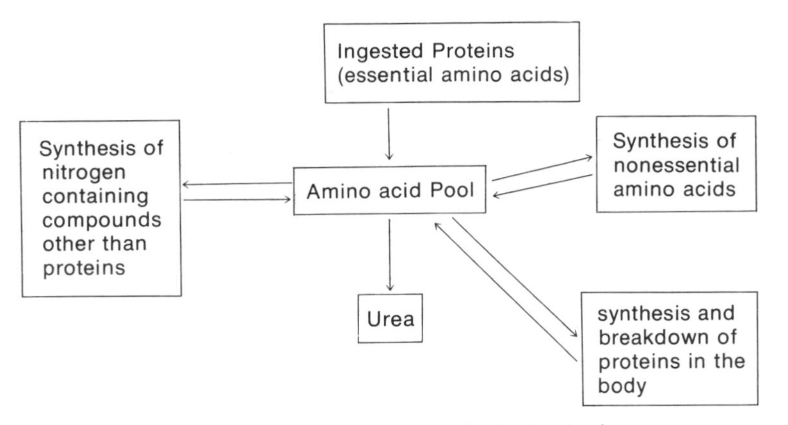

Figure 20-4. Fate of amino acids in the body.

ENERGY IN THE BODY

The energy requirements of the body are met by the breakdown of glucose to CO_2 and H_2O:

$$\underset{\text{(glucose)}}{C_6H_{12}O_6} + 6\,O_2 \longrightarrow 6\,CO_2 + 6\,H_2O + \text{energy}$$

If the reaction were to take place in one step all of the energy would be released at once. In the body the reaction takes place in a number of small steps or reactions. The energy is released in small amounts and the body can store the energy.

Each step is catalyzed by a specific enzyme, and the whole series of steps is called a metabolic pathway. Glucose is not limited to one pathway. It can enter a number of pathways which are interconnected to each other. You will see later that lipids and amino acids have their own metabolic pathways that are interconnected to the various glucose pathways. When you understand the connections among all of these pathways you will understand how a disease such as diabetic mellitus affects the body processes.

The body stores the energy released by these pathways in the form of ATP. In the body some of the metabolic reactions are exergonic and some are endergonic. The energy released from the exergonic reactions is used to power the endergonic reactions. This is called a coupled reaction. ATP is one of the compounds that passes energy from an exergonic reaction to an endergonic reaction. ATP is the basis of energy transfer in the body:

$$ATP \rightleftharpoons ADP + P_i + \text{energy}$$

The forward reaction releases energy; the backward reaction stores energy. P_i stands for inorganic phosphate.

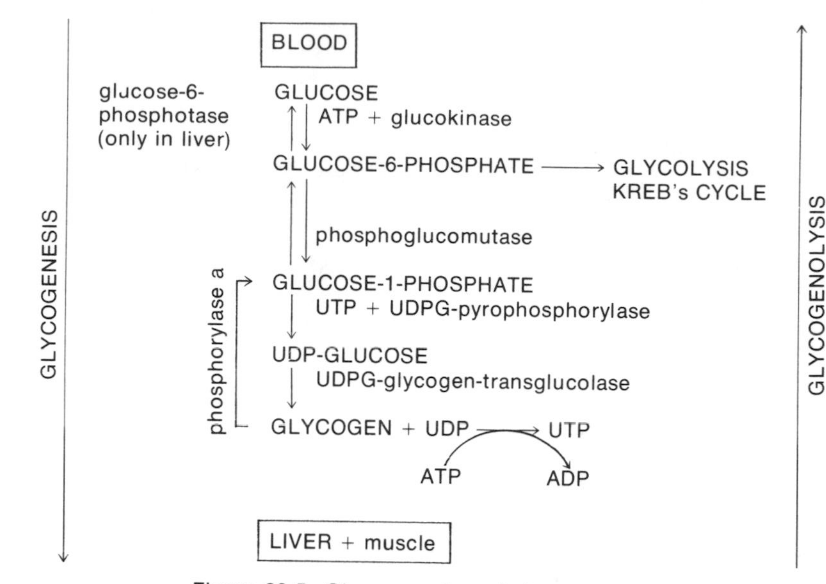

Figure 20-5. Glycogenesis and glycogenolysis.

CARBOHYDRATE METABOLISM IN THE LIVER

The liver is the major storage area for nutrients in the body. It manufactures liver glycogen, which is one way that carbohydrates are stored in the body. Glycogen is a polymer of glucose which has been brought to the liver by the portal vein after being absorbed from the small intestine. The formation of glycogen from glucose is called glycogenesis. Glucose in the bloodstream goes to the muscles where it is used to form muscle glycogen. When the stored glycogen is hydrolyzed back to glucose, the process is known as glycogenolysis (see Fig. 20-5).

Glycogenesis

Glycogenesis in liver and muscle may be summarized as follows:

1. Glucose reacts with ATP in the presence of glucokinase and Mg^2 to form glucose-6-phosphate. ATP supplies energy to power the reaction and goes to ADP.
2. Glucose-6-phosphate is transformed to glucose-1-phosphate in the presence of phosphomutase.
3. The enzyme UDPG pyrophosphorylase causes glucose-1-phosphate to react with UTP (a nucleotide similar to ATP but having a uracil molecule instead of an adenine molecule) to form activated glucose, UDP-glucose.
4. The UDP-glucose polymerizes to form glycogen. The reaction is catalyzed by UDPG-glycogen-transglucolase. UDP is released.
5. UDP reacts with ATP yielding ADP and regenerating UTP.
6. For every molecule of glucose added to a glycogen chain, 2 molecules of ATP are used.

GLYCOGENOLYSIS AND THE CASCADE REACTION

Since glycogen is formed by the polymerization of glucose, it would seem that the breakdown of glycogen to glucose would be a reversal of the steps in glycogenesis. However, such is not the case. A glucose at the end of a glycogen chain is phosphorylated and broken off as glucose-1-phosphate by phosphorylase a. Normally, this enzyme is in an inactive form called phosphorylase b. Phosphorylase b is changed to phosphorylase a by a complex series of reactions known as the cascade reaction (see Fig. 20-6). This may be started by epinephrine, glucagon or ACTH. Each step in the sequence multiplies the following reaction 10 to 100 times. Hence, 1 molecule of hormone will yield 10^8 molecules of glucose.

The muscle cells do not have the enzyme glucose-6-phosphatase, which removes the phosphate molecule from glucose-6-phosphate to yield glucose. Hence, glucose-6-phosphate in muscles can only go into the glycolysis and Kreb's cycle, which provide energy for muscle contraction. Muscle glycogen cannot be used to produce glucose in the bloodstream. But liver glycogen can produce glucose in the bloodstream.

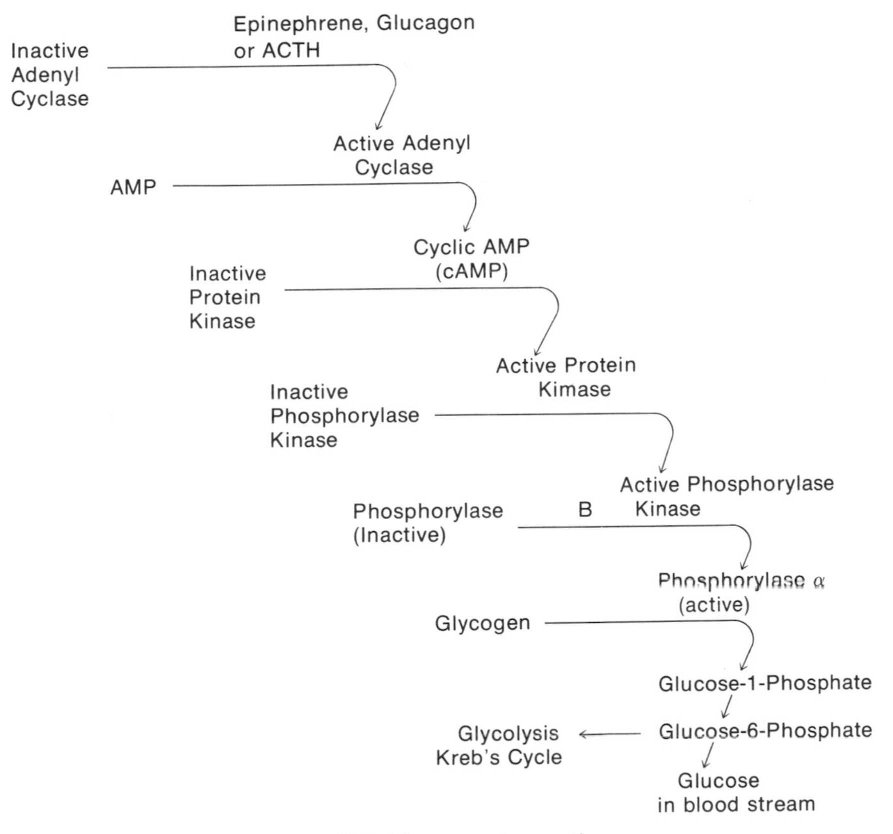

Figure 20-6. The cascade reaction.

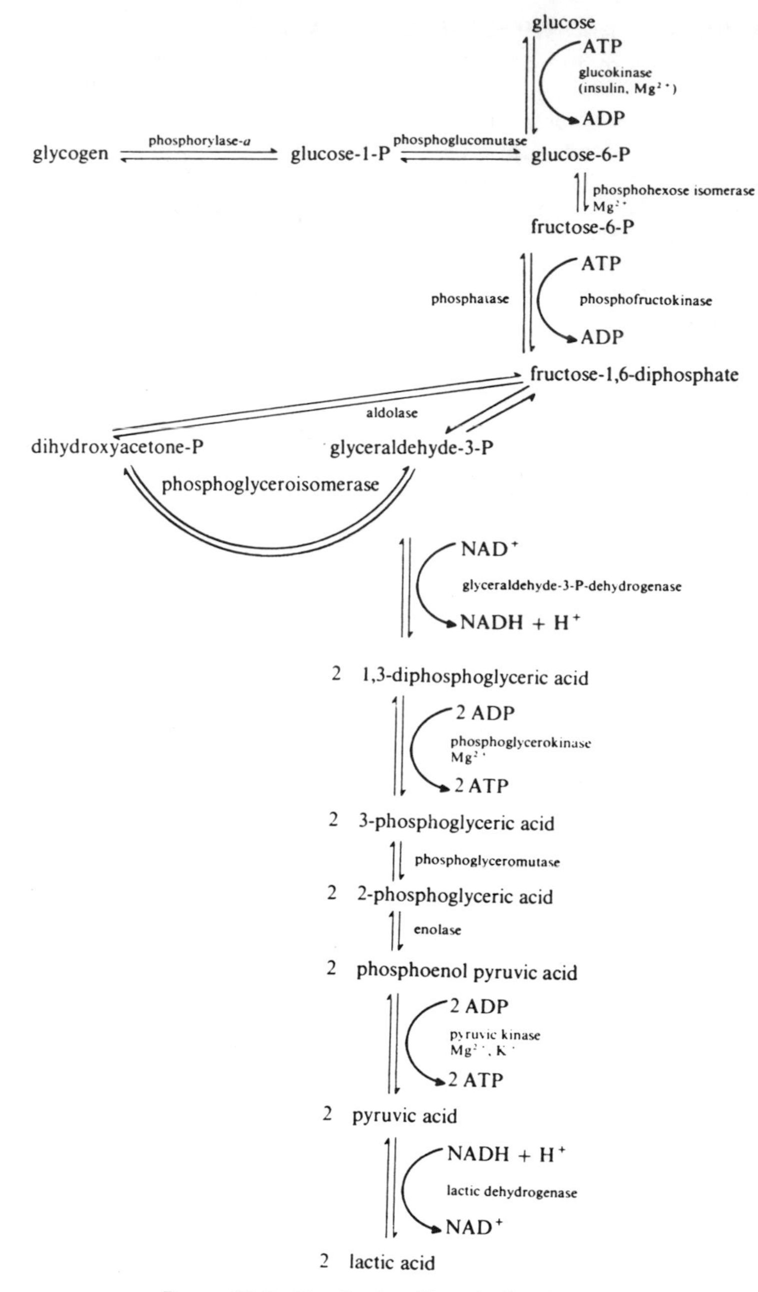

Figure 20-7. The Emden-Meyerhof pathway.

Glycolysis

Earlier you were told that glucose could use several metabolic pathways. Three such pathways are the Emden-Meyerhof pathway, the hexose monophosphate shunt, and the Kreb's cycle. The Emden-Meyerhof pathway (also called the Cori cycle, the lactic acid cycle or anaerobic glycolysis) is an anaerobic process; it takes place in the absence of oxygen. It is the principal method whereby muscles obtain energy in emergency situations. The hexose monophosphate shunt (also called the phosphoglutamate pathway) is an aerobic process. The Kreb's cycle (also called the citric acid cycle or the tricarboxylic acid cycle) is an aerobic process (it takes place in the presence of oxygen). The Emden-Meyerhof pathway can feed into the Kreb's cycle. When this occurs, the combined pathway is known as aerobic glycolysis.

EMDEN-MEYERHOF PATHWAY

The overall reaction for the Emden-Meyerhof pathway (Fig. 20-7) is:

$$1 \text{ glucose} + 2 \text{ ADP} \longrightarrow 2 \text{ lactic acid} + 2 \text{ ATP}$$

The enzymes for this pathway are found in the cytoplasm of the cell. We can summarize the pathway as follows:

1. Glycogen is converted to fructose-6-phosphate.
2. Fructose-6-phosphate splits into two 3-carbon triosephosphates.
3. Through intermediates pyruvic acid is formed.
4. Pyruvic acid is reduced to lactic acid.
5. In the anaerobic process, 2 ATP's are used to power the reactions, but 4 ATP's are generated. This gives a new yield of 2 ATP's. As we shall see later, the aerobic glycolysis yields more ATP's.

When muscles do work, they contract. This can be an anaerobic process. Lactic acid is produced. As the concentration of lactic acid increases, some of it is transported to the liver and some remains in the muscle cells. The lactic acid transported to the liver is reconverted to liver glycogen. The lactic acid in the muscle cell is oxidized back to pyruvic acid by lactic acid dehydrogenase and coenzyme NAD^+. The accumulation of lactic acid in muscle tissues leads to muscle fatigue as the glycogen reserve of the muscle tissue is used up. However, in the presence of oxygen, glycogen is regenerated from lactic acid, and the muscle regains its contractile ability.

THE HEXOSE MONOPHOSPHATE SHUNT

The overall reaction for the hexose monophosphate shunt (Fig. 20-8) is:

$$3 \text{ glucose-6-P} + 6 \text{ NADP} \longrightarrow 3 \text{ CO}_2 + 2 \text{ fructose-6-P} + \text{glyceraldehyde-3-P} + 6 \text{ NADPH}$$

where P stands for phosphate. The enzymes for this pathway are found in the cytoplasm of the cell. This pathway generates ribose and NADPH. The former is used in nucleic acid synthesis and the latter is used in the synthesis of fatty acids. Fructose-6-phosphate and glyceraldehyde-3-phosphate are junction

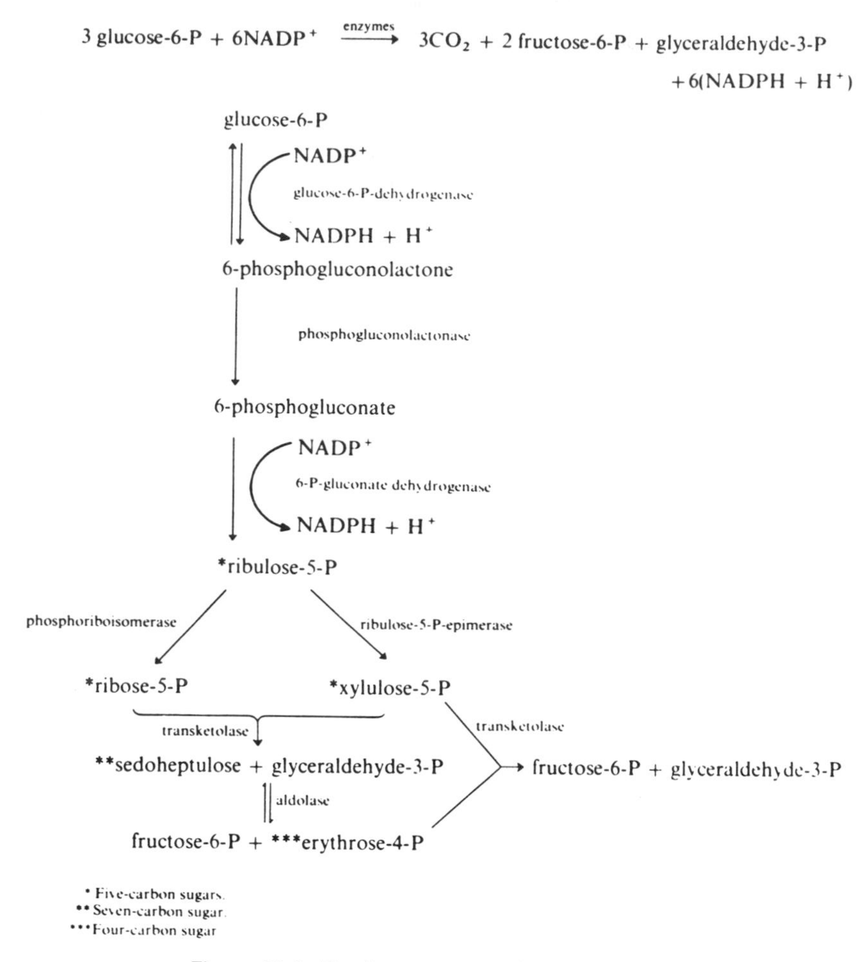

Figure 20-8. The hexose monophosphate shunt.

points between the Emden-Meyerhof pathway and the hexose monophosphate shunt.

THE KREB'S CYCLE

The principal metabolic pathway in the body is the Kreb's cycle. The enzymes for this pathway are in the cristae (foldings) of the mitochondria (Fig. 20-9). Here the enzymes are together in multienzyme complexes.

The cycle begins with pyruvic acid which combines with coenzyme A to form acetyl CoA. Each chemical reaction in the Kreb's cycle is regulated by a specific enzyme. There are five reactions in which hydrogen is lost in the form of NADH (see Fig. 20-10).

The NADH enters the electron transport chain where the hydrogens are passed from one compound to another. As the hydrogens pass these com-

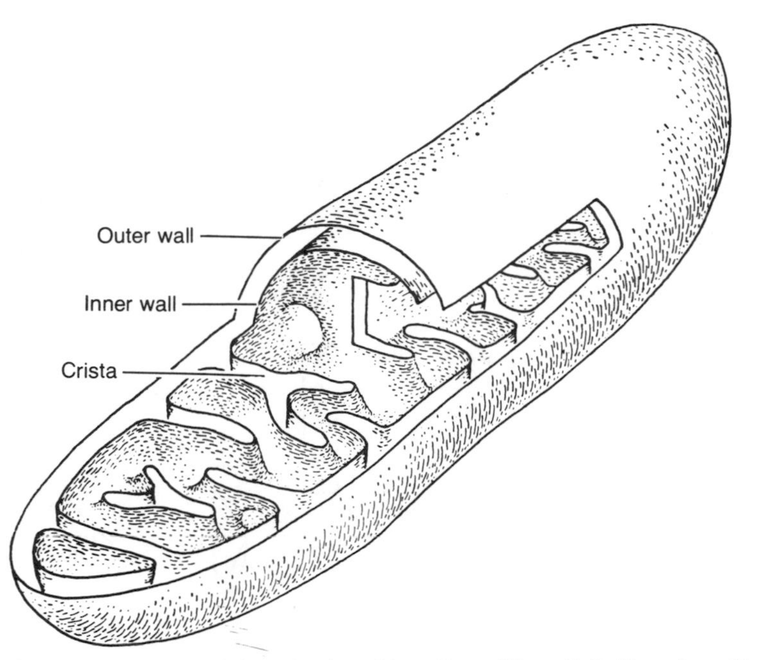

Figure 20-9. The structure of the mitochondrion. From "How Cells Transform Energy," by Albert L. Lehninger. Copyright © (1961) by Scientific American, Inc. All rights reserved.

pounds undergo oxidation-reduction reactions. At certain points in the chain, energy is stored in the form of ATP. The hydrogens finally react with oxygen to form water. Without oxygen the process would not proceed. Each molecule of NADH yields 3 molecules of ATP. A molecule of $FADH_2$ yields only 2 ATP's (Fig. 20-11).

Each glucose yields 2 pyruvic acids in the Emden-Meyerhof pathway. In the Kreb's cycle each pyruvic acid yields the equivalent of 5 NADH's or $5 \times 3 = 15$ ATP's. In aerobic glycolysis, glucose yields 38 ATP's as follows:

4 ATP	(from Emden-Meyerhof pathway)
6 ATP	(from 2 NADH release in Emden-Meyerhof pathway)
10 ATP	
− 2 ATP	(Used in the Emden-Meyerhof pathway)
8 ATP	
30 ATP	(from 2 pyruvic acids in Kreb's cycle)
38 ATP	

The overall reaction for aerobic glycolysis is:

$$\text{glucose} + 6\,O_2 \longrightarrow 6\,CO_2 + 6\,H_2O + 38\,\text{ATP}$$

Thus, lactic acid (or the oxidized form, pyruvic acid) is metabolized to give CO_2,

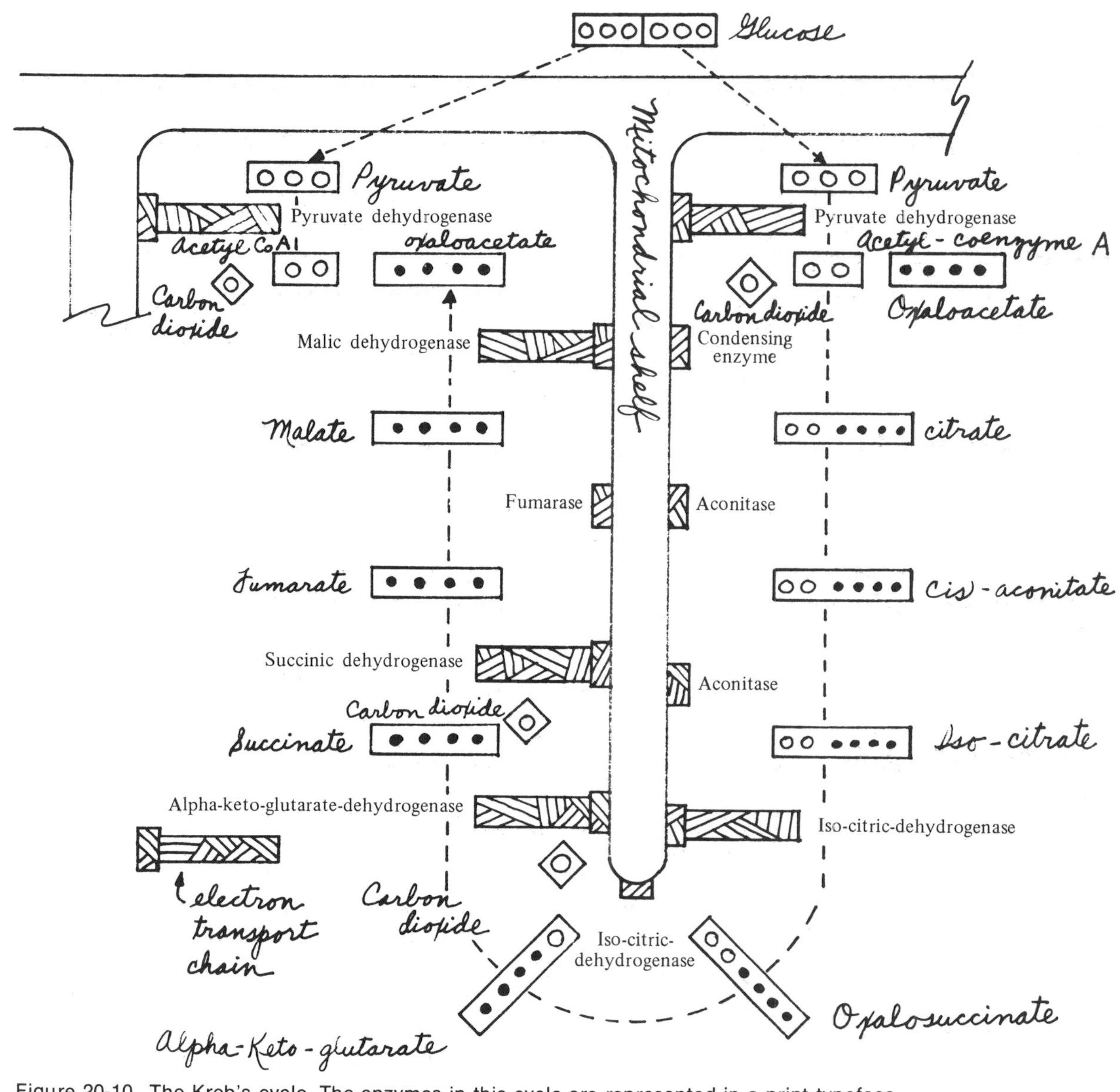

Figure 20-10. The Kreb's cycle. The enzymes in this cycle are represented in a print typeface.

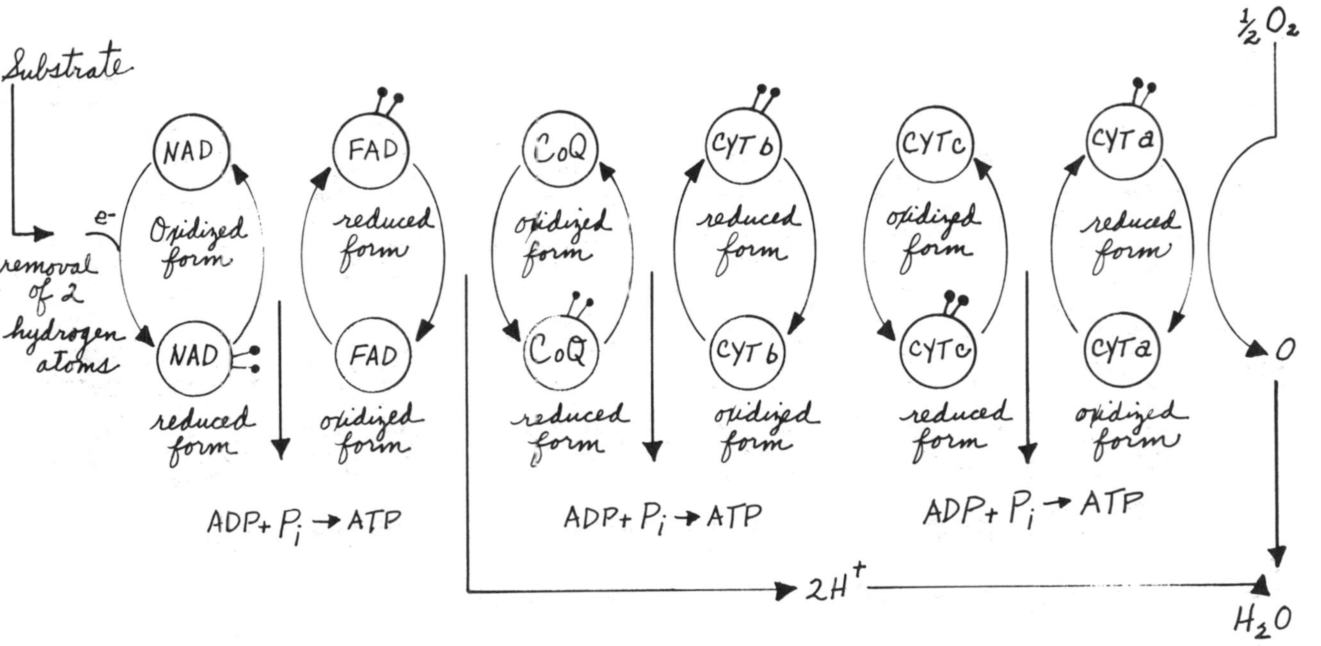

Figure 20-11. A summary of electron transport. The figure shows the removal of the 2 hydrogen atoms from an oxidizable substrate, the transport of the electrons and the final acceptance of the electrons and 2 protons by an atom of oxygen to give water. The energy released is trapped in the synthesis of ATP from ADP and inorganic phosphate (P_i).

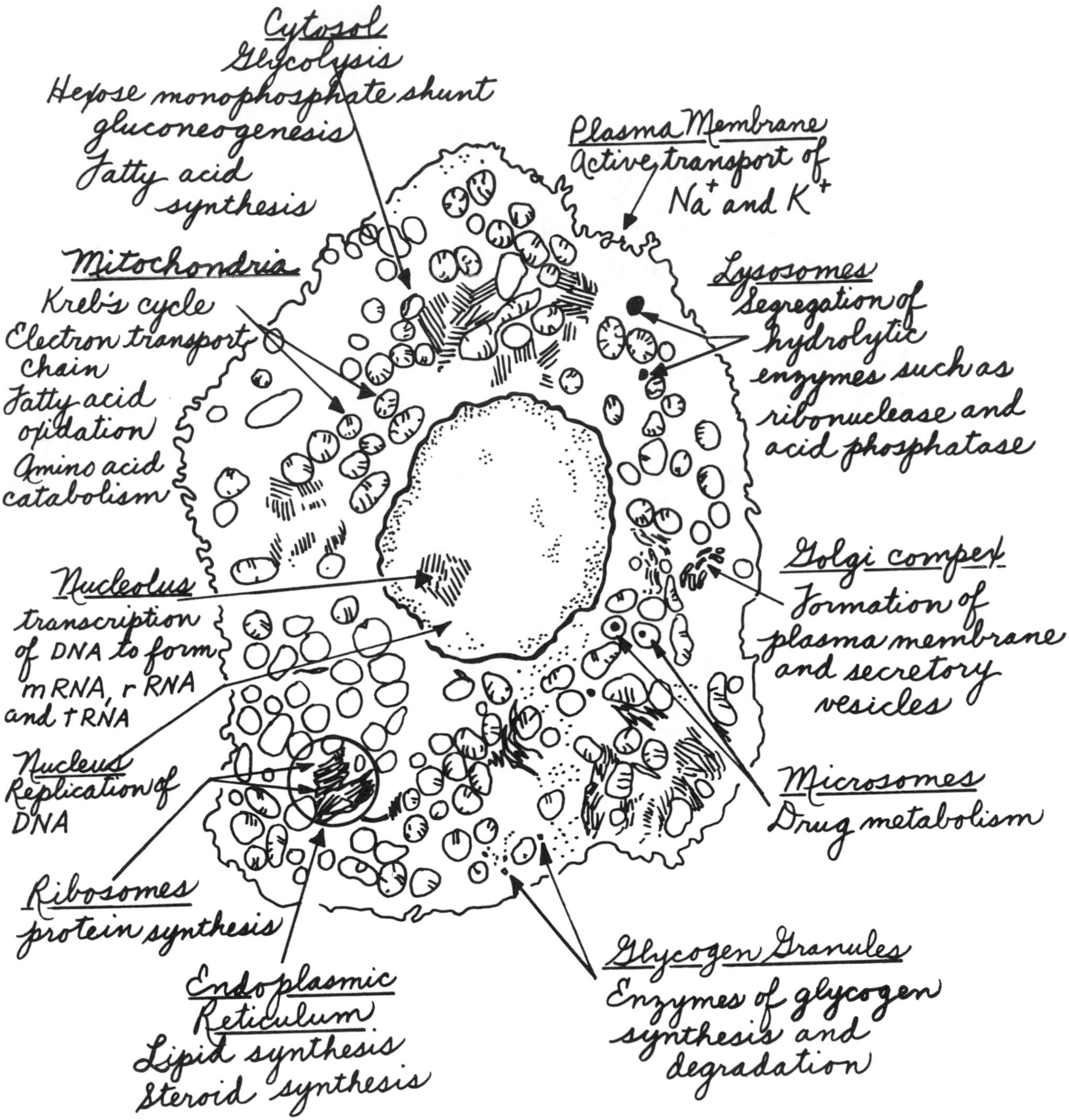

Cytosol
Glycolysis
Hexose monophosphate shunt
gluconeogenesis
Fatty acid synthesis

Plasma Membrane
Active transport of Na^+ and K^+

Mitochondria
Kreb's cycle
Electron transport chain
Fatty acid oxidation
Amino acid catabolism

Lysosomes
Segregation of hydrolytic enzymes such as ribonuclease and acid phosphatase

Golgi complex
Formation of plasma membrane and secretory vesicles

Nucleolus
transcription of DNA to form mRNA, rRNA and tRNA

Nucleus
Replication of DNA

Microsomes
Drug metabolism

Ribosomes
protein synthesis

Endoplasmic Reticulum
Lipid synthesis
Steroid synthesis

Glycogen Granules
Enzymes of glycogen synthesis and degradation

Figure 20-12. The liver cell of a rat.

H_2O, and considerable amounts of energy through the Kreb's cycle. Most biochemists agree that for each mole of glucose that is oxidized 304 kilocalories of energy is liberated, or about 45 percent of the 686 kilocalories available through the complete combustion of glucose. The locations of the various biochemical reactions within the cell are shown in Figure 20-12.

SUMMARY

The body can only metabolize monosaccharides, obtaining them from the hydrolysis of starch, polysaccharides, and disaccharides. The most common monosaccharides are glucose, fructose, and galactose. Lipids are absorbed in the digestive system as neutral fats or as the hydrolyzed products glycerol and fatty acids. These are transported to the cells where they are used as energy or stored for later use. The liver acts as a temporary storage area for fats. Proteins are hydrolyzed to amino acids in the digestive process. Since they are the major source of nitrogen in the body, the intake and output of the nitrogen containing compounds must be balanced. Glucose is converted to glycogen in the liver and muscles by the process of glycogenesis. The breakdown of glycogen to glucose is known as glycogenolysis. Liver glycogen can replenish blood sugar. Muscle glycogen can only supply energy to the muscles; it cannot replenish blood sugar. In glycolysis glucose is broken down to provide energy for the cells. In aerobic glycolysis (Emden-Meyerhof pathway), the end product is lactic acid. In aerobic glycolysis (Kreb's cycle) the end products are CO_2 and H_2O. In the hexose monophosphate shunt, the principle products are ribose and NADPH.

EXERCISE

1. Name the enzyme in the mouth that initiates carbohydrate digestion.
2. What is the mechanism for the digestion of fats?
3. Explain the nitrogen balance in the body.
4. What is the end product of protein digestion?
5. Where are fats stored in the body?
6. Name the enzymes that hydrolyze proteins in the gastric and intestinal juices.
7. What is the raw material for the synthesis of glycogen? What is the name of the process?
8. Define glycogenolysis.
9. How does muscle glycogen differ from liver glycogen? What enzyme accounts for this difference?
10. What is the difference between aerobic and anaerobic reactions?
11. What is the cascade reaction?
12. What is the end product of anaerobic glycolysis?
13. What is the overall reaction of the Emden-Meyerhof pathway?
14. What is the overall reaction of the hexose monophosphate shunt?
15. What is the purpose of the hexose monophosphate shunt in the body?
16. What is the overall reaction of the Kreb's cycle?
17. Why is the Kreb's cycle important in the body?

OBJECTIVES

When you have finished this chapter, you will be able to:

1. Describe the oxidation of fats.
2. Calculate the energy produced by a molecule of fatty acid.
3. Define lipogenesis.
4. Discuss the storage of lipids.
5. Describe the metabolism of cholesterol.
6. Explain how essential amino acids form nonessential amino acids.
7. Define transamination, amination and oxidation.
8. Diagram the urea cycle.
9. Relate the urea cycle to the Kreb's cycle.
10. Describe how the body synthesizes purines and pyrimidines.
11. Discuss the role of acetyl CoA in the body.
12. Define ketosis, ketonemia and ketonuria.
13. Discuss the metabolic effects of starvation.
14. Relate the factors that control the concentration of blood sugar.
15. Discuss the metabolic effects of diabetes mellitus.
16. Discuss the interrelationship of protein, fat, and carbohydrate metabolism.

METABOLISM II

As we mentioned in the last chapter, the Kreb's cycle is the central metabolic pathway of the body. The end products of carbohydrate, fat, and protein digestion are linked through this pathway. In the Kreb's cycle excess carbohydrates and amino acids can be transformed to fats (lipogenesis). Amino acids can also be converted to glucose (gluconeogenesis). When carbohydrates are in short supply, the Kreb's cycle can utilize fats and amino acids to produce energy.

OXIDATION OF FATS

The end products of fat digestion are glycerol and fatty acids. The glycerol can be metabolized by oxidation to dihydroxyacetone phosphate, which is transformed to pyruvic acid in a series of steps in the Emden-Meyerhof pathway. Pyruvic acid can enter the Kreb's cycle for total oxidation to CO_2, H_2O and energy.

We normally consider glucose to be the only source of energy for body processes. However, in the absence of glucose, fats can also furnish the required energy. This pathway is called beta oxidation of fatty acids or the fatty acid spiral. The enzymes for this pathway are found in the mitochondria. The fatty acids are broken down two carbons at a time. In the spiral, two carbons are broken off the fatty acid in the form of acetyl CoA. The shorter fatty acid then repeats the process until it is all broken down to acetyl CoA. The pathway requires coenzyme A and one ATP to start the process (Fig. 21-1).

Acetyl CoA enters the Kreb's cycle to yield 12 ATP's. In the electron transport chain each $FADH_2$ yields 2 ATP's and each NADH yields 3 ATP's. Stearic acid

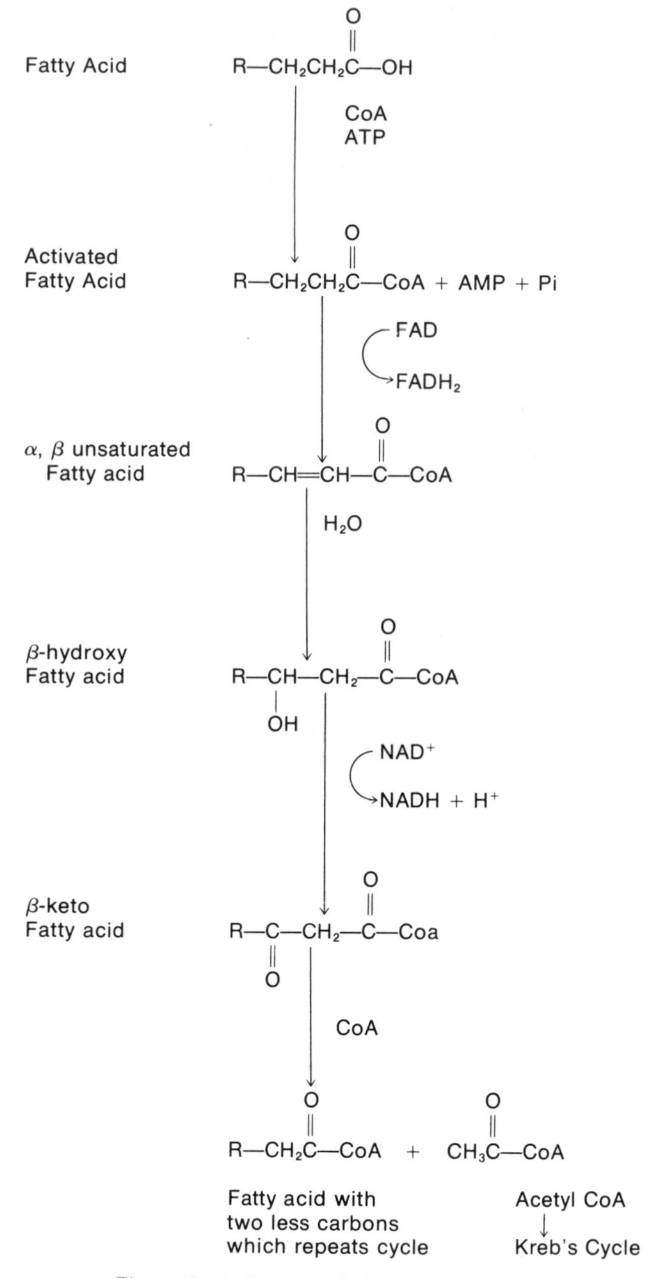

Figure 21-1. Beta oxidation of fatty acids.

($C_{18}H_{36}O_2$) will take eight turns of the fatty acid spiral to produce 9 acetyl CoA's. Thus, 1 molecule of stearic acid will yield:

108 ATP	(from nine acetyl CoA)
16 ATP	(from eight $FADH_2$)
24 ATP	(from eight NADH)
148 ATP	
− 1 ATP	(used to start the pathway)
147 ATP	(total ATP produced)

Compare this with the 38 ATP's produced by 1 molecule of glucose in the Kreb's cycle. Translating this to kcal 1 gram of fat yields 9 kcal, 1 gram of carbohydrate yields 4 kcal and 1 gram of protein yields 4 kcal. Thus fats are the most compact way of storing energy in the body.

Lipogenesis

In another sequence of reactions acetyl CoA can be converted to a fatty acid. We need not go into the exact mechanism, but acetyl CoA's are joined two by two to form a fatty acid of the desired length. The pathway needs NADPH, formed in the hexose monophosphate shunt. The acetyl CoA can come from the glycolysis of glucose or, as we shall see later, from the oxidation of amino acids. In this way, excess carbohydrates and proteins can be transformed into fats.

The triglycerides are synthesized from glycerol (obtained from the enzymatic hydrolysis of fats) and fatty acids (obtained either from the diet or the fatty acid spiral). The reaction requires energy in the form of ATP to phosphorylate the glycerol, which then reacts with the fatty acids to form triglyceride.

Storage of Lipids

The lipids are essential components of cytoplasm, cellular membranes, muscle tissues, bone marrow, and nerves. However, the fats are not stored as static lipids but as active lipids. The body constantly breaks down the stored lipids and replaces them with synthesized lipids or lipids ingested in the diet. This creates a dynamic equilibrium where lipids are constantly being replaced with new molecules of lipids. This turnover of lipid molecules in the body involves the digestion of fats, the hydrogenation of unsaturated fatty acids, the dehydrogenation of saturated fatty acids, and the synthesis of fatty acids from proteins and carbohydrates.

METABOLISM OF CHOLESTEROL

Cholesterol is absorbed in the intestinal tract along with neutral fats through the action of bile. Usually cholesterol appears in the blood as a fatty acid ester. Once cholesterol enters the blood, some of it is converted to steroid hormones, some is totally metabolized to CO_2 and H_2O, some is converted to phospholipids, some is converted to glycogen and some is excreted in the feces. Cholesterol can be totally synthesized by the body from acetyl CoA, just as the fatty acids, but by a different metabolic pathway.

METABOLISM OF AMINO ACIDS

The body has the unique ability to synthesize required building blocks from substances ingested in the diet by means of various metabolic pathways. The body has tremendous leeway to start with almost anything to synthesize needed products. All of these various transformations are performed by specific enzymes in the pathways. The body can synthesize nonessential amino acids from essential amino acids, carbohydrates, and fats. This occurs by three principal processes: transamination, amination, and oxidation.

Transamination is a process whereby one amino acid can transfer its alpha amino group to another molecule. This results in the formation of an alpha keto acid in the original molecule, which can be oxidized to CO_2 and H_2O in the Kreb's cycle. Two examples of this conversion are the transamination of alanine and aspartic acid to pyruvic acid, oxaloacetic acid, and glutamic acid (see Fig. 21-2). Glutamic acid can be oxidized to alphaketoglutaric acid. This can either react with alanine to form pyruvic acid or it can enter the Kreb's cycle (as can oxaloacetic acid, since both are molecules in the Kreb's cycle). Pyruvic acid can either go on to form glucose or can form acetyl CoA. Acetyl CoA can go into the Kreb's cycle or to the lipogenesis pathway to form fat. Amino acids can be oxidized to produce energy, transformed into glucose, or stored in the form of fat.

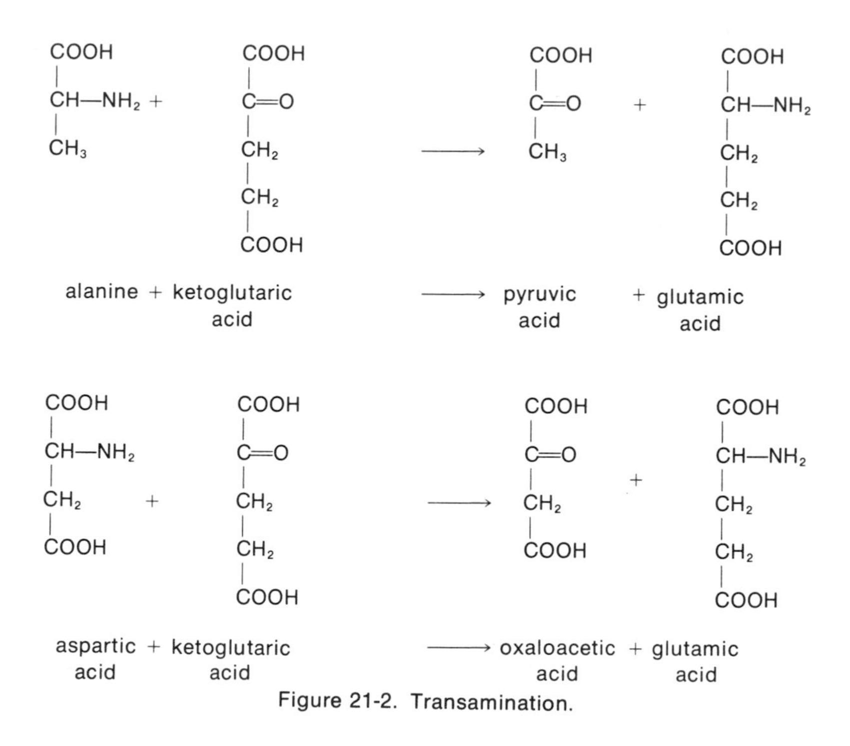

Figure 21-2. Transamination.

Amination is the process by which alpha amino acids are synthesized directly by the addition of ammonia to an alpha keto acid. Some representative reactions are:

The synthesis of glutamic acid from alphaketoglutaric acid (α-KG):

$$NH_3 + \alpha\text{-KG} + NADPH + H^+ \rightleftharpoons \text{Glutamic acid} + NADP^+ + H_2O$$

The synthesis of glutamine from glutamic acid:

$$NH_3 + \text{glutamic acid} + ATP \longrightarrow \text{glutamine} + ADP + P_i$$

Oxidation is a process whereby certain groups of amino acids are selectively oxidized to synthesize other amino acids or their intermediates. A typical example is the oxidation of phenylalanine to tyrosine:

$$\text{phenylalanine} + NADPH + H^+ + O_2 \longrightarrow \text{tyrosine} + NADP^+ + H_2O$$

As mentioned previously, the lack of the enzyme for this oxidation leads to phenylketonuria.

SYNTHESIS OF UREA

In the deamination of glutamic acid, ammonia is formed along with alphaketoglutaric acid. Ammonia is transformed in the human liver to urea, which is excreted in the urine. In this way nature provides a mechanism for the discharge of excess nitrogen to maintain a normal nitrogen balance. The ammonia is not directly converted to urea, but is converted by means of the urea cycle, in which the amino acid, ornithine, initiates the cycle. In fact, only trace amounts of ornithine are needed since it is a catalyst for the reaction (see Fig. 21-3).

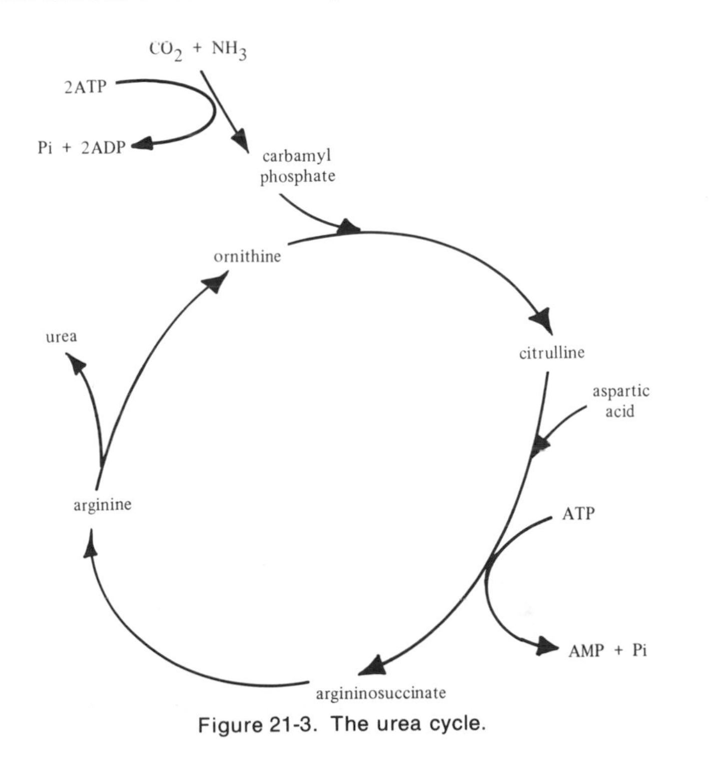

Figure 21-3. The urea cycle.

The energy needed to carry the reaction forward is provided by ATP, which causes CO_2 to react with NH_3 to form carbamyl phosphate and ADP. Then the carbamyl phosphate reacts with ornithine to form citrulline. Citrulline then goes on to produce urea.

The Kreb's Cycle and the Urea Cycle

The Kreb's cycle contains two compounds that enable it to interrelate at two junction points with the transamination products of amino acids. In this way intermediates of either the Kreb's cycle or the urea cycle are able to be metabolized through either cycle as they are needed, this being dictated by existing conditions in the body. The two common intermediates are alphaketoglutaric acid and oxaloacetic acid (Fig. 21-4).

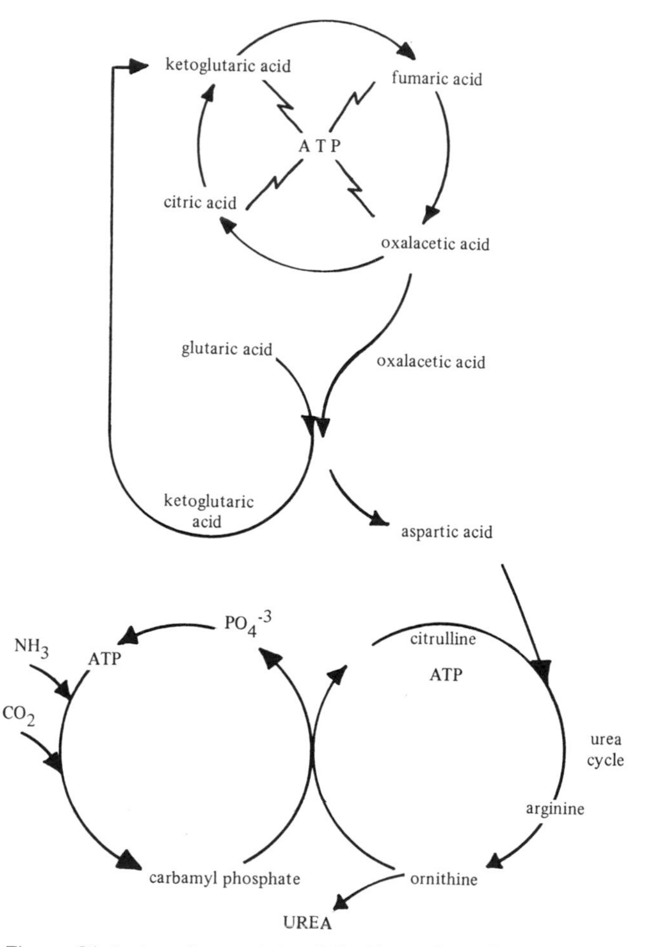

Figure 21-4. Junction points of Kreb's cycle and urea cycle.

SYNTHESIS OF PURINES AND PYRIMIDINES IN THE BODY

The body has the capability to synthesize purines and pyrimidines when these compounds are absent in the diet. If you examine the purine skeleton (Fig. 21-5), you can readily see how that ring structure was formed from the reactants and amino acids that are present in the body during normal metabolism. The pyrimidine ring structure is synthesized by the reaction of carbamyl phosphate and aspartic acid (Fig. 21-6).

THE ROLE OF ACETYL CoA

It should be obvious that acetyl CoA plays a central role in the metabolism of proteins, carbohydrates, and fats. This coenzyme can be formed from pyruvic acid and serves as the entry point in the Kreb's cycle to produce energy. It is also the main product of fatty acid oxidation and is involved in the synthesis of fatty acids and cholesterol. Finally, it is a product of amino acid breakdown. Thus acetyl CoA acts as a link between proteins, carbohydrates and fats; it is through this molecule that excess carbohydrates and proteins are converted to fat.

Another possible path for acetyl CoA is the synthesis of cholesterol. Two acetyl CoA's react to form acetoacetyl CoA, which is converted in the liver to 3-hydroxy-3-methylglutaryl CoA (HMG-CoA). This molecule goes on to form cholesterol and the steroid hormones. The amount that can be used for this purpose is generally limited. Hence HMG-CoA, in another reaction, forms acetoacetic acid, which in turn generates beta hydroxybutyric acid and to a smaller

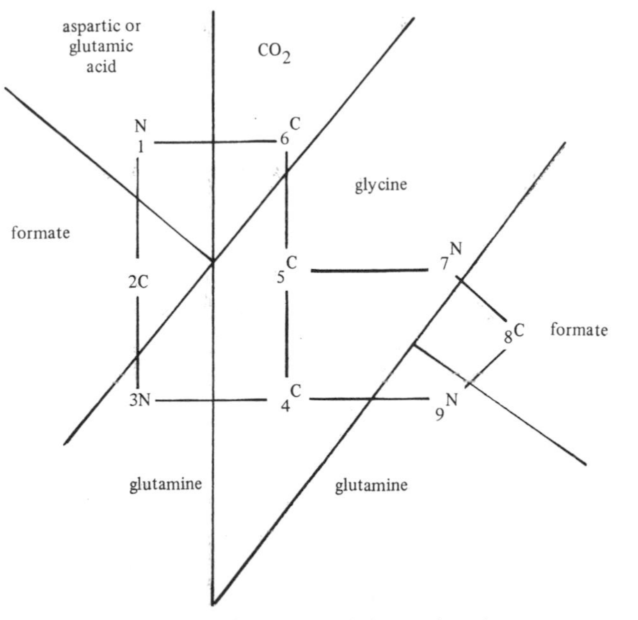

Figure 21-5. Synthesis of the purine ring.

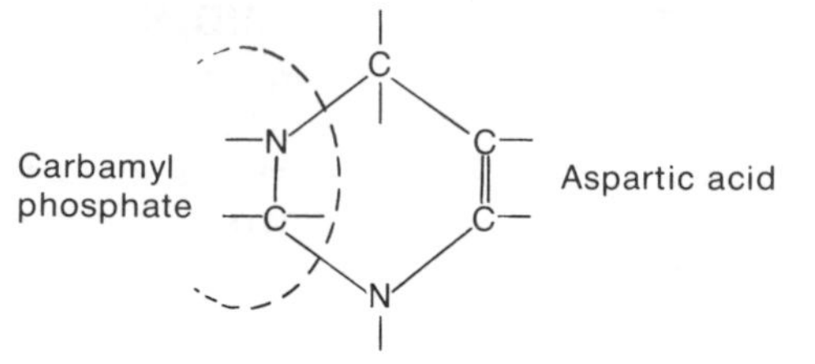

Figure 21-6. Synthesis of the pyrimidine ring.

extent, acetone. These 3 molecules, which can accumulate in the blood, are called ketone bodies.

KETOSIS

The ketone bodies are normally present in small amounts in the blood. In fact the muscles can use acetoacetic acid for a sizable fraction of their total energy needs. In this case, acetoacetic acid must be reconverted to acetyl CoA. Acetone can also be metabolized either by conversion to pyruvic acid or decomposition to a 2-carbon acetyl and a 1-carbon formyl fragment.

Certain conditions may lead to an overproduction of ketone bodies. This occurs when the other pathways of acetyl CoA are restricted or when excessive amounts of acetyl CoA are formed. When ketone bodies are produced in excessive amounts, ketosis results.

An abnormally high concentration of the three ketone bodies in the blood is called *ketonemia.* When the ketone level is high enough, the renal threshold is exceeded. Ketone bodies appear in the urine resulting in *ketonuria.* When both ketonemia and ketonuria exist, the smell of acetone will sometimes appear on the breath. The term *ketosis* describes this combination of ketonemia, ketonuria, and acetone odor on the breath.

Starvation is one possible cause of ketosis. Due to the absence of carbohydrates (the principle source of energy in the body), glycogen is quickly used up and the body must draw upon its reserve of fat. This results in a high concentration of lipids in the blood (liponemia) and an increased production of acetyl CoA. Not all of the acetyl CoA can be used in energy production. The excess acetyl CoA forms ketone bodies. A recently popular diet (Dr. Atkins' diet) causes ketosis because it is rich in lipids and low in carbohydrates.

The principal cause of ketosis is diabetes mellitus. In a diabetic individual glucose is present but cannot be used because insulin is absent. The diabetic has a high level of glucose in the blood (hyperglycemia), but the body cannot make use of it. The situation has been referred to as starvation in the midst of plenty. Storage lipids must be used for energy, generating acetyl CoA in large amounts. This will result in ketosis. In severe cases the blood ketone level may reach 90 mg per 100 ml of blood compared to the normal level of 3 mg per 100 ml.

One of the serious consequences of ketosis results from the acidic nature of two of the ketone bodies, acetoacetic acid and betahydroxybutyric acid. Their presence leads to a decrease in blood pH from the normal value of 7.4, resulting in acidosis (or in this case, ketoacidosis). This serious complication of diabetes mellitus results in diabetic coma and if left untreated, death. Blood pH falls because the ketone bodies are more acidic than carbonic acid, an important blood buffer.

Some substances tend to form ketone bodies. These substances are called ketogenic and include all of the fatty acids and these amino acids: phenylalanine, tyrosine, and leucine. Those substances that do not tend to form ketone bodies are called glucogenic. They include the carbohydrates, the polyhydric alcohols (such as glycerol), and the remaining amino acids.

BLOOD SUGAR

The concentration of blood glucose ranges from 80 to 120 mg per 100 ml of blood. Rarely does it vary from this range. The concentration depends on the following factors:

1. The formation of glycogen in the liver
2. The breakdown of glycogen in the liver to glucose
3. The formation of glycogen in the muscles
4. The utilization of glycogen in the muscles
5. The utilization of blood glucose by other tissues
6. The conversion of excess glucose to fat
7. The amount of carbohydrates ingested
8. The excretion of glucose
9. Gluconeogenesis—the conversion of amino acids to glucose

Should one or more of these regulatory mechanisms fail the blood sugar level will rise or fall appreciably (Table 21-1). Glucose in the blood is the principal means of energy supply to the brain. Any excessive reduction in blood sugar will result in coma.

In starvation the blood glucose level is maintained by the transformatlon of amino acids to glucose (gluconeogenesis). This is accomplished through the deamination of amino acids and the conversion of pyruvic acid to glucose. In starvation the body destroys muscle protein to provide glucose for the brain since fat cannot be converted to glucose.

Insulin, glucagon, and epinephrine play an important part in regulating blood sugar levels. It is sometimes impossible for the liver to completely absorb and remove all of the sugars passing through it. Glucose will enter the blood and cause a rise in the blood glucose level. After a heavy meal, the blood glucose level will rise, but several hours later the level falls back to the standard range. *Hypoglycemia* occurs when the blood glucose level falls below the normal range. *Hyperglycemia* occurs when the blood glucose level rises above the normal range. A temporary increase in the blood glucose level caused by excessive intake of carbohydrates is called alimentary hyperglycemia. When the blood glucose level is above 170 mg per 100 ml glucose appears in the urine. This point is called the renal threshold for glucose. *Glucosuria* (glucose in the urine) is a symptom of a metabolic disorder.

Table 21-1. Normal and abnormal blood sugar levels.

	170	Glycosuria
	160	
	150	Glucose synthesized into fat
	140	
	130	Hyperglycemia: glycogenesis
Blood sugar levels in mg/100 ml blood	120	
	110	
	100	Normal level
	90	Fasting: absorption of glucose
	80	glycogenolysis
	70	
	60	Hypoglycemia: shock
	50	

DIABETES MELLITUS

When a person has both glycosuria and hyperglycemia the condition is called diabetes mellitus. It is caused by the malfunctioning of the pancreas to secrete insulin.

Insulin lowers blood glucose levels by increasing the conversion of glucose into glycogen (glycogenesis) and by regulating the oxidation of glucose by the cells. Insulin is also needed to increase the rate of transport of glucose across the cellular membrane. In the liver insulin acts as a regulator in the formation of glucose-6-phosphate.

The metabolism of the fats is also adversely affected by the lack of insulin because the oxidation products of fats accumulate as ketone bodies. Acetone, detectable by its sweetish odor, is evident on the breath of a diabetic who is not under proper medical supervision. The acidosis that results from a lack of insulin causes diabetic coma. Injection of the correct amount of insulin regulates fat metabolism by increasing fatty acid synthesis and decreasing fatty acid oxidation. However, too much insulin will produce insulin shock. The excess insulin will lower the blood sugar, thus starving the brain. Coma and death will result. Immediate relief can be obtained by providing the patient with glucose to raise the blood sugar level.

AN OVERVIEW OF METABOLISM

Now that you have finished this chapter you may still be confused as to how all these pathways are related and what their relevance is. Let us consider a simple example.

Imagine that you are sitting at home, quietly reading a book and oblivious to your surroundings. Suddenly someone sneaks up behind you and bursts a paper bag. You jump out of your seat, turn around quickly, and yell. You notice that you are breathing heavily. The total elapsed time is 2 seconds.

What happened biochemically? First you hear an unexpected noise. A complex series of reactions (Fig. 21-7) releases epinephrine into your blood. Epinephrine starts the cascade reaction, which activates phosphorylase a. Phosphorylase a hydrolyzes glycogen in muscle cells to glucose-1-phosphate, which enters into the Emden-Meyerhof pathway. This pathway produces ATP, which

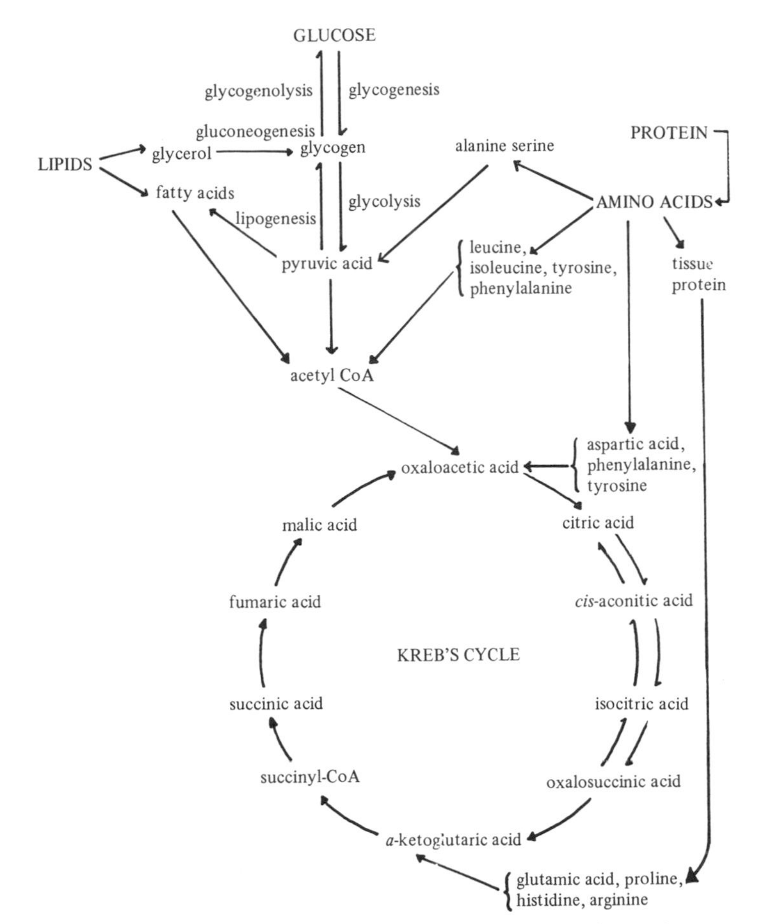

Figure 21-7. Metabolism of carbohydrates, fats and protein.

is used by the muscles for contraction. Therefore, you jump and turn around. Now the accumulated lactic acid is transformed into pyruvic acid, part of which is transported to the liver for transformation to glucose. The rest enters the Kreb's cycle, which requires energy. You gulp in air. The Kreb's cycle releases energy so that you can fight or flee.

SUMMARY

Fatty acids and glycerol can be oxidized to provide energy to the body. Fatty acids are broken down to acetyl CoA in the fatty acid spiral. Acetyl CoA can enter many of the metabolic pathways in the body; it can be used to provide energy, synthesize fatty acids, or produce cholesterol. The body can synthesize nonessential amino acids from the essential amino acids. This is done by transamination, amination, and oxidation. The body eliminates excess nitrogen in the form of urea, which is formed in the urea cycle. The urea cycle has common junction points with the Kreb's cycle. Acetyl CoA is a common interchange point for the metabolism of proteins, fats, and carbohydrates. In this way the body has the capability of substituting available nutrients when others are lacking. In starvation and diabetes mellitus amino acids and fats are used to produce energy instead of glucose. If excessive amounts of acetyl CoA are produced ketone bodies are formed. An excessive amount of ketone bodies in the blood is called ketonemia, while the presence of ketone bodies in the urine is called ketonuria. The normal concentration of glucose in the blood is about 80 to 100 mg glucose per 100 ml of blood. Lower concentrations are hypoglycemic and higher concentrations are hyperglycemic. Glucose in the urine is known as glucosuria.

EXERCISE

1. Diagram the fatty acid spiral.
2. How much ATP does the complete oxidation of 1 molecule of tristerate produce?
3. Are the depot fats in a static or a dynamic state? Explain.
4. How are the metabolism of the fats, proteins and carbohydrates related?
5. Write the reaction for the conversion of alanine to pyruvic acid.
6. Write the reaction for the conversion of glutamine from glutamic acid.
7. What is the reaction for the conversion of phenylalanine to tyrosine?
8. Discuss the relationship of the metabolism of the amino acids and the Kreb's cycle.
9. How is excess nitrogen removed from the body?
10. Diagram the urea cycle.
11. What is the relationship of the urea cycle to the Kreb's cycle?
12. How does the body synthesize purine and pyrimidine rings?
13. Describe the action of insulin on glucose metabolism.
14. What are the metabolic effects of starvation?
15. What are the metabolic effects of diabetes mellitus?
16. Define ketosis, ketonemia and ketonuria.
17. What is the difference between ketogenic and glucogenic substances?
18. Define hypoglycemia and hyperglycemia.

OBJECTIVES

When you have completed this chapter you will be able to:

1. Describe the synthesis of hemoglobin.
2. Show the breakdown products obtained by the degradation of hemoglobin.
3. Define jaundice.
4. Discuss the coagulation of blood.
5. Explain the structure of DNA.
6. Distinguish the forms of RNA.
7. Describe the synthesis of a protein.
8. Summarize the metabolism of drugs in the body.
9. Explain how a synapse works.

There are many metabolic pathways in the body. In the last two chapters we have discussed the principal pathways common to all ingested nutrients. In this chapter we will discuss a few of the more specialized pathways.

METABOLISM III

HEMOGLOBIN

Hemoglobin is a protein molecule with a molecular weight of about 68,000. It is a complex of 4 polypeptide units (globins), each having an iron containing porphyrin skeleton (a heme group) bound to itself (Fig. 22-1). It is synthesized by the following series of reactions:

1. 2 alphaketoglutaric acids + glycine ⟶ 1 pyrrole group
2. 4 pyrrole groups ⟶ protoporphyrin III
3. protoporphyrin III + Fe ⟶ 1 heme group
4. 4 heme groups + 4 globins ⟶ hemoglobin

Hemoglobin carries oxygen to the cells. It has a life span of about 125 days, after which it is decomposed into globin and heme groups. The body removes the iron from the heme group and reuses it. The heme group is decomposed to biliverdin, a bile pigment. Biliverdin then forms bilirubin and is finally excreted as urobilin in the urine, causing the yellow-orange color, and as stercobilinogen in the feces, causing the characteristic color (Fig. 22-2). The globins are hydrolyzed to their amine acid components.

If bilirubin were to accumulate in the blood the skin would take on a yellow color and the sclera (the white of the eye) would take on a yellowish tint. This is known as jaundice and may result from:

1. Excessive destruction of red blood cells (hemolytic jaundice)
2. Liver diseases such as hepatitis and cirrhosis
3. An obstruction in the bile duct causing bile to enter the bloodstream; no bile pigments enter the intestines and the feces have a clay-colored appearance

If the jaundice is not relieved permanent liver damage may result.

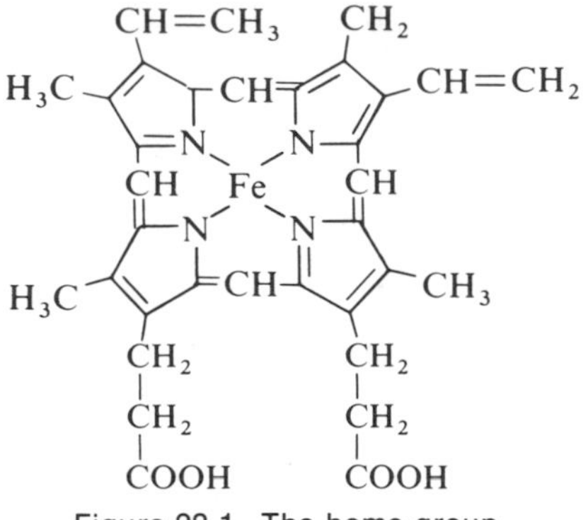

Figure 22-1. The heme group.

THE COAGULATION OF BLOOD

When a blood vessel is ruptured, severed or injured, blood will be lost. However, if the damage is not too great the body has the capability to seal the opening. On contact with injured tissue the blood changes to a jelly-like solid called a clot that blocks the injured area (see Figure 22-3). The exact mechanism of clotting is unknown, but what is known can be summarized as follows:

1. When blood is shed thromboplastin, a cephalin protein, is released from the injured cells.
2. Thromboplastin along with calcium ions activates the enzyme thrombin from its inactive proenzyme form, prothrombin.
3. Thrombin catalyzes the conversion of fibrinogen to fibrin forming a clot.

Blood does not coagulate in the blood vessels because thromboplastin only enters the plasma when tissue is damaged. Another theory involves the presence of antithrombin, an enzyme that inactivates prothrombin and prevents its conversion to thrombin. When blood flows from a cut the platelets disintegrate and neutralize the antiprothrombin. The prothrombin can then be activated and the clot forms. The disintegration of the platelets also liberates thromboplastin and other coagulative activators (such as globulin accelerator and antihemophilic globulin). Hemophiliacs are people who suffer from an inability to coagulate blood because they lack these globulins.

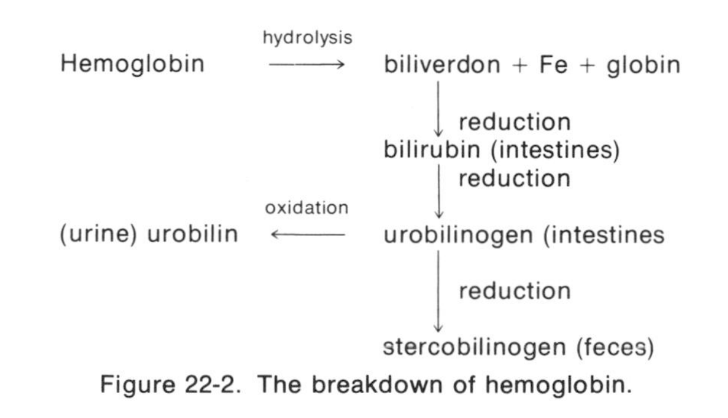

Figure 22-2. The breakdown of hemoglobin.

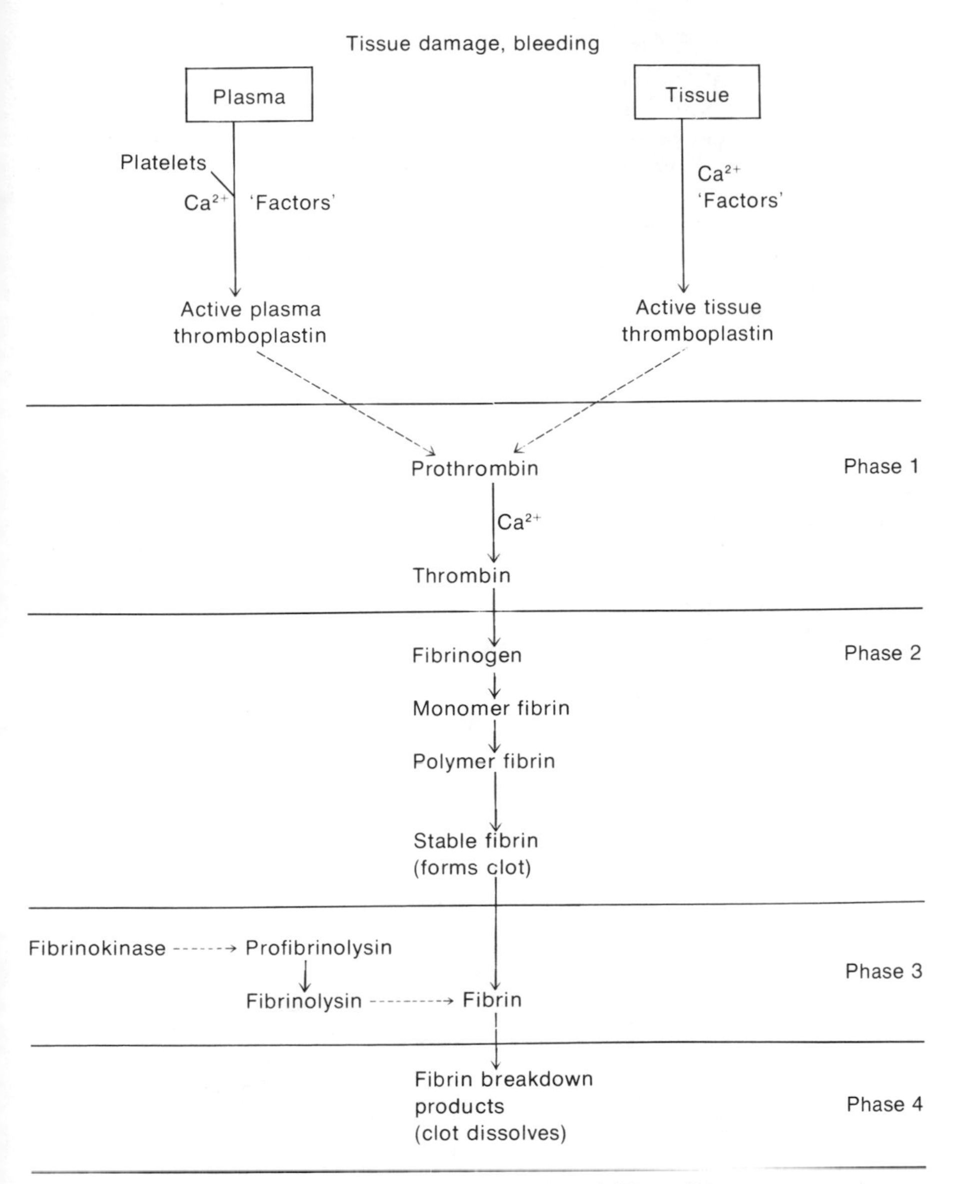

Figure 22-3. Processes involved in the clotting of blood. The solid arrow means *is converted to* and the broken arrow means *acts on*.

Anticoagulants

During surgery a clot may form in a blood vessel but it usually does no harm since blood plasma contains an inactive form of a protein splitting substance called plasmin or fibrinolysin. Upon activation plasmin removes the clot by hydrolyzing fibrin.

DNA—THE CORE OF LIFE

We now know that the basic hereditary material is a substance called deoxyribonucleic acid, DNA. It is a very large molecule containing millions of atoms. DNA dictates heredity because it has the ability of replication, and replication is the basis of life. The genes in the chromosomes are specific parts of the DNA molecule.

The sperm cell and the egg cell are composed of DNA molecules. The union of the sperm and egg results in a combination of the DNA of each cell. This combination results in a unique DNA molecule that contains the blueprint for the newly formed organism. It is the division and recombination of nuclear material containing genes from each parent that accounts for the transfer of hereditary characteristics.

In 1953, James Watson and Francis Crick proposed the structure of DNA, which enables an understanding of how the molecule contains an unlimited amount of genetic information. They proposed that DNA is a molecule composed of two strands helically wound about each other (a double helix), just as a spiral staircase is constructed (Fig. 22-4). There are pentose sugars (deoxyribose) and phosphate molecules on the outside and four types of bases on the inside: guanine, cytosine, adenine and thymine. The DNA has an almost unlimited sequence of these bases to code all of the information needed for the survival of a complex organism.

These bases are always paired on the inside of the double helix. Guanine always pairs with cytosine and adenine always pairs with thymine (see Fig. 22-5). The sequence of bases on one chain of the double helix determines the sequence on the other chain. The two helical chains fit into each other, because a given base on one chain must be opposite a particular base on the other chain. The DNA replicates itself by drawing on the pool of nucleic acids in the cell.

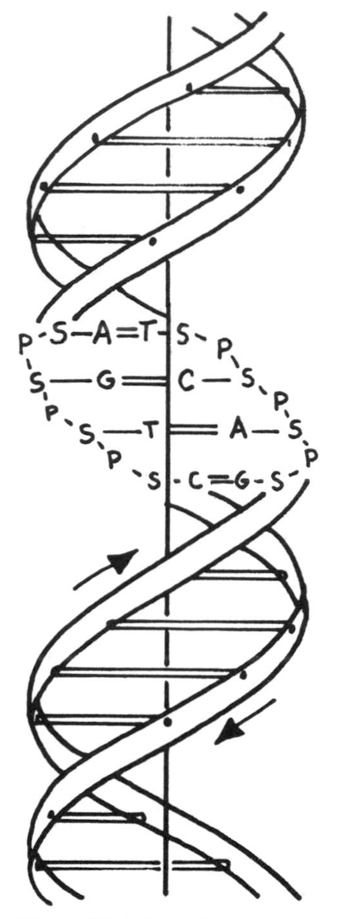

Figure 22-4. The double helix.

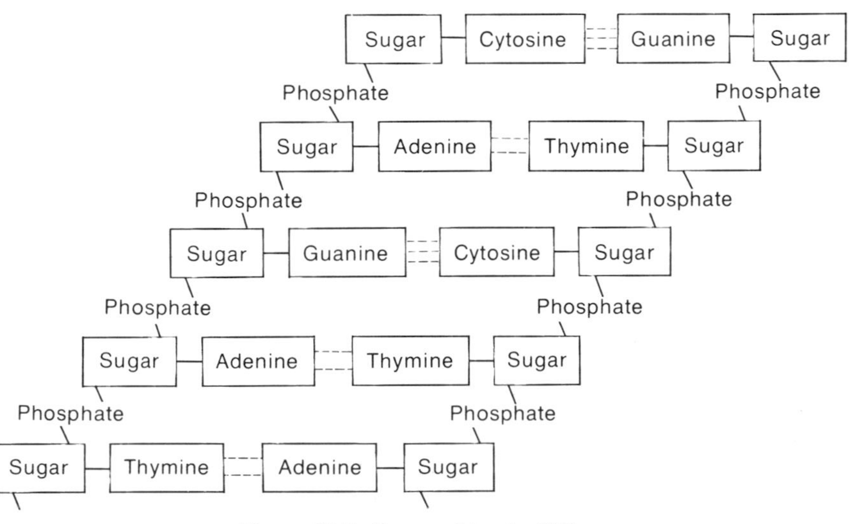

Figure 22-5. Base pairing in DNA.

Biosynthesis of Proteins

The various DNA molecules in the nucleus of a cell form the chromosomes of the cell. A sequence of nucleotides on one chain of the DNA is called a gene. One gene has a code to construct one specific protein or enzyme. The protein or enzyme controls a metabolic process in the body.

Ribonucleic acid (RNA) is needed to construct or synthesize the protein. RNA is similar to DNA, except that ribose substitutes for deoxyribose and uracil substitutes for thymine. Uracil can pair with adenine. There are three types of RNA involved in protein synthesis: messenger RNA (mRNA), transfer RNA (tRNA), and ribosomal RNA (rRNA).

The steps in protein synthesis may be summarized as:

1. The DNA molecule uncoils and forms two chains.
2. Along one of the sections of the chains, nucleotides pair up to form a small strand of mRNA.
3. The mRNA migrates out of the nucleus to the endoplasmic reticulum and attaches itself to a ribosome composed of rRNA (Fig. 22-6).
4. At the ribosome tRNA brings amino acid to the mRNA. As each amino acid is brought it is assembled to the previous amino acid to form a protein.
5. Each tRNA has a three nucleotide code which is specific for a particular amino acid. The sequence of nucleotides on the mRNA determines the order in which tRNA's are brought to the mRNA and hence, the order in which amino acids are assembled to form the primary structure of the protein (see Fig. 22-7).

We can compare the process to the assembly of a car on an assembly line. The DNA is like the blueprint of the car. The mRNA is like the manager of the factory who directs the assembly of the car. The tRNA's are like the workers who put the cars together. And the rRNA is like the assembly line upon which the cars are assembled.

Mutations

Normal replication of DNA results in the production of identical molecules of DNA. However, sometimes factors may alter the sequence of the nucleotides in the DNA. When this happens, the new molecule of DNA is different from the original molecule. This change in nucleotide sequence of DNA is called a mutation. Mutations are responsible for various hereditary diseases such as hemophilia, galactosemia, sickle cell anemia and phenylketonuria. A mutation can result from chemical agents (called mutagens) or from radiations (x-rays or gamma rays). These factors may also cause cancers.

METABOLISM OF DRUGS

Modern medicine uses ever increasing amounts of drugs. For drugs to be effective, they must reach their particular site of action. This site is the particular part of the body where the drug performs its function (see Table 22-1). Once the drug reaches the site of action, we may not know the exact mechanism of action, or how the drug affects the physiological processes in the body. The mechanism of drugs that act on the intracellular level is very complex and we only have a superficial understanding of their action.

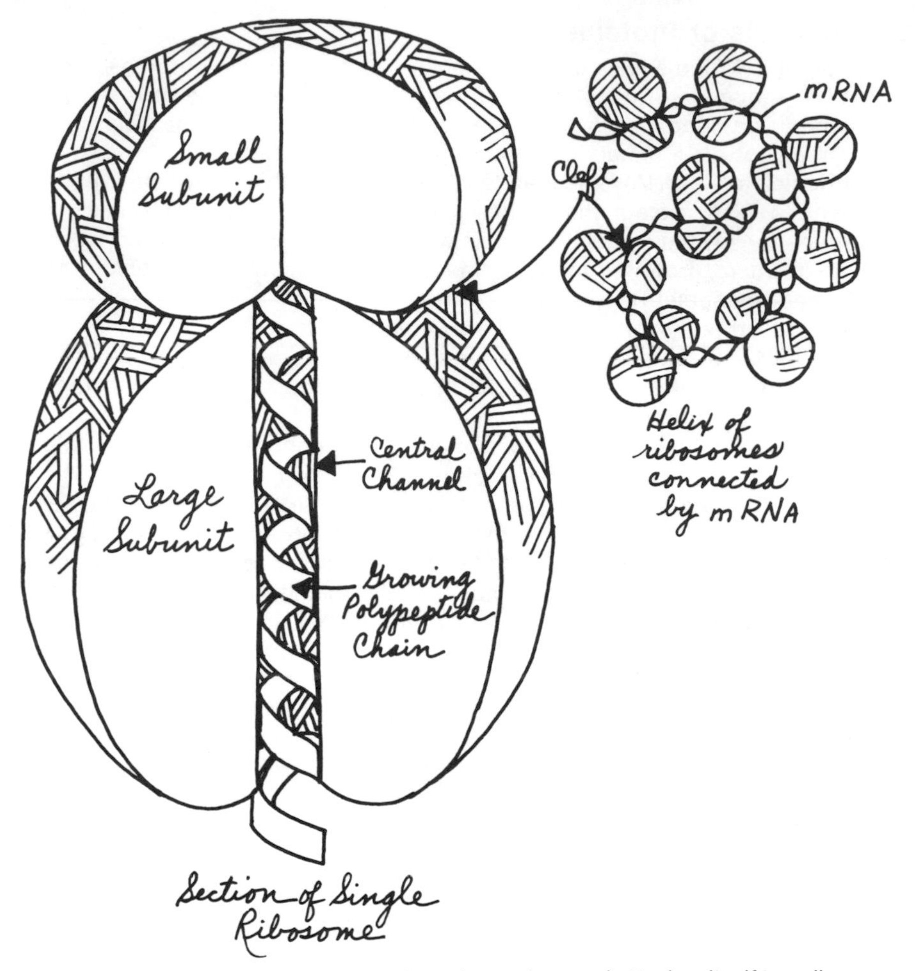

Figure 22-6. Messenger RNA migrates from the nucleus and attaches itself to a ribosome.

Everyone has ready access to drugs because we are a drug-oriented society. Drug abuse is the self administration of drugs in an excessive manner without regard to accepted medical practice. This applies to the average person who habitually takes medications such as headache pills, laxatives or vitamins. The occasional nonmedical or inappropriate use of a drug is called drug misuse. A good example is the use of antibiotics to treat the common cold. The common cold is caused by a virus and antibiotics work only on bacterial infections.

Most drugs entering the body are metabolized by the liver, which has enzyme systems (microsomes) to transform chemicals that enter the body. Some drugs can inactivate these enzyme systems. This is the reason why combinations of some drugs can be more potent than the same drugs taken individually. One such example is the alcohol barbiturate interaction. There is some evidence that alcohol blocks the enzyme system that metabolizes barbiturates. When

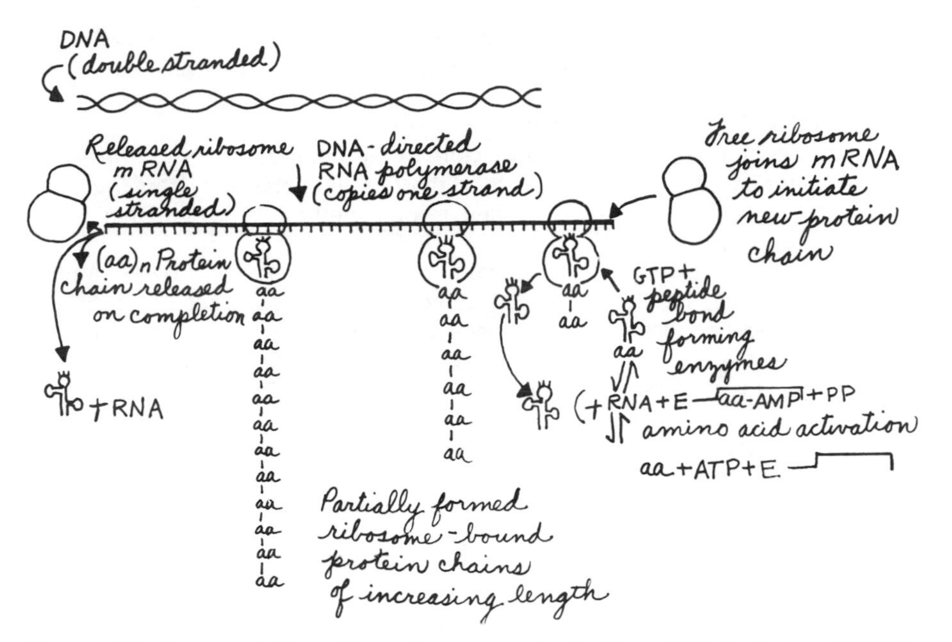

Figure 22-7. Representation of how ribosomes, attached by mRNA to form polysomes, participate in protein synthesis.

Table 22-1. Steps drugs undergo before reaching the site of action.

Step	*Amount of drug*	*Location in the body*	*Process drug undergoes*	*Effect on drug*
1	○	Mouth	Total drug in pill form is swallowed	—
2	○	Stomach	Drug dissolved by stomach fluids	Some drug is not dissolved. It is excreted. Some is degraded by fluid
3	○	Intestine	Drug is in solution	Some drug is lost by degradation or by binding to food
4	○	Villi	Drug is absorbed	Some drug is not absorbed and is excreted
5	○	Liver	Metabolism of some drug	Drug lost by secretion into bile. Some lost by biotransformation
6	○	Blood	Drug distributed to entire body	Some drug lost by biotransformation. Most drug is distributed to other tissues other than site of action. Some drug lost by excretion
7	○	Site of action	Drug affects body	—

they are taken in combination more of the barbiturate reaches the active site and in effect, the alcohol increases the potency of the barbiturate.

Drugs Affecting the Nervous System

The synapse is a gap between nerve cells in the nervous system. In order for messages (or impulses) to be transmitted through the nervous system they must bridge the synaptic gap. Acetylcholine is one method by which this gap is bridged.

Acetylcholine is stored in the presynaptic membrane. When a nerve impulse travels to the presynaptic membrane, the acetylcholine is released into the synaptic gap. The stronger the impulse the more acetylcholine is released. The acetylcholine binds to receptor sites on the postsynaptic membrane, and the nerve impulse is regenerated along the next nerve cell (Fig. 22-8). The enzyme acetylcholinesterase breaks down the acetylcholine freeing the receptor sites for the next nerve impulse.

Drugs can affect the impulse transmission in one of three ways:

1. Some drugs block the synthesis of acetylcholine; hence, the signal cannot be transmitted. Other drugs, such as heroin, block the transmission of acetylcholine. The body then compensates by using a stronger impulse to move more acetylcholine across the gap. When an addict stops taking heroin the transmission of acetylcholine is increased. This causes the withdrawal symptoms of narcotic addiction.
2. Some drugs block the receptor site to prevent the acetylcholine from binding with the receptor site. Therefore the impulse is not transmitted. Drugs that act in this manner are curare, atropine, and certain local anesthetics.
3. Some drugs inhibit the enzyme acetylcholinesterase. In this case acetylcholine stays on the receptor site and continues the transmission of the impulse. Nerve gases act in this way. They result in the loss of essential control of the nervous system, which can lead to death due to respiratory failure.

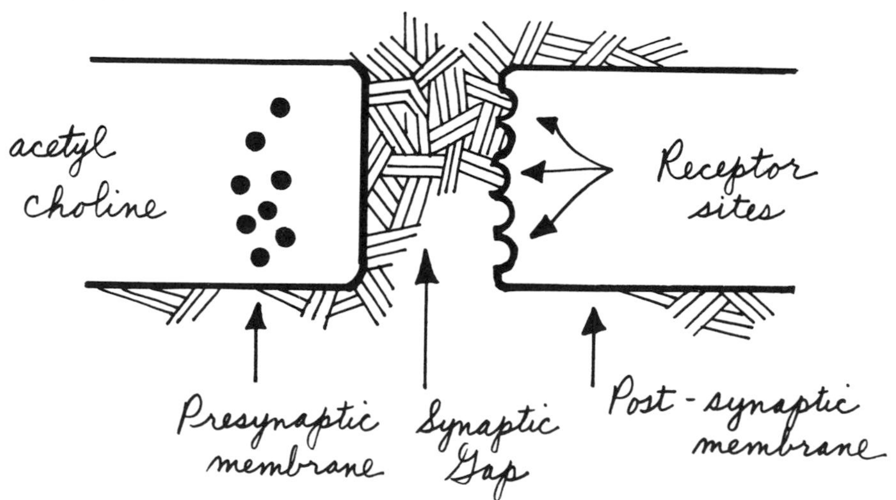

Figure 22-8. The synapse.

SUMMARY

Hemoglobin is synthesized from 4 heme groups and 4 polypeptides, called globins. It carries oxygen to the cells. Hemoglobin is broken down in the liver to bilirubin and other products which are excreted in the urine and feces. If bilirubin accumulates in the blood we have a condition known as jaundice. When blood vessels are damaged a complex series of reactions occurs that catalyzes the formation of a blood clot. DNA is the basic hereditary material of life. It has the ability to replicate itself. There are three forms of RNA: messenger RNA, transfer RNA, and ribosomal RNA. All of these are important in the synthesis of proteins. The sequence of nucleotides in RNA determines the order of amino acids in a protein. If the sequence of nucleotides is altered we have a mutation. In order to be effective drugs must reach their site of action. The synapse transmits impulses through the nervous system. Drugs that affect the nervous system alter the transmission of nerve impulses through the synapse.

EXERCISE

1. How is hemoglobin synthesized?
2. Diagram the breakdown of hemoglobin in the body.
3. What causes jaundice?
4. Define fibrin.
5. What is the function of calcium in blood coagulation?
6. What are the steps in blood coagulation?
7. What is the structure of DNA?
8. What is the difference between DNA and RNA?
9. Define mRNA, tRNA, and rRNA.
10. What are the steps in protein synthesis?
11. What is a mutation?
12. What causes mutations?
13. Define the site of action of a drug.
14. Why is alcohol and barbiturates a potent combination?
15. How does the synapse function?
16. How does heroin affect the synapse?

OBJECTIVES

When you have completed this chapter, you will be able to:

1. Describe the environment to which the mouth is subject.
2. Discuss saliva, its sources, and its physical and chemical composition.
3. Explain the action of ptylin.
4. Draw a diagram of a tooth, label its components and give the chemical composition of the components.
5. State the effects of vitamin A, C and D on the teeth.
6. Describe the organic matter of teeth.
7. Summarize the formation of calculus from plaque.
8. Write the basic reactions involved in the destruction of the apatite crystal of the enamel and the basic equation for dental caries.
9. Explain the two theories of the cause of dental caries.
10. Discuss the fluoride inhibiting theory and the pros and cons of drinking water that contains fluorides.
11. Identify the basic formulation for the silver-tin-mercury amalgam and discuss trituration, condensation and hardening of the amalgam.
12. List the functions of the components of an amalgam.
13. State why gold foil is used in dental restorations and why it is used.
14. Define dental alloys and describe the space lattice of the solute and solvent metals.
15. Enumerate the contributions of other metals to gold alloys.
16. Specify the difference between tarnish and corrosion and describe the two types of corrosion.
17. Describe the acrylic resins, and state the effect of varying their side chains on their physical properties.

CHEMISTRY OF THE MOUTH AND DENTAL MATERIALS

18. Discuss silicate cements.
19. Write the balanced chemical formula for the formation of plaster of Paris.
20. Identify the zinc oxide-eugenol reaction.
21. List the reasons why calcium hydroxide is used as a base.
22. Name the three basic sealants.
23. Explain the basic difference between composite and conventional resin filling materials.
24. Identify reversible and irreversible hydrocolloids.
25. Write the general formula of the rubber impression materials.
26. Discuss the quaternary ammonium salts as antiseptic agents.
27. Discuss (a) the sequence of events involved in developing x-ray negatives, (b) the chemical reactions of the exposure and development and (c) the fixing of the negative.

The environment of the mouth is subjected to drastic changes. The temperature can change instantaneously from that of freezing water (0°C) to hot coffee (75°C). The pH can fluctuate rapidly from acidity to alkalinity depending upon the food being eaten. The pressures being exerted on the teeth can exceed thousands of pounds per square inch. The gums, soft tissues and the teeth themselves may be easily injured by irritants. Furthermore, almost everyone at some time must go to the dentist for restorations and insertions of prosthesis or orthodontic appliances for tooth movement.

It is important to know how the environment of the mouth changes, what those changes are, and how those changes affect the biochemistry of the digestion of the food and their effect upon the teeth, tissues, and dental materials.

SALIVA

Saliva is an almost colorless secretion of the parotid, submaxillary (submandibular) and sublingual glands, containing about 99.3 percent water with a specific gravity between 1.002 to 1.008. Daily secretion is about 1,500 ml. The pH of resting saliva varies from 6.4 to 6.9 (slightly acidic), whereas during active stimulation, the pH increases slightly to the basic side, 7.0 to 7.3.

Saliva is very viscous, containing albumins, globulins, mucins, enzymes, urea, uric acid, and various inorganic salts. The primary functions of saliva are: (1) to moisten and lubricate the food so that it can be easily swallowed, (2) to dilute acids and salts to protect the mucosa and teeth, (3) to complete the iodine cycle in the deiodiation of the hormone thyroxine by releasing the iodine for absorption in the small intestine and (4) to supply the principle enzyme ptylin (which is an amylase) and also a maltase, a catalase, a lipase, and a protease.

The principle enzyme, ptylin, acts in the optimum pH range of 5.6 to 6.5, but it can be effective through the pH range of 4 to 9. Ptylin acts solely on the polysaccharides, but it does require certain ions (chlorides, bromides, iodides, nitrates, sulfates, phosphates or acetates) as activators. There are two amylases, the alpha and beta. Both act upon the alpha 1-4 glucoside linkages, splitting the polysaccharide into maltose and glucose units. The chief end products of the salivary digestion of starch are various oligosaccharides, small amounts of maltose and some glucose (hydrolyzed from maltose by the maltase present).

The buffering power of the saliva will vary due to changes in pH. The most effective buffer system is carbonate-bicarbonate, because the enzyme carbonic anhydrase catalyzes the reaction:

$$CO_2 + H_2O \rightleftharpoons H_2CO_3$$

The mucins also buffer the saliva through the salt-acid relationship:

$$\frac{\text{Salt}}{\text{acid}} = \frac{\text{Na Mucin}}{\text{H Mucin}}$$

TEETH

The teeth resemble bone. Figure 23-1 shows the names, composition and location of the various components. The upper surface of the tooth is covered by *enamel.* It is the hardest substance of the body and is used to grind and masticate the food. It is composed of 5 percent water and 95 percent inorganic material embedded in an organic matrix resembling keratin. The inorganic material is a calcium phosphate hydroxyapatite with the formula $Ca_{10}\,(PO_4)_6(OH)_2$. The remaining bulk of the tooth is *dentin*, which is chemically identical to bone structures of the body, cementum (consisting of 30 percent organic matter and water and 70 percent inorganic material), and pulp with its arteries, veins, and nerves.

To insure proper calcification of teeth the diet must contain enough calcium, phosphorus, vitamins and hormones. Developing teeth rapidly take up phosphorus until they are completely formed and calcified. The calcium of the teeth is not utilized in the same way as the calcium of the bones. Proper tooth development and calcification requires vitamins A, C and D. Vitamins A and C are required for the proper functional activities of the formative cells. Hyperplastic (imperfectly calcified) enamel results from a deficiency of vitamin A. Imperfect calcification of the dentin results from a deficiency of vitamin C. The absorption

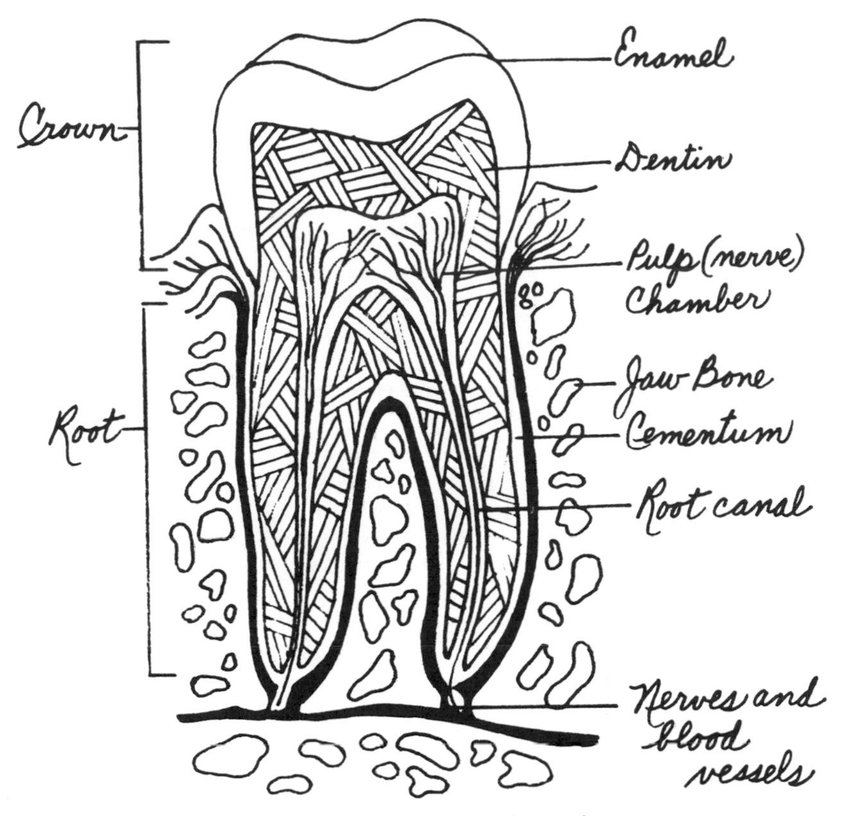

Figure 23-1. Structure of a tooth.

of calcium and the deposition of calcium and phosphorus in the teeth requires vitamin D.

Vitamin A deficiency results in a defective formation of the enamel with the consequent exposure of the more vulnerable dentin. It is believed that vitamin D indirectly prevents caries because it improves the general health and nutrition of the individual. It plays a direct part in the absorption of calcium and an indirect part in that of phosphorus in the body.

The Organic Matter of Teeth

Collagen is the main protein constituent of most connective tissues in animals, found in the form of microscopic fibers in the matrix of the dentin. When hydrolyzed completely collagen breaks down to 18 amino acids, including glycine, proline, and hydroxyproline.

The enamel also contains an organic framework which has been shown by extremely slow decalcification of specimens. Rapid decalcification will destroy organic matter due to the removal of the supporting inorganic salts. The protein of the enamel has been isolated by dialysis in acid solutions where the semipermeable membrane prevents the passage of the large protein molecules. The acid diffuses through the membrane, dissolves the mineral matter of the enamel, and then diffuses out, leaving the protein behind (see Table 23-1). The organic matter is distributed in the tooth as follows:

Enamel	1% (mainly keratin, some cholesterol and phospholipids)
Dentin	18% (collagen and lipids)
Cementum	23% (collagen)

Table 23-1. Chemical contents of enamel, dentin, cementum, and bone.

	Enamel	*Dentin*	*Cementum Compact bone*
Water	2.3%	13.2%	32%
Organic matter	1.7	17.5	22
Ash	96.0	69.3	46
In 100 gm of ash:			
Calcium	36.1 gm	35.3 gm	35.5 gm
Phosphorus	17.3	17.1	17.1
Carbon dioxide	3.0	4.0	4.4
Magnesium	0.5	1.2	0.9
Sodium	0.2	0.2	1.1
Potassium	0.3	0.07	0.1
Chloride	0.3	0.03	0.1
Fluorine	0.016	0.017	0.015
Sulfur	0.1	0.2	0.6
Copper	0.01		
Silicon	0.003		0.04
Iron	0.0025		0.09
Zinc	0.016	0.018	
	Whole teeth		Bone
Lead	0.0071 to 0.037		0.002 to 0.02

Plus small amounts of: Ce, La, Pr, Ne, Ag, Sr, Ba, Cr, Sn, Mn, Ti, Ni, V, Al, B, Cu, Li, Se.

Calculus

Calculus, also known as tartar, accumulates on the teeth and is composed mainly of calcium carbonate ($CaCO_3$) and calcium phosphate, $Ca_3(PO_4)_3$. Around the nuclei of materia alba (food particles) these salts precipitate along with calcium oxalate because of increased acidity. Calculus begins with the formation of a bacterial mucinous plaque that later becomes calcified. The calcification of this plaque is in part controlled by changes in the physiochemical properties of saliva. Changes in the salivary proteins, protein-carbohydrate complexes and a decrease in salivary pH resulting from a loss of CO_2 will cause precipitation of the calcium salts. The presence of calculus as a destructive factor in peridontal disease has been confirmed.

Dental Caries

Dental caries (cavities) result from the breaking of enamel and the exposure of the underlying dentin. No one knows the exact cause, but dental caries is one of the most widespread diseases.

The enamel and the other hard surfaces of the tooth are dissolved by chemical action and they are washed away (Fig. 23-2). Therefore, a cavity is formed that not only produces pain and discomfort, but also interferes with mastication and proper nutrition. Furthermore toxins may be absorbed or secondary infections may be initiated in other parts of the body. The enamel contains 3 vulnerable ions in the apatite. They are OH^-, PO_4^{3-} and CO_3^{2-}. These ions react with the hydrogen ions of acids formed to destroy the apatite crystal, thereby destroying the enamel. The reactions may be expressed by these basic equations:

$$OH^- + H^+ \longrightarrow H_2O$$
$$PO_4^{3-} + 3H^+ \longrightarrow H_3PO_4$$
$$CO_3^{2-} + 2H^+ \longrightarrow H_2CO_3$$

The acids are formed as a result of the hydrolysis of starches by enzymes to yield glucose. The glucose, on bacterial fermentation, is converted to lactic acid. An extremely simplified equation for dental caries is shown below:

$$Ca_{10}(PO_4)_6(OH)_2 + 8\ (H^+\ \text{lactate}^-) \longrightarrow 4\ Ca^{++}(\text{lactate}^-)_2 + 6\ CaHPO_4 + 2\ H_2O$$

There are two theories as to the possible cause of caries:

1. Caries always begin on the external enamel. It is a local action caused by imperfect enamel structure. Since the formation of the teeth begin in fetal life the nutritional intake of the mother is extremely important. The proper intake of compounds of calcium, phosphorus and fluorine, and of vitamins A, C and D by the mother determines the perfection of the tooth structure and therefore the lack of caries.
2. Carbohydrate foods, pastries, candies and cereals become lodged between teeth and are fermented by bacteria to acids. The acids destroy the enamel and lead to the formation of caries. Dentists advise cleaning teeth after eating to minimize the bacterial growth on the surface of the teeth, which disintegrate the lamellae (the flattened bands of organic protein-containing matter that extend through the enamel). Since the enamel is destroyed by acids it would seem logical that the immediate removal of the food particles by cleaning minimizes bacterial acid formation by inhibiting bacterial growth and therefore reducing caries.

In a simplified form, the process of dental caries appears to depend on an interplay between tooth composition, diet and microbial flora. Since all carious lesions are initiated under dental plaques, well controlled studies on the role of salivary factors on the formation, composition and permeability of the dental plaque are needed to clarify the process.

ACIDIC BEVERAGES

Carbonated soft drinks that contain appreciable quantities of phosphoric acid have a markedly destructive action on the tooth enamel and dentin. Fortunately the stimulated flow and the buffering action of the saliva tends to minimize the deleterious effect. Excessive intake of citrus fruits and drinks containing citric acid can cause extensive damage to the tooth.

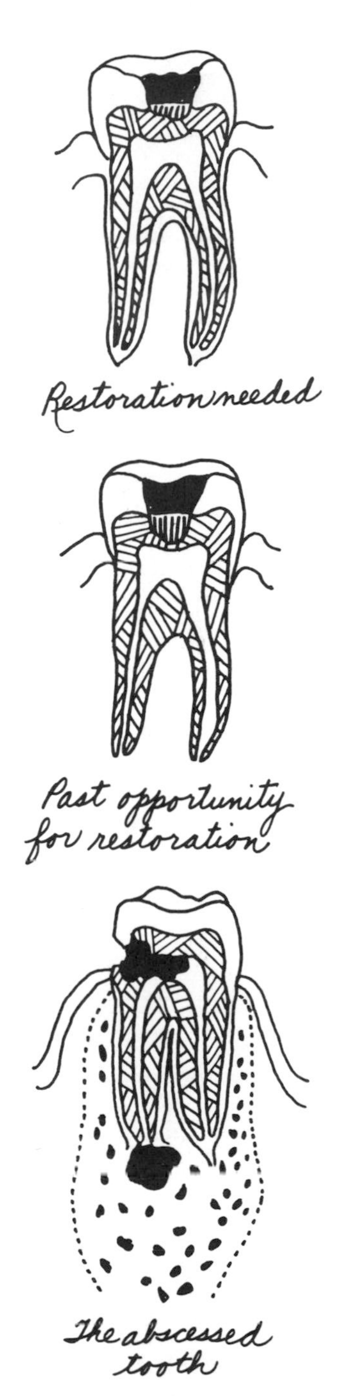

Figure 23-2. Types of dental caries.

FLUORIDES

Elemental fluorine is a gas, but in combination with alkali and alkaline earth metals it forms salts. The uptake of fluorides by the teeth is shown in Figure 23-3. In 1916 a research project initiated in Colorado Springs indicated that people drinking water containing fluoride had a significant decrease in dental caries when compared to those drinking water deficient in the fluoride ion. In 1938 the inhibitory effect of the fluoride ion on dental decay was proved by research. Since that time there has been considerable controversy over the proposals to add fluoride to the drinking water of communities.

Problems do occur when excess fluoride is added to drinking water because a condition known as mottled enamel occurs. The enamel is not only mottled but also becomes brittle, allowing caries to occur. However, the enamel seems more resistant to caries when the fluoride content of the drinking water is not sufficient to cause mottling and ranges from 0.7 to 1.2 parts per million. The dental profession advocates the topical application of fluorides to the surface enamel by professional treatment and at home in tooth pastes and cleansers. Naturally, fluorides should be ingested by the mother during fetal development and by children during the formative years for maximum development of tooth and enamel structure.

Fluorine is the most electronegative element in the periodic chart. Because of this the fluoride ion will replace the OH^- in the hydroxyapatite to form the fluoroapatite:

$$Ca_{10}(PO_4)_6OH_2 \longrightarrow Ca_{10}(PO_4)_6F_2$$

Formation of the fluoroapatite reduces the solubility of the apatite crystal and makes the hydroxyapatite less soluble in the presence of acid (H^+). Topical fluoride ions attach themselves to the surface of the tooth (see Fig. 23-4). Once the teeth have erupted, fluoridated drinking water and topical fluoride applications are significantly beneficial.

Administration of Fluorides

Fluorides can be ingested through fluoridated drinking water by home water fluoridation or fluoride-containing medication. They can be externally ingested through topical fluoride application by fluoride containing dentrifrices or fluoride prophylatic paste.

CAUTION: Excessive fluoride ingestion has a deleterious effect on the teeth. During childhood this is evidenced by calcification, where the teeth have white patches instead of the normal glistening appearance. Pitting also occurs and it is believed that the presence of excessive fluorine inhibits phosphorus and calcium metabolism, because excessive fluorine does inhibit glycolysis.

Fluoride Inhibiting Theory

Fluorine is essential to the development of the tooth structure because its presence results in the formation of less soluble complexes that are less affected by acids. The presence of fluorides may also act as an enzyme inhibitor, reducing the fermentive bacterial actions and the formation of the destructive

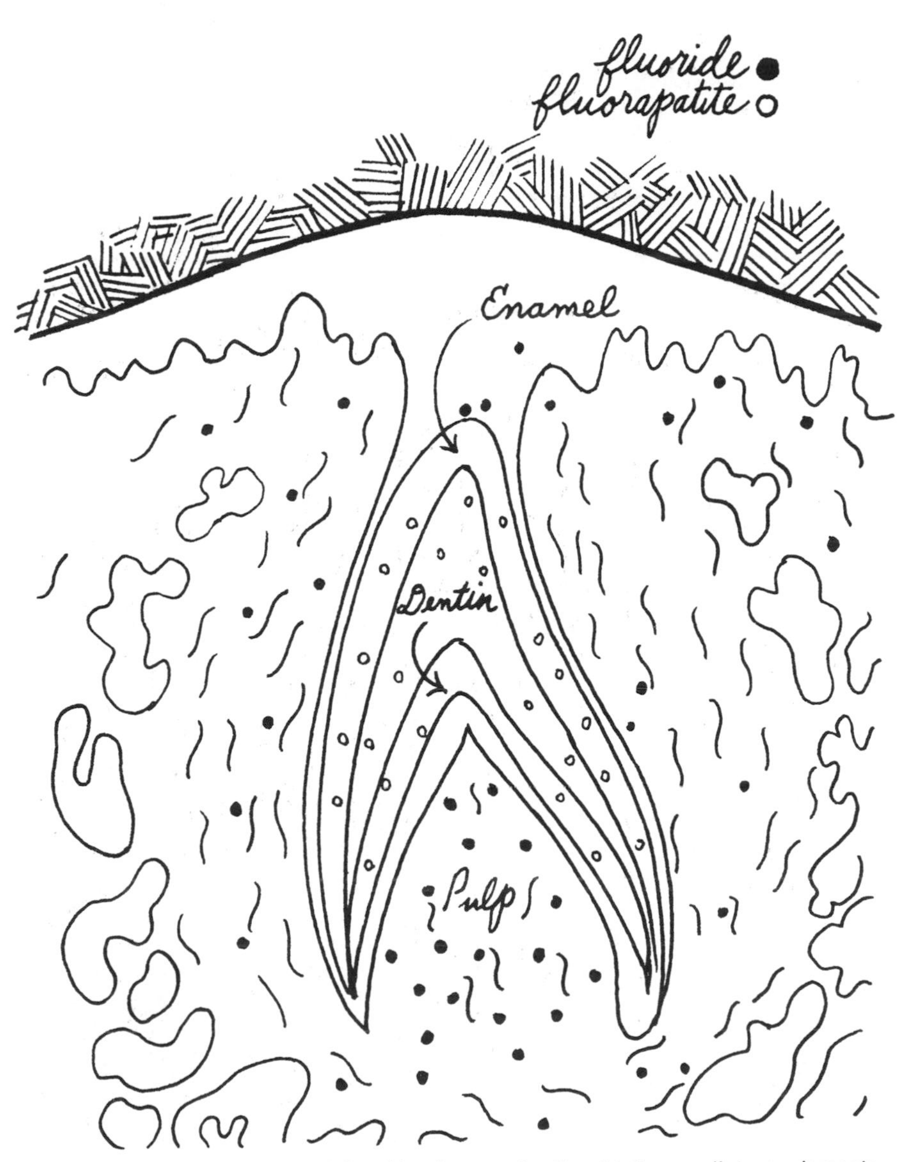

Figure 23-3. The uptake of fluorides from water fluoridation or diet supplements.

organic acids on the enamel and dentin. As a necessary component in the development of tooth structure the fluorides permanently modify the chemical structure of the enamels from the hydroxyapatites to the resistant fluoroapatite.

Apatites are double salts consisting of calcium phosphate and another calcium salt which might be the hydroxide, fluoride, chloride, oxide, sulfate or the carbonate:

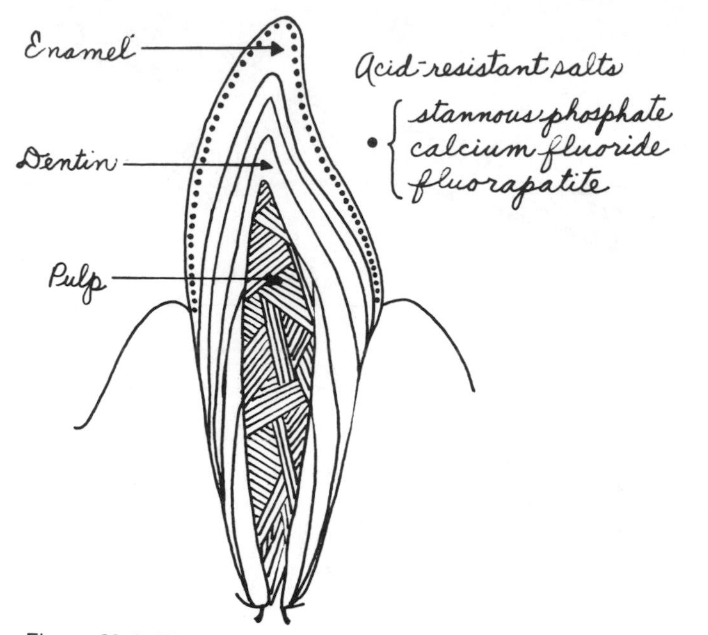

Figure 23-4. The effect of topically applied stannous fluoride.

$$Ca_3(PO_4)_2\text{—}CaX$$

where X may be the $(OH)_2$; F_2; Cl_2; O; SO_4; CO_3

Fluoride concentrations have been found to be 10 times higher on the outer surface of the enamel than in the whole enamel. On unerupted teeth that have not been in contact with fluids of the mouth the surface of the enamel is also richer in fluorides, but not as high as that of erupted teeth. These findings indicate that the increase in fluoride on the surface results from a reaction with fluoride-containing liquids in the mouth. The solubility of the enamel in mineral acids is reduced by treatment of the enamel with fluoride solution. Even when topically applied there is a chemical reaction on the surface enamel to form the protective calcium fluoride salts which are resistant to the destructive action of the fermented acids.

CONTROL OF DENTAL CARIES

The multiplicity of factors that cause dental caries make control of this disease very difficult. Our knowledge on the prevention of hyperplasia, modification of the chemical structure of the teeth to increase resistance, and modification of the protective properties of saliva is very scanty and uncertain. When you change one factor, such as reducing the carbohydrate intake, you may still have no effect on caries because the balance may already be heavily weighted in another direction by unknown factors. The discovery of the anticaries action of fluorides prompted researchers to look for a single factor that would dominate all others and control caries. If one is found that could be practically applied we would be able to control this disease.

THE CHEMISTRY OF DENTAL MATERIALS

Dental materials are those that are employed in mechanical procedures of restorative dentistry, such as crown and bridge, prosthetics, operative dentistry, and orthodontics. When you consider the destructive and corrosive effects of water, air, and everyday substances on homes, cars, buildings, monuments, bridges, metal, plastics, cement, marble, steel and minerals, you can readily appreciate the destructive environment of the mouth. And because of the drastic conditions that dental materials are subjected to, perfect restorative materials have not yet been developed, although dental research in the basic field of chemistry has produced vastly improved compounds.

Alloys

Alloys are combinations of two or more metals that are mutually soluble in the molten condition. An alloy is called a solid solution. The components may be present in varying proportions. The term alloy system is the aggregate of the two or more metals in all of their possible combinations that are being considered as a whole. The solvent metal atoms occupy more than one half of the total number in the space lattice, the solute metal being the other. If two metals are present a binary alloy exists; three metals form a ternary alloy, and so on. The atoms are miscible in the solid state, intermingling randomly in the space lattice. The alloys resemble pure metals.

Restorative Dental Materials

The silver-tin-mercury amalgam is used for about 80 percent of the restorations of lost tooth structure. An amalgam is a special type of alloy, in which the liquid molten mercury is combined or amalgamated with a silver-tin alloy. That process of alloying is known as amalgamation, and the alloy is known as dental amalgam alloy. When the alloy is mixed with mercury in a mortar or automatic electric amalgamator trituration occurs, which results in a plastic mass similar to the melt of any alloy at a temperature between the liquids and the solids. That plastic mass is forced into a prepared cavity by a process known as condensation (Fig. 23-5). In the cavity certain metalographic changes take place, with the production of new phases bringing about hardening or setting of the amalgam. The most successful amalgam alloys contain 67 to 70 percent silver, 25 to 27 percent tin, 0 to 5 percent copper and 0 to 1.7 percent zinc.

Mercury is known to be toxic, but a toxic reaction from traces of mercury penetrating the tooth is highly improbable. The mercury allows the plastic mass to be inserted and finished in the tooth before hardening to a structure that will withstand the stress and rigors of the oral environment.

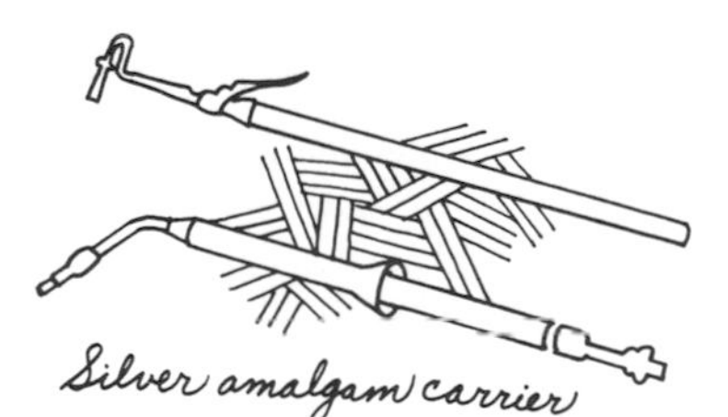

Figure 23-5. Equipment used in restorative dentistry.

Functions of Alloy Components

Each ingredient in the alloy serves a definite function.

Silver increases the strength of the final restoration, decreases the ultimate flow, resists tarnish, and hastens the time of hardening. It increases the expansion of the amalgam thereby sealing the margins of the preparation more effectively. It prevents oral fluids and contaminants from seeping into the space between the tooth margins and the restoration.

Tin reduces the expansion of the alloy. It increases the length of the hardening time so there is adequate time for amalgamation, condensation, and carving the final restoration. It has a marked affinity for mercury and therefore makes the amalgamation process easier.

Copper increases the expansion, strength, and hardness of the amalgam while reducing undesirable flow.

Zinc contributes to the cleanliness of the alloy because it is a scavenger for oxides. In addition it renders the amalgam easier to work.

Platinum increases tarnish and corrosion resistance and is a better hardener and strengthener than copper.

Palladium is used to replace platinum because it is cheaper. It is an active constituent of the white golds of dentistry.

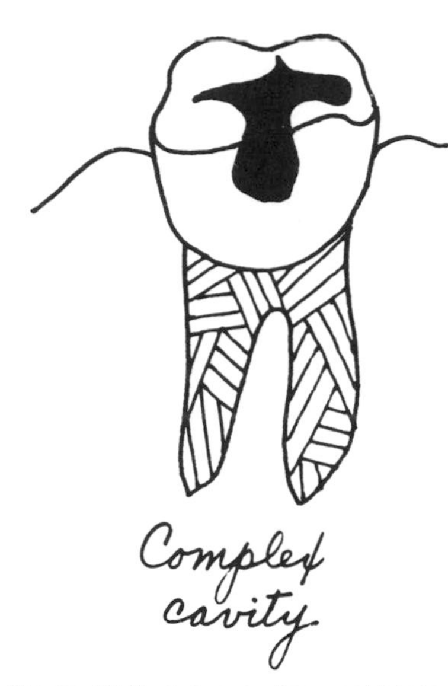

Figure 23-6. A two surface gold inlay.

GOLD

Pure gold is a noble metal. It will neither tarnish nor corrode in the mouth and when properly used it is unsurpassed in operative and restorative dentistry. In form of gold foil, it is extremely soft and very malleable. When its surface is clean and free from absorbed gases and other impurities it can be welded at room temperature because it is extremely cohesive. The gold foil used is driven into prepared cavities gradually by vibratory impaction varying from 400 to 3500 vibrations per second. The gold is therefore gradually built up in the prepared cavity into a coherent mass because the energy absorbed by the impactions causes the gold to weld. The retention is obtained by the extension of the dentin cavity walls. Cast gold restorations include inlays, crowns and partials (Fig. 23-6).

Tarnish and Corrosion

Any metal or alloy that is to be placed in the mouth should not tarnish or corrode. *Tarnish* is a surface discoloration or a slight alteration on the surface finish or luster of the metal. Tarnish is acceptable provided it is not excessive. Tarnish is usually associated with the formation and accumulation of soft and hard deposits on the surface of the restoration. Calculus is the hard deposit. When both tarnish and corrosion are present in noticeable amounts they lead to a loss of esthetic qualities and alteration of the physical properties of the alloy. *Corrosion* is not merely a surface deposit. The metal may disintegrate and dissolve by the chemical reaction of water, gases of the atmosphere, acids, bases, and other chemicals. Sulfur compounds are probably the most important destructive chemicals. The high sulfur content of foods account for marked corrosion of metal dental materials. There are two types of corrosion:

Chemical corrosion is due to the direct combination of a metal and a non-metal. Sulfurization is a good example.

Electrolytic corrosion invariably accompanies chemical corrosion and it is caused by the flow of electric currents, usually involving the replacement of hydrogen from water or acids. The electromotive force series which you studied earlier is the basis for this corrosion. Two different metals in an electrolyte solution create an electric couple and the most active metal goes into solution. Of course, the tendency for the metal to dissolve and form ions depends on the relative position of the two metals in the electromotive series, the environment and the concentration of ions present in the environment. The electro-chemical properties of saliva depend upon its composition, surface tension, pH, concentration of its components and the buffering system. Dissimilar metals produce the galvanic shock causing sharp pain, as when amalgam restorations oppose a gold foil restoration wet with electrolytic saliva.

Impurities in alloys may enhance corrosion. Accumulation of food debris as a result of poor oral hygiene may produce concentration cell corrosion because of variations in the electrolytes or composition of the electrolytes. All alloys such as the clasp for the partial denture (Fig. 23-7) must be able to withstand the corrosive environment of the oral cavity.

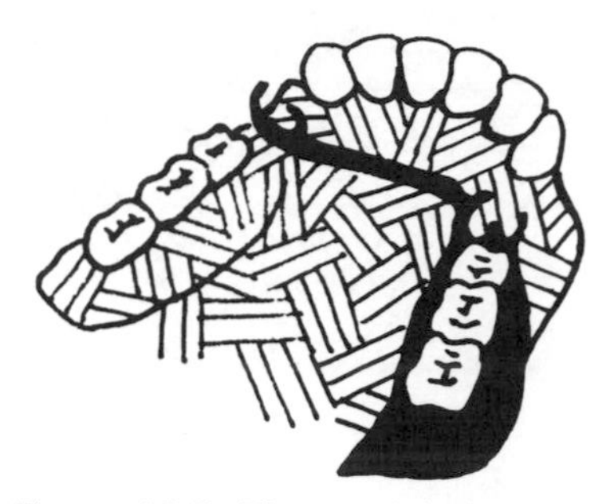

Figure 23-7. The partial denture is clasped before insertion.

Denture Base Materials—The Synthetic Resins

Although the dentist uses many forms of plastics and resins the particular synthetic resin most currently used in dentistry is the acrylic resin polymethylmethacrylate. The acrylic resins are derivatives of acrylic acid or methacrylic acid. Both polymerize to form long chains by the addition mechanism, wherein the double bond opens up to join with adjacent molecules.

Polymethylmethacrylate is a very transparent, extremely stable resin exhibiting remarkable aging properties. In practice the monomer is usually mixed with the polymer in powdered form to give a dough that is packed into the mold and then polymerized by heat.

SILICATE CEMENTS

Silicate cements are essentially acid soluble glasses composed of silica, alumina, lime, sodium fluoride, calcium fluoride and cryolite or combinations thereof. These cements in the form of powders are mixed with phosphoric acid to set to a hard translucent substance that resembles dental porcelain, yet is not a porcelain. The reaction is complex, but basically the acid reacts with the powder to form silicic acid; straight chains are then formed by eliminating water between molecules, and the long chains then cross-link by the elimination of water between the two hydroxyl groups on adjacent chains.

GYPSUM AND GYPSUM PRODUCTS

Dental plaster is the result of heating gypsum to high temperatures so that the dihydrate is converted to the hemihydrate by the loss of water. The hemihydrate is the principle constituent of dental plasters. Depending upon the method of calcination (heating) one of two forms of the hemihydrate results. When gypsum is heated in an open container in this temperature range, the beta hemihydrate results (plaster of Paris). However when gypsum is calcined under steam pressure in an autoclave, the alpha hemihydrate dental stone results. Differences between these two products exist in their crystalline structure, and irregularities and porosity of the crystals. In the setting of gypsum products both the alpha and beta hemihydrates combine with water to form gypsum and evolve heat (exothermic reaction).

Plaster of Paris is used in the preparation of a cast for an artificial denture: the plaster of Paris is mixed with water and placed in an impression tray and is impressed against the upper part of the mouth. After it has set or hardened this impression provides the dentist with a life-size negative model of the mouth. The alpha hemihydrate is then poured into the already prepared impression to

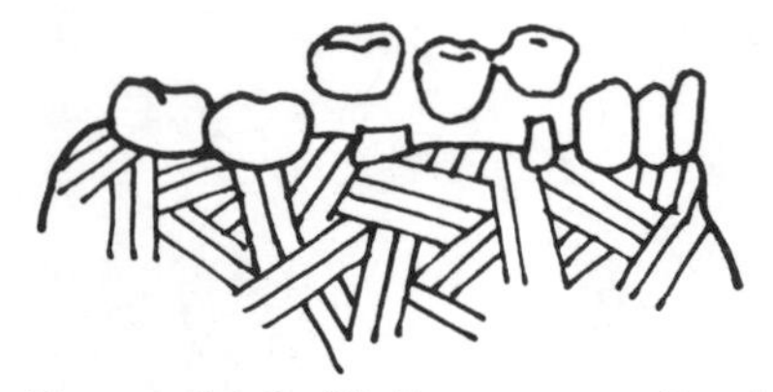

Figure 23-8. Stationary or fixed bridge.

form a *positive* model or cast on which the artificial denture is constructed by a dental technician (Fig. 23-8).

ZINC OXIDE-EUGENOL IMPRESSION PASTES

Zinc oxide reacts with eugenol to form a chelate product called zinc eugenate (Fig. 23-9). Under proper conditions a relatively hard mass is formed. The two compounds are mixed together and the impression is withdrawn after the paste has hardened.

CALCIUM HYDROXIDE

This is an odorless white powder which forms a cream paste of high pH (11-12) when mixed with water. It is used as a base, especially in deep cavities for the following reasons:

1. It neutralizes free phosphoric acid in zinc phosphate cements minimizing or avoiding irritation of the pulp.
2. It increases the density and hardness of the dentin under the base of the cavity to the pulp chamber.
3. It stimulates odontoblastic activity and the formation of secondary dentin over minute or larger pulpal exposures of healthy pulpal tissue.
4. It is a bactericide and is bacteriostatic.

SEALANTS

Sealants should act as a physical barrier to prevent oral bacteria and their nutrients from aggregating within fissures and developing the acid conditions that are considered necessary for caries initiation. There are 3 basic sealants:

1. The cyan-acrylates.
2. The BIS-GMS's, which are the result of the reaction between bisphenol A, glycidyl methacryl and methylmethacrylate.
3. The polyurethanes.

The occlusal surfaces are first treated with pumice and then with an acid conditioning agent (phosphoric or citric acid) to create *microspores* (about 25 microns in depth) so that the resin or sealant will attain maximum adhesion.

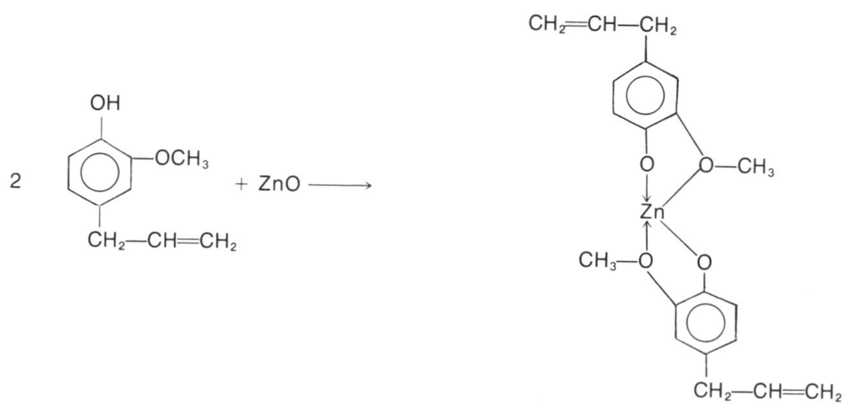

Figure 23-9. The reaction of zinc oxide and eugenol to form zinc eugenate.

Catalysts such as benzoyl peroxide or ultraviolet light (30 sec exposure) activate the polymerization.

COMPOSITE RESINS

Conventional resin filling materials are rapidly being displaced by the use of *composite* resins. The term composite is used to indicate the presence of actual reinforcing materials. These differ from a material just containing a filler without chemical bonding to the polymer matrix. Composite dental resins may contain up to 80 percent inorganic fillers such as glass, beads, rods, quartz, or lithium aluminum silicate. the filler particles themselves are coated with a coupling agent such as an appropriate vinyl silane which provides an adhesive bond between the filler and the resin. The advantages over other silicate cement restorations are: increased compression strength, higher modulus of elasticity, greater hardness, better coefficient of thermal expansion, fast settings, and with quartz fillers, translucency.

The resulting matrix is more complicated than the BIS-GMA because of side chains.

HYDROCOLLOID IMPRESSION MATERIALS

Any solution where the units of the solute are too large to dialyze through a semipermeable membrane is called a colloidal dispersion, a colloid, or a colloidal sol. Some hydrocolloids (aqueous colloids) possess the property of changing to a jelly or gel and the process is reversible under the influence of temperature. Heating the gel yields a sol, which reverts back to a gel on cooling. Agar-agar and gelatin hydrocolloids are reversible, others, such as the alginates, are irreversible. Once the gel state is formed heating will not convert it back to the sol state. The composition of a typical hydrocolloid is: potassium alginate 14 percent, calcium sulfate 8 percent, sodium phosphate 2 percent, diatomaceous earth (filler) 72 percent and sodium silico fluoride 3 percent. Potassium alginate dissolves readily in water and reacts with calcium sulfate to form the insoluble calcium alginate. Usually sodium phosphate is added as a retarder to slow the reaction. The alginates are one of the most widely used *elastic* impression materials. They are used in forming impressions for study casts in orthodontics, removable partial denture construction or in prosthetics. The agar-agar material is a polysaccharide extracted from certain types of seaweed. It is a sulfuric acid ester of a linear polymer of galactose. Soluble alginates are salts of alginic acid believed to be linear polymers of the sodium salt of anhydro-beta-d-mannuronic acid. Alginic acid itself is insoluble in water, but its sodium, potassium, ammonium and magnesium salts are soluble. Both sodium and potassium alginates are used extensively in dental impression materials.

RUBBER IMPRESSION MATERIAL

The process of changing a rubber base (a liquid polymer) to a rubber-like material is called vulcanization or curing. The basic ingredient is a polyfunctional polysulfide polymer. Lead peroxide (PbO_2) and sulfur (S) are the polymerizing agents. They are mixed outside the patient's mouth and then placed in an impression tray and impressed in the mouth. Curing occurs in the mouth with the evolution of some heat. The cured product has adequate elasticity and strength which can be readily removed over undercuts. The polysulfide polymer reacts with the lead peroxide to give the following probable formula:

$$—S(R—S—S)_{23}—R—S—S—R(S—S—R)_{23}—S + PbS$$

Silicone rubbers are based upon an organosilicone such as polydimethyl siloxane. When the cure is effected by certain organometallic compounds, large amounts of hydrogen may be evolved, which may corrode the surface of the stone (gypsum) cast.

COLD STERILIZATION SOLUTIONS

In operative dentistry many instruments are sterilized by immersion in cold sterilization solutions (Fig. 23-10). This is a water solution containing a type of bactericide called quaternary ammonium salts or merely quats. A small percentage of these substances in water sterilizes the instruments but does not kill spores. These compounds are formed by the reaction of an alkyl halide with a tertiary amine. The compound formed is a true ionic salt dissociating in water to form a cation and an ion. The substituting group on the amine may be an alkyl or aryl (aromatic ring) group. The reaction may be expressed as:

Tertiary amine + alkyl halide → quaternary ammonium halide
Trimethyl amine + methyl chloride → tetramethyl ammonium chloride

THE DEVELOPMENT OF X-RAY NEGATIVES

X-rays are electromagnetic waves of extremely short wavelengths and are similar to the gamma rays that you studied in the chapter on nuclear chemistry. They have the tremendous power of easily penetrating human body tissues but bony structures and teeth do not allow them to pass through readily. Unexposed photographic film is placed in a position so that the x-rays impact upon the film after passing through the part of the body being x-rayed. The photographic film is a sheet of plastic, such as cellulose nitrate, on which is spread a suspension (emulsion) of silver bromide and silver iodide which has been formed by the reaction of potassium bromide and potassium iodide with silver nitrate.

Exposure

When the x-rays strike the film, the silver halide grains become unstable, forming silver ions (Ag^+). Visual inspection of the exposed film shows no apparent change:

$$AgBr \text{ and } AgI + \text{x-rays} \rightarrow Ag^+$$

The greater the intensity of the x-rays on the photographic film the greater the amount of Ag^+ ions formed. This happens when the x-rays pass through soft body tissue. When the x-rays are retarded by bony structure and denser body tissues, the amount of Ag^+ is less. The number of ions formed is therefore directly proportional to the intensity of the x-rays that hit the photographic plate.

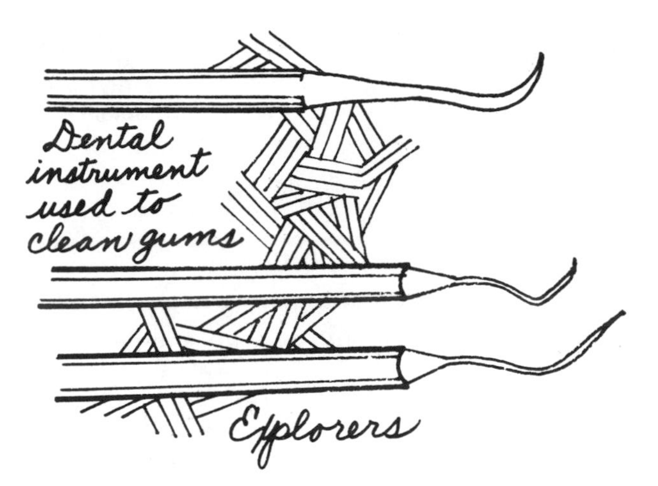

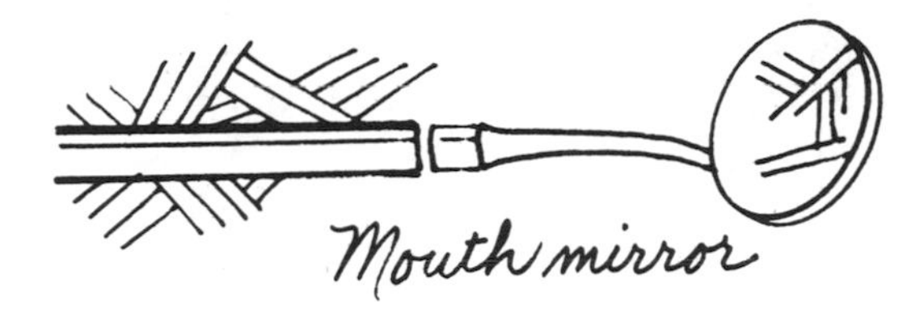

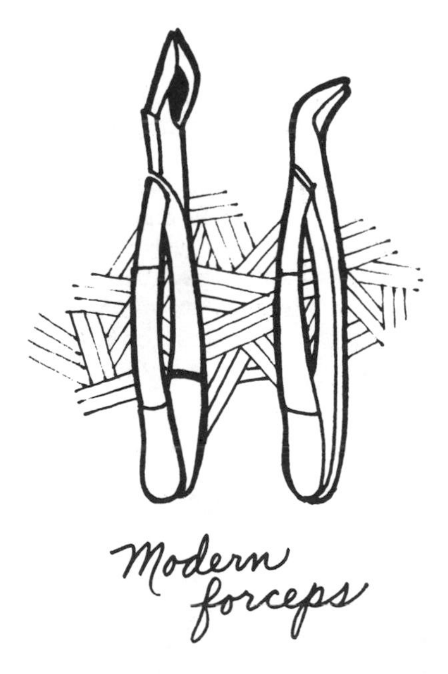

Figure 23-10. Dental instruments that are sterilized in cold solutions.

Development

The silver ions that are formed by the action of the x-rays are reduced in the developing process by chemical reducing agents to metallic silver. Typical reducing agents are pyrogallol:

$$C_6H_3(OH)_3 =$$

and hydroquinone:

$$C_6H_4(OH)_2$$

In a solution made alkaline by sodium carbonate, they reduce the Ag^+ to metallic silver. The hydroquinone is oxidized to quinone, and the silver ion is reduced to metallic silver.

The image on the photographic film is the REVERSE of what has been x-rayed, because: when the x-rays passed through easily the maximum conversion to Ag^+ ions took place, and the film is dark. Where the x-rays were impeded or stopped by bony structure, there was minimum conversion to Ag^+ ions, and the film is light. The photographic film is a NEGATIVE, just as you have when you take a black and white picture with your camera after having the film developed; the image on the film is the reversal of the picture of the object, dark where the object was light, and light where the object was dark.

The exposed x-ray film is covered so that visible light will not cause exposure just as the x-rays did. The exposed film itself must be opened in a dark room where it can be developed and fixed. The dark room is not absolutely dark but has a red electric bulb, whose radiations *do not* affect the suspension (emulsion), so that the technician can see how to process the exposed film.

Fixing

After the photographic film has been developed there may be much silver salt that was not hit by x-rays and therefore not changed to the Ag^+ ions. These unaffected salts are soluble in sodium thiosulfate, $Na_2S_2O_3$, and they may be dissolved and removed from the film by a solution of this salt. Sodium thiosulfate dissolves the silver salts AgBr and AgI. Sodium thiosulfate is commonly called hypo. The metallic silver particles that were formed in the developing process are insoluble in sodium thiosulfate. They therefore remain on the photographic film while the unaffected silver salts are dissolved and washed off. Usually some alum is added to harden the gelatin on the photographic film so that the negative can be handled and examined without being easily damaged after it has dried.

The Negative

The negative is a valuable tool for the physician, dentist and health care personnel for diagnosis and treatment.

SUMMARY

The environment of the mouth is subject to drastic changes in temperature, pH and pressure. Saliva, pH around 7, contains an enzyme that acts on polysaccharides to hydrolyze them. The teeth resemble bone covered by enamel, composed of a calcium phosphate hydroxyapatite. Proper tooth development requires vitamins, calcium and phosphorus. Collagen, a protein, is found in the matrix of dentin. Accumulations on the teeth that calcify cause calculus and tartar to form. Dental caries (cavities) result from the breaking of the enamel and the dissolving of the hard surfaces of the teeth. Acids react with the apatite material to destroy the crystal and destroy the enamel. Fluorides significantly decrease dental caries and they can be taken internally or applied to the surface of the teeth. Alloys are combinations of two or more metals which may be present in varying proportions. Certain metals used in mechanical procedures of restorative dentistry contribute desirable functions to the alloy; gold, when properly used, is unsurpassed. Tarnish is a surface discoloration; corrosion is the disintegration and dissolving of the metal. There are a number of denture base materials, but the acrylic resins are the best. Other materials used are silicate cements, zinc oxide-eugenol, calcium hydroxide, sealants, composite resins, and impression materials: the hydrocolloids, agar-agar, alginates, and rubber. Dental x-rays are used to enable the dentist to locate abnormalities.

EXERCISE

1. State the temperature range to which the mouth is exposed.
2. Give the composition, pH and functions of saliva.
3. Describe the action of the enzyme ptylin.
4. Draw a diagram of a tooth, label its components and give the basic chemical composition of each component.
5. Discuss the action of vitamins A, C, and D on proper tooth development.
6. What is the major protein constituent of most connective tissues, and what is it composed of?
7. Describe the formation of calculus from plaque.
8. Discuss the two theories of dental caries and give the possible chemical reactions involved.
9. Discuss the fluorides and their action on hydroxy apatites to form fluoro apatites.
10. Describe the silver-tin-mercury amalgam as a restorative material, including the functions of the alloy components.
11. Why does gold foil have to be clean and free from contaminants and occluded gases?
12. What is a dental alloy and name the metals that may be used.
13. Describe the difference between tarnish and corrosion.
14. What is the most widely used denture base resin?
15. What is the basic reaction involved in silicate cements?
16. Write the chemical reaction for the hydration of plaster of Paris.
17. Why is calcium hydroxide used as a base?
18. Name the three basic sealants.
19. What are the advantages of composite resins?
20. Are the following reversible or irreversible hydrocolloids?
    a) agar-agar b) gelatin c) alginic acid
21. What are the polymerizing agents for the rubber impression materials?
22. Will cold sterilization solution absolutely sterilize instruments?
23. What is the action of hydroquinone in x-ray developing solution?
24. Why is sodium thiosulfate used in x-ray fixing solution?
25. What is the action of alum in x-ray fixing solution?

INDEX